A Guide to the Zoological Literature

A Guide to the Zoological Literature

The Animal Kingdom

George H. Bell
and
Diane B. Rhodes

Illustrated by
Emily R. Rhodes

1994
Libraries Unlimited, Inc.
Englewood, Colorado

LIBRARIES UNLIMITED, INC.
P.O. Box 6633
Englewood, CO 80155-6633
1-800-237-6124

Library of Congress Cataloging-in-Publication Data

Bell, George H., 1943-
A guide to the zoological literature : the animal kingdom / George H. Bell, Diane B. Rhodes ; illustrated by Emily R. Rhodes.
xxiii, 504 p. 17x25 cm.
Includes index.
ISBN 1-56308-082-6
1. Zoology--Bibliography. I. Rhodes, Diane B. II. Title.
Z7991.B43 1994
[QL45.2]
016.591--dc20 94-12694
CIP

Contents

2 THE INVERTEBRATES (*continued*)

3 THE ARTHROPODS (Including the Spiders, Crustaceans, and Insects) . . . 77

3 THE ARTHROPODS (*continued*)

6 THE AMPHIBIANS AND REPTILES (*continued*)

7 THE BIRDS (*continued*)

ACKNOWLEDGMENTS

To accomplish a project of this magnitude, a number of people, other than the authors, are always instrumental in the endeavor. Although we cannot thank everyone involved in this effort, we would like to especially thank the following:

Foremost, we wish to thank Mara Pinckard, Head of Science Reference and Assistant Head of Noble Library, for the administrative support she has shown us during these past two years. Without her consideration and understanding, this project would never have been realized.

Special thanks are due to the Professional Development Committee, University Libraries, Arizona State University, for providing research funds.

Because many of the books we reviewed had to be borrowed, we express our heartfelt thanks to Sheila Walters and her staff at Arizona State University, for the speedy acquisition of interlibrary loan requests.

Last, but not least, we wish to show our appreciation and give thanks to Joy Bentley, Secretary of the Science Reference Department, for her expertise in WordPerfect as well as other secretarial support functions.

INTRODUCTION

This work serves as a guide to the literature of the animal kingdom by providing a useful, annotated list of traditional and electronic sources for each major animal group. The major groups include the invertebrates, fish, amphibians, reptiles, birds, and mammals. We know of no other work existing today that contains a compilation of all the major reference sources on the zoological kingdom.

Sources listed in this work contain detailed reviews, many with comparisons among similar sources, that should enable the reader to find information on a particular animal subject, or to make an intelligent choice when selecting sources for a collection.

This work does not reflect the totality of the zoological literature, rather it is intended as a major selective guide to the more important sources for each animal group. The sources are worldwide in scope, covering works for major land areas, continents, and countries. Most of the sources are in English and are in the form of monographs or monographic sets. A few classic foreign language tools are listed. Out-of-print sources that either emphasize major undertakings in the early years or are believed to be the best source available for the topics are also included. A number of electronic sources, such as online databases and CD-ROMs, can be found within the Indexes and Abstracts section of chapter 1. Internet sources can be found in the appendix.

Not included here are fossil sources, medically related sources, and geographical sources for areas smaller than country level (such as the individual states in the United States, except Alaska and Hawaii). Also, due to the large amount of published literature on the animals alone, the zoological disciplines such as animal behavior, ecology, conservation, physiology, anatomy, genetics, and so on, are not included. However, general information on many of these disciplines as they relate to specific animal groups can be found within the sources listed. Also, bibliographies, checklists, classification schemes, and zoological keys published in the journal literature are not included, but citations to these can be located using the journal indexes reviewed in the Indexes and Abstracts section in chapter 1. *Zoological Record* and *Biological Abstracts* are especially useful for this purpose.

To obtain the sources found within this work, in-house database systems, OCLC, *Books in Print*, *American Reference Books Annual*, major subject bibliographies (see References at the end of this introduction), and those bibliographies found in the more important sources for each animal group were searched. All sources were reviewed by the authors with "book-in-hand" except for the newest sources that were reviewed by use of publishers' flyers.

Prices have been included whenever possible and were taken primarily from the latest editions of standard bibliographic acquisition tools, from the most recent publisher price list, or from the publisher directly. Prices are quickly outdated, however, and may only indicate the average in an approximate range. Those without prices may indicate out-of-print publications. Out-of-print book dealers may be found in *AB Bookman's Yearbook* in the section "The O.P. Market (Directory of

Specialist & Antiquarian Booksellers)." This directory is arranged under subject headings such as birds, mammals, and so on, to help the reader find a dealer for a particular subject. Also, a special natural history issue of *Bookman's Weekly* comes out each year that features the latest natural history books and includes out-of-print dealers.

ORGANIZATION OF THIS GUIDE

This guide is organized into eight chapters and one appendix. The first chapter addresses general reference works for the entire animal kingdom. The following chapters, arranged phylogenetically, cover the major animal groups: the invertebrates (excluding arthropods), the arthropods, the vertebrates (general), the fish, the amphibians and reptiles, the birds, and the mammals. An appendix has been included to offer Internet sources of information on the animal kingdom.

Each chapter includes lists of sources for a major group of animals under the following sections: Dictionaries, Bibliographies, Indexes and Abstracts, Directories, Handbooks, Field Guides, Textbooks, Journals, Biological Keys and Classification Schemes, Checklists, Biographies, Style Manuals, Guides to the Literature, and Professional Associations. Due to the publishing record, all categories were not included in each of the eight chapters. For some of the largest categories, especially Handbooks, sources were arranged under "general" headings (for comprehensive sources on the animal group in general), geographical headings (for sources covering large land areas), and taxonomic headings. The taxonomic arrangement is based on *Synopsis and Classification of Living Organisms* (see entry 77), edited by Sybil P. Parker. This source was chosen because it covers all the animal groups and provides a single source to consult. The chapter on the arthropods, the largest group of animals, did not adhere well to this overall arrangement simply because of the magnitude of information in the arthropod literature. The introduction to that chapter explains its arrangement.

Dictionaries

These are not only sources for word definitions and English and foreign-language terms, but sources for word derivations and other sources on zoological terminology.

Encyclopedias

These are narrowly defined as sources that are truly encyclopedic in nature and that usually have the word *encyclopedia* in their titles. Works that served as a comprehensive source of information for a particular group were placed in the Handbook section.

Bibliographies

Listed here are bibliographies published as a complete volume. These bibliographies of references can be invaluable, especially for the older references that are difficult to find. We have included the major bibliographies and some specific bibliographies here. Many bibliographies on particular groups are also found in comprehensive works in the Handbooks or Checklists sections.

Indexes and Abstracts

Indexes and abstracts to the zoological literature are listed in this guide, including those found in electronic form. Significant historical indexes covering the literature of the sixteenth, seventeenth, eighteenth, and nineteenth centuries are also included.

Directories

Directories to people, organizations, collections, and other information sources are included here.

Handbooks

This is the largest category of each chapter. These sources are the primary works for each animal group, and contain the most information on each animal group. Usually the handbook covers animals of a particular geographical area or a specific order. Occasionally, if handbooks for a particular order are not available in published form, handbooks to the major family are included here. Material in handbooks may often contain checklists, bibliographies, identification keys, and so on, and so should be consulted in conjunction with sources listed under these specific categories.

Field Guides

Field guides are defined here as guides that are meant to be taken into the field as an aid in identification of a species. These guides are usually of a size that fit in pockets, jackets, or backpacks for easy retrieval. Identification guides that are not small or portable enough are, for the most part, listed in the Handbooks section.

Textbooks

Works listed here are those meant to be used as a text for a course on that particular animal group. These works are useful not only as course texts, but also as reference sources.

Journals

Journals listed here are considered core journals. We have not attempted to provide a comprehensive list.

Taxonomic Keys

The taxonomic keys listed here aid in identifying and naming a particular organism according to certain anatomical characteristics. The keys are usually dichotomous, offering choices of two characteristics (with wings or without) at a time; then, two more choices are given; this is repeated until the organism is finally identified. Many biological keys are included in handbooks, so be sure to consult the Handbooks section for that particular animal group.

Checklists and Classification Schemes

Checklists include catalogs of names of a certain animal group. Worldwide and specific geographical sources are included. These lists, in most cases, are arranged taxonomically by scientific name. Checklists are often found in handbooks for a particular animal group as well. Also, many of the more specific and most up-to-date checklists are found in the journal literature. These are not listed here, but may be found by consulting a journal index listed in the Indexes and Abstracts section. Classification schemes, which give a broad overview of the taxonomy of an animal group, are also considered in this section.

Biographies

Only collective biographies are listed here, not individual biographies.

Style Manuals

This section is only found in the first chapter on general sources. Style manuals provide a standardized form for writing papers, theses, and so on.

Guides to the Literature

For the larger animal groups, for example the insects and birds, these are guides that have already been written. They provide more sources than those covered in our work.

Associations

This category includes lists of associations and organizations with interests in the animal group for that specific chapter. Information for many of the associations and organizations was taken directly from the *Encyclopedia of Associations: National Organizations of the U.S.* (see entry 120) and *Encyclopedia of Associations: International Organizations* (see entry 119). They are arranged in alphabetical order and provide the address, telephone number, and telecommunication services. Annotations on these organizations and associations are not included here, but may be found in the above reference sources.

We hope this guide will be of use not only to the librarian, but to the student, the specialist, and the amateur needing access to information on the animal kingdom or help finding books to add to their collections. We have tried diligently to include here all of the most important works. We take full responsibility and apologize for any major source inadvertently omitted. What we anticipated as a heavy task of finding and reviewing hundreds of books was, in many ways, an enjoyable experience in discovering new books and becoming lost in especially engaging works. We hope this source will go beyond the dry reference work and prove to be an enjoyable experience for the reader as well.

REFERENCES

Guide to Reference Books, 10th ed. Sheehy, Eugene P., comp. (Chicago: American Library Association, 1986). LC 85-11208. ISBN 0-8389-0390-8.

Guide to Reference Books Covering Materials from 1985-1990: Supplement to the Tenth Edition. Balay, Robert, ed., and Eugene P. Sheehy, special editorial advisor. (Chicago: American Library Association, 1992). LC 92-6463. ISBN 0-8389-0588-9.

Guide to Sources for Agricultural and Biological Research. Blanchard, J. Richard, and Lois Farrell, eds. (Berkeley, Calif.: University of California Press, 1981). LC 76-7753. ISBN 0-520-03226-8.

Hurt, C. D. *Information Sources in Science and Technology*, 2d ed. (Englewood, Colo.: Libraries Unlimited, 1994). LC 93-43934. ISBN 1-56308-034-6.

Information Sources in the Life Sciences. Wyatt, H. V., ed. (London, Boston: Butterworths, 1987). LC 86-31731. ISBN 0-408-11472-X.

Science and Technology Annual Reference Review 1989. Malinowsky, H. Robert, ed., (Phoenix, Ariz.: Oryx Press, 1989. ISSN 1041-2557.

Smith, Roger C., and Reginald H. Painter. *Guide to the Literature of the Zoological Sciences,* 7th ed. (Minneapolis, Minn.: Burgess, 1967). LC 66-23383.

Smith, Roger C., W. Malcolm Reid, and Arlene E. Luchsinger. *Smith's Guide to the Literature of the Life Sciences*, 9th ed. (Minneapolis, Minn.: Burgess, 1980). LC 79-55580. ISBN 0-8087-3576-4.

Walford's Guide to Reference Materials, 5th ed. Walford, A. J., ed. (London: The Library Association, 1989). ISBN 0-85365-978-8.

CHAPTER 1

GENERAL REFERENCE SOURCES

The animal kingdom is incredibly diverse. It is estimated that more than 2 million animal species are alive today. They show an amazing range of diversity in both structure and mode of life, with representatives being found in every known habitat in the world, from great ocean depths to high mountain peaks, from torrid deserts to below-zero ice continents.

This chapter covers the diverse literature for the animal kingdom as a whole. These sources should be used to supplement those in the individual chapters on specific animal groups. For example, great classic indexes like *Zoological Record* (see entry 55) and *Biological Abstracts* (see entry 33) are listed in this chapter, but should be consulted in addition to those sources listed in the following individual chapters. The individual chapters, arranged by class, cover those sources specifically for that group, but need these comprehensive general sources to complete the coverage.

In this chapter on general reference sources we have included works on the broader subject of biology or science when no or few works could be found in some categories, such as biographies and dictionaries, that were specifically devoted to zoology. In these cases, the books were annotated to show their zoological significance (such as the number of pages devoted to zoology, zoological index headings, and so on). Some foreign-language sources as well as historical sources are included, and electronic databases are listed to provide additional diversity in coverage of the literature.

There is a grandeur in this view of life, with its several powers,
having been originally breathed into a few forms or into one;
and that, whilst this planet has gone cycling on according to the fixed
law of gravity, from so simple a beginning endless forms most
beautiful and most wonderful have been, and are being, evolved.
—Charles Darwin
On the Origin of Species, *1859*

DICTIONARIES

1. Allaby, Michael, ed. **The Concise Oxford Dictionary of Zoology**. Oxford, New York: Oxford University Press, 1991. 508p. $35.60. LC 90-21596. ISBN 0-19-866162-2.

Based on the entries contained within the *Oxford Dictionary of Natural History* (see entry 2), this source represents approximately 6,000 terms. The length of the definitions vary from a few sentences to one column or more of text. The average length of a definition is one paragraph. A large number of the terms are taxonomic, representing phylums, classes, orders, families, and their respective subgroups. These are defined morphologically. In addition, larger taxonomic groups are incorporated within the definition when the smaller taxonomic groups are defined. Common names are also listed, referring the reader to the scientific classification. For example: "prairie dog." See Sciuridae.

Other subject areas considered are ecology, evolution, ethology, genetics, cytology, physiology, and zoogeography. Physiological and biochemical terminology are not as nearly represented as that of taxonomy.

2. Allaby, Michael, ed. **The Oxford Dictionary of Natural History**. Oxford, New York: Oxford University Press, 1985. 688p. $45.00. LC 85-13758. ISBN 0-19-217720-6.

Taxonomic animal, plant, and microbial terms make up the majority of this work. Definitions range from a few lines to several paragraphs. In some cases an entire column is devoted to a particular entry, as in the case of Araneae (spiders; class Arachnida). Common names are also listed for taxonomic entries in most cases. In addition, common names are interfiled in the work, which lead to the taxonomic entry. Terminology used in other biological disciplines can also be found.

3. Chinery, Michael, consultant ed. **The Kingfisher Illustrated Encyclopedia of Animals: From Aardvark to Zorille, and 2,000 Other Animals**. New York: Kingfisher Books, 1992. 379p. $19.95. LC 92-53113. ISBN 1-85697-801-X.

This is the revised edition of *Dictionary of Animals* (New York: Arco, 1984), and so is placed here in the dictionary section instead of the encyclopedia section as the title suggests. Although this work is not comprehensive in coverage, it does define and describe 2,000 of the world's animals. Each brief description (one paragraph to one column in length) includes information on the size, habitat, and behavior of the animal. At the end of each description, the names of the order, family, genus, and species of the animal are provided. Full-color photos and drawings of selected animals are featured on every page. A glossary of terms used in the work and an index to common and scientific names complete the book. This is a useful work for public, school, and home libraries.

4. Henderson, I. F. **Henderson's Dictionary of Biological Terms**. 10th ed. Eleanor Lawrence, ed. New York: Wiley-Interscience, 1989. 637p. $49.95. LC 89-171806. ISBN 0-470-21446-5.

Terminology represented in this work consists of biology, botany, and zoology, as well as molecular and cellular biology, and electron microscopy. Definitions range from a few words to several sentences. Containing 6,000 more terms than *Chambers Biology Dictionary* (see entry 13), this dictionary does not expand on certain entries as does *Chambers*.

In addition to the definitions, supplementary types of material can be found. Among them are units and conversion tables, SI prefixes, the greek alphabet, and common latin and greek noun endings. The appendices consist of structural formulas of important biological molecules, an outline of the plant, animal, protista, and monera kingdoms, and a list of the common chemical elements listing symbol, atomic number, and atomic mass.

5. Lapedes, Daniel N., ed. **McGraw-Hill Dictionary of the Life Sciences**. New York: McGraw-Hill, 1976. 907p. LC 76-17817. ISBN 0-07-045262-8.

Over 800 illustrations of plants, animals, cells, structural formulas, and charts can be found interspersed among the over 20,000 definitions in this text. Definitions usually range from one to several sentences. Many of the newer terms in molecular and cellular biology will not be found; however, the work abounds in morphological, chemical, and taxonomic terminology. Many of the entries listed have been taken from the *McGraw-Hill Dictionary of Scientific and Technical Terms* (1974).

For each word listed, the area of the biological sciences in which the term is associated can be found. For example, the term Antilopinae has next to it, in brackets, VERT ZOO. The definition then follows.

Appendices consist of various forms of measurement systems, fundamental constants, a list of antibiotics, values in clinical pathology (e.g., erythrocyte blood count in humans), as well as animal, plant, and bacterial taxonomy.

6. Leftwich, A. W. **A Dictionary of Zoology**. 3d ed. London: Constable; distr., New York: Crane, Russak, 1973. 478p. LC 73-161027. ISBN 0-09-454972-9.

Approximately 6,700 terms are defined in this third edition. In addition to the taxonomic terminology, words dealing with DNA, RNA, and the genetic code are included. Concise definitions are usually the case for each entry. However, paragraphs are usually written when discussing taxonomic terms, such as Families.

Besides the major portion of the dictionary there is what might be called a mini dictionary. In this part of the work can be found a listing of the common names of the animals, with cross references to the scientific terms found in the main listing.

Some general information on classification and nomenclature, transliteration of greek words, a simplified animal classification scheme, and a brief bibliography complete the volume.

7. Lincoln, R. J., and G. A. Boxshall. **The Cambridge Illustrated Dictionary of Natural History**. Cambridge, N.Y.: Cambridge University Press, 1987. 413p. $27.95; $14.95pa. LC 87-8018. ISBN 0-521-30551-9; 0-521-39941-6pa.

Unlike the *Oxford Dictionary of Natural History* (see entry 2), this volume abounds with over 700 animal and plant line-drawings. The majority of the terms listed represent all the major animal, plant, and microbial taxonomic groups. Entries are usually described employing one to several sentences. Common names are interfiled in the dictionary and cross-referenced back to the main entry, which is the scientific taxonomic term.

8. Martin, Elizabeth, ed. **Concise Dictionary of Biology**. New York: Oxford University Press, 1985. 272p. $17.95. LC 86-195762. ISBN 0-19-866144-4; Paperback ed. **Concise Dictionary of Biology: New Edition**. New York: Oxford University Press, 1990. 272p. (Oxford Paperback Reference Series). $8.95. LC 90-38629. ISBN 0-19-28609-3.

Covering the very basic zoological terms, this work was derived from the *Concise Science Dictionary*, which was published by Oxford University Press in 1984. From that work, some terms in chemistry and geology that relate to zoological phenomena are included. The length of the definitions vary from one sentence to several paragraphs. Tables, graphs, and labeled drawings are used from time to time in order to enhance the meaning of the word. Although the major animal groups are defined and explained quite adequately, the smaller taxonomic groups are not present. Appendices consist of SI unit tables, an animal and a plant classification chart, and a geological time scale.

9. Martin, E. A., ed. **Dictionary of Life Sciences**. 2d ed. New York: Pica, 1983. 396p. LC 83-13258. ISBN 0-87663-740-3.

Although not as comprehensive as the *McGraw-Hill Dictionary of the Life Sciences* (see entry 5) or *Henderson's Dictionary of Biological Terms* (see entry 4), the definitions given for most of the entries are longer; some even border on a semi-encyclopedic discussion. There are a number of illustrations to aid in the definitions, many of them being structural formulas. For example, when meiosis is discussed, diagrams showing the various stages of the meiotic process are depicted. According to the editor, in order for this edition to absorb some 300 new entries (genetics, molecular biology, microbiology, and immunology), a small reduction in the coverage of terms associated with more traditional aspects of biology had to be realized.

10. **Nature in America**. Pleasantville, N.Y.: Reader's Digest Books, 1991. 456p. $32.50. LC 90-46146. ISBN 0-89577-376-7.

Containing over 1,200 entries, this dictionary-like volume considers the more familiar terms that deal with animals, plants, landforms, and other natural features found in the United States. Arranged alphabetically, the majority of the terms are defined using easy-to-understand language. They are defined by relating general and basic information to the reader. The length of the definitions vary from usually one paragraph to several pages. In addition, a two-page spread can be found for certain ecological niches, such as the cypress swamp, deciduous forest, desert, and so on. Along with a brief description of the niche, a variety of colored photographs depicting certain animals and plants endemic to the region are present.

This volume abounds with excellent colored photographs; most entries are accompanied by a photograph that depicts the animal, plant, or landform. Colored tabs representing the letters A through Z aid the reader. A combined subject and common name index concludes this mini-encyclopedic source.

11. Pennak, Robert W. **Collegiate Dictionary of Zoology**. New York: Ronald Press, 1964. 583p. $32.50. LC 64-13331. ISBN 0-471-06790-3. Reprint, Malabar, Fla.: Robert E. Krieger, 1988. $26.50pa. LC 85-23983. ISBN 0-89874-921-2.

Containing about 19,000 entries, all zoological disciplines are covered in this text. Most definitions range in length from one to several sentences. Common and taxonomic names are interfiled. In some cases, a number of different definitions may be given for the same term. Because of its publication date, the majority of the terms are taxonomic and not biochemical. Family and species names are included in many cases. An appendix, consisting of a taxonomic outline of the animal kingdom, completes the work.

12. Tootill, Elizabeth, ed. **The Facts on File Dictionary of Biology**. Rev. and exp. ed. New York: Facts on File, 1988. 336p. $19.95. LC 88-45476. ISBN 0-8160-1865-0.

This compact dictionary includes major terms from all areas of biology. Some definitions are concise, while others are two to three paragraphs long. A number of excellent illustrations can be found throughout the text. For example, in defining the heart in a semi-encyclopedic manner, a labeled picture of the heart, depicting all its primary parts, can be found. This volume is not as comprehensive as many of the other biology dictionaries mentioned. However, it does a fine job defining the major terms that are listed.

13. Walker, Peter M. B., ed. **Chambers Biology Dictionary**. Cambridge, N.Y.: W & R Chambers, 1989. 324p. $34.50; $14.50pa. LC 90-113402. ISBN 1-85296-152-X; 1-85296-153-8pa.

The length of each of the approximately 10,000 terms defined range from one to five or more sentences. In the case of Rodentia, one-third of a column is used to explain this class. Over 3,000 of these words are devoted exclusively to the field of zoology. The remainder deal with the other areas in biology, as well as chemistry, medicine, radiology, and statistics.

A unique feature of this work is the essay-type treatment given to over one hundred terms. For example, in describing oxidative phosphorylation, the better part of a page is used in discussing this process, along with a diagram depicting the chemical mechanism. Many prefixes of terms are also included.

ENCYCLOPEDIAS

14. Buchsbaum, Ralph, and others. **The Audubon Society Encyclopedia of Animal Life.** A Chanticleer Press ed. New York: Clarkson N. Potter; distr., New York: Crown Publishers, 1982. 606p. $45.00. LC 82-81466. ISBN 0-517-54657-4.

This one-volume encyclopedia is arranged in taxonomic sequence from mammals to invertebrates. It is geared more for school and public library use with brief, less technical descriptions than the Larousse encyclopedia (see entry 20). It is heavily illustrated with striking full-color photos. Based on the seven-volume *World of Nature* series, it has been updated by several authorities to make it a fine reference source as well as a source for enjoyable reading.

15. Burton, Maurice, Robert Burton, and Mark Lambert, eds. **The Marshall Cavendish International Wildlife Encyclopedia**. New York: Marshall Cavendish, 1989. 24 vols. $449.95 set. ISBN 0-86307-734-X set.

This twenty-four volume encyclopedia is intended as a guide to the wildlife of the world. It is the newest edition, first published in 1969 as *The International Wildlife Encyclopedia* (Marshall Cavendish). Some updating, primarily in the indexing and illustrations, has been done since the 1969 edition. But some entries, like the one on Pika, are word-for-word from the 1969 edition. Entries are arranged alphabetically by common name of an animal, but also include articles on general topics such as migration and endangered animals. Each article is one to three pages in length beginning with a brief physical description and usually including a small map showing the location of the species. The article is enhanced with several full-color photos. Indeed, the illustrations and easy reading of the articles are the strengths of this set. Volume 25 contains the index and must be consulted to get the most use of this encyclopedia because entries are alphabetical by common name. For example, there is no entry for penguin, but there are articles under Emperor Penguin, King Penguin, and so on. The index volume also includes "A Selected List of Endangered Species" as well as a list of major wildlife refuges, parks, and sanctuaries. A 13-page bibliography is also included with entries as recent as 1989. *The Illustrated Encyclopedia of Wildlife* (see entry 19), is comparable, except for its systematic arrangement by class or phylum. This encyclopedia is suitable for public and school libraries, but may also be suitable for academic libraries where good illustrations of animals are desired.

16. **Cambridge Natural History**. Harmer, S. F., and A. E. Shipley, eds. Reprint, New York: Hafner, 1959. 10 vols.

Originally published 1895-1906, it is included here as an English language work similar to *Handbuch der Zoologie* (see entry 68), and *Traité de Zoologie* (see entry 69), which are still continuing today. Arranged systematically from protozoa to mammals; includes black-and-white line-drawings and indexes at the end of each volume. Interesting for historical work. Footnotes are used in place of bibliographies.

17. Friday, Adrian, and David S. Ingram, eds. **The Cambridge Encyclopedia of Life Sciences**. Cambridge, N.Y.: Cambridge University Press, 1985. 432p. $45.00. LC 84-1829. ISBN 0-521-25696-8.

Although this one-volume encyclopedia covers life sciences in general, it is included here as an excellent introduction and background for zoological study. Part one covers the basic processes and organization of the cell and the organism and involves chapters on genetics, behavior, and ecology. Part two covers the marine, coastal, terrestrial, freshwater, and wetland environments, and how animals live and interact with other organisms. Part three covers the evolutionary process and paleontology and finishes with a review of the classification of living organisms.

It is a scholarly publication developed at Cambridge University with more than 30 scientists contributing to the effort. The whole is illustrated with both full-color and black-and-white photos and illustrations. Species and subject indexes complete the volume.

18. Grzimek, Bernhard, ed. **Grzimek's Animal Life Encyclopedia**. New York: Van Nostrand Reinhold, 1972-1975. 13 vols. LC 79-183178.

This is a classic set of comprehensive knowledge on animals. Translated from the 1968 German edition, this set provides the most thorough information in one place for animals. It is arranged by class with lower animals in volume 1, insects in volume 2, mollusks and echinoderms in volume 3, fish in volume 4, and fish and amphibia in volume 5, reptiles in volume 6, birds in volumes 7-9, and mammals in volumes 10-13. The mammal volumes have been revised and published separately in a new edition (see entry 1404).

The text is authoritative, having been prepared by nearly two hundred international scholars, primarily European. It is profusely illustrated with full-page color photos and drawings as well as black-and-white drawings and maps in the margins. Each volume has its own index (but there is no comprehensive index to the set); a systematic classification; a four-way animal dictionary in English, German, French, and Russian; and a list of supplementary readings. Although dated (except for the new Mammal edition), this set is still the most complete source of information on the animal kingdom in the English language.

19. **The Illustrated Encyclopedia of Wildlife**. Lakeville, Conn.: Grey Castle Press; distr., Chicago: Encyclopaedia Britannica Educational, 1991. 15 vols. (3053p.) $495.00 set. LC 90-3750. ISBN 1-55905-052-7.

This beautifully illustrated new reference source for school and public libraries features easy, informative reading and striking full-color illustrations that fill almost every page. Originally published in 1989 in London, it is organized into 15 continuously paged volumes with American spellings, weights, and measures substituted for the British. Mary Corliss Pearl, Assistant Director, Wildlife Conservation International, served as the consultant on this American edition. Although the introduction to the set states that wildlife experts from many nations cooperated to make this publication available, no other information as to who they are is included. The set is arranged by class with mammals comprising volumes 1-5, birds in volumes 6-8, reptiles and amphibians in volume 9, fishes in volume 10, and invertebrate classes in volumes 11-15. The 53-page index to the set is in the final volume after the text for part 5 of the invertebrates.

The illustrations in the set are exceptional. There are 4,000 color photos and 1,400 remarkable color artworks.

20. **The New Larousse Encyclopedia of Animal Life**. rev. ed. New York: Larousse, 1980. 640p. LC 79-91865. ISBN 0-88332-132-7.

This is a new edition of *The Larousse Encyclopedia of Animal Life* published in 1967 by McGraw-Hill. It has the same arrangement by taxonomic sequence from Protozoa to Mammalia and so is a handy, authoritative one-volume work to consult for basic information on these groups. The main improvement in this edition is the new full-color photos and illustrations. The text itself has not been greatly changed since the first edition. The encyclopedia also includes a classification list, glossary, bibliography, and index by scientific and common name.

21. Pfeffer, Pierre, ed. **Predators and Predation: The Struggle for Life in the Animal World**. New York: Facts on File, 1989. 419p. $50.00. LC 88-3880. ISBN 0-8160-1618-6.

Compiled by a team of French scientists, this readable work provides general information on more than 500 predators. Arranged alphabetically by common name in an encyclopedic format, this source describes the way in which species preserve themselves by feeding off one another.

Each predatory entry (well-balanced between invertebrates and vertebrates) includes common and scientific names; the name of the family, suborder and/or order, class, and subphylum to which it belongs; physical description of the animal and its habitat; a discussion of its life cycle, mating, and offspring; what the predator preys upon and how, and what preys upon it and how (such information as how predators hunt and trap prey, how they kill and eat their prey, and whether or not they themselves become prey to yet other predators). The length of an entry is usually between one-half to one full page. There are no illustrations and there is no bibliography in this reference source. The index, which completes the book, is a guide to all prey, and to the natural habitats of both prey and predators. For example, if one wishes to know what predators mentioned in the work prey on ants, the index will provide a list of these predators under the heading "Ants," thereby providing an additional access point to the predatory/prey relationship.

BIBLIOGRAPHIES

22. Agassiz, Louis. **Bibliographia Zoologiae et Geologiae. A General Catalogue of All Books, Tracts, and Memoirs on Zoology and Geology**. New York: Johnson. Reprint, 1968. 4 vols. (The Sources of Science, no. 20). $275.00 set. LC 67-29077. ISBN 0-384-00404-0.

Originally compiled by Louis Agassiz as an in-house tool, this tome attempts to cover zoological and geological literature of the seventeenth, eighteenth, and early nineteenth centuries. The first part of this work, which encompasses 85 pages, consists of serial and miscellaneous publications dealing with zoology and geology; it is arranged by European countries. The second part, which fills the remainder of volume 1 and runs through the end of volume 4, is arranged by author. It lists the articles that were found in the publications listed in part one, as well as other articles written by the scientists. Materials cited are in the major European languages, as well as English. The citations to the literature are adequate in identifying the sources.

Because there is no index, the only major access is by author. The book would be more valuable if there was a subject approach. In any case, it serves as a good source for retrospective information.

23. Besterman, Theodore. **Biological Sciences: A Bibliography of Bibliographies**. Totowa, N.J.: Rowman and Littlefield, 1971. 471p. (The Besterman World Bibliographies) $20.00. LC 72-177947. ISBN 0-87471-055-3.

Arranged by broad subject headings, the volume attempts to include all of the most significant bibliographies that exist on a particular biological topic: examples are zoology, ornithology, animals, biology, entomology, and the like. Subheadings can be found under each of the main subject headings. As in the case of animals, phylum, classes, and common names of selected animals are considered.

For each entry listed, the author, title, pagination, and number of citations contained within the bibliography are given. Partial imprints and series statements are entered when appropriate. This is a world list and as such contains many bibliographic entries in languages other than English. Many of the citations to bibliographies were published during the first half of the twentieth century, with a few in the eighteenth, nineteenth, and latter half of the twentieth century.

Although computer searching can compile bibliographies at record speed, they are limited by the number of years that can be searched. This book provides a useful tool in examining retrospective citations.

24. Blacker-Wood Library of Zoology and Ornithology. **A Dictionary Catalogue of the Blacker-Wood Library of Zoology and Ornithology**. Boston: G. K. Hall, 1966. 9 vols. LC 75-15898. ISBN 0-8161-0719-X.

Reproduced catalog cards arranged by author, title, and subject in a dictionary format comprise this nine-volume set. The set represents the collection of the Blacker-Wood Library of Zoology and Ornithology found at McGill University, Montreal. Much of this material is older and represents in many cases unique titles. Government document publications are also included. As of the publication of this set, the Blacker-Wood Library contained 60,000 volumes.

25. Carus, J. Victor, and Wilhelm Engelmann, eds. **Bibliotheca Zoologica I**. Leipzig: W. Engelmann, 1861. 2 vols.

This work, which is continued by entry number 30, represents a continuation of *Bibliotheca Historico-Naturalis*. It covers the years 1846-1860 and includes periodical articles from 1700. Its arrangement is similar to that of *Bibliotheca Historico-Naturalis* (see entry 26).

26. Engelmann, Wilhelm. **Bibliotheca Historico-Naturalis**. Weinheim, Germany: H. R. Engelmann (J. Cramer); New York: Hafner, 1960. 786p.

A reprint of the 1846 work, this bibliography contains citations in the field of natural history covering the period 1700-1846. The citations are arranged alphabetically by author under their respective subject categories, such as anatomy and physiology, bibliographies, natural history, and so on. Likewise, the categories themselves are divided into works produced in Germany (inlandische Werke) and those produced in other parts of the world (auslandische Werke). The citations listed in these categories as well are arranged alphabetically by author. There is an author and subject index at the end of the bibliography to access the citations.

27. Harvard University. Library of the Museum of Comparative Zoology. **Catalogue of the Library of the Museum of Comparative Zoology, Harvard University**. Boston: G. K. Hall, 1968. 8 vols. First supplement, 1976. LC 77-359046. ISBN 0-8161-0811-0.

Arranged alphabetically by main entries (author or title), this book catalog (reproduction of catalog cards) represents the holdings of the Library of the Museum of Comparative Zoology found at Harvard University. It represents over 250,000 items, as of 1976. Much of this material is older and represents many unique titles. Government document publications are also included. Unfortunately there is no subject approach.

28. Reuss, Jeremias David. **Repertorium Commentationum a Societatibus Litterariis Editarum. Secundum Disciplinarum Ordinem Digessit. T. 1, Historia Naturalis, Generalis et Zoologia**. New York: B. Franklin, 1961. 574p. (Burt Franklin Bibliography and Reference Series, vol. 29). $550.00 16-volume set. LC 63-6359. ISBN 0-8337-2966-7.

A reprint of a bibliography published from 1801 to 1821 in Gottingen by Dieterich, this volume of a 16-volume set represents a list of zoological publications of learned societies. Citations are included up to 1800 and are comprised mostly of eighteenth century entries. The majority of the listings are arranged under species name or animal group by author of the work. The authors are listed first name first and then last name. There does not seem to be an alphabetical arrangement. Citations appear in the language in which they were written. Unfortunately, there is no author or subject index at the end of this volume. There is, however, an author index to volume 1 and volume 2 at the end of volume 2. Volume 2 deals with botany and mineralogy. There is a table of contents in the beginning of volume 1 that lists in order all the animal groups and species names found within the body of the work.

29. Sims, R. W., ed. **Animal Identification: A Reference Guide**. Chichester, N.Y.: John Wiley, 1980. 3 vols. $59.95 vol. LC 80-40006. ISBN 0-471-27765-7.

Unlike Besterman's *Biological Sciences: A Bibliography of Bibliographies* (see entry 23), this work enters citations to bibliographies that relate specifically to taxonomic literature.

Volume 1 is devoted to marine and brackish water animals, volume 2 to land and fresh-water animals, and volume 3 to the insects. The books are basically arranged by phylum and class; the insect volume subdivides by families. Other subdivisions include geographical considerations.

Covering both monographic and journal literature, the author, title, journal title, imprint information, and pagination are considered. In some cases a sentence explains the entry. The majority of the citations cover twentieth-century literature in English, German, Russian, and other languages. The greatest percentage of the works are in English. The 1970s is the last decade included. A phylogenetic index is present at the end of the insect volume and indexes only that volume.

As was mentioned previously, these are sources that provide access to much of the information not attainable through computer searching due to the limited number of years one can search. The arrangement by phylum, class, and family makes this useful work more useful.

30. Taschenberg, O., comp. **Bibliotheca Zoologica II**. Leipzig: W. Engelmann, 1887-1923. 19 vols. in 9.

This is a continuation of *Bibliotheca Zoologica I* (see entry 25) and covers the years 1861-1880. Together with *Bibliotheca Historico-Naturalis* (see entry 26), and *Bibliotheca Zoologica I*, these three works represent much of the writing that was done in the zoological sciences from 1700 to 1880.

INDEXES AND ABSTRACTS

31. **Aquatic Science and Fisheries Abstracts**. Vols. 1-7. London: Information Retrieval, 1971-1977. Monthly. **Aquatic Science and Fisheries Abstracts. Part 1. Biological Sciences & Living Resources**. Vol. 8- . Bethesda, Md.: Cambridge Scientific Abstracts, 1978- . monthly. $785.00. ISSN 0140-5373.

The initial publication of *Aquatic Science and Fisheries Abstracts* (vols. 1-7) represented a merger of *Aquatic Biology Abstracts* (1969-1971) by Information Retrieval Limited and *Aquatic Sciences and Fisheries* (1959-1971) by the Food and Agricultural Organization. Presently this work has been divided into three parts. The title of the three parts are as follows:

Part 1: "Biological Sciences and Living Resources" (1978-).
Part 2: "Ocean Technology, Policy and Non-Living Resources" (1978-).
Part 3: "Aquatic Pollution and Environmental Quality" (1990-).

For the purposes of this book only part 1 will be discussed.

Part 1 is an indexing and abstracting service that basically indexes articles from journals, books, reports, and conference proceedings. It considers those articles that represent all biological facets of organisms found in fresh, salt, and brackish waters. Such animal groups as the invertebrates, fish, birds, and mammals are included. In addition, materials discussing aquatic ecology, population sudies, ecosystems, and the like are also considered. There is an annual index that provides access to the abstracts by author, subject, taxonomy (scientific names), and geography.

This service is also available online. The online vendors are DIALOG (File 44; 1978- .; all three parts included; 375,000 records as of 1992); CISTI; DIMDI; and the European Space Agency. Approximate cost for searching this database is $100.00 per hour.

In addition, Cambridge Scientific, which had initially produced this work on CD-ROM, handed over all rights to the production of this CD-ROM product to SilverPlatter in 1993. Coverage begins with 1978 entries and includes all three parts; part 3 began in 1990. The start-up cost from 1978 to the present is $5,690.00; from 1982 to the present is $4,995.00; from 1988 to the present is $2,495.00; and the 1978 to 1982 backfiles is $695.00. The CD-ROM is updated quarterly. The yearly renewal cost is $1,495.00. SilverPlatter software that is used in this CD-ROM product offers a user friendly interface along with a sophisticated command interface.

32. **Bibliography of Agriculture**. Vols. 1- . Phoenix, Ariz.: Oryx Press, 1942- . Monthly, cumulated yearly. $650.00/yr.; cum. $595.00. ISSN 0006-1530.

This index is compiled by the National Agricultural Library (NAL) and up to 1984 included only the holdings of that library. It now also contains references to materials held outside the NAL. It is actually based on the records found in AGRICOLA, mentioned below. It includes the world's literature, journals, and monographs on agriculture and related studies. For the purposes of this book, it involves entomology, especially insects as pests. More than 2,000 journals are scanned for this publication. It also includes references to pamphlets, government documents, reports, proceedings, books, theses, patents, software, and audio-visuals. Abstracts or annotations are provided for about 400 indexed publications. It is divided into nine sections: a main entry section, five main entry subsections (USDA Publications; State Agricultural Experiment Stations; State Agricultural Extension Service Publications; FAO Publications; and Translated Publications), a corporate author index, a personal author index, and a subject index.

This is also available online. The database from which the *Bibliography of Agriculture* is now drawn is called AGRICOLA (AGRICultural On-Line Access). Vendors are DIALOG (File 110, 1970-1978; and File 10, 1979-present) and BRS (File AGRI). Approximate cost for searching the database is a reasonable $45.00 per hour, plus an additional cost per citation viewed or printed.

This is also available on compact disk from SilverPlatter. It is titled *AGRICOLA on SilverPlatter* and is available in three archival disks and one current disk, which is updated quarterly. The three archival disks are from 1970 to 1984 and cost $950.00 for single-user; $1,425.00 for multi-user. The current disk that includes materials from 1984 is $825.00 per year for single-user; $1,238.00 per year for multi-user. Purchase of current disk and archival disks are available at a discount rate.

33. **Biological Abstracts**. Vols. 1- . Philadelphia: BIOSIS, 1926- . Semi-monthly, cumulated yearly. $4,720.00/yr. (universities $4,225.00/yr.). ISSN 0006-3169.

This is the most comprehensive index to the biological and biomedical literature in the world. It is a two-part index. The four main indexes—author, biosystematic, generic, and subject—refer to an abstract number. The researcher first looks in the indexes under the author or subject term, for example, where an abstract number will be found. The researcher then looks in the abstract section to find the citation and abstract of the article. The biosystematic index is used to find entries by taxonomic category (i.e., phylum, class, order, family). The generic index is used to find entries according to genus or genus-species names. The subject index is by keywords that appear in the title or added by BIOSIS.

The abstract section is arranged in broad subject categories for browsing. The broad subject categories are arranged alphabetically from aerospace to virology. Detailed abstracts are in English.

This service is also available online and on CD-ROM (see *Biological Abstracts/RRM*, entry 34).

34. **Biological Abstracts/RRM** (Reports, Reviews, Meetings). Vols. 18- . Philadelphia: BIOSIS, 1980- . Semi-monthly, cumulated yearly. $2,400.00/yr. (universities $2135.00/yr.). ISSN 0192-6985.

This complements *Biological Abstracts* by covering research reports, reviews, books on biology and biomedicine, and conference literature. No abstracts. Uses the same indexes of author, biosystematic, generic, and subject as BA. Formerly titled *BioResearch Index* until 1980.

This service is also available online. *BIOSIS Previews*. Vendors: DIALOG (File 5, 1969 - ; and File 55, 1985-), BRS (File BIOL, 1978-; BIOB, 1970-1977; BIOZ, 1970-), CISTI, DIMDI, Data-Star, European Space Agency, and STN International (File BIOSIS/RN, 1969 -).

Database includes citations from *Biological Abstracts, Biological Abstracts/RRM*, and *BioResearch Index*, making this one-stop searching a time-saver, although an expensive one at approximately $85.00 per connect hour (DIALOG) with citations printed at approximately 78¢ each.

File size in 1990 was 7,196,209 records. Coverage begins in 1969 with abstracts from 1976 (no abstracts from *Biological Abstracts/RRM* or *BioResearch Index*) and book synopses in *Biological Abstracts/RRM* starting in 1985.

This service is also available on CD-ROM by SilverPlatter. *Biological Abstracts on Compact Disc* and *Biological Abstracts/RRM on Compact Disc*. Coverage begins in 1985. Updated quarterly. BA and BA/RRM issued on separate CD-ROMS, so searching takes more time than online database, but is still faster than paper index. For BA on CD, one year's database takes up two disks. Includes all searching features of online database (e.g., biosystematic and concept codes). Prices differ according to whether or not paper indexes are subscribed to. Minimum is approximately $2,000.00 per year for each.

35. **Biological & Agricultural Index.** Vols. 1- . Bronx, N.Y.: H. W. Wilson, 1964- . Monthly, except August. Quarterly, annual cumulation. $175.00/yr. minimum, based on subscriber's periodical holdings. ISSN 0006-3177.

Covers the core English language journals in biology, agriculture, and related sciences. The selection of journals covered are voted on by subscribers. It also indexes experimental station publications and has a separate section for book review citations. Because this continues *Agricultural Reviews, 1919-1964*, this is a good source for retrospective searches. Easy-to-use index, subject index only, with subject headings in bold print and citations listed under the headings. Suitable for high school and undergraduate use as well as a good beginning for graduate research.

This service is also available online. Vendor: Wilsonline. At $45.00 per connect hour, this is the lower cost alternative biology database. Those who subscribe to the printed versions are given a discount of $10.00 per hour. Although not nearly as comprehensive as *BIOSIS Previews*, nor as retrospective (database only goes back to 1983), it covers the core journals in biology that are readily available in most four-year colleges and universities and larger public libraries, and at about half the cost of *BIOSIS Previews*.

Also available on BRS (File WBAI, 1983- . $60.00 per connect hour plus a charge per citation viewed or printed.

This database is also available on CD-ROM from H. W. Wilson Co. It is very easy to use. There are three modes of searching from straightforward subject searching to combining terms with Boolean operators. The cost is $1,495.00 per year and the disks are updated monthly. Coverage begins in 1983.

Also available on magnetic tape for loading into an online catalog. Pricing is based on several variables; contact publisher for pricing information.

36. **Biology Digest.** Vols. 1- . Medford, N.J.: Plexus, 1974- . Monthly, September to May. $95.00/yr. ISSN 0095-2958.

Aimed for use in high school and junior college libraries, this index covers approximately 200 periodicals and includes lengthy abstracts. Divided into topical subject sections for browsing, including a section for the animal kingdom. Has author and keyword indexes.

37. **CAB Abstracts.** Database. Farnham Royal, Slough, U.K.: CAB International, 1972- . Updated monthly.

This database contains the records of all 26 abstract journals published by the Commonwealth Agricultural Bureaux (CAB). It contains references to agricultural and biological information from over 8,500 journals in 37 different languages. Books are also covered as well as reports, theses, conference proceedings, patents, annual reports, and guides. See the following individual chapters for the zoological areas covered by this database. Vendors: DIALOG (File 53, 1972-1983; and File 50, 1984-); BRS (File CABA); STN International (File CABA).

This database is also available on compact disk from SilverPlatter and is titled CABCD. Coverage is 1984 to the present. It includes English abstracts. There are three archival disks: vol. 1, 1984-1986, $3,600.00 for single-user, $7,200.00 for multi-user; vol. 2, 1987-1989, $7,200.00 for single-user, $14,400.00 multi-user; vol. 3, 1990-1992, $9,675.00 for single-user, $19,350.00 for multi-user. The current disk is updated annually and is $11,400.00 for single-user, $22,800.00 for multi-user.

38. **Chemical Abstracts.** Vols. 1- . Columbus, Ohio: Chemical Abstracts Service, 1907- . Weekly. $13,900.00, including indexes. ISSN 0009-2258.

Although this is the world's major index to chemical literature, it is included here for its comprehensive coverage of the many biochemical aspects of zoology such as DNA and RNA chemistry, mammalian hormones, enzymes, animal nutrition, and toxicology. Weekly indexes include only those for author's names, patent numbers, and keywords. The six-month indexes are more comprehensive and include those for general subjects, chemical substances, formulas, ring systems, authors, and patents. An index guide is indispensable for using the general subject and chemical substance indexes, and should be consulted first. After finding the abstract number from using the indexes, the researcher finds this number in the abstracts section. Abstracts are arranged under 91 different subject areas for browsing.

This service is also available online. Vendors: STN (File CA, 1967-) ; BRS (File CHEM, 1977- , File CHEB 1970-1976). More than 10.5 million records in the complete database (1967-) as of 1993. Updated monthly. BRS database does not contain abstracts or images, such as the chemical structure. Approximate cost to search STN database is $139.00/hr. and BRS database $120.00/hr. plus a separate charge per citation displayed or printed. STN charges subject to 80% discount to college and university accounts, 90% to small colleges.

39. **Current Awareness in Biological Sciences.** Elmsford, N.Y.: Oxford, Pergamon, 1954- . 144/yr. $3,715.00/yr. ISSN 0733-4443.

A current awareness tool that lists periodical articles under broad subject categories such as biochemistry, cell and developmental biology, ecological sciences, physiology. Includes an author index but no subject index. Formerly titled *International Abstracts of Biological Sciences* until 1982, which included signed abstracts. This database is available online. Vendor: BRS (CURB). Coverage is 1983 to present. The approximate cost for searching per hour is $77.00, plus a separate charge per citation displayed or printed.

40. **Current Contents: Agriculture, Biology & Environmental Sciences**. Vols. 1- . Philadelphia: Institute of Scientific Information, 1970- . Weekly. $360.00/yr. ISSN 0090-0508.

Intended as a current awareness tool, this includes the table of contents pages from approximately 1,200 journals and book series under sections such as biology, aquatic sciences, environment ecology, and animal sciences. Journal issues are indexed within two weeks of their receipt at ISI. Has author and title indexes with an author address directory to request reprints. Current book contents is another section that lists the table of contents of selected new books.

This service is also available online. Vendor: BRS (File AGRI). This database covers the print edition and is a 12-month file updated weekly. New features include abstracts and enhanced keywords. Approximate cost for searching this database is $95.00 per hour, plus a separate charge per citation displayed or printed.

Also available in magnetic tape for loading into online catalogs. Costs for magnetic tape subject to many variables; contact the Institute for Scientific Information for pricing.

Also available on CD-ROM. A one-year subscription of $1,995.00 provides 51 weekly issues, plus a year-end archival disk. Abstracts are included.

41. **Current Contents: Life Sciences**. Philadelphia: Institute for Scientific Information, 1958- . Weekly. $360.00/yr. ISSN 0011-3409.

Same format as above, but has sections that show the multidisciplinary nature of life-sciences research (e.g., neurosciences and behavior, animal and plant science, chemistry). Has same indexes and sections as *Current Contents: Agriculture, Biology & Environmental Sciences* (see entry 40).

Also available online. Vendor: BRS (LIFE). Covers the print edition and is a 12-month file, updated weekly. New features include abstracts and enhanced keywords. Approximate cost for searching this database is $95.00 per hour, plus a separate charge per citation displayed or printed.

Also available in magnetic tape for loading into online catalogs. Costs for purchase of magnetic tape varies; contact Institute for Scientific Information for pricing.

Also available on CD-ROM. A one-year subscription of $2,495.00 provides 51 weekly issues, plus a year-end archival disk. Abstracts are included.

42. **Fish and Wildlife Reference Service** (FWRS). 5430 Grosvenor Lane, Suite 110, Bethesda, Md. 20814 Phone: 1-800-582-3421.

Operated by MAXIMA Corporation under contract with the U.S. Fish and Wildlife Service. FWRS maintains a full-text database on fish and wildlife management. It serves state or federal agencies (no charge) and others outside these groups at minimal charge (literature search $30.00 each, 10¢/page and photocopy of technical reports (10¢/page). It will send a free quarterly newsletter that highlights the publications indexed in their database. Toll free number allows requests for database searches.

43. **Government Reports Announcement and Index.** Vols. 75- . Springfield, Va.: National Technical Information Services, U.S. Dept. of Commerce, 1975- . Biweekly. $495.00/yr. ISSN 0097-9007.

Another index not primarily for zoology but includes indexing of technical reports from NTIS on life sciences, subjects such as ecology, fisheries and aquaculture, parasitology, natural resource management, limnology, and cytology. The citations with abstracts are arranged in 38 broad subject categories with keyword, personal and corporate author, and the NTIS order/report number indexes. GRAI indexes U.S. government sponsored research as well as foreign technical reports and others prepared by the government agencies or from grants or contracts. NTIS is the source for sale of these reports from many government agencies and some state and local government agencies as well. These reports can be purchased individually from NTIS or through subscription.

This database is also available online: Vendors: DIALOG (File 6, 1964-); ORBIT (NTIS, 1964-); STN (NTIS, 1964-); BRS (NTIS, 1964-). Coverage is from 1964 with more than 1.5 million records in the database in 1993. Updated biweekly. Approximate cost for searching this database is $85.00 per hour, plus a separate charge per citation displayed or printed.

Also available on CD-ROM from SilverPlatter. Covers 1983 to present on two disks. Updated quarterly. Minimum subscription approximately $2,700.00 per year.

44. **Index to Scientific and Technical Proceedings.** Vols. 1- . Philadelphia: Institute for Scientific Information, 1978-. Monthly, with one annual cumulation a year. $1225.00/yr. ISSN 0149-8088.

Of the estimated 10,000 scientific meetings each year, three quarters of them publish the results of those meetings. ISTP indexes about half of the published proceedings and papers from these conferences, conventions, workshops, seminars, symposia, and colloquia from throughout the world and in a broad range of scientific disciplines. These appear in published form as books, reports, or in the periodical literature. Its coverage in zoology is seen in the category index section in such areas as biology, ecology, entomology, environmental sciences, fisheries, limnology and water resources, ornithology, parasitology, and zoology. Besides the category index, there are indexes by permuterm subject, sponsor, author/editor, meeting location, and corporate index with geographic and organization sections. *Biological Abstracts/RRM* (see entry 34), is another good source for conference literature.

45. **Index to Scientific Reviews**. Vols. 1- . Philadelphia: Institute for Scientific Reviews, 1974- . Semiannual. $780.00/yr. to main libraries, $250.00 for departmental libraries. ISSN 0360-0661.

This is an index to the scientific review literature. Review articles are helpful in familiarizing a researcher on a particular topic and also for additional sources of articles on the topic. ISR is a subset of *Science Citation Index* (see entry 51), selected from the magnetic tapes of SCI by the following criteria: 1) articles with keywords such as "advances," "progresses," or "review"; 2) articles with 50 or more references; and 3) articles coded "R" (for review or bibliography). The Research Front Specialty Index was created by computer selection of ISR's database for articles on a particular topic and allows the researcher to find other similar articles in a particular topic. The other indexes are similar to those in *Science Citation Index* (i.e., Corporate Index, Source Index, and Permuterm Subject Index). *Biological Abstracts* (see entry 33) is also a source of review literature.

46. **Key-Word-Index of Wildlife Research**. Vols. 1- . Zurich, Switzerland: Swiss Wildlife Information Service, 1974- . Annual. SM 70.00. ISSN 0250-3859.

Indexes approximately 600 international periodicals as well as some reports and conference proceedings on wildlife and conservation. The keyword index is based on the thesaurus included after the author index. Also listed in the keyword index are the common names of animals, which can be found in the Systematic Species List that lists the English, German, and scientific names of organisms in three separate columns. Although not comprehensive, it is a handy source specifically for wildlife literature that complements *Wildlife Review* (see entry 54).

47. **Life Sciences Collection**. Database. Bethesda, Md.: Cambridge Scientific Abstracts, 1978- . Updated monthly.

This is an online index that corresponds to 17 abstracting journals published by Cambridge Scientific such as *Marine Bio-technology Abstracts, Oncogenes and Growth Factors,* and *Animal Behavior*. In 1990 the file size was 1,254,581 records. It contains abstracts of literature in the multidisciplinary life sciences such as animal behavior, ecology, and endocrinology. Database vendor is DIALOG (File 76). Coverage begins in 1978 and is updated monthly. Approximate cost to search the database is $90.00 per hour, plus a separate charge per citation displayed or printed.

This is also available on CD-ROM from SilverPlatter. The disks for 1982 to present are $2,995.00 per year; 1987 to present are $2,495.00 per year; and 1991 to present are $1,795.00 per year. The database is updated quarterly.

This database is also available on magnetic tape for loading into an online catalog. See vendor for pricing information.

48. **Monthly Catalog of United States Government Publications**. Vols. 1- . Washington, D.C.: Government Printing Office, 1895- . Monthly. $166.00/yr. ISSN 0362-6830.

This is the most extensively used index to U.S. government publications. Publications by the Dept. of Interior, Dept. of Agriculture, the Smithsonian, Environmental Protection Agency, and others that deal with zoological subjects are all indexed here. The publications include reports, fact sheets, maps, pamphlets, and conference proceedings as well as Senate and House hearings. It is arranged by government department and includes indexes by subject, author, title, and series/report.

Also available online. Vendor: DIALOG (File 66, 1976-). Contained more than 300,000 records in 1991. Updated monthly. Approximate cost to search this database is $48.00 per hour, plus a separate charge per citation displayed or printed.

This database is also available on magnetic tape for loading into online catalogs. Pricing varies.

49. **PASCAL Folio. F 53: Anatomie et Physiologie des Vertébrés**. Vols. 1- . Vandoeuvre-Les-Nancy Cedes, France, Centre National de la Recherche Scientifique, Institut de l'Information Scientifique et Technique, 1984- . 10/yr. Fr. 1,870/yr. ISSN 0761-1900.

Title has changed several times. Formerly, *PASCAL Folio. Part 64: Anatomie et Physiologie des Vertébrés*; supersedes in part (1961-1984); *Bulletin Signaletique. Part 365: Zoologie des Vertébrés. Ecologie Animale. Physiologie Appliqué Humaine*; formerly, *Bulletin Signaletique. Part 365; Zoologie des Vertébrés*; supersedes in part *Bulletin Signaletique. Part 360. Biologie et Physiologie Animales*. This indexes international literature in the biological sciences covering journals, masters and doctoral theses, conference proceedings, reports, and books. One of the many parts of the PASCAL index, similar to the many-parted *Referativnyi zhurnal* (see entry 50).

This database is available online. Vendor: DIALOG (File 144, 1973-). It covers all 79 PASCAL journals. The file is in French and English and about 50% of records have abstracts. International coverage. Updated monthly. Approximate cost to search this database is $60.00 per contact hour, plus a separate charge per citation displayed or printed.

50. **Referativnyi zhurnal:Biologiya**. Moscow, Akademiia Nauk SSSR: Institut Nauchnoi Informatsii. 1958- . Monthly. 619.50 Rub. (861.00 Rub. including index). ISSN 0034-2300.

One of many subject series of the *Referativnyi zhurnal* that covers most science and technology subjects, this biology section covers the world's literature in this area. Includes abstracts.

51. **Science Citation Index**. Vols. 1- . Philadelphia: Institute for Scientific Information, 1961- . Bimonthly, annual cumulations, and 5- and 10-year cumulations. $9,500.00/yr. ISSN 0036-827X.

The uniqueness of this index is in its Citation Index that tracks who is citing who in the literature of science, medicine, technology, behavioral sciences, and agriculture. This is especially useful when one has a pertinent paper on a particular subject and wants to find other papers similar to it. By finding this paper in the citation index, one can find a list of works that cited that paper and therefore may be on similar research. Searching can also be done in the Permuterm Subject Index, where keywords from the title are placed in an alphabetical arrangement with every other significant word in the same title to form a type of subject access to articles. The Corporate Index is used to find out what research has been done in a particular institution for that year. Each index refers to a particular author, and the full bibliographic citation is found in the author or Source Index. A very useful index, but limited to the 3,000+ journals it covers. Coverage in zoology is good but not comprehensive. Ten-year cumulative indexes and five-year cumulative indexes through 1984 are available.

Also available online: Vendors: DIMDI, DataStar, DIALOG (Files 34, 432, 433, 434 SciSearch). This database covers 1974 to present and in 1990 included more than 10 million records, updated weekly. Includes records from various *Current Contents* (see entries 40 and 41) to make this database more comprehensive than the print index. Approximate cost to search this database is $162.00 per connect hour, plus an additional charge per citation displayed or printed. Subscribers to the print index receive a reduced rate of approximately $66.00 per connect hour in DIALOG.

52. **Wildlife Abstracts**. Washington, D.C.: Dept. of the Interior, U.S. Fish and Wildlife Service; printed by United States Government Printing Office, 1957-1983.

This indexed *Wildlife Review* (see entry 54), nos. from 1 to 179, 1935-1980. It was compiled by various biologists with the U.S. Fish and Wildlife Service and published in six cumulations: 1935-1951, 1952-1955, 1956-1960, 1961-1970, 1971-1975, and 1976-1980. Combines the indexes in *Wildlife Review* and includes a systematic index. Continued by annual indexes of *Wildlife Review*, 1981-1984. After 1984 no cumulative indexes were published.

53. **Wildlife Information Service**. 9956 North Hwy. 85, Las Cruses, N. Mex. 88005 Phone: (505) 527-2547.

This biological and ecological information service produces the H.E.R.M.A.N. database, which covers research literature on wildlife, including marine birds and mammals. The database is gleaned from *Biological Abstracts,* AGRICOLA, *Zoological Record, Current Contents*, and others (see entries 33, 32, 55, 40, and 41). In 1991 the database held approximately 120,000 citations.

This is also available on CD-ROM from National Information Services Corporation, titled *Wildlife Worldwide*. This database is combined with *Wildlife Review* and BIODOC from the Biological Documentation Center of the National University of Costa Rica to provide a broad coverage of wildlife literature. The database is on a single disk. Approximate price is $700.00 per year. Coverage is 1971 to present for full coverage, and 1935-1970 partial coverage. It contains approximately 320,000 citations.

54. **Wildlife Review**. Vols. 1- . Ft. Collins, Colo.: U.S. Fish and Wildlife Service, 1935- . 6/yr. Free to some government agencies. ISSN 0043-5511.

Selectively indexes worldwide wildlife and natural resource literature from over 1300 journals as well as more than 500 books and symposia proceedings. Separate indexes for author, geography, and subject; and systematic indexes for mammals, birds, and reptiles/amphibians. Indexed from 1935 to 1980 by *Wildlife Abstracts* (see entry 52), then annual indexes until 1984. However, no annual index since 1984, so it is necessary to check index in each quarterly issue back to that time. Excellent coverage of wildlife and conservation literature, though some references, especially state wildlife department reports, may be difficult to obtain.

This database is available on two CD-ROMs from National Information Services Corporation, Suite 6, Wyman Towers, 3100 St. Paul St., Baltimore, Md. 21218. These CD-ROMs use RomWright software. One CD-ROM, titled *Wildlife Review/Fisheries Review*, is combined with the *Fisheries Review* (see entry 722) database. This single disk covers the years 1971 to the present for both databases with quarterly updates. The price is approximately $700.00 per year. The other CD-ROM is titled *Wildlife Worldwide* and not only includes the entire *Wildlife Review* database but also the HERMAN database (see entry 53), and the BIODOC database of the National University of Costa Rica. This CD-ROM goes back to 1930 and is more comprehensive for wildlife studies than the *Wildlife Review/Fisheries Review* CD-ROM. The approximate cost of the CD-ROM is $700.00 per year. Both disks are updated semiannually.

55. **Zoological Record.** Vols. 1- , nos. 1- . Philadelphia: BIOSIS, Zoological Society of London, 1865- . Annual. $2,200.00/yr. ISSN 0084-5604.

Indexing about 5,500 scientific journals, serials, books, dissertations, and conference proceedings annually, *Zoological Record* is one of the most comprehensive indexes of systematic zoology and has been for more than 100 years. At the end of each year, a 20-section set (in 27 parts) of *Zoological Record* is published. Each section records all the literature published in that year pertaining to a certain class or phylum in the animal kingdom. Section 20 lists new generic and subgeneric names. Each of these sections contains an author, subject, systematic, geographic, and paleontological index. Full bibliographical information is given in the author index. No overall index to all 20 sections.

This database is also available online, titled *Zoological Record Online*. Vendor: DIALOG (File no. 185). Coverage begins in 1978, updated monthly. Database of approximately 850,000 records in 1991, includes all 27 sections of the paper version. Each record includes basic bibliographic information and systematic classification of up to six levels for the organism addressed. Approximate cost for searching the database is $87.00 per hour.

This database is also available in CD-ROM from SilverPlatter. Retrospective disks cover 1978 to 1992. The current disk is updated semiannually. This allows searching through all 27 sections of the print index with one search. It also includes its own searchable thesaurus, which includes changes in vocabulary that may have occurred over the years. Prices for disks are dependent on whether or not the purchaser subscribes to the print index. For volume 129 (1993) and continuing, the price is $2,900.00 per year, but only $800.00 per year if purchaser subscribes to the print index. For volumes 115-128 (1978-1992) the price is $7,200.00, but $4,800.00 if purchaser subscribes to 9 or more volumes of the print indexes.

HISTORICAL PERIODICAL INDEXES

56. **Bibliographia Zoologica**. Zurich: Sumptibus Concilii Bibliographici, 1896-1934. 43 vols.

Covers the zoological periodical literature from 1895-1934. Before this title, *Bibliotheca Zoologica I* and *Bibliotheca Zoologica II* (see entries 25 and 30), covered the zoological literature between 1846 and 1860 and Wilhelm Englemann's *Bibliotheca Historico-Naturalis* (see entry 26), covered the literature from 1700 to 1846.

57. **International Catalogue of Scientific Literature. N: Zoology**. London: published for the International Council by the Royal Society of London, 1901-1914. 14 vols.

This is part of a 17-section work that was one of the most important bibliographies of its time covering all sciences. It indexes books and periodical articles from the major scientific periodicals. It continues the coverage of the *Catalogue of Scientific Papers* (see entry 58), also published by the Royal Society of London.

The work is divided systematically into 20 sections similar to those of *Zoological Record* (see entry 55); indeed, volumes 6-14 of this index are identical to volumes 43-51 of *Zoological Record*, which continues the coverage. This arrangement was made at an international convention that met in 1905 where the Zoological Society of London agreed to an amalgamation of the two publications for five years. This continued until 1914 when the Zoological Society of London was responsible for the entire volume, because, due to the international situation, the Royal Society of London was unable to financially support the project.

58. Royal Society of London. **Catalogue of Scientific Papers**, 1800-1900. London: H. M. Stationery Office, 1867-1872; London: Clay, 1877-1902; Cambridge University Press, 1914-1925. 19 vols. Reprint. New York: Johnson, 1968.

This is the main index to nineteenth-century scientific papers. The arrangement is alphabetical by author. However, only three subject indexes were published (Pure Mathematics, Mechanics, and Physics), so zoological subject access to this set is not available.

Researchers can, however, find articles by the known zoological scientists of this period. Coverage is continued by the *International Catalogue of Scientific Literature* (see entry 57).

59. **Zentralblatt fur Zoologie, allgemeine und experimentelle Biologie**. Bd. 1-6. Leipzig: Teubner, 1912-1918. 6 vols.

Although short-lived, this index covered the zoological periodical literature during its time. Includes abstracts for more important papers.

60. **Zoologischer Bericht; im Auftrage der Deutschen Zoologischen Gesellschaft**. Bd. 1-55. Jena: G. Fischer, 1922/1943-1944. 55 vols.

Indexes both books and periodical articles and includes abstracts.

61. **Zoologischer Jahresbericht**. Berlin: R. Friedlander, 1879-1913. 35 vols.

Indexed the entire field of zoology through 1884. Did not cover systematic zoology after this time because *Zoological Record* (see entry 55), covered this area.

DIRECTORIES

62. Cichonski, Thomas J., ed. **Research Centers Directory**. 18th ed. Detroit: Gale Research, 1994. 2 vols. $435.00. LC 60-14807. ISBN 0-8103-8159-1 (set).

More than 13,000 university-related and other nonprofit research organizations in all disciplines can be found throughout this two-volume set. For each organization entered, the address, phone number, director, general and historical information, research activities, and publications and services are given. Volume 2 contains a subject index, geographic index, and master index (names of the centers) to volumes 1 and 2. In addition to the many research centers listed in biology, there are numerous centers relating specifically to zoology. Some of the subject headings include animal behavior, ornithology, ichthyology, entomology, evolution, and the like.

Many of the research centers listed in this work will also be found in the *Life Sciences Organizations and Agencies Directory* (see entry 63).

63. Darnay, Brigitte T., and Margaret Labash Young, eds. **Life Sciences Organizations and Agencies Directory**. Detroit: Gale, 1988. 864p. $155.00. LC 87-29090. ISBN 0-8103-1826-1.

This reference work lists approximately 7,600 organizations and agencies related to the life sciences. The names, addresses, phone numbers, directors, and pertinent information of these bodies fall under broad categories. Among these categories are U.S. and international associations; botanic gardens; national and international computer information services; consulting firms; educational institutions; libraries and information centers; U.S. and worldwide research centers; U.S. and international organizations involved in creating standards; state and federal government agencies; federal government research centers; federal grants and domestic assistance programs; and U.S. foreign agricultural officers.

Information about these organizations may include history, list of publications, library holdings, research activities, and subjects that are endemic to that body. A master name and keyword index completes the work.

This is an extremely useful work and it is recommended; however, some of the material found in this directory can be found in many of the other directories published by Gale.

64. **Directory of Research Grants**. Phoenix, Ariz.: Oryx Press, 1993. 1216p. $125.00pa. LC 76-47074. ISBN 0-89774-699-6.

This work represents a listing of research grants and programs in all major areas of knowledge. For each listing, the name of the body granting the funding, a description of what type of research is supported by the grant, the amount of the grant, and the sponsor are given. U.S. government programs are included. The end of the work features a subject index, a sponsoring organization index, and a sponsoring organization (by type) index. Some of the biological headings appearing in the subject index are biology, zoology, zoos, and so on. For additional up-to-date information on grants, one should search the Grants database found on DIALOG.

65. Montney, Charles B., ed. **Directories in Print**. 11th ed. Detroit: Gale Research, 1994. 2 vols. $285.00. LC 90-32021. ISBN 0-8103-8197-4.

Over 14,000 listings of directories in all areas of knowledge can be found among this two-volume set. For each directory entered, the address of the publisher, a descriptive annotation of the directory, arrangement, pagination, frequency, editor, price, and fax number are given. Volume 2 contains an alternative formats index, a subject index, and a title and keyword index. Subject headings found in the subject index and pertaining to the biological sciences are biology, biotechnology, birds, biological research, zoological societies, zoology, zoological parks and aquariums, and so on.

HANDBOOKS

66. Altman, Philip L., and Dorothy S. Dittmer, eds. **Biology Data Book**. 2d ed. Bethesda, Md.: Federation of American Societies for Experimental Biology, 1972. 3 vols. $250.00. LC 72-87738. ISBN 0-08-030071-5.

Although this tabular, three-volume work is devoted to all areas of biology, much of the material contained therein is zoological in nature. Life-spans, body weights, reproductive information, and development and growth of animals are some of the many examples contained within these books. References, providing citations to the original papers, are given for each piece of tabular information presented.

Each volume has its own theme and its own index. There is no subject index to the complete set. Volume 1 deals with all phases of reproduction, from genetics to development and growth; volume 2 considers biological regulators, the environment, parasitism, and neurobiology; while volume 3 depicts physiological processes having to do with digestion, metabolism, circulation, respiration, and the blood and body fluids.

This set is an absolute must for colleges and universities. It abounds with hard-to-get-at information. Its only draw back is that it has not been updated since 1972.

67. Halstead, Bruce W. **Dangerous Aquatic Animals of the World: A Color Atlas: With Prevention, First Aid, and Emergency Treatment Procedures**. Princeton, N.J.: Darwin Press, 1992. 264p. $60.00. LC 91-13746. ISBN 0-87850-045-6.

Consisting of invertebrate and vertebrate animals, this source provides factual information on these creatures that can be found in aquatic environments. The work, by means of chapter headings, divides the organisms into invertebrates and vertebrates. Futhermore, the sections are then refined by considering those groups of animals that sting versus those that are poisonous. In addition, sections dealing with wound-producing aquatic animals, electric aquatic animals, and human-parasitic catfish are also considered.

The particular animal groups listed are discussed, in many cases, in some depth. Such areas as habitat, size, external anatomy, behavioral patterns, and coloration are provided. Information on the species includes common and scientific name, size, coloration, habitat, geographical distribution, and mode of attack. An emphasis, as one would imagine, is placed on information dealing with the mechanics of how the animal injures as well as its stinging apparatus and/or venom produced.

Cross references to the plates accompany the species accounts. Over 470 colored photographs, a number of black-and-white photographs and labeled line-drawings provide the illustrative material. Pictures showing the animals' effects on humans are also provided. Included are a list of geographical range maps; a section on prevention, first aid, and emergency treatment procedures; and a glossary; a one-page bibliography concludes the handbook. Although the volume is lacking an index, the table of contents aids in finding major common-name animal groups.

68. **Handbuch der Zoologie: Eine Naturgeschichte der Stämme des Tierreiches**. 2d ed. Gegrundet von Willy Kukenthal. Hrsg. von J. G. Helmcke, and others. Aufl. Berlin: W. de Gruyter, 1968- . Price vol. varies.

The second edition of this work begins with volume 4 (Arthropoda: Insecta) to the first edition, which began in 1923. Arranged systematically from protozoa to mammalia, this is a comprehensive work comparable to *Traité de Zoologie* (see entry 69), and *The Cambridge Natural History* (see entry 16). A great deal of information on each of the animal groups is contained within this set. A large number of detailed black-and-white drawings and photos provide the illustrative material. Long bibliographies and indexes by scientific name and subject are at the end of the text. Most of the volumes are in German; a few of the newer ones are in English. A description of the major parts of this set can be found in the relevant chapters.

69. **Traité de Zoologie: Anatomie, Systématique, Biologie**. Pierre-Paul Grassé, ed. Paris: Masson, 1948- . 17 vols. (various fascicules). Many fascicules still in print. Price fascicule varies. LC 49-2833.

This classic, comprehensive and authoritative multivolume treatise is arranged by systematic groups. Vol. 1—Phylogenie: Protozoaires; vols. 8-10—Insects; vol. 11—Echinodermes, Stomocordes, Procordes; vol. 12—Vertebres; vol. 13—Agnathes et Poissons; vol. 14—Reptiles; vol. 15—Oiseaux; vols. 16-17—Mammiferes. This work is written entirely in French and includes detailed information regarding living and fossil species.

Fascicules are published irregularly to update and expand the set. Each volume (fascicules) is illustrated by black-and-white photos and detailed drawings with some color plates. Each section includes a limited bibliography. A subject and organism index completes each volume.

Comparable sets are the German work *Handbuch der Zoologie* (see entry 68), and the older, but standard British work *The Cambridge Natural History* (see entry 16).

A description of each main part of this set can be found in the relevant chapters.

69a. Shuker, Karl. **The Lost Ark: New and Rediscovered Animals of the Twentieth Century**. London, HarperCollins, 1993. 287p. $30.00. LC gb93-59214. ISBN 0-00-219943-2.

The coelacanth, known to be extinct for millions of years, was caught December 22, 1938 off the coast of South Africa; another was captured in December 1952 northwest of Madagascar. A new genus was created when a new type of pig was discovered that exceeded 7 ft. in length. It was first formally detected in 1904 and became known as the giant forest hog.

These are just a few of a myriad of facts that can be found within this interesting and informative source. Arranged phylogenetically (Mammalia through Invertebrates), this work considers mammals, birds, reptiles, amphibians, fishes, and invertebrates that had escaped scientific discovery until the twentieth century. Each of the animal entries provides a brief history as to how they were detected. Black-and-white as well as colored plates, depicting the animals, can be found throughout the book. A 17-page bibliography and a combined common, scientific-name index concludes the volume.

FIELD GUIDES

Listed below, in addition to the main entries, are several series that comprise field guides. Many of the field guides contained within these series can be found under their respective chapters throughout this book. The main entries listed in this section are guides dealing with all types of animals collectively.

Audubon Society Field Guide Series. New York: Knopf.

Golden Field Guide Series. New York: Golden Press.

Macmillan Field Guides. New York: Macmillan.

Quick Reference Field Guide Series. New York: Macmillan.

Peterson Field Guide Series. Boston: Houghton Mifflin.

70. Murie, Olaus J. **A Field Guide to Animal Tracks**. 2d ed. Boston: Houghton Mifflin, 1975. 375p. (The Peterson Field Guide Series). $17.95; $13.95pa. LC 74-6294. ISBN 0-395-19978-6; ISBN 0-395-18323-5pa.

Although most of this work encompasses mammalian tracks from North America, Mexico, and Central America, there are brief sections on the common birds, amphibians, reptiles, and a number of insects. For each organism mentioned there are line-drawings of the tracks and a drawing of the organism, as well as a morphologic description of the tracks. The animals are arranged by their respective families. In addition, there is a very brief section of gnawed twigs and limbs, and one on bone and horn chewing. A concise bibliography and an index, consisting of both common and scientific names, complete the work.

This is considered to be one of the more comprehensive works identifying animal tracks. Because of its size, it can easily be carried in the field.

71. Palmer, E. Laurence, revised by H. Seymour Fowler. **Fieldbook of Natural History**. 2d ed. New York: McGraw-Hill, 1975. 779p. $42.95. LC 73-18290. ISBN 0-07048425-2.

Although this work does not lend itself to be taken readily into the field (hard cover, and 24 cm in height), it nevertheless provides the user with the information necessary to identify common plants and animals. Approximately 50% of the book is devoted to the animal kingdom.

The organisms are arranged by phylum, and then class, order, family, and species of the specimen in question. The common name is listed in bold print, with the scientific name underneath it. There are excellent line-drawings that will aid in identifying the fauna in question. In addition, for each species considered, there is a description that discusses its morphology, behavioral patterns, habitat, diet, and so on. In order to use this work for identification, it will be necessary to have some knowledge of the type of animal being identified; at the very least, its phylum and class.

In addition to the plants and animals listed, the volume contains star charts of the constellations, and a small section devoted to the identification of minerals. An index to both common and scientific names of plants, animals, and minerals can be found completing the guide.This is an excellent work and contains a great percentage of the types of plants, animals, and minerals inhabiting the planet earth.

ILLUSTRATIONS

This section includes works about zoological illustrations as well as indexes that enable one to find a picture or drawing of a particular animal.

72. Clewis, Beth, comp. **Index to Illustrations of Animals and Plants**. New York: Neal-Schuman, 1991. 225p. $49.95 LC 90-23816. ISBN 1-55570-072-1.

This is intended as an update and supplement to *Index to Illustrations of the Natural World* (see entry 75), and *Index to Illustrations of Living Things Outside North America* (see entry 73).

This index covers books published in the 1980s and includes plants and animals from around the world. Illustrations to 6,200 species of animals and plants located in 142 books are included here. Anyone wishing to find a picture or drawing of a certain animal can use either the common name index or the scientific-name index. Under the name, one or more sources are listed by a three-letter code and page number. Also included are the abbreviations for color plate, black-and-white plate, or black-and-white illustration to aid the researcher in finding specifically that kind of illustration. The key to the abbreviations and codes are found at the beginning of the book.

73. Munz, Lucile Thompson, and Nedra G. Slauson, comps. **Index to Illustrations of Living Things Outside North America: Where to Find Pictures of Flora and Fauna**. Hamden, Conn.: Archon Book, 1981. 441p. $55.00 LC 81-8037. ISBN 0-208-01857-3.

This index complements that of the *Index to Illustrations of the Natural World* (see entry 75), in that it includes plants and animals found outside North America. It indexes illustrations to more than 9,000 species of plants and animals found in 206 books. Entries are listed by common name in boldface followed by the scientific name in parenthesis. A separate scientific-name index is found at the back of the book.

74. Nissen, Claus. **Die Zoologische Buchillustration. Ihre Bibliographie und Geschichte.** Stuttgart: Hiersemann, 1969-1978. 2 vols.

Arranged by author and illustrator, this bibliography (v. 1) attempts to list all pertinent book and journal citations that deal with illustrations. Works listed either contain animal illustrations or instructions on how to draw animals. Citation titles are listed in their respective languages. Much of the material is drawn from the eighteenth and nineteenth centuries. Some citations from the early twentieth century can also be found. There are a number of indexes at the end of volume one that aid in accessing the material. Among these indexes are geographical, subject, animal groups, and author. A total of 4,826 citations are included.

Volume two, written in narrative form, discusses the history of animal book illustrations. Black-and-white plates and drawings enhance the text. As one may expect, however, from the title of this work, the text is in German. Indexes include geographic, subject, animal groups, and author. The last section represents a list of illustrators that are responsible for the illustrations in the cited works found in both volume one and two. Entries in bold type indicate that the illustrator is also the author.

75. Thompson, John W., comp. **Index to Illustrations of the Natural World: Where to Find Pictures of the Living Things of North America**. Ed. Nedra Slauson. Syracuse, N.Y.: Gaylord Professional, 1977. 265p. Reprint. Hamden, Conn.: Shoe String Press, 1981. $49.50. LC-77-4143. ISBN 0-915794-12-8.

This is the best of the indexes to illustrations, followed closely by Lucille Thompson Munz's work (see entry 73). This index covers approximately 6,200 entries found in 178 books and covers those species found primarily in North America. This unique index is the result of a lifetime of keeping an index card file by Mr. Thompson, a self-educated man who went on to become a noted botanist and owner of a company that provides color slides of animal and plant species to science teachers around the world.

CHECKLISTS AND CLASSIFICATION SCHEMES

75a. International Commission on Zoological Nomenclature. **International Code of Zoological Nomenclature [Code international de nomenclature zoologique]**. 3d ed. Adopted by the XX General Assembly of the International Union of Biological Sciences. London, International Trust for Zoological Nomenclature, in association with British Museum (Natural History); Berkeley: University of California Press, 1985. 338p. LC 84-40785. ISBN 0-520-05546-X.

In order to stabilize zoological nomenclature, or the scientific naming of animals, an international commission was established and in 1961 the first edition of the code was published. The aim of this third edition "is to provide the maximum universality and continuity in the scientific names of animals compatible with the freedom of scientists to classify animals according to taxonomic judgments" (p. xiii).

It is arranged under 88 articles, with French and English text on opposite pages. Several appendices include a code of ethics, transliteration and latinization of greek words, latinization of geographical and proper names, recommendations on the formation of names, and general recommendations. There is an index to scientific names as well as a subject index in English and French. A necessary source for all academic, large public libraries and specialized zoological libraries.

76. Margulis, Lynn, and Karlene V. Schwartz. **Five Kingdoms: An Illustrated Guide to the Phyla of Life on Earth**. 2d ed. New York: W. H. Freeman, 1988. 376p. $37.50. LC 87-210. ISBN 0-7167-1885-5.

Bacterial, plant, and animal phyla are considered in a succinct and informative manner. Eighty-eight pages are devoted to all the phyla that represent the animal kingdom. Each phyla is discussed in an encyclopedic manner. The morphology, size, classes, movement, number of species, reproductive methods, and other interesting facts are included. In addition, excellent, labeled line-drawings, representing internal and/or external anatomy, enhance the text. In most cases, photographs depicting a representative animal of the phyla can also be found. A brief bibliography for each of the phyla is listed at the end of each kingdom section.

Near the end of the book is a list of genera, alphabetically arranged, giving phylum and common name; a glossary of major terminology; and an alphabetically arranged subject and taxonomic index.

Although this work is not as comprehensive as *Synopsis and Classification of Living Organisms* (see entry 77), it serves as an excellent introduction to the various groups of organisms that make up the microbial, plant, and animal kingdoms. This work is not only useful as a reference but also as a teaching tool for high school and college freshman biology.

76a. Minelli, Alessandro. **Biological Systematics: The State of the Art**. London: Chapman & Hall, 1993. 387p. $69.00. LC 93-3578. ISBN 0-412-36440-9.

True to its title, this book presents a review of the state of the art in biological systematics. This is an especially interesting work in light of the new project to create a Biological Diversity Inventory, an inventory that would list all presently known plants and animals in existence today. This work discusses the present thoughts and practices in systematics for the major groups of plants and animals. Part one discusses the problems and methods used in the present classification systems and comparative biology. Part two covers the systematics and classification systems of 26 of the major groups, both animal and plant. Part three discusses dangerous trends in classification and what the future may hold. The appendix includes excerpts from 22 of the major recent systematic works for groups such as birds, fish, chordates, mammals, crustacea, and so on. A lengthy bibliography can be used to find papers and books on classification systems discussed in the text. An author, subject and taxonomic index (including scientific and common names) complete this work.

77. Parker, Sybil P., ed. **Synopsis and Classification of Living Organisms**. New York: McGraw-Hill, 1982. 2 vols. $280.00. LC 81-13653. ISBN 0-07-079031-0.

In addition to a general discussion of each phyla, this two-volume set considers morphologic and habitat information on each of the orders, and their respective families. References are listed throughout the taxonomic descriptions. In many cases there is usually one or more references cited for each of the families described. Line-drawings and photographs are interspersed among the textural material, enhancing the morphological descriptions. The length of narrative describing each family varies, but, in most cases, one-third of a column is the norm. The animal kingdom encompasses one and one half of the two volumes. The work concludes with a brief history of nomenclature in the taxonomy and classification of organisms; an essay on biological classification; a taxonomic table listing superkingdom, kingdom, division, class, order, and family for all living species; and an index that includes, alphabetically, both common and taxonomic names.

This reference represents an extremely comprehensive look at all the major taxonomic groups of viruses, bacteria, plants, and animals and should be on the shelves of college and university libraries offering programs in the life sciences.

78. Sherborn, Charles Davies. **Index Animalium: Sive, Index Nominum Quae ab A.D. MDCCLVIII Generibus et Speciebus Animalium Imposita Sunt, Societatibus Eruditorum Adiuvantibus**. Bath, England: Chivers, 1969. 1195p.

First published by Typographio Academico (1902-1933), this source provides a complete list of those animals that have been named by zoologists from 1758 to 1800. The animal list is a combined alphabetical arrangement by genus and species. For each of the genera listed, the person who first named it, date of the naming, and the pertinent citation to the taxonomic literature is given. For each of the species listed, the person who first named it, date of the naming, and pertinent citation to the taxonomic literature is given. In addition, the genus is also provided.

A 44-page bibliography, listing those works referred to in the compilation of this checklist, can be found at the beginning of the work. An index to generic names showing the species names associated with each from 1758 to 1800 completes the checklist.

This volume may very well represent one of the first checklists ever published.

79. Sims, R. W., ed. **Animal Identification: A Reference Guide**. Chichester, N.Y.: John Wiley, 1980. 3 vols. $59.95/vol. LC 80-40006. ISBN 0-471-27765-7.

See entry 29 for an annotation of this work.

79a. World Conservation Monitoring Centre. **1990 IUCN Red List of Threatened Animals**. Gland, Switzerland: Cambridge, 1990. 192p. (IUCN Red Data Book Series). $15.00. ISBN 2-8317-0031-0.

Arranged taxonomically, this work represents a list of those species of animals known by IUCN to be threatened with extinction. The list is divided by large taxonomic groups. Among these groups are mammals, birds, reptiles, amphibians, fishes, invertebrates (noninsects) and insects. The species are arranged under their respective families, which in turn is entered under order and class within these large taxonomic groups. For each species listed, the scientific name, common name, status (extinct, endangered, vulnerable, rare, indeterminate, insufficiently known, threatened, commercially threatened), and geographical area is given. A scientific-name index completes the book.

TEXTBOOKS

80. Alexander, R. McNeill. **Animals**. Cambridge, England, New York: Cambridge University Press, 1990. 509p. $89.50; $34.50pa. LC 89-9766. ISBN 0-521-34391-7; 0-521-34865-x.

After a brief introduction to a variety of laboratory procedures, this volume follows the typical phylogenetic approach to the invertebrates and vertebrates with some variations. For example, many of the chapters are not titled by the name of the phylum but rather by a class, a common name, or a morphological distinction. Chapter titles such as "Single-Celled Animals," "Animals with Mesogloea," "Flatworms, and Segmented Animals," and "Crustaceans" can be found. In addition, although anatomy is considered in each of these sections, much of the emphasis is placed on physiological concepts; mathematical formulas can be found from time to time as part of the discussion as in the case of the shell of molluscs and in the flight of insects. Furthermore, the number of animal groups and animals included are less than in other zoological texts. There are a number of labeled line-drawings to help explain the internal anatomy, but not to the same degree as in the other texts.

Each chapter concludes with a list of selected readings. A subject index completes the textbook; page numbers in italics refer to illustrations.

This work is more rigorous than other texts in zoology in that it presupposes some basic knowledge in calculus and in physiology. Its purpose is to have the student acquire a solid foundation in basic zoological principles that deal with anatomy, physiology, and biochemistry.

81. Boolootian, Richard A., and Karl A. Stiles. **College Zoology**. 10th ed. New York: Macmillan, 1981. 803p. $47.25. LC 80-22446. ISBN 0-02-311990-x.

After a brief introduction to cytology, histology, and taxonomy, much of this work is arranged phylogenetically. Each phylum is introduced and by means of diagrams, line-drawings, photomicrographs, and photographs; the major features of each group are considered. Many of the labeled line-drawings are excellent. Organisms are expounded upon as representative examples of the phyla under discussion.

The remaining third of the volume presents chapters on animal behavior, physiology, genetics, zoogeography and ecology, and organic evolution. Humans are used as the representative organism in the sections on physiology and, to some degree, genetics.

At the end of each chapter is a brief set of review questions and a brief list of selected references. The appendices consist of a biochemical flow chart depicting glycolysis, Krebs Cycle, and the electron transport chain; a classification scheme for animals, describing morphology for phylums and classes; and a review of some elementary mathematics. A glossary and subject index complete the work.

82. Hickman, Cleveland P., Jr., Larry S. Roberts, and Frances M. Hickman. **Biology of Animals**. 5th ed. St. Louis: Time Mirror/Mosby College, 1990. 720p. $46.95. LC 89-5197. ISBN 0-8016-5481-5.

Approximately one-half of this work is devoted to zoological principles in evolution, ecology, ethology, and morphology and physiology. It is divided into two parts. Examples of some areas in part 1 are evolution of animal diversity and genetic basis of evolution. Part 2 considers chapters dealing with animal form and function such as animal behavior, homeostasis, internal fluids, and nervous coordination. Part 3, representing the other half of the text, portrays the parade of the animals. Chapters are presented in a phylogenetic arrangement. Illustrations, which are excellent, include black-and-white and colored, labeled diagrams; color and black-and-white photographs; and, to some degree, charts and tables.

At the end of each chapter is a series of review questions and a few selected references. A glossary that provides definitions and, in some cases, pronunciations, and derivations can be found near the end of the volume. A subject index completes the work.

83. Kershaw, Diana R. **Animal Diversity**. London, Boston: Unwin Hyman, 1988. 428p. $29.95pa. LC 88-5674. ISBN 0-04-445177-6.

Divided equally between the invertebrates and the chordates/vertebrates, this text parades through the myriad of animal types encompassing the animal kingdom. As one browses through the pages, the most obvious focus is on the excellent, labeled line-drawings of the internal anatomy of representative organisms. As a former high school biology teacher, I find the diagrams alone are worth the price of the volume.

After a very brief introduction to the internal anatomy of the cell and the mitotic and meiotic divisions, the text follows the standard presentation. Chapters are arranged phylogenetically from the protozoans to the mammals. Each chapter discusses the salient features of the phylum, using a representative animal to elucidate the characteristics of the group. The evolutionary relationships between the phyla are considered.

A subject index completes the work. Index terms in italics refer to illustrations.

84. Villee, Claude A., Warren F. Walker, Jr., and Robert D. Barnes. **General Zoology**. 6th ed. Philadelphia: Saunders College, 1984. 856p. $40.00. LC 83-20253. ISBN 0-03-062451-7.

Although only one-third of this work is devoted to the phylogenetic representation of the phylums of invertebrates and vertebrates, the remaining two-thirds is devoted to a physiological/biochemical, genetics/evolution, and ecological approach. Representative organisms are used to illustrate phylum characteristics as well as those principles found in physiology, ecology, genetics, and evolution. Comparative vertebrate physiology is stressed, from dogfish sharks to humans.

Diagrams (some in color), graphs, drawings, photomicrographs, and an occasional color photo can be found throughout the book. Each chapter ends with a summary, listing the most important points, as well as a brief list of references and selected readings. A combined subject index/glossary completes the volume.

JOURNALS

85. **American Zoologist: A Journal of Integrative and Comparative Biology**. Vols. 1- , nos. 1- . Thousand Oaks, Calif.: American Society of Zoologists, 1961- . Bimonthly. $127.00/yr.(institutions). ISSN 0003-1569.

Covering all areas in zoology, this journal usually consists of two themes per issue. The themes originate from symposia sponsored by the Society. A series of articles, which combine review and original research, are then published that emphasize that particular theme. The number of papers contributed to a particular subject varies. Book reviews appear at the end of each issue; obituaries may also be found here.

86. **Bulletin of the British Museum of Natural History: Zoology Series**. Vols. 1- , nos. 1- . London: British Museum of Natural History, 1949- . Semiannually. £60.00/yr. ISSN 0007-1498.

Issues of this serial are devoted to taxonomic, phylogenetic, and anatomical articles of both vertebrate and invertebrate animals. Excellent line-drawings and photographs usually accompany the descriptive treatises. Some articles are 20 or more pages in length.

87. **Bulletin of the Museum of Comparative Zoology**. Vols. 1- , nos. 1- Cambridge, Mass.: Museum of Comparative Zoology, Harvard University, 1863- . Published irregularly. Price varies. ISSN 0027-4100.

Each issue is devoted to a specific taxonomic article. Both vertebrate and invertebrate species are considered. For example, some of the six issues comprising volume 152(1989-1991) consists of articles on lepidopterans, arachnids, lizards, and mammals. Issues vary in length from approximately 50 to 150 pages or more.

88. **Bulletin of Zoological Nomenclature**. Vols. 1- , nos. 1- . London: International Commission on Zoological Nomenclature, 1943- . Quarterly. $135.00/yr. ISSN 0007-5167.

Representing the official periodical of the International Commission on Zoological Nomenclature, this journal consists of brief articles (one to several pages) on scientific names of animal species. Living species as well as fossils are included.

Approximately 40 brief articles and technical communications appear in each issue. These are categorized under headings such as applications, comments, and rulings of the commission (opinions). In most issues a general taxonomic article appears. Illustrations are lacking.

89. **Canadian Journal of Zoology**. Vols. 1- , nos. 1- . Ottawa: National Research Council of Canada, 1929- . Monthly. $310.00/yr. (institutions). ISSN 0008-4301.

Representing all areas of zoology, this journal usually includes approximately 40 articles ranging from three to ten pages in length. Zoological fields such as behavior, biochemistry and physiology, developmental biology, ecology, genetics, morphology and ultrastructure, parasitology and pathology, systematics, and evolution are considered. Abstracts to the papers are written in English and French. Photomicrographs, line-drawings, charts, and graphs accompany the research articles.

90. **Journal of Experimental Zoology**. Vols. 1- , nos. 1- . New York: Wiley-Liss, 1904- . Monthly. $1,058.00/yr. ISSN 0022-104X.

Published under the auspices of the American Society of Zoologists and the Division of Comparative Physiology and Biochemistry, this work, as the name implies, publishes original research articles that are considered experimental. Both invertebrates and vertebrates are included. For example, one can find articles dealing with the regeneration of planaria, molluscan shell matrix phosphoproteins, and gonadotropin receptors in salmon, to name a few. Charts, graphs, line-drawings, and photomicrographs accompany the text.

The articles found in the table of contents are arranged by broad subject categories. Examples are comparative physiology and biochemistry, endocrinology, general developmental biology, reproductive biology, and genetics.

91. **Journal of Zoology**. Vols. 1- , nos. 1- . London: Zoological Society of London, 1833- . Monthly. $685.00/yr. ISSN 0952-8369.

Each issue of this serial consists of papers that describe original research in all zoological disciplines. Both vertebrates and invertebrates are included. Charts, graphs, drawings, and photographs are present. From the semicircular canals of hagfishes to mink populations in Ireland, this journal has a variety of articles that would appeal to all types of zoologists. In addition, brief reviews and communications from different societies, in the form of short papers, can be found from time to time.

Important to note is the fact that this journal incorporates the *Proceedings of the Zoological Society of London* (founded in 1830), and the *Transactions of the Zoological Society of London* (founded in 1833).

92. **Physiological Zoology**. Vols. 1- , nos. 1- . Chicago: University of Chicago Press, 1928- . Bimonthly. $160.00/yr. (institutions). ISSN 0031-935x.

As the name implies, this work is devoted to original research articles in the comparative and ecological physiological fields. Both vertebrate and invertebrate species are used in this type of experimentation. From 10 to 20 articles can be found in a single issue varying in length from 10 to 20 pages. Graphs and tables are used to enhance the text. Some examples of the papers include metamorphosis in salamanders, evolution of mammalian blood parameters, and nitrogenous excretion in terrestrial crabs.

93. **Systematic Zoology**. Vols. 1- , nos. 1- . Washington, D.C.: Society of Systematic Biologists, 1952- . Quarterly. $40.00/yr. ISSN 0039-7989.

As one may surmise, this serial devotes its papers to both micro and macro areas of systematics. Line-drawings, graphs, and charts can be found throughout the articles. The individual papers usually range from 10 to 20 pages in length. Some examples of the research are cause of geographic variation, DNA hybridization and marsupial phylogeny, and phylogeny of osmeroid fishes. A number of book reviews can be located at the end of each issue.

94. **Zoo Biology**. Vols. 1- , nos. 1- . New York: Wiley-Liss, 1982- . Bimonthly. $320.00/yr. ISSN 0733-3188.

Contained within these issues are original research articles dealing with the conservation and preservation of threatened and endangered species in zoos and other preservation centers. Additionally, it covers not only care, feeding, and health, but such topics as captive breeding programs for threatened and endangered species. The papers include most of the zoological disciplines such as anatomy, ethology, genetics, reproduction, nutrition, and physiology. Graphs, charts, and photographs accompany the articles when appropriate.

95. **Zoological Journal of the Linnean Society**. Vols. 1- , nos. 1- . London: Academic Press, 1855- . Monthly. $534.00/yr. ISSN 0024-4082.

From articles dealing with onychophorans in Singapore to the ecomorphology of Bornean tree frogs, this serial publication includes original papers in the fields of taxonomy, systematics, ecology, and comparative zoological studies. Both vertebrate and invertebrates are covered. Usually no more than three or four articles can be found in an issue. Many of the articles are 50 or more pages in length. In some instances an entire issue may be devoted to a particular treatise. Occasionally book reviews can be found in particular issues. Plates, line-drawings, maps, and other forms of illustrations are present, enhancing the textural material.

96. **Zoomorphology: An International Journal of Comparative and Functional Morphology**. Vols. 1- , nos. 1- . Heidelberg: Springer International, 1924- . Quarterly. $502.00/yr. ISSN 0720-213X.

In addition to embryological studies, this publication considers morphological research at the macro and micro levels. Vertebrates and invertebrate species are included. Phylogenetic and structure/function relationships are considered. Between four and six articles are usually present in each issue. The articles range from 10 to 30 pages in length. Photomicrographs, line-drawings, graphs, tables, and charts can be found throughout the research papers.

TAXONOMIC KEYS

97. Winchester, A. M., and H. E. Jaques. **How to Know the Living Things**. 2d ed. Dubuque, Iowa: William C. Brown, 1981. 173p. (The Pictured Key Nature Series). $12.80. LC 81-65442. ISBN 0-697-04780-6.

Encompassing the plants and animals, this reference is a key in identifying organisms down to orders. Morphological descriptions are used in the identification process. Line-drawings also aid the user in determining the type of plant or animal in question. Each chapter represents a kingdom, and the person using the key would have to know, at the very least, the kingdom to which the organism belongs. A particular species (scientific name included), is noted for each of the line-drawings, along with a brief, general description of the type of organism. For example, by using the line-drawings and scientific names, one can make the distinction between an African and an Indian elephant.

This is a simple key to use and only the major types of plant and animal organisms are included. A brief glossary and index complete the work. Index includes biological subjects, and common and scientific names, alphabetically arranged.

BIOGRAPHIES

98. Abbott, David, ed. **The Biographical Dictionary of Scientists: Biologists**. London: Blond Educational, 1983. 182p. (Biographical Dictionary of Scientists Series). $28.00. LC 84-10972. ISBN 0-911745-82-3.

Consisting of biologists from the ancient through the modern world, this biographic reference depicts the significant work that made each scientist famous in their own right. Biologists from the fields of taxonomy, palaeontology, evolution, microbiology, genetics, medicine, biochemistry, biophysics, and so on, are included. Many of the entries include Nobel Prize winners.

Each biographical description is one or two columns in length. The years of birth and death are given along with a description of scientific contributions. Occasionally, a number of key papers are mentioned in the textural account. A few diagrams are scattered throughout the text to enhance the explanation. An adequate glossary and subject index complete the work.

This compilation does not come close to the exhaustive descriptions found in *Dictionary of Scientific Biography* (see entry 106). Except for bibliographical information given in the biography, there is no formal bibliography included.

99. **American Men & Women of Science**. 18th ed. New York: Bowker, 1992. 8 vols. $750.00/set. LC 06-7326. ISBN 0-8352-3074-0. ISSN 0192-8570.

Men and women in the physical, biological, and related sciences who are considered leaders in their fields are included in this current edition. Related sciences encompass public health, engineering, mathematics, statistics, and computer science. Unlike many of the other biographical compilations, persons listed are still living and active in their respective disciplines. For each scientist mentioned, their birth data, scientific field, education, work locations, memberships, research activities, and mailing address are given. Approximately one-eighth of a column is employed in presenting the biographical information. It is also understood that these scientists have distinguished themselves in research and hold important positions in their place of employment.

The last volume of this set is devoted to indexing the work. The researchers included are arranged by broad scientific disciplines. Among these are agricultural and forest sciences, biological sciences, chemistry, computer sciences, engineering, environmental, earth and marine sciences, mathematics, medical and health sciences, physics and astronomy, and other professional fields, such as history and philosophy of science. Approximately 26% of the index is devoted to the listing of biological scientists.

AMWS, now in its 18th edition, began publications in 1921 and so is an excellent reference source for those wishing to locate information on leaders in science in the twentieth century.

100. Asimov, Isaac. **Asimov's Biographical Encyclopedia of Science and Technology: The Lives and Achievements of 1,510 Great Scientists from Ancient Times to the Present Chronologically Arranged**. 2d ed., rev. New York: Doubleday, 1982. 941p. $29.95pa. LC 81-47861. ISBN 0-385-17771-2.

In addition to the information contained within the subtitle, birth and death data, a thumbnail biographical sketch, and pronunciation information for names is given for each person listed. Most of the biographical data is contained within one column, a few are several pages in length. Portraits of a few of the scientists are intermingled with text. A few of the biologists mentioned are Darwin, De Duve, Pasteur, Linnaeus, Schleiden, and Schwann.

Because this is a chronological arrangement, the alphabetical list of biographical entries found near the front of the work is invaluable; referrals are to the biographical entry number, not the page. The index at the end of the volume allows access to subject matter and individuals not included as a biographical entry.

101. **Biographical Memoirs.** Vols. 1- . Washington, D.C.: National Academy Press, 1877- . Published irregularly. $22.50/vol. LC 05-26629. ISSN 0077-2933.

In order to be included in this multivolume work, the scientist must be deceased, must have been a member of the National Academy of Sciences, and must have made significant contributions to the scientific world.

For each entry, a portrait of the scientist is included along with birth and death dates and a 10- to 20-page biographical essay. The essay includes a brief life history along with major contributions to the respective field. A list of honors and distinctions, as well as a selected bibliography of writings, arranged by year, complete each of the monographic biographies. The most recent volume includes an alphabetical biographee index to all previously published volumes. Biologists such as Dobzhansky, Romer, and DuBois can be found in these pages.

This work, because of the limitations for inclusion, is not as comprehensive as *Dictionary of Scientific Biography* (see entry 106), and should only be considered for purchase after more comprehensive biographies have been acquired. It should also be noted, however, that a number of entries found in this series would not be comprehensively covered in other biographical works.

102. **Biographical Memoirs of Fellows of the Royal Society.** Vols. 1- . London: Royal Society, 1955- . Annual. Price varies/vol. $160.78 for 1990 vol. LC 56-26605. ISSN 0080-4606.

As in the case of *Biographical Memoirs* (see entry 101), one must be deceased and have been a member of the Royal Society, and one must have contributed significantly to scientific endeavors, to be mentioned here.

For each inclusion, a portrait, a 15- to 20-page biographical sketch, a list of honors and distinctions, and a bibliography of said person is given. The biography includes a brief life history and scientific contributions to the respective fields. Volume 36 (1990) contains a biographee index to volumes 27-36 (1981-1990), volume 26 lists the biographee index to volumes 1-26 (1955-1980). Biologists such as Krebs, Frisch, and Urey are included.

Similar to *Biographical Memoirs*, libraries may wish to purchase this set with the understanding that it is more limited than other comprehensive biographies mentioned in this section.

103. Daintith, John, Sarah Mitchell, and Elizabeth Tootill, eds. **A Biographical Encyclopedia of Scientists**. New York: Facts on File, 1981. 2 vols. $125.00. LC 80-23529. ISBN 0-87196-396-5.

Published in a larger type than most reference sources, this biography consists of some 2,000 scientists. For each one mentioned, birth and death data and a biographical sketch depicting contributions to science can be found. On average, the sketches are usually no longer than one column. Entries are based upon the significance of the scientific work. There are several appendices that close this two-volume set. Among them is a chronological listing of scientific events beginning with 590 B.C. and ending with 1981; a list of books and papers that influenced and developed scientific knowledge; an index of the scientists' names; and a subject index. Biologists such as Darwin, Szent-Gyorgi, Spemann, Ponnamperuma, and Mendel, have been included.

104. Debus, Allen G., ed. **World Who's Who in Science: A Biographical Dictionary of Notable Scientists from Antiquity to the Present**. Chicago: Marquis, 1968. 1855p. LC 68-56149. ISBN 0-8379-1001-3.

Covering all areas of science, and spanning two centuries, this work presents thumbnail sketches of those persons who have made contributions to scientific knowledge. Many of the entries would not be found in other biographical sources listed in this section. For each inclusion, scientific field, birth and death data, mate, number of children, memberships, and contributions are described. The dates of these events are also considered. The length of information (in small type) varies from several lines to one-sixth of a column.

105. Elliott, Clark A., comp. **Biographical Index to American Science: The Seventeenth Century to 1920**. New York: Greenwood Press, 1990. 300p. (Bibliographies and Indexes in American History, no. 16). $55.00. LC 90-31735. ISBN 0-313-26566-6.

This volume provides indexing for some 2,850 American scientists found in approximately 310 sources. Inclusion consists of researchers from the seventeenth century, those who died before 1921. Besides the classical biographical works, entries were also taken from a number of journals. For each person listed, the birth and death data, occupation, and the citations to the reference materials are given. The index provides access to the biographees by broad scientific discipline and year of birth. Some of the biological disciplines found are anatomy, biochemistry, zoology, physiology, ornithology, entomology, and biology. This is an excellent companion to *Index to Scientists of the World from Ancient to Modern Times: Biographies and Portraits* (see entry 107), (coverage is continued by *American Men of Science*, then *American Men & Women of Science*) as well as *Prominent Scientists: An Index to Collective Biographies* (see entry 109).

106. Gillispie, Charles Coulston, ed. **Dictionary of Scientific Biography**. New York: Scribner (published under the auspices of the American Council of Learned Societies), 1970-1990. 18 vols. $750.00/16-vol. set; $80.00/vol. LC 69-18090. ISBN 0-684-16962-2.

This classical biographical set includes deceased scientists who have made significant contributions to the world of science; the more significant, the better chance for inclusion. It considers persons from ancient Greece through and including the twentieth century. Biologists such as Charles Darwin, Carl Linnaeus, Antoni Van Leeuwenhoek, Joseph Lister, and the like have found a place in this work.

For each person listed, a biographical sketch discussing the scientist's contribution to his/her specific field is noted. These biographical depictions can run from several to 20 or more pages. Diagrams and mathematical formulas are used in order to supplement and explain the text. A lengthy bibliography, listing works by and about the biographee, can be found at the end of each entry. Volume 15, besides containing additional entries, also presents a number of extended essays on the history of mathematics and astronomy.

Volume 16 is the index to the main set. The main part of the index contains an alphabetical arrangement of the scientists, subject matter, societies, periodicals, societies, organizations, and countries. The subject approach, for example, allows one to create a list of people associated with zoological taxonomy. In addition, there is a separate alphabetical listing of contributors, societies, and periodicals at the end of the index volume, as well as a list of scientists arranged under broad scientific disciplines, such as life sciences, earth sciences, and physics.

Volumes 17 and 18 consist of scientific biographies that were not included in the original 16 volumes. The same format is followed as with the original set. Biologists such as Karl Frisch, Theodosius Dobzhansky, and the like are included.

This is a classic work and should be found on the shelves in every major college and university library.

107. Ireland, Norma Olin. **Index to Scientists of the World from Ancient to Modern Times: Biographies and Portraits**. Boston: F. W. Faxon, 1962. 662p. (Useful Reference Series No. 90). $13.00. LC 62-13662. ISBN 0-87305-090-8.

Over 7,475 scientists, from the ancient world through modern times, can be found in this collected work. The biographees have been indexed from 338 biographical collections. The vast majority of names are from the seventeenth, eighteenth, and nineteenth centuries; the twentieth century is not covered very well, as one might expect considering the imprint date.

For each person listed, the birth and death dates, name and page of the collection, and page on which a portrait can be found are given. Portraits are not listed for every scientist.

A useful work for finding information on biologists, chemists, physicists, and the like that may not necessarily appear in standard reference sources. It would be useful for libraries to own many of the selected works that have been indexed in this volume.

108. **McGraw-Hill Modern Scientists and Engineers**. New York: McGraw-Hill, 1980. 3 vols. $180.00. LC 79-24383. ISBN 0-07-045266-0.

The majority of the entries found in this compendium are autobiographical. Each of the biographies/autobiographies listed consists of significant contributions to the scientist's field of endeavor. Approximately one page is devoted to each of the researchers. A portrait and birth/death data are also given. These people are leaders of the twentieth century. They were considered for inclusion based on awards and prizes presented by societies, organizations, and institutions. It is international in scope. Approximately 40% of those listed are biologists. An index by specialty along with a separate subject index can be found in volume 3 of the set. Biologists such as Charles Best, Bernardo Houssay, Nikolaas Tinbergen, Francis Crick, and James Watson are included.

109. Pelletier, Paul A., ed. **Prominent Scientists: An Index to Collective Biographies**. 2d ed. New York: Neal-Schuman, 1985. 356p. $45.00. LC 85-3079. ISBN 0-918212-78-2.

Two hundred sixty-two collected biographical works representing 12,211 scientists have been indexed in this source. *Who's Who* books are not represented. In many respects this volume begins where *Index to Scientists* (see entry 107), leaves off. Books published from 1960 through 1983 are included whereas *Index to Scientists* covers the time period from the 1930s to 1960. As one would suspect, many of the scientists listed appear in more than one source. Birth and death dates can be found next to each entry. Near the end of the volume, the names of scientists are arranged under their respective field; the fields listed are broad. Among some of the headings are ornithology, ecology, ethology, evolution, embryology, and ichthyology.

As with any of these indexes to collected works, not having most of the indexed works depreciates the value of the index.

110. **Who's Who in Science in Europe: A Biographical Guide in Science, Technology, Agriculture, and Medicine**. 7th ed. London: Longman; dist. U.S., Detroit, Mich.: Gale Research, 1991. 4 vols. $895.00. LC gb92-4735. ISBN 0-582-08658-2.

Arranged by the European country and then alphabetically by name of the person, this biographical set covers scientists and engineers, academics, book authors, and society officers residing and working in western and eastern Europe. For each one listed, the birth date, job title, scientific interests, degrees held, appointments, societies, publications, telephone number, and address are given. As one can surmise, this is not a retrospective work; researchers listed are actively engaged in their job responsibilities.

Volume 1 represents the United Kingdom, volumes 2 and 3, the European Community countries, and volume 4 is devoted to the non-EC countries and eastern Europe.

For each of the four volumes there is an index allowing access to that volume. The index, which is a subject approach, lists the scientists' names under their areas of specialty. Some of the types of biological headings used are anatomy and physiology, biochemistry, biophysics, microbiology, molecular biology, and zoology.

STYLE MANUALS

111. CBE Style Manual Committee, comps. **CBE Style Manual: A Guide for Authors, Editors, and Publishers in the Biological Sciences**. 5th ed., rev. and exp. Bethesda, Md.: Council of Biology Editors, 1983. 324p. $27.95. LC 83-7172. ISBN 0-914340-04-2.

This work, similar to style manuals in other fields, discusses the various rules and procedures necessary to publish biological articles and monographs. It is a step-by-step guide for how to write an article or a book, how to cite references, the use of illustrative material, copyright protection, the art of indexing a manuscript, and so on. In addition, it considers certain styles that are endemic to specific fields in the biological sciences. Among the fields included are zoology, botany, microbiology, chemistry, and biochemistry. A subject index, which allows easy access to the particular rule or discipline, completes the volume.

GUIDES TO THE LITERATURE

112. Blanchard, J. Richard, and Lois Farrell, eds. **Guide to Sources for Agricultural and Biological Research**. Berkeley, Calif.: University of California Press, 1981. 735p. $50.00. LC 76-7753. ISBN 0-520-03226-8.

This bibliographic guide represents a listing of the major works that are useful in researching topics in biology and agriculture. In addition to general works in agriculture and biology, there are areas in plant sciences, crop protection, animal sciences, physical sciences, food science and nutrition, environmental sciences, and social sciences, and computerized data bases for bibliographic research. Types of works include, but are not limited to, indexes and abstracts, library catalogs, government publications, directories, dictionaries, encyclopedias, handbooks, and dissertations. Each of the reference sources listed are descriptively annotated. They range in length from one sentence to one paragraph. Citations include author/editor, title, imprint, pagination, series, and NTIS numbers when appropriate. They do not include LC or ISBN numbers, nor do they include the price. This volume contains 5,720 books and 59 databases, and covers the time period from 1959 to 1980.

There is an excellent introduction in the researching of these fields, along with an appendix consisting of acronyms and abbreviations. A separate subject and title index completes the guide. This work represents a vast undertaking; it covers the literature in these areas extremely well.

113. Davis, Elisabeth B. **Using the Biological Literature: A Practical Guide**. New York: Marcel Dekker, 1981. 286p. $69.75. LC 81-7832. ISBN 0-8247-7209-1.

Some of the most important reference sources in the life sciences are listed in this useful work. Chapters reflect the major botanical, zoological, and microbiological disciplines. The different types of tools are considered separately under each chapter. For example, the section on entomology lists abstracts, bibliographies, indexes, dictionaries, directories, guides, handbooks, periodicals, histories, and so on. The annotations range from one sentence to one paragraph. The title, author/editor, edition, and imprint are given for each of the books mentioned. In addition, there is a chapter describing the history of the biological literature, as well as an author/editor and title index.

This is a very useful work for those interested in a core set of reference works in all areas of the biological sciences.

114. Dronick, David A. **The Literature of the Life Sciences: Reading, Writing, Research**. Philadelphia: ISI, 1985. 219p. $29.95. LC 85-4283. ISBN 0-89495-045-2.

Written in a more narrative style than *Information Sources in the Life Sciences* (see entry 118), this treatise attempts to discuss the historical, philosophical, and research elements of the reference tools contained therein. Chapters consider such aspects as primary and secondary literature, writing and publishing, citation indexing, and scientific communication, to name a few. Throughout these sections, reference materials are mentioned; however, complete bibliographic information is absent. The appendix consists of a selected list of guides to the literature. A subject and title index provides access to the text.

This is a readable guide that will provide insight to the reader in using and applying this type of literature.

115. Sheehy, Eugene P., ed. **Guide to Reference Books**. 10th ed. Chicago: American Library Association, 1986. 1560p. $65.00. LC 85-11208. ISBN 0-8389-0390-8.

The granddaddy of all bibliographic guides, this classic attempts to list and annotate every major reference source in all areas of knowledge. The citations to the work include author/editor, title, imprint, and pagination. LC and ISBN numbers, including the cost, have not been listed. Annotations vary in length from one sentence to several paragraphs. The biological section, which includes general works, ecology, natural history, botany, zoology, bacteriology, biochemistry, and entomology, consists of 22 pages. At the end of the work can be found an author/title/subject index.

Although this volume is a classic, its coverage is limited to only the very major works in any subject area. If a more comprehensive listing of reference works in the biological/zoological sciences is needed, consider *Guide to Sources for Agricultural and Biological Research* (see entry 112).

116. Smith, Roger C., and Reginald H. Painter. **Guide to the Literature of the Zoological Sciences**. 7th ed. Minneapolis, Minn.: Burgess, 1966. 238p. LC 66-23383.

Although this work is outdated it represents the last published guide dealing exclusively with the zoological literature. In addition to introductory chapters discussing literature problems of the scientists and the mechanics of the library and book classifications, this guide lists and annotates bibliographies, abstract journals, taxonomic indexes, and a selected list of world periodicals covering all phases of zoology. The annotations describing many of the historical bibliographies/indexes are especially useful. The description, however, to titles such as *Biological Abstracts* are outdated. In addition, chapters considering bibliographic entries, different types of zoological literature, the preparation of a scientific paper, and library assignments are included.

117. Walford, A. J., ed. **Walford's Guide to Reference Material. Volume 1, Science and Technology**. 5th ed. London: Library Association, 1989. 791p. £60.00. LC GB89-53868. ISBN 0-85365-978-8.

A listing with, in most cases, one-paragraph annotations of the major reference works in the sciences, medicine, and engineering can be found in this bibliographic tool. Chapters are devoted to each of the sciences. In addition, there are chapters that consider patents, medicine, engineering, household management, chemical industry, building industry, agriculture and livestock, and so on. A total of 48 pages, encompassing 469 citations, are devoted to biology works in general, natural history, and zoology. Citations contain author/editor, title, imprint, collation, and in many cases the ISBN/ISSN numbers and price. There are 5,995 entries making up the complete work. Besides a brief introduction and list of abbreviations, there is a separate author/title and subject index.

This work represents a major undertaking in compiling all the basic reference sources in all areas of science, medicine, and engineering.

118. Wyatt, H. V., ed. **Information Sources in the Life Sciences**. 3d ed. London, Boston: Butterworths, 1987. 191p. (Butterworths Guides to Information Sources). $75.00. LC 86-31731. ISBN 0-408-11472-x.

Written in a textural style, this work walks the user through many of the important reference works in the biological sciences. In addition, basic texts, journals, newsletters, and databases are covered. The main portion of the guide is divided into a series of chapters representing biological disciplines such as zoology, microbiology, genetics, ecology, and botany. There are chapters that include databanks, guides to the literature, computer searching, secondary sources, and current awareness. For each reference listed, the title, author/editor, edition, publisher, and date are given. There is a brief bibliography at the end of some of the chapters. A title and broad subject-area index can be found completing the book.

Although not nearly as comprehensive as *Guide to Sources for Agricultural and Biological Research* (see entry 112), this guide offers the reader a more interesting narrative.

ASSOCIATIONS

Listed below are the directories of associations. Following these is a list of associations pertinent to zoology in general.

Directories

119. **Encyclopedia of Associations: International Organizations**. Detroit, Mich.: Gale Reseasch, 1989- . 2 vols. Annual. $490.00. ISSN 1041-0023.

This work represents a compilation of over 13,500 international organizations as well as those national organizations found outside the United States. Associations listed represent all areas of knowledge. Over 1,354 organizations are listed in the fields of engineering, technology, natural sciences, and social sciences. For a complete write up see *Encyclopedia of Associations. National Organizations of the U.S.* (entry 120).

120. **Encyclopedia of Associations: National Organizations of the U.S.** Detroit, Mich.: Gale Research, 1961- . 3 vols. Annual. $375.00. ISSN 0071-0202.

Making its debut in 1961, this classic and comprehensive directory lists over 23,000 national as well as some international organizations. All areas of knowledge are represented in this listing. The associations are listed under broad subject headings such as environmental and agricultural organizations; engineering, technological, and natural and social science organizations, and cultural organizations. For each entry found, the name, address, telephone number, executive director, founding date, number of members, number of staff, history, purpose, telecommunications services, location, time, dates, and so on of the conventions/meetings, and committees, and a list of its publications are presented.

Part 3 of this three-part work is devoted to a name and keyword index. Not only does this index lead to entries in this particular work but also to associations found in the *Encyclopedia of Associations. International Organizations* (see entry 119). This is an indispensable reference work for all libraries.

Associations

121. **Academy of Zoology**. Khandari Rd., Agra 2, Uttar Pradesh, India.

122. **American Association for the Advancement of Science**. 1333 H St. NW, Washington, DC 20005. (202) 326-6400. FAX (202) 289-4021.

123. **American Association of Zoological Parks and Aquariums**. Oglebay Park, Rte. 88, Wheeling, WV 26003. (304) 242-2160. FAX (304) 242-2283.

124. **American Association for Zoological Nomenclature**. c/o NHB 163, Smithsonian Institution, Washington, DC 20560. (202) 357-4668.

125. **American Institute of Biological Sciences**. 730 11th St. NW, Washington, DC 20001-4521. (202) 628-1500. FAX (202) 628-1509. TELEX 209061. TOLL FREE (800) 992-AIBS.

126. **American Philosophical Society**. 104 S. Fifth St., Philadelphia, PA 19106. (215) 440-3400. FAX (215) 440-3436.

127. **American Society of Zoologists**. 104 Sirius Circle, Thousand Oaks, CA 91360. (805) 492-3585.

128. **Association of Systematics Collections**. 730 11th St. NW, 2nd Fl., Washington, DC 20001. (202) 347-2850.

129. **Biological Council**. The Inst. of Biology, 20 Queensberry Pl., London SW7, 2DZ, England. 71 5818333. FAX 71 8239409.

130. **British Association for the Advancement of Science**. Fortress House, 23 Savile Row, London W1X 1AB, England. 71 4943326. FAX 71 7341658. TELEX 092091.

131. **Council of Biology Editors**. 9650 Rockville Pike, Bethesda, MD 20814. (301) 530-7036. FAX (301) 571-1848.

132. **European Science Foundation**. 1, quai Lezay-Marnesia, F-67000 Strasbourg, France. 88 353063. FAX 88 370532. TELEX 890440 F.

133. **Federation of American Societies for Experimental Biology**. 9650 Rockville Pike, Bethesda, MD 20814. (301) 530-7000. FAX (301) 571-1855.

134. **International Commission on Zoological Nomenclature**. Natural History Museum, Cromwell Rd., London SW7, 5BD, England. 71 9389387.

135. **International Union of Biological Sciences**. 51, boulevard de Montmorency, F-75016, Paris, France. 1 45250009. FAX 1 42889431. TELEX c/o ICSU 630533 F.

136. **National Academy of Sciences**. Office of News and Public Information, 2101 Constitution Ave. NW, Washington, DC 20418. (202) 334-2138.

137. **National Research Council**. 2101 Constitution Ave. NW, Washington, DC 20418. (202) 334-2000. FAX (202) 334-2158.

138. **National Science Foundation**. 1800 G St. NW, Rm. 527, Washington, DC 20550. (202) 357-9498.

139. **New York Academy of Sciences**. 2 E. 63rd St., New York, NY 10021. (212) 838-0230. FAX (212) 888-2894.

140. **Nordic Council for Wildlife Research**. c/o J.O. Pettersson, Svenska Jagareforbundet, P.O. Box 26091, S-100, 41 Stockholm, Sweden. 8 236340. FAX 9 7912303.

141. **Royal Zoological Society of Scotland**. Corstorphine Rd., Edinburgh, EH12, 6TS, Scotland. 31 3349171.

142. **Society for Conservation Biology**. Univ. of Florida, Florida Museum of Natural History, Gainesville, FL 32611. (904) 392-1721.

143. **Society for Experimental Biology**. Burlington House, Piccadilly, London WLV OLQ, England. 71 4398732. FAX 71 2874786.

144. **Society of Systematic Zoology**. Smithsonian Institution (NHB), No. 163, Washington, DC 20560. (202) 357-4757.

145. **Zoological Society of Israel**. Dept. of Zoology, Tel-Aviv Univ., 69978 Tel-Aviv, Israel. 3 5459818. FAX 3 6425518.

146. **Zoological Society of London**. Regent's Park, London, NW1 4RY. (071) 722-3333. TELEX 265247.

147. **Zoological Society of Southern Africa**. Univ. of Port Elizabeth, Dept. of Zoology, Box 1600, Port Elizabeth 6000, South Africa. 41 5311308. TELEX 747342.

CHAPTER 2

THE INVERTEBRATES (Excluding the Arthropods)

You may have heard from time to time the derogatory remark about some people being described as not having a backbone. Fortunately or unfortunately, as the case may be, all of the animals set forth in this chapter, alas, have no backbone. They are the invertebrates and are described as such. They do not seem to have suffered due to a lack of a vertebral column. These animals are some of the most successful creatures on the planet Earth. They have been around for eons and are still thriving. According to Brusca and Brusca, in their book entitled *Invertebrates* (see entry 213), there are approximately 220,751 extant species of invertebrates excluding the arthropods, but including the hemichordates, tunicates, and cephalachordates. Compare this number to approximately 47,000 extant species of vertebrates. It should be noted that although Brusca's work *Invertebrates* considers the number of vertebrate species to be approximately 47,000, other sources provide different numbers. For example, Willson's text *Vertebrate Natural History* (see entry 696), suggests approximately 42,460 vertebrates (not including hemichordates, tunicates, and cephalachordates). In any event, it is safe to say that the invertebrate species now living far outnumber the vertebrate species that are in existence today.

These organisms come in all shapes and sizes and exhibit a wide array of lifestyles. From the microscopic protozoans to the sessile sponges, to the reproductive machines known as the parasitic worms, to the hard-shell clams, to the spiny-skinned starfish, these animals show an amazing array of external and internal characteristics. Many exhibit a parasitic way of life, others provide food for humans, such as many of the mollusks, while others are useful toil tillers, such as the earthworms. These lowly creatures are interesting to study, not only from an ecological or economic perspective, but from an evolutionary standpoint. They show relationships among each other and that of their vertebrate neighbors.

If Aristotle was the "father of invertebrate zoology," then Libbie Hyman deserves the title "mother of invertebrate zoology" for what may be the last great individual attempt to comprehensively monograph the invertebrates in a work that saw six volumes completed before her death.
—Engemann and Hegner
Invertebrate Zoology, *3d ed., 1981*

DICTIONARIES

148. Martignoni, Mauro E., and others. **Terms Used in Invertebrate Pathology in Five Languages: English, French, German, Italian, Spanish**. Portland, Oreg.: U.S. Dept. of Agriculture, Forest Service, Pacific Northwest Forest and Range Experiment Station, 1984. 195p. (General Technical Report PNW, vol. 169).

Divided into five parts, this polyglot dictionary considers a multitude of terms associated with diseases found in the invertebrate groups. Each part presents alphabetically the terms in English, French, German, Italian, and Spanish, along with the translation of the word into the other languages.

149. Stachowitsch, Michael. **The Invertebrates: An Illustrated Glossary**. New York: John Wiley, 1992. 676p. $175.00; $79.95pa. LC 91-21129. ISBN 0-471-83294-4; 0-471-56192-4pa.

Arranged phylogenetically by phylum and containing more than 10,00 entries, this work includes anatomical terms that are specific for the group. In many cases, due to the size of the phylum, the terms are entered under their respective class. Each term is defined in usually five lines or less. The German equivalent of the term is also given. Each phylum and/or class is introduced by a one- or two-page illustration showing various anatomical forms, including a labeled drawing. Minor phyla are also included in the work. The arrangement of this work is extremely interesting: it conforms to the thinking of a zoologist and so could provide a useful supplement to a course in invertebrate zoology.

ENCYCLOPEDIAS

150. George, J. David, and Jennifer J. George. **Marine Life: An Illustrated Encyclopedia of Invertebrates in the Sea**. New York: John Wiley, 1979. 288p. LC 79-10976. ISBN 0-471-05675-8.

Teeming with invertebrate life, this source is divided basically into two sections: one of textural material and another of colored plates. The text is arranged by phyla, followed by class, order, family, and species. A brief description of external and internal anatomical structure, along with line-drawings, depicts each of these groups. Reproductive processes are also mentioned in many cases. The colored plates are also arranged by phyla, followed by colored pictures of the species in their natural environments. These organisms have been identified within the text. Phyla represented are Porifera, Coelenterata (Cnidaria), Ctenophora, Platyhelminthes, Nemertea, Rotifera, Gastrotricha, Kinorhyncha, Nematoda, Nematomorpha, Entoprocta, Priapulida, Sipuncula, Echiura, Pogonophora, Annelida, Crustacea,

Chelicerata, Tardigrada, Mollusca, Bryozoa, Phoronida, Brachiopoda, Chaetognatha, Echinodermata, Hemichordata, and Chordata. A bibliography arranged by phyla, a list of identification guides, a brief glossary, photographic credits, and a scientific/common-name index completes the encyclopedia. The numbers in bold print within the index refer to the plates.

151. Grzimek, Bernhard, ed. **Grzimek's Animal Life Encyclopedia: Lower Animals, Volume 1, 1974**. New York: Van Nostrand, 1974. 599p. LC 79-183178.

Employing a myriad of full-page color photographs and diagrams, this volume (of a thirteen-volume set) walks the reader through many of the invertebrates found in nature. From the unicellular animals (protozoans) through the tracheates (centipedes, millipedes, etc.), it describes in very clear language the morphology and physiology of the groups. In addition, behavioral patterns, ecology, and evolution are interspersed among the pages. Other animal groups considered are the sponges, coelenterates (jellyfish, coral, hydra, etc.), flatworms (planaria, flukes, and tapeworms), ribbon worms, roundworms, sipunculids, annelids (earthworms, sandworms, etc.), water bears, arachnids (spiders, etc.), and the crustaceans (crayfish, lobsters, etc.). The insects are devoted to volume 2.

In addition, there are chapters devoted to forms and life processes of the animal world, and animal behavior. The closing sections of the work encompass a classification scheme, and an English-German-French-Russian dictionary of the animals considered in this volume. Conversion tables of metric to U.S. and British measurement systems, a bibliography, and an excellent index encompassing scientific and common names, as well as subjects, end the work.

This volume contains many interesting facts and some of the information could be used as a supplement to lectures at the high school and college levels.

152. Grzimek, Bernhard, ed. **Grzimek's Animal Life Encyclopedia: Mollusks and Echinoderms, Volume 3, 1974**. New York: Van Nostrand, 1974. 541p. LC 79-183178.

Following the same format as the other volumes in this set, this book contains a wealth of information on the variety of mollusks and echinoderms that inhabit the land and sea. The mollusks considered are the chitons, limpets, gastropods (snails), scaphopods and bivalves (tooth shells, clams, oysters, etc.), cephalopods (squid and octopus), lophophores (bryozoans, phoronids, etc.), and the arrow worms. In like fashion, the echinoderms featured consist of crinoids (sea lilies), holothuroids (sea cucumbers), echinoids (sea urchins), asteroids (starfish), and the ophiuroids (brittle stars). Chapters on the hemichordates (acorn worms) and tunicates (sea squirts) are also present. These last two groups are neither mollusks nor echinoderms; rather, they seem to be transitional organisms between the invertebrates and the vertebrates. It is also useful to note that the lophophores and arrow worms are other animals that are not usually considered to be mollusks or echinoderms.

As in the case of the other volumes in this set, the color plates are excellent; the animals are shown in all their glory. None of the volumes in this encyclopedic set, however, considers the detailed morphology present in Libbie Hyman's work (see entry 166).

A systematic classification scheme, an English-German-French-Russian dictionary, and a bibliography of these animals can be found near the end of the work. A conversion table of metric to U.S. and British measurement systems is also included. The index is composed of scientific names, common names, and subjects.

This volume contains many interesting facts and some of the information could be used as a supplement to lectures at the high school and college levels.

INDEXES AND ABSTRACTS

Invertebrates (General)

153. **Zoological Record**. Vols. 1- , nos. 1- . Philadelphia: BIOSIS, Zoological Society of London, 1865- . Annual. $2,200.00/yr. ISSN 0084-5604.

For a description of this index, see entry 55. The various sections relating to this chapter are:

Section 2:	Protozoa
Section 3:	Porifera & Archaeocyatha
Section 4:	Coelenterata & Ctenophora
Section 5:	Echinodermata
Section 6A:	Platyhelminthes & Nematoda; together with Nemertinea, Mesozoa, Nematomorpha, Acanthocephala, Placozoa
Section 6B:	Annelida; together with Rotifera, Chaetognatha, Echiura, Sipuncula, Gastrotricha, Kinorhyncha, Priapulida, Gnathostomulida, Pogonophora
Section 7:	Brachiopoda
Section 8:	Bryozoa (Polyzoa) & Entoprocta
Section 9:	Mollusca

Protozoa (Unicellular Organisms)

154. **Protozoological Abstracts**. Vols. 1- . Wallingford, Oxon, England: CAB International, 1977- . Monthly. $338.00/yr. ISSN 0309-1287.

Published monthly in hard copy or as part of the CAB database, this indexing and abstracting service attempts to provide comprehensive, worldwide information on all protozoan diseases occuring in animals and humans. Journals, technical reports, books, and other publications are covered. The abstracts contained within this service are sufficient for a good understanding of the article.

This work can also be searched via DIALOG employing the *CAB International* database (File 53, 1972-1983; and File 50, 1984-). This particular abstract title in the CAB database can be searched by employing the subfile field (CAB contains over 50 abstract journals in agricultural and related science fields). Likewise, this title can be searched on BRS (File CABA) using the hardcopy field as well as STN International (File CABA) using the file segment field.

Helminths (Parasitic Worms)

155. **CAB Abstracts on CD (CABCD)**. Wallingford, Oxon, England: CAB International, 1984- . $3,000.00, archival disc (1984-1986); $6,000.00, archival disc (1987-1989); $3,000.00, archival disc (1990); $7000.00, latest disc (1990-1992).

Contains the entire CAB database from 1984 to the present. *Protozoological Abstracts* and *Helminthological Abstracts* are included.

156. **Helminthological Abstracts**. Vols. 1- . Wallingford, Oxon, England: CAB International, 1932- . Monthly. $376.00/yr. ISSN 0957-6789.

This indexing and abstracting service covers literature comprehensively and on a worldwide basis. Subjects deal with all of the biological disciplines necessary for an understanding of the parasitic worms and their effect on plants, animals, and humans. Types of literature covered include journals, technical reports, and books. This title is divided into two sections, which are entitled "Series A: Animal and Human Helminthology," and "Series B: Plant Nematology."

This work can also be searched via DIALOG employing the *CAB International* database (File 53, 1972-1983; and File 50, 1984-). This particular abstract title in the CAB database can be searched by using the subfile field (CAB contains over 50 abstract journals in agricultural and related science fields). Likewise, this title can be searched on BRS (File CABA) using the hardcopy field, as well as on *STN International* (File CABA) using the file segment field.

157. **VETCD: Veterinary Science and Animal Health Information.** Wallingford, Oxon, England: CAB International, 1973- . $9,000.00, archival disc (1939-1990); $1,200.00, annual updates.

This CD-ROM, produced by CAB International and employing SilverPlatter Software (SPIRS), contains a subset of index/abstracting journals contained with the CAB database. The titles are *Index Veterinarius*; *Veterinary Bulletin*; *Review of Medical and Veterinary Entomology*; *Review of Medical and Veterinary Mycology*; as well as *Protozoological Abstracts* and *Helminthological Abstracts*. SilverPlatter's software allows for sophisticated field searching.

HANDBOOKS

Invertebrates (General)

158. Adiyodi, K. G., and Rita Adiyodi, eds. **Reproductive Biology of Invertebrates.** Chichester, N.Y.: John Wiley, 1983-(1988-). 5 vols. to date. LC 81-16355.

Each of these volumes presents a reproductive theme, such as oogenesis, fertilization, and so on. The chapters within the volumes are arranged phylogenetically by phylum and class. Therefore, the theme is discussed as it relates to each major group of invertebrates. All invertebrate groups are considered except the protozoans. References to the literature are cited throughout the text. Photomicrographs, labeled line-drawings, and photographs supplement and help explain the textural material. A bibliography consisting of those references cited can be found at the end of each chapter. A scientific-name index, and a subject index conclude each of the volumes.

The writing of this set is scholarly and assumes a biological background. However, it is quite readable. Listed below are the volumes that have been published thus far. According to volume 1, a total of six volumes will eventually make up the set.

Volume 1. *Oogenesis, Oviposition, and Oosorption.* $160.00pa. ISBN 0-8357-6637-3.
Volume 2. *Spermatogenesis and Sperm Function.* $215.00. ISBN 0-471-90071-0.
Volume 3. *Accessory Sex Glands.* $215.00. ISBN 0-471-91466-5.
Volume 4. *Fertilization, Development, and Parental Care.* Part A. $166.00. ISBN 0-471-92269-2.
Volume 4. *Fertilization, Development, and Parental Care.* Part B. $166.00. ISBN 0-471-92271-4.
Volume 5. *Sexual Differentiation and Behaviour.* $168.00. ISBN 0-471-93410-0.

159. Brusca, Richard C. **Common Intertidal Invertebrates of the Gulf of California.** 2d ed., rev. and exp. Tucson, Ariz.: University of Arizona Press, 1980. 513p. $29.95pa. LC 79-19894. ISBN 0-8165-0682-5.

Covering descriptive and taxonomic information, this volume considers all invertebrate groups found in the geographical area suggested by the title. For each phylum under consideration, preservation techniques, general taxonomic information of the group as it relates to the area, a key to the species, a description of the species and the person who first identified it are given. The order and family are also listed before the description of the species. Numerous line-drawings and a number of colored plates enhance the text. Phylums considered are Porifera, Cnidaria, Ctenophora, Platyhelminthes, Nemertea, Annelida, Sipuncula, Echiura, Mollusca, Arthropoda (makes up the largest group), Bryozoa, Brachiopoda, Echinodermata, Hemichordata, and Chordata. An introductory section includes physical evolution, water temperatures, habitats, and tides of the Gulf of California. A glossary, an extensive bibliography (citations arranged under phylum names), and an index to scientific names as well as a separate general index (common names, geographical locations, etc.) can be found at the end of the source.

160. Conn, David Bruce. **Atlas of Invertebrate Reproduction and Development**. New York: Wiley-Liss, 1991. 252p. $79.95. LC 90-19577. ISBN 0-471-56079-0.

Considering all major phyla of invertebrates, this atlas considers the reproductive mechanisms using text and plates. Invertebrates included are the Proifera, Cnidaria, Platyhelminthes, Rhynchocoela, Nematoda, Mollusca, Annelida, Arthropoda, Echinodermata, the primitive chordates, and some miscellaneous phyla. The plates depict stages of reproduction and embryology as seen under the light microscope. All photographs of plates (black-and-white) that make up the greater part of the chapter are labeled. A list of selected references can be found within each chapter. In addition to a glossary, there are appendices that discuss materials and methods of microscope slide preparation, and a list of companies that supply prepared microslides and preserved specimens. An index of subjects and common and scientific names concludes the atlas.

The photographs are produced extremely well and are good for study. This is a highly specialized work and should be found only in those collections supporting invertebrate programs.

161. Corning, W. C., J. A. Dyal, and A. O. D. Willows, eds. **Invertebrate Learning**. New York: Plenum, 1973-1975. 3 vols. LC 72-90335.

Considering all of the major invertebrate groups, this three-volume set presents reviews that discuss the behavioral aspects of learning in invertebrates. Such areas as conduction, tactile behavior, feeding responses, habituation, sense receptors, social behavior, conditioning, reflexes, and the like are considered. Such animal groups as the cephalopods, echinoderms, chelicerates, crustaceans, insects, snails, protozoans, cnidarians, turbellarians, and annelids are present. The review articles contained within these books are scholarly; citations to the primary literature are scattered throughout the text. Bibliographies appear at the end of each section. Line-drawings, graphs, and charts provide the illustrative material. A subject index can be found at the end of each volume.

Volume 1. *Protozoans through Annelids*. ISBN 0-306-37671-7.
Volume 2. *Arthropods and Gastropod Mollusks*. ISBN 0-306-37672-5.
Volume 3. *Cephalopods and Echinoderms*. ISBN 0-306-37673-3.

162. Freeman, W. H., and Brian Bracegirdle. **An Atlas of Invertebrate Structure**. London: Heinemann, 1971. Reprint, 1973; 1976. 129p. LC 72-182149. ISBN 0-435-60315-9.

This unique source provides photomicrographs of the external and internal anatomy of many of the invertebrate groups, along with a labeled drawing of the photograph. In this way, one can make the comparison of preserved material with that of an illustrative diagram. Cross sections and sagital sections of material are also present. A number of representative animals are included within their respective phyla. Phylums included are Protozoa, Porifera, Coelenterata (Cnidaria), Platyhelminthes, Nematoda, Annelida, Arthropoda, Mollusca, Echinodermata, Rotifera, Endoprocta, Ectoprocta, Chaetognatha, and Hemichordata. Some of the organisms include amoeba, euglena, paramecium, hydra, obelia, aurelia, planaria, flukes, tapeworms, ascaris, the earthworm, locust, snails, and the starfish. The work concludes with photographs and labeled drawings of the mitotic process in ascaris and the meiotic process in a grasshopper. There is no index.

163. Giese, Arthur C., and John S. Pearse, eds. **Reproduction of Marine Invertebrates.** New York: Academic Press, 1974. 7 vols. (1-6, 9). LC 72-84365.

This set of books considers in a review format all of the various aspects of reproduction found in marine invertebrates. It covers asexual and sexual reproduction as well as embryonic development. The articles are written in a scholarly manner with many references to the primary literature cited within the text. Bibliographies appear at the end of each section. Labeled line-drawings, charts, and graphs comprise the majority of the illustrations. A separate author, subject, and taxonomic index can be found at the end of each volume.

Volume 1. *Acoelomate and Pseudocoelomate Metazoans.* (Out of print.)

Volume 2. *Entoprocts and Lesser Coelomates.* $116.00. ISBN 0-12-282502-0.

Volume 3. *Annelids and Echiurans.* ISBN 0-12-282503-9. (Out of print.)

Volume 4. *Molluscs: Gastropods and Cephalopods.* $116.00. ISBN 0-12-282504-7.

Volume 5. *Molluscs: Pelecypods and Lesser Classes.* $116.00. ISBN 0-12-282505-5.

Volume 6. *Echinoderms and Lophophorates.* Pacific Grove, Calif.: Boxwood Press. $65.00. ISBN 0-940168-09-X.

Volume 9. *General Aspects: Seeking Unity in Diversity.* Palo Alto, Calif.: L. Blackwell Scientific. $65.00. ISBN 0-940168-08-1.

164. **Handbuch der Zoologie: Eine Naturgeschichte der Stämme des Tierreiches.** Gegrundet von Willy Kükenthal. Hrsg. von J. G. Helmcke and others. Berlin: W. de Gruyter, 1923- .

Beginning in 1923 with the first edition, and in 1968 with the second, this work now consists of over 15 volumes. It is still in progress. A general description of this work appears in chapter 1 (see entry 68). Similar to the *Traité de Zoologie* (see entry 69), it attempts to cover significant information on all the animals. Arranged taxonomically by phylum, followed by order, suborder, class, and so on, it concentrates mainly on morphology, physiology, and taxonomy with some information dealing with ecology and geographical distribution. In addition to the text, the pages abound with line-drawings showing internal and external anatomy (many cross sections and sagital sections); the majority of these are labeled. Map diagrams, showing geographical distribution, can also be found. Brief bibliographies are at the end of each section. These citations are arranged under zoological headings, such as Morphologie and Physiologie und Okologie. A scientific-name/common-name/subject index appears at the end of each volume. Except for the last few publications, the work is written in German. All works listed below depicting the invertebrate part of the source are in German.

Erster Band. *Protozoa, Porifera, Coelenterata, Mesozoa.* 1925. Various forms of protozoans, sponges, cnidarians (jellyfish, sea anemones, hydra, coral, etc.), and the mesozoans are contained herein.

Zweiter Band, Erste Halfte. *Plathelminthes: Turbellaria, Trematoda, Cestoidea, Nemertini. Nemathelminthes: Rotatoria, Gastrotricha, Kinorhyncha, Nematodes, Nematomorpha, Acanthocephala. Kamptozoa (Bryozoa entoprocta).* 1933. Considers the flatworms (planaria, flukes, tapeworms), ribbon worms, roundworms, and a number of animals belonging to minor phyla.

Zweiter Band, Zweite Halfte. *Vermes Polymera: Arachiannelida, Polychaeta, Clitellata, Priapulida, Sipunculida, Echiurida.* 1934. Encompasses the earthworms, sandworms, and some minor phyla organisms.

Dritter Band, Erste Halfte. *Tardigrada, Pentastomida, Myzostomida, Arthropoda, Allgemeines, Crustacea.* 1927. Some minor phyla, general considerations on arthropods, and detailed information on the arthropods, belonging to the class crustacea (lobsters, crayfish, barnacles, etc.) are described.

Dritter Band, Zweite Halfte. *Chelicerata, Pantopoda, Onychophora, Vermes Oligomera.* 1932-1937. Discusses some minor phyla, beginning with the chelicerates.

Dritter Band, 2. Halfte. *Pogonophora.* 1968. Volume totally devoted to the pogonophorans.

Funfter Band. *Solenogastres, Mollusca, Echinodermata*. 1926. Contains animals such as the clam, oyster, squid, octopus, starfish, sand dollar, and sea urchin.

Funfter Band. *Tunicata*. 1956. Volume totally devoted to the tunicates, otherwise known as the sea squirts. These organisms are considered to be prochordates.

Sechster Band, Erste Halfte Erster Teil. *Acrania (Cephalochorda), Cyclostoma, Pisces*. 1962. Much of this work considers the vertebrates, such as the lamprey eel and the fish. However, it also depicts the Cephalochordata group, which takes into account organisms such as amphioxus. This creature is a chordate. It is one of the transitional animals between the invertebrates and the vertebrates.

165. Harrison, Frederick W., and John O. Corliss. **Microscopic Anatomy of Invertebrates**. New York: John Wiley, 1991- . 15 vols. (projected). $150.00/vol. LC 89-12117.

As of this writing there are four chapters in print of a projected 15-volume set. Although this work is being listed under handbooks, it could just as easily serve as an encyclopedia. When completed it should serve as one of the major comprehensive works on the functional anatomy of invertebrates.

The volumes are arranged by phylum, followed by class, which constitues chapters. For each class of organism, gross anatomy, microscopic anatomy, and cellular anatomy is considered. Species within these classes are used to illustrate differences. Biological systems are included in the higher animal groups. The work makes excellent use of scanning and transmission electron microscopy photomicrographs. In addition to the photomicrographs, a large number of labeled line-drawings of external and internal anatomy are included. A comprehensive list of references, which have been cited within the chapter, are listed at the end of each chapter.

There is no doubt that once this multivolume set is completed it will become a definitive work in the field. The volumes constituting the set are as follows:

Volume 1. *Protozoa*. ISBN 0-471-56842-2.
Volume 2. *Placozoa, Porifera, Cnidaria, and Ctenophora*. ISBN 0-471-56224-6.
Volume 3. *Platyhelminthes and Nemertinea*. ISBN 0-471-56843-0.
Volume 4. *Aschelminthes*. ISBN 0-471-56103-7.
Volume 5. *Mollusca: Monoplacophora, Aplacophora, Polyplacophora, and Gastropoda*.
Volume 6. *Mollusca: Bivalvia, Scaphopoda, and Cephalopoda*.
Volume 7. *Annelida*.
Volume 8. *Chelicerate Arthropoda*.
Volume 9. *Crustacea*.
Volume 10. *Decapod Crustacea*.
Volume 11. *Insecta*.
Volume 12. *Onychophora, Myriapod Arthropoda, and Lesser Protostomata*.
Volume 13. *Lophophorates and Entoprocta*.
Volume 14. *Echinodermata*. ISBN 0-471-56121-5.
Volume 15. *Hemichordata, Chaetognatha, and the Invertebrate Chordates*.

166. Hyman, Libbie Henrietta. **The Invertebrates**. New York: McGraw-Hill, 1940-1967. 6 vols. (McGraw-Hill Series in the Invertebrates). LC 40-5368.

It is unfortunate that the life span of a human being is so short, especially in the case of Libbie Hyman. It was her life's ambition to describe, as completely as possible, all of the invertebrate groups. Following a phylogenetic approach she managed to publish six volumes, ending with the first half of the phylum Mollusca. She never reached the Arthropoda.

What we have, however, and what she did contribute to the zoological sciences, however, is one of the major classics in the field of invertebrate zoology. Abounding with invaluable labeled pen-and-ink drawings of the gross morphology of the organisms as well as their internal parts (mostly from living or prepared material), this source presents many of the biological aspects of these creatures. This famous, classic work should be on the shelves of every college and university library offering programs in zoology. Libbie proceeds to describe each one's morphology, taxonomy, physiology, embryology, and life cycle. The morphological and taxonomical considerations make up the greater part of these volumes. Each of the six volumes is arranged by phylum, followed by class, order, and/or subclass. Each work is followed by an extensive bibliography and index. Bibliographic entries are referred to in the text. The index is an alphabetical arrangement of common names, scientific names, and subjects. The title of the volumes are as follows:

Volume 1. *Protozoa through Ctenophora*. Contains organisms such as the amoeba, paramecium, euglena, Sporozoa, sponge, jellyfish, hydra, coral, and comb jellies.

Volume 2. *Platyhelminthes and Rhynchocoela: The Acoelomate Bilateria*. Contains organisms such as planaria, flukes, tapeworms, and nemertines.

Volume 3. *Acanthocephala, Aschelminthes, and Entoprocta: The Pseudocoelomate Bilateria*. Contains organisms such as the nematodes (e.g., Ascaris), rotifers, Kinorhyncha, Priapulida, Gastrotricha, and the like.

Volume 4. *Echinodermata: The Coelomate Bilateria*. Contains organisms such as the sea cucumber, starfish, seastar, sea urchin, sea lily, and brittle stars.

Volume 5. *Smaller Coelomate Groups: Chaetognatha, Hemichordata, Pogonophora, Phoronida, Ectoprocta, Brachiopoda, Sipunculida, the Coelomate Bilateria*. These types of organisms are usually included in most works as the minor phyla. Groups such as the hemichordates contain animals that have a dorsal notochord. It is this group of organisms that may have given rise to the vertebrates.

Volume 6. *Mollusca I: Aplacophora, Polyplacophora, Monoplacophora, Gastropoda, the Coelomate Bilateria*. Contains organisms such as the snails, limpets, Chitons, and Solenogasters.

167. Kaestner, Alfred. **Invertebrate Zoology**. New York: Wiley Interscience, 1967- . $32.50 (vol. 2); Reprint. LC 67-13947. ISBN 0-470-45416 (vol. 2); 0-471-45417-6 (vol. 3).

This work represents a translation and adaptation of Kaestner's *Lehrbuch der Speziellen Zoologie*. According to volume 1, a total of four volumes are to make up the set. However, as of this writing, only three volumes have been published. Each of the chapters represents a large group of invertebrates at the phylum and/or the class level. Such subject areas as classification, general anatomy, reproduction, habitats, behavior, ecology, embryology, and physiological parameters are considered for each of the groups. Labled and unlabeled line-drawings depicting external and internal anatomy aid the explanations found in the text. A list of references can be found at the end of each chapter. A combined subject and scientific-name index concludes each of the volumes.

According to Herbert Levi and Lorna Levi, translators and adaptors, this work is more advanced than Barne's *Invertebrate Zoology* (see entry 212), and less technical than Hyman's *The Invertebrates* (see entry 166). In addition, it should be noted that the section on protozoans found in the original German work has been omitted in this set.

Volume 1. *Porifera, Cnidaria, Platyhelminthes, Aschelminthes, Mollusca, Annelida, and Related Phyla*.

Volume 2. *Arthropod Relatives, Chelicerata, Myriapoda*.

Volume 3. *Crustacea*.

Volume 4. *Lophophorates and Deuterostomes*. (Not yet published as of this writing.)

168. Pennak, Robert W. **Fresh-Water Invertebrates of the United States: Protozoa to Mollusca**. 3d ed. New York: John Wiley, 1989. 628p. $59.00. LC 88-18570. ISBN 0-471-63118-3.

Arranged phylogenetically, this book presents a comprehensive picture of those invertebrates that inhabit freshwater climes. Each chapter either represents a phylum, a class, an order, or a family. Among them are the Protozoa, Porifera, Coelenterata, Turbellaria, Nemertea, Gastrotricha, Rotifera, Nematoda, Nematomorpha, Tardigrada, Bryozoa, Annelida, Eubranchiopoda, Cladocera, Copepoda, Ostracoda, Mysidacea, Isopoda, Amphipoda, Decapoda, Hydracarina, Gastropoda, and Pelecypoda. For each animal group listed, general information, internal and external anatomy, locomotion, feeding, physiology, reproduction, development, ecology, geographical distribution, collecting, culturing, preparing, preserving, and taxonomy is given. The taxonomy information consists of keys to the group. There are fine line-drawings (many of them labeled) that enhance the discussion of the invertebrate organisms. In addition, there is an introductory chapter discussing freshwater invertebrates in general as well as an appendix consisting of reagents, solutions, and laboratory items that were mentioned throughout the text. An adequate bibliography can be found at the end of each chapter. An index of subjects, scientific names and common names completes the work. This is an excellent volume. It can just as easily serve as a textbook or encyclopedia.

169. Shaw, A. C., S. K. Lazell, and G. N. Foster. **Photomicrographs of Invertebrates**. London: Longman, 1974. 96p. LC 75-305410. ISBN 0-582-32279-0.

Following the same format as *An Atlas of Invertebrate Structure* (see entry 162), this work provides photomicrographs of the internal and external anatomy of invertebrates, along with labeled diagrams of the photomicrographs. In comparing this source with *An Atlas of Invertebrate Structure*, there is very little discernable difference, save for the fact that some photomicrographs may occur in one and not in the other. For example, cystercercoids of tapeworms are absent in this source as well as the minor phyla groups. Phylums include Protozoa, Coelenterata, Platyhelminthes, Nematoda, Annelida, Mollusca, Arthropoda, and Chordata. Representative animals are included to show anatomical features of the various phyla and classes. The work concludes with a one-page scientific/common-name index.

All in all, this source is not quite as comprehensive as *An Atlas of Invertebrate Structure*, but it nevertheless serves the same basic function.

170. Thorp, James H., and Alan P. Covich, eds. **Ecology and Classification of North American Freshwater Invertebrates**. San Diego, Calif.: Academic Press, 1991. 911p. $59.95. LC 90-46694. ISBN 0-12-690645-9.

Encyclopedic in scope, this exhaustive source provides much more than taxonomic schemes, and therefore has been placed under the handbook section, rather than under taxonomic keys. In addition to briefly describing the types of freshwater invertebrates and giving an overview of their habitats and geographic ranges, this source includes in-depth information on the creatures, following a phylogenetic approach. For each of the groups listed, anatomy, physiology, ecology, evolution, collecting, rearing, preparation, and taxonomic keys are included. An adequate bibliography is listed at the end of each chapter. The animal groups considered are Protozoa, Porifera, Cnidaria, Platyhelminthes, Gastotricha, Rotifera, Nematoda, Mollusca (gastropods and bivalves), Annelida, bryozoans, Tardigrada, water mites, aquatic insects, Crustacea, Ostracoda, Cladocera, and other Branchiopoda, Copepoda, and Decapoda. This work abounds with illustrations, among which are line- and labeled drawings of external and internal anatomy, photographs, photomicrographs, tables, and graphs. A glossary, a taxonomic index, and a subject index conclude the source.

This is a first-rate publication and will more than likely become a major reference on the freshwater invertebrates of North America.

171. Grassé, Pierre-Paul, ed. **Traité de Zoologie: Anatomie, Systématique, Biologie**. Paris: Masson, 1948. 17 vols. (various fascicules). LC 49-2833.

A general description of this work appears in chapter 1 (see entry 69). Published out of sequence, this major, classic work devotes 7 of its 17 volumes to the invertebrates, excluding the insects. The insects can be found in 3 of its other volumes. Arranged by class and subclass, each of these 7 volumes gives the reader an extremely comprehensive description of the morphology, physiology, and taxonomy of all major and minor groups of invertebrates. Excellent drawings of the external and internal anatomy (many in color; many cross and sagital sections) enhance the written contents. In addition, there are a number of transmission electron microscopy plates scattered throughout the volumes. A bibliography can be found at the end of each section, usually after a treatise on a class or subclass. The works in the bibliography have been cited within the text. There is a common-name/scientific-name/subject index found at the end of each volume and/or section (fascicule).

It is unfortunate that this work has not been translated, as the entire work is written in French. Regardless, the diagrams in these volumes are useful. Many of the scientific words are easy to translate into their English equivalents.

Listed below are the invertebrate volumes/sections that are pertinent to this chapter.

Tome I (Premier Fascicule). *Phylogenie. Protozaires: generalites, flagelles*. 1952. Protozoans in general as well as the protozoans that move by means of flagella (e.g., Euglena).

Tome I (Fascicule II). *Protozoaires: rhizopodes, actinopodes, sporozoaires, cnidosporidies*. 1953. Fr. 1,482. ISBN 2-225-58223-8. Contains protozoans such as amoebas, foraminferans, heliozoans (having fine pseudopods), and parasitic protozoans that cause malaria.

Tome II (Fascicule I). *Infusoires ciliés*. 1984. Fr. 1,477. ISBN 2-225-65263-5. Protozoans that move by means of cilia (e.g., Paramecium).

Tome III (Fascicule I). *Spongiaires: anatomie, physiologie, systématique, écologie*. 1973. Fr. 1,133. ISBN 2-225-38742-7. Describes the anatomy, physiology, taxonomy, and ecology of the sponges.

Tome III (Fascicule III). *Cnidaires: anthozoaires*. 1987. Fr. 1,415. ISBN 2-225-74668-0. Contains the sea anemones and the coral.

Tome IV (Premier Fascicule). *Plathelminthes, mésozoaires, acanthocéphales, némertiens*. 1965. Encompasses the flatworms, mesozoans, spiny-headed worms, and the ribbon worms.

Tome IV (Deuxieme Fascicule). *Némathelminthes. rotiféres, gastrotriches, kinorhynques*. 1965. Fr. 2,058. ISBN 2-225-58267-X. Consists of the free-living (e.g., vinegar eel) and the parasitic (e.g., ascaris) roundworms, the remainder of the roundworms, and a number of minor phyla.

Tome V (Premiere Fascicule). *Annélides, myzostomides, sipunculiens, echiuriens, priapuliens, endoproctes, phoronidiens*. 1960. Fr. 1,482. ISBN 2-225-59510-0. Contains the earthworms, sandworms, and a number of minor phyla.

Tome V (Deuxieme Fascicule). *Broyzoaires, brachiopodes, chetognathes, pogonophores, mollusques (generalities, aplacophores, polyplacophores, monoplacophores, bivalves)*. 1960. Fr. 1,482. ISBN 2-225-58300-5. Describes a number of minor phyla; description of the molluscs, in general: chitons, clams, mussels, oysters, and so on.

Tome V (Fascicule III). *Mollusques: gastéropodes etc. scaphopodes*. 1968. Fr. 1,482. ISBN 2-225-59510-0. Encompasses the snails and the tooth shells.

Tome V (Fascicule IV). *Cephalopodes*. 1989. Fr. 1,298. ISBN 2-225-80419-2. Describes the squids and octopi.

Tome VI. *Onychophores*, ... 1949. Fr.1,376. ISBN 2-225-58311-0. In addition to some minor phyla, this volume considers the trilobites, scorpions, and spiders.

Tome XI. *Echinodermes, Stomocordes, Procordes*. 1948. Takes into account the echinoderms (e.g., starfish, brittle stars, sea cucumbers, etc.), as well as the protochordates, such as the tunicates and amphioxus. It is believed that the protochordates gave rise to the vertebrates.

172. Wells, Susan M., Robert M. Pyle, and N. Mark Collins, comps. **The IUCN Invertebrate Red Data Book**. Gland, Switzerland: IUCN, 1983. 632p. $32.00. LC 83-222770. ISBN 2-88032-602-X.

Arranged phylogenetically, this compilation attempts to list and describe all invertebrate species that are threatened to some extent. A useful description of each phylum preceeds the species listed. Terms that are used to describe their vulnerability are *extinct*, *endangered*, *vulnerable*, *rare*, *indeterminate* (not sure of which three of the preceeding categories they fall into), *out of danger*, *insufficiently known* (insufficient information to place them in a category), *commercially threatened*, *threatened community*, and *threatened phenomenon*.

For each species mentioned, the common name, scientific name, name of person who first identified it, phylum, class, order, family, and category of vulnerability are given. In addition, textural material on the organisms contain a summary, description, distribution, population, habitat and ecology, scientific interest and potential value, threats to survival, conservation measures taken, conservation measures proposed, and a bibliography.

Because of the phylogenetic approach, the index will be used in most cases. It contains alphabetically the scientific and common names of the species mentioned.

Considering the date of imprint (1983), some of the assigned vulnerability categories may no longer be valid.

Invertebrates (Systematic Section)

PROTOZOA (Unicellular Organisms)

173. Margulis, Lynn, and others, eds. **Handbook of Protoctista: The Structure, Cultivation, Habitats and Life Histories of the Eukaryotic Microorganisms and Their Descendants Exclusive of Animals, Plants and Fungi**. Boston: Jones and Bartlett, 1990. 914p. $195.00. LC 88-8530. ISBN 0-86720-052-9.

Describing the algae, ciliates, flagellates, foraminifera, sporozoa, water molds, slime molds, and other protoctists, this is an extremely comprehensive work. It considers the structure, cultivation, habitats, and life histories of these eukaryotic organisms and their descendants, exclusive of animals, plants, and fungi. The arrangement of organisms in this volume lends itself well as a textbook and reference work.

The phyla of these species are discussed in separate chapters. Each chapter is broken down into five sections. The first of these includes the estimated numbers of genera and species, localization of the organisms in nature or in culture collections, the major sources of literature, and medical or agricultural importance. The second and third sections deal with habitat, life cycles, and taxonomic considerations. The fourth describes the maintenance and cultivation of organisms within the phylum, and the fifth considers the evolutionary history and fossil record of the group, whenever feasible. A bibliography follows each chapter. Pen-and-ink drawings, as well as transmission and scanning electron microscopy plates of the various organisms, can be found throughout each chapter. The illustrations are excellent.

A glossary, consisting of etymological roots and definitions of appropriate terms, can be found near the end of the book. An extensive organism, author, and general (subject) index completes the volume.

This is an extremely well-researched work. The contributors have given a great deal of themselves in completing their respective chapters. High praise should be given to the editors.

PORIFERA (Sponges)

174. Vacelet, Jean, and Nicole Boury-Esnault, eds. **Taxonomy of Porifera: From the N.E. Atlantic and Mediterranean Sea**. Berlin, New York: Springer-Verlag, 1986. 332p. (NATO ASI Series. Series G, Ecological Sciences; no. 13). $104.00. LC 87-13129. ISBN 0-387-16091-4.

Arranged by research-type articles written by many experts in the field, this work considers the systematics of those sponges from the North East Atlantic and the Mediterranean. Considerations given in virtually all of the chapters consist of species from different sponge groups. They are discussed in terms of their taxonomic description, spicules, distribution, and so on. Line-drawings and few photographs provide the illustrations. A scientific-name index completes the work.

HELMINTHS (Parasitic Worms)

175. Yamaguti, Satyu. **Systema Helminthum**. New York: Interscience, 1958-1963. 5 vols. in 7. LC 57-10544.

Helminthology is the study of parasitic worms. With this in mind, the treatise describes in considerable detail, the internal morphology that distinguishes one genera from another. Keys to the genera of families are also employed. Because of the precarious nature of parasites, the majority of their internal organs are male and female reproductive systems; they are reproductive machines. Therefore, great emphasis has been placed in describing slight differences in the testes and ovaries of each group of genera. Excellent sets of plates can be found at the end of each volume depicting the internal organs of these creatures. In most cases the sections of the volumes are arranged by the vertebrate host that is parasitized upon.

In addition to the plates, a massive bibliography and index can be found at the end of each volume. Works in the bibliography have been cited throughout the descriptive text. The index consists of the scientific names of the organisms. The person who first identified and named the parasite is also included within the index.

Volume 1. *Digenetic Trematodes.*
Volume 2. *Cestodes.*
Volume 3. *Nematodes.*
Volume 4. *Monogenea and Aspidocotylea.*
Volume 5. *Acanthocephala.*

PLATYHELMINTHES (Free-living and Parasitic Flatworms)

176. Yamaguti, Satyu. **Synopsis of Digenetic Trematodes of Vertebrates**. Tokyo: Keigaku, 1971. 2 vols. LC 72-880644.

This two-volume tome represents a revision of *Systema Helminthum, Volume I, Digenetic Trematodes, Parts I and II* (see entry 175). The general arrangement is the same. Major changes are an updated bibliography, a redrawing of the specimens, and the incorporation of newer classification schemes for the digenetic trematodes.

177. Yamaguti, Satyu. **Synoptical Review of Life Histories of Digenetic Trematodes of Vertebrates with Special Reference to the Morphology of Their Larval Forms**. Tokyo: Keigaku, 1975. 590p. 219 leaves of plates. LC 76-376411.

In essence, this work supplements *Synopsis of Digenetic Trematodes of Vertebrates* (see entry 176), as it concentrates specifically on the larval trematodes regarding their morphologies and life histories. The source is arranged by the vertebrate class upon which the larval trematode can be found. Examples are Digenea of Fishes, and Digenea of Reptiles. For each species listed, in most cases, the size of eggs, description of the miracidium, sporocyst, cercaria, and adult fluke are given. An extensive bibliography, a scientific-name index, and

219 plates can be found ending the volume. The plates, which consist of finely detailed drawings, depict the various larval stages of the trematodes included in the book.

NEMATA (Free-living and Parasitic Roundworms)

178. Anderson, R. C. **Nematode Parasites of Vertebrates: Their Development and Transmission**. Wallingford, Oxon: CAB International, 1992. 578p. £75.00. LC gb92-45857. ISBN 0-85198-799-0.

The species (561) are arranged under their respective genus, which in turn is entered under subfamily, family, superfamily, order, and subclass. Information regarding internal anatomy, eggs, larvae, penetration, hosts, and so on is considered for the subfamily, family, and superfamily. For each of the species listed, the scientific name, person who first named it, date of naming, parasitic mode of existence, hosts, eggs, larvae, and disease caused by same is given. Line-drawings depicting the internal anatomy of these worms, their eggs, and larvae, as well as drawings depicting a few life cycles, provide the illustrative matter. The descriptive accounts differ in length from one-half to several pages. References that were cited in the description can be found at the end of each taxonomic section. A scientific-name index closes the volume.

178a. Hunt, D. J. **Aphelenchida, Longidoridae and Trichodoridae: Their Systematics and Bionomics**. Wallingford: CAB International, 1993. 352p. $95.00. LC gb93-62174. ISBN 0-85198-758-3.

Consisting of one order and two families of roundworms, this work presents taxonomic information on all species contained within the above groups. Approximately 1,000 species contained within some 40 genera are represented. For each of the taxonomic groups listed (order, suborder, superfamily, family and genus), name of person who named the group as well as the synonyms, detailed external and internal anatomical information, and the names of the species within the genera are given. In addition to the names of the species and their synonyms the person who first named them and date of naming is considered. Taxonomic keys to the families are also presented throughout the text. A large number of black-and-white drawings and some photomicrographs depicting internal and external anatomical structures enhance the descriptive material. General information on the general morphology and bionomics of these groups can be found in the opening chapters. A 48-page bibliography consisting of citations cited within the text and a scientific-name index concludes the work.

MOLLUSCA
(Snails, Slugs, Clams, Oysters, Squid, Octopi, etc.)

179. Abbott, R. Tucker. **American Seashells: The Marine Mollusca of the Atlantic and Pacific Coasts of North America**. 2d ed. New York: Van Nostrand Reinhold, 1974. 663p. LC 74-7267. ISBN 0-442-20228-8.

This work represents a taxonomic description of 2,000 mollusks, along with 4,400 recorded species. It is more detailed and systematically oriented than Abbott and Tuckers' *Compendium of Seashells: A Color Guide to More than 4,200 of the World's Marine Shells* (see entry 180).

The chapter headings are broad. The organisms contained within these chapters are arranged by families. Descriptions contain the scientific name, name of person who first identified it, common name, geographical distribution, size, and shell anatomy. The source is illustrated with small black-and-white photographs and drawings. The center of the book consists of 24 plates of colored photographs of the shells. Common names, scientific names, and subjects make up the index.

This is a very detailed taxonomic work and should not be confused with the coffee table-like book called *Shells* (see entry 183).

180. Abbott, R. Tucker, and S. Peter Dance. **Compendium of Seashells: A Color Guide to More Than 4,200 of the World's Marine Shells**. New York: Dutton, 1982. Reprint, Melbourne, Fla.: American Malacologists, 1990. 411p. $49.95. LC 81-67757. ISBN 0-915826-17-8.

Richly illustrated with colored photographs, this handsome compendium parades through the world of shelled mollusks inhabiting the globe. The shells are arranged under their common-name grouping along with their family name. Examples are moon shells: Naticidae, dove shells: Columbellidae and spindles: Fasciolariidae. For each shell, included are the common name, scientific name, person who first named it, date of naming, size of shell (in inches and centimeters), geographical distribution, status (rare, uncommon, etc.), and synonyms when appropriate. Each shell description is accompanied by a colored photograph. In addition, a very brief description of the family is considered.

General information on seashells, classification of mollusks, habitats, shell features and identification, conservation, geographical ranges, measurements, the art of obtaining shells, care of shells, and tips on how to catalog a shell collection can be found in the introductory pages. A taxonomic classification of mollusks, along with bibliographic references for each group and a separate common-name index and scientific-name index can be found closing the volume. This is an impressive book.

181. Cattaneo-Vietti, R., R. Chemello, and R. Giannuzzi-Savelli, eds. **Atlas of Mediterranean Nudibranchs. Atlante dei Nudibranchi del Mediterraneo**. Roma: Editrice La Conchiglia, 1990. 264p. DM 135.00.

The species are arranged under their respective family, which in turn is entered under family, superfamily, and suborder. General anatomical features, scientific name, person who first named it, date of naming, and, in some cases, the generalized size of the organism is presented for the suborder, superfamily, and family. For each of the species mentioned (approximately 230), the scientific name, person who first named it, date of naming, citation to the taxonomic literature, main synonyms, references, general anatomical description, coloration, radula formula, size, habitat, geographical distribution, and a black-and-white drawing of the specimen are given. Each descriptive account can be found on two pages, one in Italian and the other in English. A systematic checklist of the Mediterranean nudibranchs, based on the *Annotated Check-List of Mediterranean Marine Mollusks* (see entry 255), can be found preceeding the accounts. A series of 14 colored plates containing four to eight specimens per plate can be found near the end of the source. A brief glossary in English and one in Italian, a six-page bibliography, and a scientific-name index concludes the handbook.

182. Damme, Dirk Van. **The Freshwater Mollusca of Northern Africa: Distribution, Biogeography and Palaeoecology**. Dodrecht, Boston: W. Junk, 1984. 164p. (Developments in Hydrobiology; 25). $115.00. LC 84-9999. ISBN 90-6193-502-4.

Divided into four major parts, this taxonomic work considers those molluscan species endemic to North Africa. Part one, which is about taxonomic synopsis and quaternary distribution, arranges the organisms under their respective class, subclass, family, and genus. The species themselves, along with a taxonomic description and distribution, are found under the genera. Interspersed among this section, which makes up the largest portion of the work, are line-drawings of shells, and distribution maps.

Part two considers biogeography of the region (name of region and organisms endemic to it), part three discusses the cretaceous and tertiary history, and part four includes the quaternary malacofauna of North Africa. A comprehensive bibliography, a list of localities, a general subject index, and a species index complete the volume.

183. Feininger, Andreas, and William K. Emerson. **Shells**. New York: Viking Press, 1972. 295p. LC 73-157616. ISBN 0-500-54008-X.

Written in nontechnical language by William K. Emerson, photographs taken by Andreas Feininger, this beautifully illustrated source documents the incredible variety of shells that exist in nature. The black-and-white and colored photographs of the shells are art forms in themselves. Most of the photos take up an entire page.

The text is not arranged in classic taxonomic order; rather, the chapters consider the forms and variety of shells. Examples are "An Infinite Variety of Shape," "The Spiral," "The Structure of Shells," "The Color of Shells," and "Knobs and Spines." General information is given that depicts each of the above themes. There are as many photographs as there are text.

The book closes with brief descriptions of the species. Photographs accompany this material as well. The classes Gastropoda (snails), Polyplacophora (chitons), Scaphopoda (tusk shells), and Bivalvia (oysters, clams, etc.) are considered. Some very general information on the classes of mollusks can be found in the introduction. A brief bibliography and an index containing subjects, common names and scientific names completes the volume.

This is not a typical zoological work. Rather, it is a source that can be enjoyed by all shell enthusiasts.

184. Keen, A. Myra. **Sea Shells of Tropical West America: Marine Mollusks from Baja California to Peru**. 2d ed. Stanford, Calif.: Stanford University Press, 1971. 1064p. LC 70-143786. ISBN 0-8047-0736-7.

This is a major taxonomic tome that covers the Pelecypoda, Gastropoda, Monoplacophora, Aplacophora, Polyplacophora, Scaphopoda, and Cephalopoda. The arrangement is by class, followed by subclass, order, superfamily, family, genus, and subgenus. Species are listed under genera/subgenera. For virtually each species mentioned, the scientific name, person who first named it, date of naming, synonym, range, habitat, and taxonomic description are given. Black-and-white as well as colored plates aid in the identification. Family keys can be found throughout the text.

In addition to the major portion of the volume, a glossary (terms used in morphological description), a geographic index, sources of illustrations, a very comprehensive bibliography, and a scientific-name index complete the source. Bibliographic entries are cited in the taxonomic portion of the book.

185. Nesis, Kir N. **Cephalopods of the World: Squids, Cuttlefishes, Octopuses, and Allies**. Moscow: V.A.A.P. Copyright Agency of the USSR for Light and Food Industry Publishing House, 1982. Reprint, Neptune City, N.J.: T. F. H. Publications, 1987. 351p. $69.96. LC 87-402052. ISBN 0-86622-051-8.

Translated from the Russian by B. S. Levitov and edited by Lourdes A. Burgess, this work contains an additional amount of text and photographs not in the original Russian edition.

Cephalopods are molluscs that are known, when loosely translated, as the head/foot animals. The squids, octopi, and cuttlefishes belong to this group. The main portion of the volume is divided into three parts. Part I considers external structure (tentacles, suckers, mantle, fins, funnels, etc.), part II includes internal structure (coelom, circulation, gills, excretion, ink sac, nerves, etc.), and part III is a key (taxonomic considerations) that considers the cephalopods of the world. Of these three parts, part III makes up the majority of the book (244 pages), as compared to part I (23 pages) and part II (49 pages).

There are a large number of fine pen-and-ink drawings that enhance these sections. In addition, there are a number of colored photographs showing the external anatomy of many of these creatures. There is an adequate bibliography. Citations are arranged under "Literature to General Section," and "Literature to Taxonomic Section." A taxonomic index can be found completing the work. Pages indicated in bold print refer to the illustrations.

186. Parkinson, Brian, Jens Hemmen, and Klaus Groh. **Tropical Landshells of the World**. Wiesbaden, Germany: Verlag Christa Hemmen, 1987. 279p. ISBN 3-925919-00-7.

Most works on the mollusks usually consider the marine varieties and not those endemic to the land. In this volume, however, the material is totally devoted to all types of land snails (phylum: Mollusca; class: Gastropoda) found throughout the world. The book is arranged by geographical regions and covers Florida and the Caribbean, Central and South America, Africa and adjacent islands, tropical Asia, and the tropical Pacific.

This source is divided basically into three sections. The first of the sections describes in a textural format, along with colored photographs, the geographic region and the types of snails found therein. The second lists a bibliography of the snails found in that region, while the third consists of magnificiently colored plates of the shells of the snails found in each of the geographical regions of the world. For each shell represented in the color-plate section, the family, the scientific name, person who first named it, date it was named, and geographical location is given. The source concludes with a genera, subgenera, species, and subspecies index.

It is useful to note that this book could just as easily serve as a taxonomic work as well as a handbook.

187. Pilsbry, Henry Augustus. **Land Mollusca of North America (North of Mexico)**. Philadelphia: Printed by the George W. Carpenter Fund for the Encouragement of Original Scientific Research, 1939-1948. 2 vols. in 4. (Academy of Natural Sciences of Philadelphia, Monographs, no. 3). $200.00. LC 40-5753. ISBN 0-685-08428-0.

Although this multivolume set was written many years back, it is still in print; it is a classic work. Arranged by families, and followed by subfamilies, genera, and subgenera, this is an extremely detailed taxonomic tome. Species are listed under their respective genera/subgenera and are discussed in some length. For each organism listed, the name of the person who first identified it, date of identification, bibliographic citations, color, size, detailed distribution, detailed taxonomic description, and in many cases internal anatomy, and ecology are given. The illustrations are small and consist of line-drawings and photographic plates. They are interspersed among the text. All families containing species of land mollusks in the geographical area are covered; these are the land snails of North America. Each part of each volume has its own separate index to scientific names.

188. Roper, Clyde F. E. **Cephalopods of the World: An Annotated and Illustrated Catalogue of Species of Interest to Fisheries**. Rome: Food and Agricultural Organization of the United Nations, 1984. 277p. (FAO Species Catalogue, vol. 3; FAO Fisheries Synopsis, no. 125, vol. 3).

Representing 173 species of cephalopods, this work provides the reader with general information on each of the organisms listed. The species are arranged under their respective families. For each of the families listed, the scientific name, person who named it, date of naming, citations to the literature, FAO names, basic external anatomy with appropriately labeled line-drawings, coloration, geographical distribution, habitat, biology, feeding behavior, interest to fisheries, and illustrated keys to the genera are considered. For each of the species mentioned, the scientific name, person who named it, date of naming, citations to the literature, synonymy, FAO names, basic external anatomy, geographical distribution, habitat and biology, size, interest to fisheries, local names, black-and-white line-drawings depicting the organism as a whole as well its tentacles, and a geographical range map are given. The amount of information varies for each species. Descriptive accounts are usually one page in length. An illustrated glossary of technical terms and a key to orders and families can be found in the introductory section. A classification of the genera of recent cephalopods, a list of species by major fishing areas, a 15-page bibliography, and a combined index of scientific names and international FAO names conclude the work.

189. Sweeney, Michael J., and others, eds. **"Larval" and Juvenile Cephalopods: A Manual for Their Identification**. Washington, D.C.: Smithsonian Institution Press, 1992. 282p. (Smithsonian Contributions to Zoology, no. 513). LC 91-22098.

Representing the results of a workshop organized by the Cephalopod International Advisory Council, this work provides the taxonomic information necessary to identify cephalopod species in their larval and juvenile growth stages. Species are arranged under their respective families. For each of the families listed, basic external identifying characteristics for the adult, young, and in some cases egg information, and geographical distributions are given. In addition, keys to the families, subfamilies, genera, and species are also present. Information on the species consists of the person who named it, date of naming, external identifying characteristics (young and adults), geographical distribution, and relations to similar species. A series of black-and-white line-drawings showing the body forms of these creatures aids in the identification process. A 10-page bibliography completes the volume.

189a. Thiele, Johannes. **Handbook of Systematic Malacology**. Washington, D.C.: Smithsonian Institution Libraries and The National Science Foundation, 1992. 2 vols. 198,000 DM. LC 87-600185. ISBN 1-56081-390-3.

Representing a translation of the *Handbuch der Systematischen Weichtierkunde* (1929-1935), this work forms the basis of molluscan systematics. According to the foreword it has become the basis for the arrangement of scientific collections throughout the world. This two-volume set represents two parts of a four-part set. The third part consists of the scaphopods, bivalves and cephalopods, while the last part contains general and comparative aspects. As of this writing only the first two parts have been translated. The first two parts are represented within these two volumes and consists of the classes Loricata and Gastropoda. The species are arranged under their respective genus, which in turn is entered under family, subfamily, family, and order. General external anatomical descriptions of the above groups are present. For each species listed, the scientific name, person who first named it, date of naming, synonyms, external anatomical description, and geographical distribution is given. Black-and-white drawings enhance the descriptive information. An alphabetical scientific-name index (genera, subgenera) can be found at the end of each volume.

Part I: Loricata/Gastropoda I: Prosobranchia.

Part II: Gastropoda: Opisthobranchia and Pulmonata.

190. Wilbur, Karl M., ed. **The Mollusca**. New York: Academic Press, 1983-1988. 12 vols. LC 82-24442.

Written as review articles, this set of books discusses the various aspects of molluscan biology. Citations to the primary literature are referenced throughout the text. Bibliographies appear at the end of each section. Graphs, charts, line-drawings (many of them labeled), photographs, and some electron microscopic photomicrographs provide the illustrative material. Separate taxonomic and subject indexes appear at the end of each volume.

Volume 1. *Metabolic Biochemistry and Molecular Biomechanics.* ISBN 0-12-751401-5.

Volume 2. *Environmental Biochemistry and Physiology*. ISBN 0-12-751402-3.

Volume 3. *Development*. $74.00. ISBN 0-12-751403-1.

Volume 4. *Physiology, pt. 1*. $105.00. ISBN 0-12-751404-X.

Volume 5. *Physiology, pt. 2*. $116.00. ISBN 0-12-751405-8.

Volume 6. *Ecology*. ISBN 0-12-751406-6.

Volume 7. *Reproduction*. $105.00. ISBN 0-12-751407-4.

Volume 8. *Neurobiology and Behavior, pt. 1*. $114.00. ISBN 0-12-751408-2.

Volume 9. *Neurobiology and Behavior, pt. 2*. $116.00. ISBN 0-12-751409-0.

Volume 10. *Evolution*. $113.00. ISBN 0-12-751410-4.

Volume 11. *Form and Function*. $133.00. ISBN 0-12-751411-2.

Volume 12. *Paleontology and Neontology of Cephalopods*. $127.00. ISBN 0-12-751412-0.

ECHINODERMATA (Starfish, Sea Urchins, Sea Cucumbers, etc.)

191. Clark, Ailsa McGown. **The Echinoderms of Southern Africa**. London: British Museum (Natural History), 1976. 277p. (British Museum [Natural History]; no. 776). $45.00. LC 77-357182. ISBN 0-565-00776-9.

This taxonomic work consists of approximately 280 species of echinoderms, of which 48% are endemic to South Africa and nearly 30% can be found in either the whole tropical Indo-West Pacific or the western part of the Indian Ocean. These 30% are from various parts of the Atlantic, Indian, and Pacific oceans. All classes of echinoderms are represented except for the holothurians (sea cucumbers). For each class discussed, distribution tables and keys are given for each of the species within that class. The person who first named the organism is included, as well as a description of the class, family, and species. Bibliographic references accompany the descriptions.

192. Clark, Ailsa M., and Maureen E. Downey. **Starfishes of the Atlantic**. London, New York: Chapman & Hall, 1992. 794p. $250.00. LC 91-15257. ISBN 0-412-43280-3.

Consisting of 21 families, 140 genera, and 374 species/subspecies, this source provides information on all starfish found throughout the Atlantic Ocean. The species are arranged under their respective genera, which in turn are entered under family, order, and class. Information such as citations to the taxonomic literature, history of the nomenclature, external and internal anatomical features, a systematic outline, and biological keys are given for the class Asteroidea. For each of the orders listed, the person who first named it, date of naming, citations to the taxonomic literature, identifying anatomical features, and biological keys to the families are presented. Family information consists of person who first named it, date of naming, citations to the taxonomic literature and anatomical identifying features, while person who first named it, date of naming, citations to the taxonomic literature, biological keys to the species, a taxonomic discussion, and the biology of the group is presented for the genus. For each of the species/subspecies presented, the person who first named it, date of naming, citations to the taxonomic literature, size, identifying anatomical features, color, taxonomic discussion, type (neotype, etc.), and geographical distribution and habitat are considered.

Descriptive accounts usually range from one-half to one full page in length. A chart depicting the geographical distribution of these organisms, a summary of taxonomic changes, and a 39-page bibliography can be found in the last third of the volume. The bibliography consists of those references cited within the text; it is extensive. The volume concludes with 107 pages devoted to black-and-white line-drawings depicting both external and internal anatomical features of the starfish (seastar), a series of 113 black-and-white plates (photographs) depicting gross morphology of the species, and a scientific-name index.

This work represents a scholarly approach to the starfish of the Atlantic; it is a first-class publication.

193. Downey, Maureen E. **Starfishes from the Caribbean and the Gulf of Mexico**. Washington, D.C.: Smithsonian Institution Press, 1973. 158p. (Smithsonian Contributions to Zoology, no. 126). LC 72-10019.

Consisting of a large number of black-and-white plates depicting these creatures, this work considers those species of starfish (seastars) found throughout the waters of the Caribbean and the Gulf of Mexico. The species are arranged under their respective genera, which in turn are entered under families, and orders. Biological keys to the orders, families, genera, and species are present. For each of the species listed, the scientific name, person who first named it, bibliographic citations to the taxonomic literature, detailed external anatomical characteristics, and geographical distribution are given. Some general external anatomy is also provided for the genera, families, and orders. An 11-page bibliography listing those citations cited in the descriptive accounts, and 48 pages of black-and-white plates of these echinoderms complete the volume.

194. Fisher, Walter K. **Starfishes of the Hawaiian Islands**. Washington, D.C.: U.S. Government Printing Office, 1906. 987-1130p. (U.S. Fish Commission Bulletin, part III, 1903). LC f11-386.

Based on a collection made by the United States Fisheries steamer *Albatross* in 1902, this work provides information on the starfish (seastars) obtained among the Hawaiian Islands. The species (60) are arranged under their respective genera, which in turn are entered under families and orders. For each of the species listed, the scientific name, person who first named it, date of naming, reference to the literature, detailed anatomical measurements, anatomical relationship to other species, type locality, coloration, and geographical distribution, in many cases, are given. Biological keys to family, genus, and species can be found throughout the work. Biological descriptions vary in length from one to three pages. The introductory information consists of the relationships of the Hawaiian starfish to that of the Eastern Archipelago, Indian Ocean, Japan, Polynesia, Australia, and the west coast of the United States. A series of 49 black-and-white plates depicting the animal as a whole as well as its anatomical parts can be found near the rear of the volume. Cross references to the plates appear in the descriptive accounts of the species. Also present is a two-page bibliography that preceeds the plates. A scientific-name index consisting of the families, genera, and species contained within this work completes the reference source.

195. Fisher, Walter K. **Starfishes of the Philippine Seas and Adjacent Waters**. Washington, D.C.: U.S. Government Printing Office, 1919. 712p. (Smithsonian Institution. United States National Museum. Bulletin 100, vol. 3). LC 19-26691.

Based on a collection made by the United States Fisheries steamer *Albatross* from 1907 to 1910, this work provides information on the starfish (seastars) obtained in the region of the Philippine Islands, Celebes, and Molucca islands. The species (192) are arranged under their respective genera, which in turn are entered under subfamilies and families. For each of the species listed, the scientific name, person who first named it, date of naming, detailed anatomical measurements, anatomical relationship to other species, type locality, geographical distribution, and the number of specimens examined are included. Biological keys to species level can be found throughout the descriptions. The descriptions vary in length from one to five or more pages. The introductory information consists of the distribution of species, a list of species recorded from Celebes and the Moluccas, including Bouro, Amboina, and Ceram, and a list of dredging stations at which the starfishes were secured, with the number of species taken. A series of 156 black-and-white plates depicting the animal as a whole as well as its anatomical parts can be found near the rear of the volume. Cross references to the plates appear in the descriptive accounts of the species. A six-page bibliography, and a scientific-name index, consisting of the families, genera, species, and subspecies of Asteroidea contained within this source, complete the work.

FIELD GUIDES

Invertebrates (General)

196. Gosner, Kenneth L. **A Field Guide to the Atlantic Seashore: Invertebrates and Seaweeds of the Atlantic Coast from the Bay of Fundy to Cape Hatteras**. Boston: Houghton Mifflin, 1979. 329p. (The Peterson Field Guide Series, 24). $17.95; $12.95pa. LC 78-14784. ISBN 0-395-24379-3; 0-395-31828-9pa.

Although some portions of this guide consider the identification of algae and seed plants (27 pages), arthropods (45 pages), and primitive chordates (10 pages), the majority of the work is devoted to the remainder of the invertebrate groups (77 pages). The field guide follows a phylogenetic approach. The organisms are arranged under broad phylum or class groups. For each of the species presented, the common name, scienitific name, brief anatomical structure, coloration, similar species, geographical distribution, and family names are given.

Some general information about the animal group as a whole preceeds the descriptive accounts. Introductory information consists of collecting and preserving the specimens as well as their distribution and habitats. In addition, a series of 64 plates can be found in the center of the source. Each plate represents a two-page spread. On one page is a list of organisms giving common name, scientific name, brief anatomical structure, size, and coloration; the opposing page presents pictures of the species either in black and white or in color. A brief glossary, a five-page bibliography, and a combined index of common and scientific names concludes the volume.

197. Gotshall, Daniel W., and Laurence L. Laurent. **Pacific Coast Subtidal Marine Invertebrates: A Fishwatcher's Guide**. Los Osos, Calif.: Sea Challengers, 1979. 112p. $16.95pa. LC 79-64128. ISBN 0-930118-02-2.

Abounding with a large number of colored photographs, this source attempts to identify the most common invertebrates in this geographical area. The species contained within this source are arranged under their respective phyla. For each organism considered, the common name, scientific name, identification information, habitat, size, range, and a colored photograph are given. The photographs show the species in their natural habitats. Encompassing 161 organisms, this work includes representatives from Porifera, Cnidaria, Annelida, Arthropoda, Mollusca, Ectoprocta, Echinodermata, and Chordata. A brief bibliography, a common-name index, a scientific-name index, and maps of the area can be found at the end of the guide.

198. Ruppert, Edward, and Richard Fox. **Seashore Animals of the Southeast: A Guide to Common Shallow-Water Invertebrates of the Southeastern Atlantic Coast**. Columbia, S.C.: University of South Carolina Press, 1988. 429p. $49.95; $24.95pa. LC 87-27349. ISBN 0-87249-534-5; 0-87249-535-3pa.

Arranged by phyla, this guide serves as a source of information regarding many of the invertebrates found off the southeastern Atlantic coast. For each species listed under its respective phylum, the common name, scientific name, identifying characteristics, habitat, and interesting modes of behavior are described. Colored photographic plates are included for many of the species. In addition, a number of black-and-white photographs as well as line-drawings are presented within the textural material. Phylums represented are Porifera, Cnidaria, Ctenophora, Hemichordata, Chordata, Echinodermata, Phoronida, Brachiopoda, Bryozoa, Kamptozoa, Sipuncula, Mollusca, Nemertea, Platyhelminthes, Echiura, Annelida, and Arthropoda.

This field guide also includes an illustrated key to seashore animals; a description of the major groups of marine animals (phylums, subphylums, orders, and classes considered); a section on water, currents, waves and tides; a chapter on marine ecology; a glossary; a list of references; an animal classification scheme down to family; and a combined index of common and scientific names.

Besides acting as an excellent field guide, this volume contains a large number of facts concerning the invertebrate groups.

199. Sefton, Nancy, and Steven K. Webster. **A Field Guide to Caribbean Reef Invertebrates**. Monterey, Calif.: Sea Challengers, 1986. 112p. $19.95pa. LC 85-050789. ISBN 0-930118-12-X.

After a brief description of the coral reefs and a description of the major phyla of invertebrate animals, the invertebrate species follow, arranged by common name. Each species entry, found under its respective phylum, consists of common name, scientific name, anatomical description (including color and size), and habitat. In addition, a colored photograph of each organism included can be found on the opposing page. The photographs are well produced and show the invertebrates in their natural habitats. One should be aware of the general invertebrate phyla in order to make good use of the guide. A combined index of common and scientific names can be found at the end of the volume.

Invertebrates (Systematic Section)

MOLLUSCA
(Snails, Slugs, Clams, Oysters, Squid, Octopi, etc.)

200. Abbott, R. Tucker. **Seashells of North America: A Guide to Field Identification**. New York: Golden Press, 1968. 280p. (Golden Field Guide Series). Price not reported. LC 68-10083. ISBN 0-307-13657-4.

Written for use by all ages (nine and up), this guide contains the common types of shells likely to be encountered on the North American shores. After a brief introduction describing mollusk physiology, anatomy, and taxonomy, along with collecting, cleaning, and preserving them, this source walks you through the various mollusk classes. Organisms are arranged under their respective classes and accompanied by many colored drawings of the shells. A brief description of the species is also present. In the identification process, it helps to know the class; otherwise, it will be necessary to browse through the entire guide to determine a find. A brief bibliography appears at the end of the work, along with an index of common and scientific names.

201. Clarke, Arthur H. **The Freshwater Molluscs of Canada**. Ottowa, Canada: National Museum of Natural Sciences, National Museums of Canada, 1981. 446p. $39.95. LC 87-27349. ISBN 0-660-00022-9.

Serving as both a taxonomic work and a field guide, this work presents the molluscs of Canada according to their respective classes. For the novice there is a key to the families of Canadian freshwater molluscs that leads to the proper family within the text. Organisms are arranged by class, followed by subclass, order, superfamily, and then family. For each species listed, the scientific name, common name, person who first identified it, date of naming, size, external shell anatomy, color, distribution, and ecology are considered. Black-and-white photographs as well as range maps are included for each of the species. A series of colored plates representing many of the bivalves are present. Classes include Gastropoda and Pelecypoda.

Introductory material consists of the why and how of collecting freshwater shells, arrangement and care of the collection, molluscan classification and features, distribution patterns in Canada, and molluscs as pollution indicators. A brief glossary, a list of references, and a combined index to scientific and common names conclude the guide.

202. Douglass, Jackie Leatherbury. **Peterson First Guide to Shells of North America**. Boston: Houghton Mifflin, 1989. 128p. $4.95 pa. LC 88-32884. ISBN 0-395-48297-6.

Written with the novice in mind, this easy-to-understand field guide contains a sample of the 70,000 or more marine mollusks found along the shores of North America. It is arranged by class and consists of the gastropods, chitons, bivalves, and scaphopods. There is a general description of each of the classes, followed by a select number of species arranged by common name. For each species included, an anatomical description including size and color is given. In addition, a colored picture of the shell can be found on the opposing page. A brief description of the group in general is also presented, as well as the species description. For example, there is a general description of gastropods, followed by a general description of conchs, followed by a brief description of a number of species that are conchs. A common-name index completes the guide.

This guide is easy to use and carry about. It presents some of the more familiar shelled mollusks. This volume is not an extensive guide to this group.

203. Eisenberg, Jerome M. **A Collector's Guide to Seashells of the World**. New York: McGraw-Hill, 1981. Reprint. New York: Crescent Books; distr., Crown, 1989. 239p. $15.99. LC 89-32303. ISBN 0-517-69096-9.

Abounding with 158 colored plates, this volume depicts over 2,600 species of shells, arranged by families, found throughout the world. Approximately half of the work consists of the plates and the other half describes the families. For each of the families listed, the class, subclass, common name(s), principal genera, shell morphology, and geographical distribution are given. Unlike *Seashells of North America* (see entry 200), this source is physically larger and one may have difficulty carrying it while collecting shells along the shore; this guide makes for good identification of shells at home. In addition, the pictures are larger than in *Seashells of North America*. A brief discussion of the classes, along with pointers on how to collect and clean shells, can be found in the introduction. The index consists of common and scientific names.

204. Emerson, William K., and Morris K. Jacobson. **The American Museum of Natural History Guide to Shells: Land, Freshwater, and Marine, from Nova Scotia to Florida**. New York: Knopf, 1976. 500p. LC 74-21304. ISBN 0-394-73048-8.

Arranged by class, followed by subclass, order, and family, this work provides descriptive and visual information on shells found from Nova Scotia to Florida. For each family included, the common name, key to genera, genus, etymology, size, distribution, characteristics and an adquate description of the species within the group are given. The middle portion of the work contains colored and black-and-white photographs of the shells for aid in identification.

In addition to an introduction to molluscs and a glossary, which appear at the beginning of the work, a bibliography arranged by authors, and an index of common and scientific names complete the volume.

The format of the guide does not lend itself as readily to shell identification as *The Audubon Society Field Guide to North American Seashells* (see entry 208), due to the absence of a shape index and due to its hard-cover format. This work, however, provides adequate information for identification as well as descriptive reference sources.

205. **The Macdonald Encyclopedia of Shells**. London: Macdonald, 1982. 512p. LC gb82-15097. ISBN 0-356-08575-9.

Flexible in format, this work can be easily carried during walks along the shore. Although titled as an encyclopedia (encompasses the coasts of the world), this volume serves very well as an identification guide. The book is divided into soft-surface mollusks (occurring in sand, mud, and aquatic vegetation); firm-surface mollusks (occurring in rocks, gravel, and oyster beds); coral dwellers; other marine mollusks; and land and freshwater mollusks. For each organism included, a world-range map, common name, scientific name, family, anatomical description (including size and color), body, habitat, and distribution are presented. In addition, a colored photograph of the shell appears on the opposing page. A general description of mollusks, including external and internal anatomy, interactions with the environment, methods of reproduction and dispersion, classification, habitat, collection methods, steps involved in preserving the animal as a whole, cleaning, and classifying shells can be found in the introductory pages. A comprehensive classification table of species included, a brief glossary, and a combined index of scientific and common names can be found closing the volume. This is a useful work and can be used as an encyclopedia as well as a field guide.

206. Morris, Percy A. **A Field Guide to Shells of the Atlantic and Gulf Coasts and the West Indies**. 3d ed. Boston: Houghton Mifflin, 1973. 330p. (The Peterson Field Guide Series). LC 72-75612. ISBN 0-395-16809-0.

Consisting of 75 pages of plates of shells, in both color and black and white, this guide encompasses those shells found in the Atlantic, Gulf Coasts, and the West Indies. The species included on each page of plate can be identified by common and scientific name on the proceeding page. Plates are arranged by type of shell and, in some cases, location. For example, there are pages of plates that include New England shells, Caribbean shells, whelks and conchs, and so on.

In addition to the plates, found in the center of the guide, descriptions of the organisms are given. Descriptions include taxonomic information, common and scientific names, person who first identified the species, range, habitat, and gross morphology. The molluscs are arranged under their respective families.

The plates seem to be the key in the initial identification process. These will lead to the page of text where the description is located. The end of the volume provides a glossary, a list of authors, a brief bibliography, a list of eastern shell clubs, and an index of common and scientific names.

The guide has been published for easy transportation to the shore; however, some knowledge of shells would be helpful for using this source.

207. Oliver, A. P. H. **The Henry Holt Guide to Shells of the World**. New York: Henry Holt, c. 1975, 1989. 320p. $12.95pa. LC 89-1716. ISBN 0-8050-119-6.

Covering a wide range of saltwater shells from all over the world, this identification guide encompasses shells found in all seas from polar to tropical and in shallow and deep water. About 1,200 species are described and over 100 are illustrated in full color. The description of a species falls on the same two-page spread as its illustration and gives the size, distribution, color, characteristic features of the shell, scientific name, and the date and person of the first description. The illustrations are life-size whenever possible. In many cases the scale is altered; however, shells on any one page are in proportion to each other.

The majority of this work is devoted to the Gastropoda (snails; 283 pages). The author has also included some of the more interesting Bivalvia (12 pages) as well as short references to the Cephalopoda, Scaphopoda, Amphineura, and Monoplacophora. The basis for inclusion is that a shell must be fairly common, although some rare ones are found in the volume. An introductory section includes information about mollusks and their classification, hints for collecting shells, and a short bibliography and glossary. An alphabetical index of scientific and common names can be found at the close of the guide.

Although this is not a comprehensive work, what is included is extremely well done. The color paintings are excellent.

208. Rehder, Harald A. **The Audubon Society Field Guide to North American Seashells**. New York: Knopf, 1981. 894p. $17.95. LC 80-84239. ISBN 0-394-51913-2.

Consisting of 705 colored plates of shells, this field guide enables the collector to identify seashells of North America. The plates are arranged taxonomically, and are given such headings as chitons, moon snails, limpets, and clams. In addition, at the bottom of the plate, the page number is listed in which one can find a description of the mollusc. Descriptions give common and scientific name, size, color, gross morphology, habitat, and range. In using this guide it is best to start out browsing the colored plates, which will then lead to the description. There is a shape index at the front of the guide that will give access to those colored plates that would be most useful in the identification process.

A glossary and an index of common and scientific names can be located at the end of the volume along with a four-page explanation on how to collect and preserve shells. The book has a soft and flexible cover that makes for easy transport in the field, or, in this case, at the shore.

209. Wye, Kenneth R. **The Simon & Schuster Pocket Guide to Shells of the World**. New York: Simon & Schuster, 1989. 192p. $11.95pa. ISBN 0-671-68263-6.

The shells are arranged by their class, followed by superfamily. These headings appear at the top of each page. Each colored plate of shells is presented on a black background. A number of shells are found on each page. For each shell included, the scientific name, common name, person who named it, date of naming, size (in cm. and in.), geographical location, coloration, shell thickness (in some cases), and status are given. Status considers availability, such as abundant, common, uncommon, and rare. In addition, a brief description of the family is also considered. The classes Gastropoda and Bivalvia are the main representatives contained within this field guide, with a few pages devoted to the Cephalopoda, Scaphopoda, and Polyplacophora. A two-page glossary and a combined index of common and scientific names can be found at the end of this volume. Within the introduction are shell outlines that lead to the proper pages within the text. For the novice, these shell outlines (identification charts) should be used first.

The book is published in a thin, long format with a soft flexible cover that makes for easy carrying along the shore.

TEXTBOOKS

Invertebrates (General)

210. Alexander, R. McNeill. **The Invertebrates**. Cambridge, London: Cambridge University Press, 1979. 562p. $82.00; $32.50pa. LC 78-6275. ISBN 0-521-22120-X; ISBN 0-521-29361-8pa.

Although arranged in a phylogenetic manner, this work tends to emphasize physiological characteristics of invertebrates rather than the classic anatomical, morphological approach. For example, when discussing one of the flatworm chapters, a rigorous presentation of the respiration and diffusion of gases within the group is given, along with the mathematical formulas to support it. Some calculus would be useful in fully understanding the discussion. Another example can be found in the chapter entitled "Insects in General": the physiologies of flight, wing muscles, respiration, water balance, and so on are presented. It assumes some knowledge of physiological concepts. It is also useful to note that in a few cases the biochemistry of a group of organisms is discussed.

The text also differs from many of the other texts in that the names of the chapters do not necessarily reflect the names of the phylums. For example, when considering the Cnidaria, the chapters are entitled the "Sea Anemones and Corals," and "Hydroids and Jellyfishes."

Illustrative matter consists of labeled and unlabeled line-drawings, molecular structures, graphs, and mathematical formulas. A brief bibliography can be found at the end of each chapter. The index consists of common and scientific names as well as subjects. Page numbers in italics refer to the illustrations.

This volume represents an interesting approach to the invertebrates. It assumes, however, that one has some background in higher mathematics, physiology, and biochemistry.

211. Barnes, Richard, Stephen Kent, P. Calow, and P. J. W. Olive. **The Invertebrates: A New Synthesis**. Oxford, Boston: Blackwell Scientific, 1993. 488p. £47.50; £19.95pa. LC 92-32501. ISBN 0-632-03125-5; 0-632-03127-1pa.

Although there are a few chapters discussing the invertebrate phyla, such as worms, molluscs, lophophorates, deuterostomes, and invertebrates with legs, this text emphasizes invertebrate functional biology rather than the classic phylogenetic approach. Chapters about feeding, mechanics and movement, defence, reproduction and life cycles, and control systems, to name a few, make up half the text. Specific invertebrate animals are used to represent these types of functions. For example, in the chapter dealing with mechanics and movement, protozoans, molluscs, worms, and arthropods are used to illustrate the concepts in question. In addition, physics principles are taken into account, such as kinetic energy and friction, resolution of forces, and principles of a lever. Brief taxonomic keys enhance some chapters.

The majority of the illustrations are labeled line-drawings of the internal and external anatomy of the organisms. A number of graphs depicting physiological concepts are also present. A brief bibliography is present at the end of each chapter. A glossary, a list of illustration sources, and a common/scientific-name and subject index complete the text.

As in the case of *The Invertebrates,* by Alexander (see entry 210), this work does not follow the traditional scheme; still, this volume may be easier to digest than the former.

212. Barnes, Robert D. **Invertebrate Zoology**. 5th ed. Philadelphia: Saunders College, 1987. 864p. $40.00. LC 86-10023. ISBN 0-03-008914-X.

Following the traditional format as found in other texts, this work presents the invertebrate animals in a phylogenetic fashion. Each chapter represents a phylum. Representative animals are then discussed showing features of the phylum and their respective classes. For each group of organisms (many of the minor phyla are also considered), their external and internal anatomy, physiology, life cycles, feeding habits, reproduction, and the like are considered. Lined and labeled drawings, as well as photographs, charts, and graphs, enhance the explanations of each group. Labeled drawings include external and internal anatomy.

For each class of animals mentioned within the chapter, a summary of the most salient features of the group and the systematics of the class are presented. In addition, a list of references can be found at the end of each chapter. The index includes access by subjects, common names, and scientific names.

This is a comprehensive work on invertebrates and would serve well for a one- or two-semester course.

213. Brusca, Richard C., and Gary J. Brusca. **Invertebrates**. Sunderland, Mass.: Sinauer Associates, 1990. 922p. $49.95. LC 90-30061. ISBN 0-87893-098-1.

Arranged in a phylogenetic manner, this text takes the reader through the major and minor groups of invertebrates, as well as the prochordates. Phylums, subphylums, and classes are represented within each section to show the variety of organisms within the group. Photographs and labeled line-drawings enhance the discussion. The emphasis seems to be on taxonomy, morphology, evolution, reproduction, and feeding mechanisms of the invertebrate groups. Physiology is also discussed to some extent, but not to the depth that was found in *The Invertebrates* (see entry 210), and *The Invertebrates: A New Synthesis* (see entry 211).

A number of introductory chapters, such as "Classification, Systematics, and Phylogeny"; "Animal Architecture"; and general discussions on the metazoa, are present and set the stage for the main body of the volume. A list of selected references is found at the end of each chapter and the references are arranged under their respective taxonomic groups. A subject, common name, and scientific-name index completes the work. Numbers in boldface signify definitions, while those in italics signify illustrations.

This is a very comprehensive treatise on all the invertebrate groups and may be a bit much for a one-semester course.

214. Buchsbaum, Ralph, and others. **Animals Without Backbones**. 3d ed. Chicago: University of Chicago Press, 1987. 572p. $29.95, $19.00pa. LC 86-7046. ISBN 0-226-07873-6; 0-226-07874-4pa.

Initially appearing in 1938, this work is one of the classic texts in invertebrate zoology. Following the typical phylogenetic approach, this work is an interesting and informative read. The large number of photographs and excellent, labeled line-drawings enhances the concepts and factual information on these invertebrate creatures. The drawings, showing external and internal anatomy (sagital and cross sections), are large and simplified in order to gain a better appreciation of the invertebrates. All major groups of invertebrates are covered, using representative organisms to illustrate a phylum or class. The chapters, themselves, are not identified by the name of the phylum, but rather by the common name of the group. There are also conceptual chapters introduced that allow the reader a better understanding of the animals that follow. For example, there is a chapter entitled "Two Layers of Cells" that is introduced before the chapter called "Stinging-Celled Animals." In addition to the major groups, there is a short chapter entitled "Lesser Lights" that incorporates some of the minor phyla. A selected list of references, a general classification scheme, and a subject and common/scientific-name index can be found at the close of the volume.

Although slanted more toward morphological information, life cycles, ecology, evolutionary trends, and some physiology are present. This is an excellent work that seems to get better with time.

215. Crawford, Clifford S. **Biology of Desert Invertebrates**. Berlin, New York: Springer-Verlag, 1981. 314p. $79.00. LC 81-9024. ISBN 0-387-10807-6.

Written in a textbook format, this work could just as easily serve as a handbook, or an encyclopedia. All various types invertebrates are considered, such as the protozoans, nematodes, annelids, mollusks, crustaceans, arachnids, and insects. Part 1, broken up into two chapters, consists of desert ecology and the various forms of invertebrates found in the desert. Part 2 considers adaptations to xeric environments and considers the use of light and timing of activity, water relations, temperature relations, and energetics. Part 3 delves into life-history patterns and encompasses multivoltine and univoltine species, herbivores, detritivores, and carnivores. Part 4 discusses the ecology of deserts and the invertebrates living therein, while part 5 summarizes the invertebrates in desert ecosystems. Most of the illustrations are black-and-white photographs of particular organisms. A 33-page bibliography and a combined subject, common-name, and scientific-name index concludes the text. This volume represents an excellent introduction to those invertebrates endemic to the deserts as well as their adaptive mechanisms to this type of environment.

216. Engemann, Joseph G., and Robert W. Hegner. **Invertebrate Zoology**. 3d ed. New York: Macmillan, 1981. 746p. LC 80-12063. ISBN 0-02-333780-X.

Arranged along traditional lines, this work follows the typical phylogenetic approach. Each of the chapters represent a phylum, and the chapter itself is broken up into classes and superclasses. Habitat, locomotion, reproduction, nutrition, internal and external anatomy, taxonomy, and some physiology is covered for each of the groups. A representative animal is used in order to discuss the principles of the phylum. The majority of the illustrations are in the form of labeled and unlabeled diagrams. Occasional transmission electron microscopy photomicrographs, photographs, and graphs are present.

At the end of each chapter is a summary taking into account the salient features of the group. The summaries are useful and serve as a minitext. In addition, in most cases, a classification scheme of the phylum can be found. A selected list of references completes each chapter. The index follows along traditional lines and combines subject, common names, and scientific names together in an alphabetical arrangement.

217. Meglitsch, Paul A., and Frederick R. Schram. **Invertebrate Zoology**. 3d ed. New York: Oxford University Press, 1991. 623p. $47.50. LC 89-26592. ISBN 0-19-504900-4.

Phylogenetically arranged, this work not only devotes chapters to each of the major phyla, but also devotes space to chapters discussing the minor phyla. In most other texts, the minor phyla are usually lumped together. For example, besides chapters dealing with Protozoa, Mollusca, Platyhelminthes, and so on, there are chapters about Gastrotricha, Rotifera, Priapulida, Kinorhyncha, Chaetognatha, Sipuncula, and Echiura, to name a few.

The chapters are divided into classes and, using representative animals, all major zoological concepts are discussed. Labeled and unlabeled line-drawings provide the illustrations. These drawings, as in the case of the invertebrate texts in general, consider external and internal anatomy. Each chapter concludes with a thumbnail taxonomic sketch as well as a brief listing of selected readings. There is, at the end of the volume, a taxonomic index and a subject index.

This work provides a good balance between morphology, physiology, and the other zoological disciplines. No one area seems to dominate.

218. Morris, S. Conway, ed., and others. **The Origins and Relationships of Lower Invertebrates**. Oxford: Published for the Systematics Association by Clarendon Press, 1985. 397p. (The Systematics Association, Special Volume; No. 28). $125.00. LC gb84-43831. ISBN 0-19-857181-X.

Although this is not a textbook in the traditional sense, it can provide supplementary material in the fields of lower invertebrates and minor-phyla zoology. This volume attempts to review all the work to date concerning the evolution of these organisms and their relationships to each other. The phyla considered are Porifera, Cnidaria, Ctenophora,

Platyhelminthes, Gnathostomulida, Annulonemertes, Rotifera, Gastrotricha, Priapulida, Sipuncula, Annelida, and the Pogonophora. The work is enchanced by line-drawings and transmission electron microscopy photomicrographs. A list of references can be found at the end of each chapter. The work concludes with a separate author index, a subject index, and an index of genera and species.

Because so many of the texts cover the minor phlya in a superficial way, this work may aid in a more complete understanding of these interesting organisms.

219. Pearse, Vicki, and others. **Living Invertebrates**. Palo Alto, Calif.: Blackwell Scientific, 1987. 848p. $49.95. LC 86-10790. ISBN 0-86542-312-1.

A broadened out version of *Animals Without Backbones* (see entry 214), which was first published in 1938, this textbook provides the reader with a phylogenetic approach to the invertebrates. In some cases, several chapters have been written for a particular phylum. For example, there is a chapter entitled "Cnidarian Body Plan: Medusas and Polyps," as well as "The Cnidarian Array: Hydrozoans, Scyphozoans, Anthozoans." The former discusses the group of stinging-cell animals in general, while the later considers many of the types and varieties of organisms making up the group. Much of the material making up the chapters includes internal and external morphology. Zoological disciplines, such as locomotion, feeding, regeneration, and reproduction are covered in a painless manner. It is an interesting and informative read. A brief classification scheme is offered at the end of each chapter.

The volume abounds with labeled diagrams and photographs, at both the microscopic and the macroscopic levels. The labeled diagrams depict external and internal anatomy, while the photographs show the diversity of the many types of invertebrates found in each of the phylums. Many of these illustrations have been taken from *Animals Without Backbones*. The work ends with a number of general chapters, including "Animal Relationships" (embryology, larval morphology, fossil evidence, synthesis) and "Colors of Invertebrates," which includes a large number of colored plates depicting the variety of colors found in the many invertebrate groups. In addition to the index of subjects, common names, and scientific names, a brief listing of references arranged by chapter headings can be found at the end of the work.

This is a very readable text. The multitude of diagrams and photographs keep the interest of the reader. When I was a biology teacher (and a terrible artist), this text provided me with diagrams that I was able to reproduce easily on the chalk board.

Protozoa (Unicellular Organisms)

220. Kudo, Richard R. **Protozoology**. 5th ed. Springfield, Ill.: Charles C. Thomas, 1977. Reprint. Springfield, Ill.: Charles C. Thomas, 1966. 2 vols. $145.25. LC 65-23316. ISBN 0-398-01058-7.

Consisting of 388 sections of diagrams, this work is a classic in its field. Although it has been placed under the textbooks, it could just as easily serve as an encyclopedia or a handbook, or, for that matter, a sophisticated field guide.

It is divided into two parts. Part I considers general biology and consists of the ecology, morphology, physiology, reproduction, variation, and heredity of the protozoans. Part II is strictly taxonomic (morphological and anatomical considerations), and includes the subphylums, classes, subclasses, and orders of the one-celled creatures.

The illustrations consist mainly of line-drawings of the beasts showing external and internal morphology. A set of colored plates are used to illustrate the various stages and species of plasmodium in RBC's. Drawings are used to illustrate points in the text and to make identification easier. A chapter entitled "Collection, Cultivation, and Observation of Protozoa," and an author and subject/scientific-name index complete the tome. A list of references can be found at the end of each chapter.

This is a classic work in the field of protozoology; however, some of the material in part I may be outdated.

Mollusca (Snails, Slugs, Clams, Oysters, Squid, Octopi, etc.)

221. Purchon, R. D. **The Biology of the Mollusca**. Oxford, New York: Pergamon, 1977. 560p. (International Series in Pure and Applied Biology: Division, Zoology; vol. 57). $105.00. LC 76-10804. ISBN 0-08-021028-7.

Arranged under physiological concepts, this source delves into external and internal anatomy, feeding methods and adaptive radiation in the Gastropoda and Bivalvia; digestion, reproduction, distribution, and the nervous system in the Dibranchiate Cephalopoda. Photographs and labeled line-drawings provide the illustrations. In addition to the main portion of the book, there are a number of appendices. Appendix A discusses the anatomy and physiology of four minor classes (Monoplacophora, Polyplacophora, Aplacophora, and the Scaphopoda), while appendix B considers the anatomy and physiology of the three major classes (Bivalvia, Gastropoda, and Cephalopoda.). An index of subjects, common names, and scientific names concludes the volume.

Echinodermata (Starfish, Sea Urchins, Sea Cucumbers, etc.)

222. Lawrence, John. **A Functional Biology of Echinoderms**. London: Croom Helm, 1987. 340p. $60.00. LC 87-2843. ISBN 0-7099-1642-6.

Arranged by function rather than class, this book considers the life processes of the echinoderms. Sections include characteristics of the phylum and its classes, feeding and nutrients, respiration, circulation, locomotion, taxonomic and evolutionary position, defense mechanisms, and reproduction. Examples from various classes are used to illustrate these concepts. Photographs and drawings provide the many illustrations. An appendix discussing the classification of extant echinoderms, a comprehensive bibliography, and an index of subjects, common names and scientific names can be found at the close of the work. Bibliographic entries have been cited within the text.

JOURNALS

Invertebrates (General)

223. **Invertebrate Taxonomy**. Vols. 1- , nos. 1- . East Melbourne, Victoria, Australia: Commonwealth Scientific and Industrial Research Organisation (CSIRO), 1987- . Bimonthly. $220.00/yr. ISSN 0818-0164.

This serial includes research work involved with the taxonomy and systematics of invertebrates. Many of the articles are concerned with the Indo-Pacific region of the world. Most of the papers are enhanced by line-drawings, maps, scanning electron microscopy photomicrographs, tables, charts, or any combination thereof. The number of articles contained within each issue varies from one to eight or more. All invertebrate types are considered, including the insects. For example, in 5(6), 1991, the entire issue (239 pages) is devoted to a revision of the family Raspailiidae, a family belonging to the sponges (Porifera). The issues, for the most part, are over 200 pages in length.

The articles represent finely detailed work and are extremely useful for the taxonomist involved in one of these specific invertebrate groups.

Protozoa (Unicellular Organisms)

224. **Journal of Foraminiferal Research**. Vols. 1- , nos. 1- . Washington, D.C.: Cushman Foundation for Foraminiferal Research, 1971- . Quarterly. $80.00/yr. (institutions). ISSN 0096-1191.

This is a highly specialized work dealing with a specific group of protozoans, known as the Foraminifera and related groups. Ecology, evolution, and taxonomy of both living and fossil forms are considered. The articles usually number 10 or less for each issue. Most of the research papers range from seven to fifteen pages. Scanning electron microscopy photomicrographs, graphs, tables, and charts encompass the majority of the illustrations. The Cushman Foundation Membership Directory is included in the final issue of the volume. It should be noted that this journal continues *Contributions from the Cushman Laboratory for Foraminiferal Research* (1925-1950) and *Contributions from the Cushman Foundation for Foraminiferal Research* (1950-1970).

225. **The Journal of Protozoology**. Vols. 1- , nos. 1- . Lawrence, Kans.: Society of Protozoologists, 1954- . Bimonthly. $116.00/yr. (institutions). ISSN 0022-3921.

As the name implies, this journal is devoted exclusively to the protozoans, including the parasitic ones. Original research articles deal with the biochemical, physiological, growth, medical, veterinary, systematics, evolution, ecological, and ethological aspects of the group. Excellent photomicrographs (transmission electron microscopy and scanning electron microscopy), colored and black-and-white diagrams, tables, graphs, and charts aid in the research discussion. The table of contents lists the papers under broad subject headings such as cell biology.

Supplements can be found within certain issues encompassing a large number of short communication papers. For example, 38(6), 1991 contains, as well as the main articles, the "Second International Workshop on Pneumocystitis, Cryptosporidium and Microsporidia." This supplement covers 243 pages of a 345-page issue. The main articles vary in length from 20 to around 50 pages.

Book reviews, memoriums, and abstracts of meetings of protozoology associations occur from time to time in the issues. This is a major journal for protozoologists.

Nemata (Free-living and Parasitic Roundworms)

226. **Journal of Nematology**. Vols. 1- , nos. 1- . Lake Alfred, Fla.: Society of Nematologists, 1969- . Quarterly. $70.00/yr. (institutions). ISSN 0022-300x.

This publication considers research articles dealing with free-living and parasitic roundworms. Population studies, growth, physiology, morphology, parasitism, ecology, and so on are areas included. Many of the papers encompass the effect of nematodes on economic plants. Illustrative material includes transmission and scanning electron microscopy photomicrographs, line-drawings, charts, and graphs.

The number of research articles contained within each issue are between 15 and 20 and range in length from approximately four to twelve pages. Memoriums are also included in some issues. Supplements to the journal are published and titled *Annals of Applied Nematology*. Articles consider nematode problems as they relate to agricultural crops.

This periodical not only serves the invertebrate zoologist, but the agriculturalist involved in plant production.

227. **Nematologica: International Journal of Nematological Research**. Leiden, The Netherlands: E. J. Brill, 1956- . Quarterly. $150.00/yr. ISSN 0028-2596.

Unlike the *Journal of Nematology* (see entry 226), whose slant is more towards the applied end of nematodes, this periodical considers research articles dealing with taxonomy and systematics, ultrastructure, functional morphology, ecology, ethology, life histories, control of nematodes, physiology, biochemistry, and genetics. Scanning and transmission electron microscopy photomicrographs, drawings, charts, and graphs provide much of the illustrative work.

Each issue contains between seven and ten articles ranging between four and ten pages in length. All disciplines are not covered in every issue. The table of contents arranges articles under broad biological disciplines. Conference proceedings and short communications appear in selected issues.

Parasites

Because the majority of parasites are invertebrates, a number of core parasitology journals are included here. Those dealing specifically with the medical aspects are not considered in this listing.

228. **Experimental Parasitology**. Vols. 1- , nos. 1- . 1951- . San Diego, Calif.: Academic Press. Published irregularly. $342.00/yr. ISSN 0014-4894.

Physiology, biochemistry, molecular biology, and immunology provide the slant for these research articles dealing with animal parasites. In addition to the main research articles, research briefs and minireviews are also present in many issues. Subject areas such as cloned ribosomal RNA gene probes, regulated eggshell formation by pH and calcium, and identification of parasites by in situ hybridization with total and recombinant DNA probes are examples of the types of papers included.

Transmission and scanning electron microscopy photomicrographs, charts, and tables enhance the research discussions. Approximately 10 to 15 articles appear regularly in each issue, each of which is approximately 10 to 20 pages in length.

229. **International Journal for Parasitology**. Vols. 1- , nos. 1- . Oxford, England: Published for the Australian Society for Parasitology by Pergamon Press, 1971- . 8/yr. $575.00/yr. ISSN 0020-7519.

This serial provides research papers on all types of parasitic invertebrates in all biological disciplines. Subject examples include nuclear magnetic resonance studies, genetic variations, vaccinations, ultrastructure, and regenerative processes. Line-drawings, tables, charts, and scanning and transmission electron microscopy photomicrographs can be found throughout the pages. Approximately 15 to 20 articles appear in each issue, ranging in length from approximately 5 to 20 pages. In addition to the research papers, book reviews; a parasitological calendar listing meetings, symposia, and so on; obituaries; and centenary biographical notes can be found.

230. **The Journal of Parasitology**. Vols. 1- , nos. 1- . Lawrence, Kans.: American Society of Parasitologists, 1914- . Bimonthly. $90.00/yr. (institutions). ISSN 0022-3395.

Dealing with animal parasites, such as the protozoans and helminths, this serial publishes articles in biological areas such as behavior, biochemistry, biosystematics and surveys, ecology, diagnostics, epidemiology, drug action, in vitro cultivation, life cycles, morphology, therapeutics, pathogenesis, and host response. Citations found in the table of contents are arranged under their respective headings. Book reviews, memoriums, obituary notices, and general announcements appear in most issues. Charts, graphs, tables, and transmission and scanning electron microscopy photomicrographs provide the illustrations. Approximately 25 to 30 articles, ranging in length from 5 to 20 pages, appear in each issue.

231. **Molecular and Biochemical Parasitology**. Vols. 1- , nos. 1- . Amsterdam: Elsevier, 1986- . Published irregularly. $1,207.00/yr. ISSN 0166-6851.

Covering parasitic protozoans and helminths, this journal attempts to publish papers in the fields of molecular biology and immunology as well as biochemistry. In addition, the emphasis is on how the parasites affect the intermediate and definitive hosts at the molecular level. Such subjects as molecular cloning, tyrosine phosphorylation, mitochondrial development, and so on are examples of the research papers included. Charts, graphs, drawings, and photomicrographs enhance the textural contents. There are approximately 12 to 15 articles per issue ranging in length from 4 to 20 pages.

232. **Parasite Immunology**. Vols. 1- , nos. 1- . Oxford, England: Blackwell Scientific, 1979- . Bimonthly. $337.50/yr. ISSN 0141-9838.

Research papers discussing immunological aspects of parasites can be found throughout these issues. In addition to animal parasites such as helminths and protozoans, the bacteria, fungi, and viruses are also included. However, in reviewing a number of issues, the emphasis seems to be on the protozoans and helminths. Articles consider how hosts immunologically react to parasites as well as the pathology that occurs in the infected hosts. Charts, graphs, tables, and photomicrographs provide the illustrations. Approximately nine articles can be found in each issue, along with a brief communication paper (most issues). The research papers usually range in length from approximately 10 to 15 or more pages.

233. **Parasitology**. Vols. 1- , nos. 1- . Cambridge, England: Cambridge University Press, 1908- . Bimonthly. $336.00/yr. ISSN 0031-1820.

Considering all types of invertebrate parasites in all biological disciplines, this work provides a forum for parasitic research being undertaken. In addition to research articles, reviews of particular interest are also included. Examples of representative subjects include serological activity, culture characteristics, larval development, penetration and migration routes, and distribution patterns. Line-drawings, tables, charts, and scanning and transmission electron microscopy photomicrographs enhance the text. Approximately 15 to 20 articles appear in each issue ranging in length from 5 to 20 pages.

In addition to the six issues per year, a supplement is published with the first volume that contains the "Proceedings of the Symposia of the British Society for Parasitology."

234. **Systematic Parasitology**. Vols. 1- , nos. 1- . The Netherlands: Kluwer Academic, 1979- . Published irregularly. $388.50/yr. ISSN 0165-5752.

As the name implies, this serial publication accepts research papers in the fields of systematics, taxonomy, and nomenclature dealing with parasites. It considers invertebrate organisms such as the roundworms, flukes, tapeworms, Acanthocephala, protozoans, and molluscs. Line-drawings, tables, charts, and photomicrographs can be found complementing the papers. The drawings are exceedingly well done, some taking up an entire page. There are approximately 4 to 10 articles per issue ranging in length from approximately 10 to 20 pages. The articles are published in English, and in French (with English summaries).

Mollusca (Snails, Slugs, Clams, Oysters, Squid, Octopi, etc.)

235. **Journal of Molluscan Studies**. Vols. 42- , nos. 1- . Oxford, England: Oxford University Press, 1976- . Quarterly. $130.00/yr. ISSN 0260-1230.

Continuing the *Proceedings of the Malacological Society of London*, this periodical covers all zoological disciplines related to molluscs. From the feeding mechanism of the toxoglossan to the enzyme analysis of DNA from species of Bulinus, this work publishes original research papers in the field of malacology. Graphs, charts, transmission and scanning photomicrographs, and drawings enhance the text.

There are approximately 10 articles per issue ranging in length from 10 to 12 pages. Issues also contain research notes, and an occasional obituary.

A supplement issue makes up part of one volume per year and usually contains 20 to 25 articles.

236. **Malacologia**. Vols. 1- , nos. 1- . Philadelphia: Institute of Malacology, 1962- . Published irregularly. $35.00/yr. (institutions). ISSN 0076-2997.

Very similar in many ways to the *Journal of Molluscan Studies* (see entry 235), this serial publication offers research articles that consider all aspects of molluscs pertaining to all zoological disciplines. Papers contained within each issue range from 10 to 15 pages in length. The length of articles vary considerably, from 10 to over 100 pages. Illustrative material includes photographs, transmission and scanning micrographs, labeled line-drawings, graphs, charts, and tables.

In addition to the contributed papers, the research presented at the International Malacological Congress is also presented. Each of these congresses has a theme and the papers reflect the same. For example, the 9th International Malacological Congress was devoted to the evolutionary biology of opisthobranchs.

237. **The Nautilus**. Vols. 1- , nos. 1- . Silver Spring, Md.: Trophon Corporation, 1889- . Quarterly. $35.00/yr. (institutions). ISSN 0028-1344.

Mostly devoted to research articles dealing with the systematics of molluscs, this periodical usually contains less than 10 articles per issue, ranging in length from 3 to 10 pages. Labeled line-drawings, photographs, charts, graphs, and scanning and transmission photomicrographs provide the illustrations.

238. **The Veliger**. Vols. 1- , nos. 1- . Berkeley, Calif.: California Malacozoological Society, 1958- . Quarterly. $56.00/yr. (institutions). ISSN 0042-3211.

Although the word *veliger* is used to describe a mollusc larval stage following the trochophore larval stage, this serial is devoted to all aspects of molluscan biology covering all zoological disciplines. However, in examining volume 34 of this publication, a good number of the papers were devoted to systematics and taxonomy.

Illustrated with photographs (colored and black and white), charts, graphs, drawings, and scanning electron micrographs, each issue contains approximately 7 to 10 articles ranging in length from 10 or less to 30 or more pages.

A section entitled "Notes, Information and News" can be found at the end of each issue. It contains very brief articles and general molluscan information. Some issues also contain a few pages that consider reviews of new books, periodicals, and pamphlets in the science of malacology.

Echinodermata (Starfish, Sea Urchins, Sea Cucumbers, etc.)

239. **Echinoderm Studies**. Vols. 1- . Rotterdam, The Netherlands: A.A. Balkema, 1983- . Biennially. $54.00/yr. ISSN 0168-6100.

Published as a hard-copy book, this serial is devoted to all aspects of echinoderm biology. Contributions to the fields of ecology, evolution, paleontology, developmental biology, molecular biology, reproduction, behavior, and physiology are covered. The articles reflect reviews of these topics. Extensive bibliographies follow each of the review papers. Illustrations consist of drawings and photographs. Because these are reviews, the work is not as heavily illustrated as in many of the journals listed in this section. Examples of some of the review papers include influence of echinoderms on coral-reef communities, calcification in echinoderms, and effects of salinity on echinoderms.

Each issue contains approximately four to seven articles, varying in length from 50 to over 100 pages.

TAXONOMIC KEYS

Invertebrates (General)

240. Edmonson, W. T., ed. **Fresh-Water Biology**. 2d ed. New York: Wiley, 1959. 1248p. LC 59-6781.

Originally published in 1918 by Ward and Whipple, this source essentially contains illustrated keys to freshwater animals and plants of North America (southern boundary being the Rio Grande). The animal portion of the volume deals with the invertebrates and comprises the majority of the information. The keys to the animal groups are introduced under either their phylum, class, or order. For each of these groups, brief information dealing specifically with characteristics and identification preceeds the keys themselves. The keys key down to the species level. Line-drawings provide the illustrations. All major invertebrate groups are covered, including the insects. The insects account for 215 pages, the other invertebrate groups take up 763 pages, and the remainder of the source is devoted to the nonvascular and vascular plants. A section on techniques and materials useful for collecting and preserving these organisms can be found near the end of the work. A combined index of subjects, common names, and scientific names concludes this taxonomic key. Although outdated, it is still a very useful source and has been around for a long time.

241. Kozloff, Eugene N. **Marine Invertebrates of the Pacific Northwest**. Seattle, Wash.: University of Washington Press, 1987. 511p. $35.00. LC 87-10550. ISBN 0-295-96530-4.

Covering all invertebrate groups found in this area, this work, arranged phylogenetically, provides the keys necessary to identify the organisms down to their species. The phyla represented in this taxonomic work are Porifera, Cnidara, Ctenophora, Platyhelminthes, Orthonectida, Dicyemida, Nemertea, Rotifera, Gastrotricha, Kinorhyncha, Nematoda, Entoprocta, Mollusca (makes up the largest group), Annelida, Priapulida, Sipuncula, Echiura, Tardigrada, Arthropoda, Phoronida, Bryozoa, Brachiopoda, Echinodermata, Chaetognatha, Hemichordata, and Urochordata. Each phlyum section, in addition to its keys, briefly describes the class and/or phylum as it relates to the Puget Sound area. Brief bibliographies occur as part of the description. A large number of line-drawings aid in the identification of the animal groups. A five-page glossary and a scientific-name index complete the work. It should be noted that this work represents a revised and expanded edition of Eugene N. Kozloff's *Keys to the Marine Invertebrates of Puget Sound, the San Juan Archipelago, and Adjacent Regions* (University of Washington Press, 1974).

242. Peckarsky, Barbara L., and others. **Freshwater Macroinvertebrates of Northeastern North America**. Ithaca, N.Y.: Cornell University Press, 1990. 442p. $57.50; $26.50pa. LC 89-17468. ISBN 0-8014-2076-8; 0-8014-9688-8pa.

Although presenting itself as a textbook, the chapters in this volume mainly comprise, among other pieces of information, lengthy taxonomic keys. Therefore, this work has been placed in the taxonomic key category. For each group of organisms considered, general information on classification, life history, habitat, feeding, respiration, collection and preservation, and identification is given. In addition, there are keys to the orders of the taxonomic group, along with line-drawings of the species. An adequate bibliography closes each of the chapters. Approximately 330 pages deal with the arthropods, specifically the insects. Other freshwater invertebrates considered include the Mollusca, Oligochaeta, and Hirudinea. There are a total of 18 chapters that describe certain groups of invertebrates. Many of these are insect orders. A glossary and a scientific-name index conclude the book.

Protozoa (Unicellular Organisms)

243. Carey, Philip G. **Marine Interstitial Ciliates: An Illustrated Key**. London, New York: Chapman & Hall, 1992. 351p. $169.50. LC gb92-6616. ISBN 0-412-40610-1.

Consisting of those ciliated protozoans found between the sand grains in marine sediments, this source provides the biological keys necessary to ascertain identification of the same. Keys to the class, subclass, order, suborder, family, genus, and species are present. For each of the species listed (937 species) within the key, the scientific name, person who first named it, date of naming, size, shape, and external and internal anatomical structures are given. In addition, drawings of each of the species depicting shape and organelles can be found near the end of the volume. Reference numbers to the drawings are located within the main portion of the work (biological keys). The drawings span a total of 106 pages. Information in the introduction consists of practical methods leading to the collection, transportation, and laboratory handling of these protozoans. A 12-page bibliography, a 7-page glossary, a methods index, and a scientific-name index conclude the reference source.

244. Jahn, Theodore Louis, Eugene Cleveland Bovee, and Frances Floed Jahn. **How to Know the Protozoa**. 2d ed. Dubuque, IA: Wm. C. Brown, 1979. 279p. (The Pictured Key Nature Series). $12.80 spiralbound. LC 78-52716. ISBN 0-697-04759-8.

Arranged by phyla, this guide enables one to identify the various types of protozoans either by the written key (morphological descriptions from the broad to the narrow) or by the pen and ink drawings accompanying the descriptions. Phyla included are Mastigophora (movement by flagella), Sarcodina (movement by pseudopods), Sporozoa (no major means of movement; all parasitic), and the Ciliophora (movement by cilia). The Protozoa is considered to be a Subkingdom. If one can identify the phylum of the organism in question, then the identification process is less taxing.

In addition to the main portion of the work, there are brief chapters on what is a protozoan, where to find them, equipment needed for study, collecting procedures, sizes of the one celled creatures, staining procedures, reproduction, movement, and how to employ the key. A brief bibliography and an index can be found at the end of the volume. The index is an alphabetical arrangement of common and scientific names as well as subjects. Some superficial drawings of the beasts can be found interspersed among the pages of the index.

The drawings in this key are more than adequate in identifying the protozoans in question. A good microscope also helps.

Platyhelminthes (Free-Living and Parasitic Flatworms)

245. Schmidt, Gerald D. **How to Know the Tapeworms**. Dubuque, Iowa: Wm. C. Brown, 1970. 266p. (The Pictured-Key Nature Series). LC 71-129602. ISBN 0-697-04860-8 pa.

This volume provides a means to identify a large number of tapeworms found in nature. It is obviously for the lab and not the field. As in the case of the other books found in this series, this source provides the keys to zero in on the genus and species of the unknown organism. The scolex or head of the tapeworm, along with the internal anatomy of the proglottid (tapeworm section), plays an important role in the identification process. Therefore, one can find drawings of the above interspersed among the textural content. A brief glossary and bibliography, as well as a scientific-name index, can be found at the end of the volume.

Because preparation of the specimens play a vital role in the identification, an introductory chapter discusses how to obtain, fix, stain, mount, and label the worms in question.

Nemata (Free-living and Parasitic Roundworms)

246. Skryabin, K. I. **Key to Parasitic Nematodes**. Leiden, New York: E. J. Brill, 1991. 2 vols. $82.00 vol. 1; $114.10 vol. 2. LC 89-9737. ISBN 90-04-09133-5 vol. 1; 90-04-09134-3 vol. 2; 90-04-09132-7/set.

These two volumes of a four-volume (projected) set consider the taxonomy of all parasitic nematodes found in all types of animals. By means of taxonomic keys, lined drawings, and detailed anatomical descriptions, any parasitic nematode can be identified down to genus. Anatomical descriptions are given for class, subclass, order, suborder, superfamily, family, subfamily, tribe, and genus.

In addition, species within the genera are listed. For each organism included, the scientific name, person(s) who first identified it, date of identification, and scientific name of the host are recorded.

A comprehensive bibliography of the taxonomic literature can be found at the end of each volume. Volume 2 has two sets of indices, one listing the genera and the other the species of the suborders Oxyurata and Ascaridata. These are the two suborders considered in volume 2. Volume 1 contains the suborders Spirurata and Filariata. Unfortunately, there is no index to volume 1. These two volumes represent a finely detailed taxonomic work. They are extremely scholarly.

Mollusca (Snails, Slugs, Clams, Oysters, Squid, Octopi, etc.)

247. Keen, A. Myra, and Eugene Coan. **Marine Molluscan Genera of Western North America: An Illustrated Key**. 2d ed. Stanford, Calif.: Stanford University Press, 1974. 208p. $29.50. LC 73-80625. ISBN 0-8047-0839-8.

Arranged by classes, this work considers those genera having shells. For each class included, a key is listed in order to identify the organism down to genus. As a further aid, black-and-white line-drawings of shells can be found interspersed among the keys. Classes include Gastropoda, Pelecypoda, Polyplacophora, Aplacophora, Scaphopoda, and Cephalopoda. In addition to the keys, which comprise the greatest part of the work, a classification scheme, ranges and habitats, a list of species representative of the genera found in the keys, a glossary, an adequate bibliography, and a scientific-name index can be found completing the volume.

CHECKLISTS AND CLASSIFICATION SCHEMES

Invertebrates (General)

248. Seymour, Paul R., comp. **Invertebrates of Economic Importance in Britain: Common and Scientific Names**. 4th ed. London: Her Majesty's Stationery Office; distr., Lanham, Md.: UNIPUB, 1989. 147p. $21.00pa. LC 90-127224. ISBN 0-11-242829-0.

First published under the title *Common Names of British Insects and Other Pests* by the Association of Applied Biologists, and including 1,838 species, this work attempts to list all invertebrates of economic importance in Britain. The species are arranged by phylum and consist of nematodes and platyhelminths, annelids, arthropods, and molluscs. Each of the organisms is entered twice; therefore, there is a scientific name to common name arrangement as well as a common name to scientific name arrangement. The arthropods comprise the greater portion of the checklist. Along with the scientific-name listing is the person who first named the organism.

Mollusca (Snails, Slugs, Clams, Oysters, Squid, Octopi, etc.)

249. Abbott, R. Tucker, and Kenneth J. Boss, eds. **A Classification of the Living Mollusca**. Melbourne, Fla.: American Malacologists, 1989. 189p. $21.00; $17.00pa. LC 88-83463. ISBN 0-915826-22-4; 0-915826-6pa.

This work represents a listing of all the various taxonomic groups comprising the Mollusca phylum. Within the first few pages a generalized schematic arrangement is presented. The groups considered within this abbreviated section are class, subclass, order, and suborder. Classes listed are Aplacophora, Polyplacophora, Monoplacophora, Gastropoda, Cephalopoda, Bivalvia, and Scaphopoda. The main portion of the volume refines these groups and presents class, subclass, order, family, and genus. For each of the genera considered, the person who first named it and the date of naming are given. A brief list of references appears after a listing of particular taxonomic groups. A scientific-name index can be found closing the volume.

250. Committee on Scientific and Vernacular Names of Mollusks of the Council of Systematic Malacologists. **Common and Scientific Names of Aquatic Invertebrates from the United States and Canada: Mollusks**. Bethesda, Md.: American Fisheries Society, 1988. 277p. (American Fisheries Society Special Publication; 16). LC 88-70617. ISBN 0-913235-47-4; ISBN 0-913235-48-2pa.

Scientific names are arranged alphabetically under their respective families, which in turn are entered under classes and orders. For each scientific-name entry, the person who first named it, date of naming, occurrence (Atlantic, Pacific, Arctic, freshwater, terrestrial, restricted estuarine, introduced and established, and extinct or thought to be extinct), and common names are included. Appendix 1 contains a list of endangered and threatened mollusks of North America and appendix 2 considers a list of possibly extinct mollusks of the United States. A combined common/scientific-name index, and a series of colored photographs, conclude this checklist. An alphabetical list of families, as well as a phylogenetic list by class, order, and family, can be found on the opening pages.

251. Conchological Club of Southern California. **Check List of West North American Marine Mollusks**. San Diego, Calif.: Conchological Club of Southern California, 1946. 30p.

This source provides a list of 2,208 species of marine mollusks found in the waters from San Diego to the Polar Sea. The species are arranged under their respective subgenus, which in turn is entered under genus and family. For each of the organisms presented, the scientific name, person who named it, and date of naming are given. An index to the genera concludes the list.

252. Lea, Isaac, and others. **Check Lists of the Shells of North America**. Washington, D.C.: Smithsonian Institution, 1860. [52p.], (Smithsonian Institution, Smithsonian Miscellaneous Collections, vol. 2, art 6). LC 16-8202.

The lists of species have been organized under eight geographical sections (section 4 that considers Florida and the Gulf of Mexico was not published in this checklist) representing all of North America. The shells are listed, in most cases, under their respective families. For each of the species listed, the scientific name and person who first named it (abbreviated form of name) are given. There is no index.

253. May, W. L. **A Check-List of the Mollusca of Tasmania**. Tasmania: J. Vail, government printer, 1921. 114p. LC 24-8116.

The species (1,058 of them) are arranged under their respective genera, which in turn are entered under families, orders, and classes. A taxonomic citation to the literature accompanies the genus name. For each of the species listed, the scientific name, person who first named it, and citations to the taxonomic literature are given. A scientific-name index to the genera completes the checklist.

254. Smith, Shelagh Mary, and David Heppell. **Checklist of British Marine Mollusca**. Edinburgh: National Museums of Scotland, 1991. 114p. LC gb91-34019.

The species are arranged under their respective genera, which in turn are entered under families, orders, and classes. For each of the genera and species listed, the scientific name, person who first named it, and date of naming are given. All mollusks found within the region extending from latitude 47 degrees, 30 minutes north to 62 degrees, 40 minutes north and from longitude 2 degrees, 40 minutes east to 15 degrees west have been included. Therefore, land, pelagic, and deep-water species are presented within this checklist.

A systematic list of families, superfamilies, and orders appearing in the checklist, along with appropriate page numbers, can be found preceeding the listing of the species. A series of notes on the taxonomy and nomenclature of many of the species, a six-page bibliography, and an index to the genera and species complete the volume.

255. Societ'a Italiana di Malacologia. **Annotated Check-List of Mediterranean Marine Mollusks. Catalogo Annotato dei Molluschi Marini del Mediterraneo**. Bolognese, Italy: Libreria Naturalistica Bolognese, 1990- . 2 vols. DM 99.00 vol. 1.

Each species is arranged under its respective genus and/or subgenus, which in turn is entered under family, superfamily, order, superorder, subclass, and class. The person who named it, and date of naming are given for all groups mentioned above, including the species. This checklist is divided into two volumes. Volume 1 includes the taxonomic list, while volume 2 is comprised of notes, bibliography (biology, anatomy, and ecology references with emphasis on publication since 1979), and an analytical index (consisting of synonyms, obsolete names, etc.). Volume 2 is expected to be published during the autumn of 1992.

ASSOCIATIONS

Invertebrates (General)

256. **International Society of Invertebrate Reproduction**. Univ. of Newcastle-upon Tyne, Dove Marine Laboratory, Cullercoats, Tyne and Wear NE30 4PZ, England. 91-2226000.

257. **Society for Invertebrate Pathology**. P.O. Box 38, Univ. of Maryland, Chesapeake Biological Laboratory, Solomons, MD 20688. (301) 326-4281. FAX (301) 326-6342.

PROTOZOA (Unicellular Organisms)

258. **Society of Protozoologists**. Amer. Type Culture Collections, 12301 Parklawn Dr., Rockville, MD. 20852-1776. (301) 231-5516.

Cnidaria (Hydra, Jellyfish, Coral, Sea Anemones, etc.)

259. **International Association for Biological Oceanography**. (Association Internationale pour l'Oceanographie Biologique - AIOB). Station Biologique, Place Georges Teissier, F-29211 Roscoff, France. 98-292323. FAX 98-292324.

Responsible, among other things, for holding the annual International Coelenterate (Cnidaria) Conference.

Parasites

Parasitic organizations are listed due to the fact that the greatest majority of parasites are invertebrates.

260. **American Association of Veterinary Parasitologists**. The Upjohn Company, Dept. 0690-190-40, Kalamazoo, MI 49001. (616) 385-6523. FAX (616) 385-6707.

261. **American Society of Parasitologists**. 1041 New Hampshire St., Lawrence, KS 66044.

262. **Asian Parasite Control Organization**. 1-2, Sadohara-cho, Ichigaya, Shinjuku-ku, Tokyo 162, Japan.

263. **Australian Society for Parasitology**. Secretariat, ACTS, GPO Box 2200, Canberra, ACT 2601. (062)-573256.

64. **British Society for Parasitology**. Dept. of Zoology, Univ. of Nottingham, Nottingham NG7 2RD, England.

65. **European Federation of Parasitologists**. Natl, Swedish Environment Protection Bd., Marine Section, Postfack 584, S-740 71 Oregrund, Sweden. 173-31305. FAX 173-30949.

66. **International Commission on Trichinellosis**. Inst. of Zoology, Univ. of Warsaw, Zwirki i Wiguri 93, PL-02-089 Warsaw, Poland.

67. **Japan Association of Parasite Control**. (Nihon Kiseichu Yobokai - NKY), Hoken Kaikan, 1-2, Sadohara-cho, Ichigaya, Shinjuku-ku, Tokyo 162, Japan. 3-32681800. FAX -32668767.

68. **Parasitological Society of Southern Africa**. (Parasitologiese Vereniging van Suide-ke Afrika). Med. Univ. of Southern Africa, Dept. of Microbiology, Parasitology Div. Medunsa 0204, South Africa. 12-5294117. FAX 12-582323. TELEX 320580.

69. **World Association for the Advancement of Veterinary Parasitology**. SmithKline Beecham, AHP, Applebrook Res. Center, 1600 Paoli Pike, West Chester, PA 19380. (215) 51-7416. FAX (215) 647-4563.

70. **World Federation of Parasitologists**. (Federation Mondiale des Parasitologues), Natl. Inst. of Public Health and Environmental Protection, Postbus 1, NL-3720 BA Bilthoven, Netherlands. 30-742106. FAX 30-742971.

Mollusca (Snails, Slugs, Clams, Oysters, Squid, Octopi, etc.)

271. **American Malacological Union**. P.O. Box 30, North Myrtle Beach, SC 29582. (803) 249-1651.

272. **Malacological Society of Australia**. Dept. of Invertebrate Zoology, Museum of Victoria, 285-321 Russell St., Melbourne, Vic, 3000. (03)-6699880.

273. **Malacological Society of London**. British Antarctic Survey, High Cross, Madingley Rd, Cambridge, CB3 0ET.

274. **Malacological Society of the Philippines**. (Samahang Malakolohiya ng Pilipinas - SMP). Natural Sciences Research Inst., Univ. of the Philippines, Diliman, Quezon City 1101, Philippines. 2-975736.

275. **National Shellfisheries Association**. Dept. of Biological Sciences, Univ. of New Orleans - Lakefront, New Orleans, LA 70148. (504) 286-7042.

276. **Society for Experimental and Descriptive Malacology**. P.O. Box 3037, Ann Arbor, MI 48106. (313) 747-2189.

277. **Unitas Malacologica**. Rkysmuseum van Natuurlyke Historie, Postbus 9517, NL-2300 RA Leiden, Netherlands. 71-143844. FAX 71-274900.

278. **Western Society of Malacologists**. 1633 Posilipo Ln., Santa Barbara, CA 93108. (805) 969-1434.

279. (This entry number not used.)

CHAPTER 3

THE ARTHROPODS

(Including the Spiders, Crustaceans, and Insects)

The arthropods make up by far the largest group of animals in the animal kingdom. It contains more than 1 million species. The insects alone, the largest subgroup of the arthropods, account for three-fourths of all the animal species. The name *arthropod* comes from the Greek *arthro* (joint), and *pod* (foot), meaning the group is distinguished as having jointed feet or appendages. Members of this enormous group can be found worldwide and are adapted to all habitats and foods. They range in size from the microscopic gall mites to the giant spider crabs. They are also of great economic importance, being beneficial as the primary pollinators of plants and providers of fine seafood, and being detrimental as one of the greatest destructive forces to plants and crops.

There are three large subphyla of the arthropods: Chelicerata, Crustacea, and Uniramia. The Chelicerata include the scorpions, spiders, mites, and ticks. The Crustacea are primarily aquatic, gill-breathing animals that include the shrimp, lobsters and crabs, barnacles, ostracods, copepods, and isopods. One of the common isopods, however, is the common pill bug that lives on land. The Uniramia include the millipedes, centipedes, and insects, which is the largest group of all.

Due to the enormous size of this group and the corresponding volume of literature, this chapter covers only the most important sources for each group. Often a source listed in the Handbooks section will include checklists or bibliographies or taxonomic keys, so if a checklist or key is not listed for a particular group, it is important to check in the Handbooks section to see if it may be included there. This chapter does not go into as much depth of coverage (especially geographic) as found in other chapters. It also does not cover the medical, veterinary, or agricultural aspects, such as parasitology or pest control. To do so would make this chapter larger than the book! Indeed there are separately published books, mentioned in this chapter, that are guides to the entomology (insect) literature alone and may be used to supplement sources found in this chapter.

Ants are so much like human beings as to be an embarrassment. They farm fungi, raise aphids as livestock, launch armies into war, use chemical sprays to alarm and confuse enemies, capture slaves. The families of weaver ants engage in child labor, holding their larvae like shuttles to spin out the thread that sews the leaves together for their fungus gardens. They exchange information ceaselessly. They do everything but watch television.
—Lewis Thomas
The Lives of a Cell, *1974*

DICTIONARIES

280. Stachowitsch, Michael. **The Invertebrates: An Illustrated Glossary**. New York: John Wiley, 1992. 676p. $175.00; $79.95pa. LC 91-21129. ISBN 0-471-83294-4; 0-471-56192-4pa.

See entry 149 in chapter 2, "The Invertebrates," for a full description of this work. More than 100 pages of this dictionary are devoted to the arthropods. A very useful and needed book.

281. **Thesaurus of Agricultural Organisms: Pests, Weeds, and Diseases**. Derwent Publications with the assistance of CIBA-Geigy SA. London, New York: Chapman and Hall, 1990. 2 vols. $395.00. LC 90-2185. ISBN 0-442-30422-6 set.

This thesaurus was devised as an aid in searching the database Pestdoc, a database compiled by Derwent Publications Ltd., which covers the literature on agricultural pest, weed, and disease control. This thesaurus is included here because it is a recent (1990) attempt to standardize the names of insects and other arthropods. It provides a cross-referenced source of Latin names, synonyms, and common names in English, French, and German. More than 16,500 main entries are arranged in alphabetical order. The main entry is a Latin name, followed by the higher taxa of which it is part, usually the family name. Synonyms of the Latin name and the common names are listed under the main Latin-name entry. Common names are in English, French, and German. Volume two includes an index of inverted species names (i.e., the species name is listed first, then the genus name). This is a useful tool in finding the standardized name of a particular arthropod, at least those judged here to be pests.

Insecta

282. Ericson, Ruth O. **A Glossary of Some Foreign-Language Terms in Entomology.** Washington, D.C.: U.S. Dept. of Agriculture, Entomology Research Division; for sale by the Supt. of Docs., U.S.G.P.O., 1961. 59p. (Agriculture Handbook, no. 218).

This glossary is only 59 pages long but is a useful first source in translating foreign-language papers on entomology. The foreign terms are listed in alphabetical order with the designation of the language indicated in parentheses, followed by the English equivalent of the term. It includes Czech, Danish, Dutch, French, German, Polish, Russian, and Swedish terms. The Russian terms are transliterated using the Library of Congress cataloging rules. For those terms not covered here, the author supplied a list of foreign-language dictionaries that may be consulted. Line-drawings of the anatomical parts of insects are included at the end of the text. The parts are labeled with the English and foreign-language names for each part.

283. Foote, Richard H. **Thesaurus of Entomology.** College Park, Md.: Entomological Society of America, 1977. 188p. LC 77-153953.

This thesaurus is intended to help standardize the terminology in entomology and to aid in the indexing and retrieval of information from the entomological literature. The author, along with 75 other professional entomologists, prepared this listing of standard entomological terms. The thesaurus is organized in two sections. First is a hierarchical section listing the major concepts of entomology. For example, one part is on genetics with general terms, then cytogenetics, molecular genetics, microbial genetics, and population genetics. Within each part the specific terms are listed in alphabetical order with cross-references to other terms. The second section is an index with terms arranged in alphabetical order that refer to entries in the hierarchical section. Although dated, this thesaurus is still useful for standardization of entomological terms. However, its use in retrieval of information in the entomological literature is eclipsed by other thesauri designed specifically for use in specific databases; for example, *CAB Thesaurus* of the Commonwealth Agricultural Bureau is the basis of nomenclature in *The Bibliography of Agriculture* and the AGRICOLA database (see entry 32).

284. Greiff, Margaret, comp. **Spanish-English-Spanish Lexicon of Entomological and Related Terms, with Indexes of Spanish Common Names of Arthropods and their Latin and English Equivalents.** London: Commonwealth Institute of Entomology, 1985. 158p. $30.00pa. ISBN 0-85198-5602.

This useful dictionary first lists Spanish entomological terms and their English equivalents followed by a section on English entomological terms and their Spanish equivalents. Two indexes are included after this dictionary to help in identifying specific arthropod names. One index lists the Latin scientific names of arthropods with their corresponding Spanish and English common names. The other index lists the Spanish common names with the corresponding scientific names. Although this dictionary calls itself a list of arthropod names, it does not include the crustacea, and it is specifically for entomological research.

285. Harbach, Ralph E., and Kenneth L. Knight. **Taxonomists' Glossary of Mosquito Anatomy**. Marlton, N.J.: Plexus, 1980. 415p. $24.95. LC 80-83112. ISBN 0-685-04089-5.

This glossary, ten years in the making, is an example of excellence in glossaries for specific groups of animals. It is comprehensive, detailed and carefully researched and reviewed. The glossary is separated into five sections, one each for terms describing adult, egg, larva, and pupa stages of the mosquito and a section on the vestiture, or surface covering, of the mosquito.

Not only is a glossary of terms listed in each section, but labeled, detailed line-drawings and electron micrographs are included after each text. A 24-page bibliography of references cited in the glossaries and an index to the entire glossary are included at the end. Corrections and additions to this text are included in *Mosquito Systematics*, 1982, v.13, pages 201-17.

286. Leftwich, A. W. **A Dictionary of Entomology**. New York: Crane Russak, 1976. 360p. $95.70pa. LC-83-26504. ISBN 0-318-347393,2031996. Bks on Demand UMI.

This dictionary contains more than 4,000 definitions on the anatomy and physiology of insects as well as definitions of orders, suborders, and families. The author has also defined individual species with agricultural, medical, or veterinary importance. These species are mainly British and European insects, but American species are included. The dictionary ends with a section on the classification of insects and a two-page bibliography of sources consulted in compiling the dictionary. This dictionary is primarily for the amateur entomologist or student; the professional entomologist may want to consult *The Torre-Bueno Glossary of Entomology* (see entry 287).

287. Nichols, Stephen W., comp. **The Torre-Bueno Glossary of Entomology.** Edited by Randall T. Schuh. New York: Entomological Society in cooperation with the American Museum of Natural History, 1989. 840p. LC 89-12095. ISBN 0-913424-13-7.

This is a revised edition of Jose Rollin de la Torre-Bueno's *A Glossary of Entomology* (1937) and *Supplement A* by George S. Tulloch (1962). This revision includes definitions compiled from more than 200 texts, glossaries, and journal articles. A group of 49 professional entomologists from around the world edited and supplied definitions from their own areas of specialty. Each definition lists the source where the definition was found and often includes more than one definition of the term when professional opinions differ or when a term has both an archaic and a current meaning. Extensive cross-referencing is also used, to connect related meanings or subjects. More than 16,000 terms are defined here with the length of the definitions going from one line to an entire paragraph. In addition to the 16-page bibliography, a listing of non-English-language dictionaries, sources of common names of insects, and other reference sources that may be useful to entomologists are included at the end. This is the most complete and up-to-date glossary of entomological terminology. It is more technical than *A Dictionary of Entomology* (see entry 286) and is therefore of more use to professional entomologists; however, students and serious amateurs will also find this glossary very useful.

288. Séguy, Eugène. **Dictionnaire des termes techniques d'entomologie élémentaire.** Paris: Lechevalier, 1967. 465p. (Encyclopédie entomologique, no. 41). $175.00. LC 68-71485. ISBN 2-7205-04661, M-6512.

This is the basic French dictionary for entomological terms. It covers approximately 4,000 terms and includes 200 figures with black-and-white line-drawings to illustrate selected terms. The terms were derived from 16 sources listed in the bibliography at the beginning of the text. Antonyms and synonyms are given at the end of some definitions.

289. Tuxen, S. L., ed. **Taxonomist's Glossary of Genitalia in Insects**. 2d ed., rev. and enl. Copenhagen, Denmark: Munksgaard, 1970. 359p. $35.00. LC 71-509798. ISBN 0-934454-76-0.

This specialized glossary is intended as an aid to the systematist when trying to understand terms the taxonomists and morphologists used in the literature. The text is divided into two sections. The first section lists each insect order with a two- to sixteen-page description of the genitalia with detailed line-drawings accompanying the text. These descriptions are authored by 34 authorities in this particular field. Important references are included at the end of each description. The second section is an extensive glossary of terms used in the descriptive section. Although specialized, this kind of glossary is invaluable for the correct description and identification of insects.

290. von Kéler, Stefan. **Entomologisches Wörterbuch mit besonderer Berucksichtigung der morphologischen Terminologie**. 3. durchgesehene und erw. Aufl. Berlin: Akademie Verlag, 1963. 774p. LC 63-50195.

This is a comprehensive entomology dictionary for the German language. Line-drawings accompany some definitions and 33 plates of labeled anatomical parts of insects and their corresponding German names are included at the end. A 13-page alphabetized list of the authors of literature cited in the dictionary is also included. An appendix attempting to standardize the morphological terminology of 230 important insect muscles follows the dictionary of terms. This excellent dictionary is equivalent to the *Torre-Bueno Glossary of Entomology* (see entry 287).

ENCYCLOPEDIAS

291. Grzimek, Bernhard. **Grzimek's Animal Life Encyclopedia**. Vol. 1, **Lower Animals** and Vol. 2, **Insects**. New York: Van Nostrand Reinhold, 1974. LC 79-183178.

This set is reviewed in chapter 1 (see entry 18). Two volumes of this set cover the arthropods. Volume 1, *Lower Animals*, includes a general section on the arthropods and the arachnids and crustaceans. Volume 2 covers the insects. Each group is introduced with general descriptions of the major families. They are described in nontechnical terms and illustrated with full-color photos. Information covered includes behavior, reproduction, distribution, and classification, as well as information on the major families and a few representative species. Black-and-white drawings and diagrams are found in the margins illustrating information from the text. Very readable and interesting, for the general public and the student.

292. Klots, Alexander B., and Elsie B. Klots. **Living Insects of the World**. Garden City, N.Y.: Doubleday, 1975. 303p. LC 59-9100. ISBN 0-385-06873-5.

Written in nontechnical language, this work covers the major groups of insects in taxonomic order. Full-color photographs (151 of them) with as many black-and-white photos and drawings enhance this interesting and easy-to-read text. Each chapter covers a major group, from the springtails to the wasps, ants, and bees, in a few to 60 pages. The material is dated but still interesting for general information and the illustrations.

293. Line, Les, Lorus Line, and Margery Milne. **The Audubon Society Book of Insects**. New York: Harry N. Abrams, 1983. $35.00. 260p. LC 82-16457. ISBN 0-8109-1806-4.

Stunning full-color photos, often covering a full page or two, highlight this beautiful book on common and unusual insects. The photos help illustrate the essays written by Lorus Line and Margery Milne under such titles as "Bombardiers and Borers" and "Two-Winged Aviators." The essays are entertaining reading containing interesting facts about particular groups of insects. The book serves as a wonderful introduction to insects and an enticement to read other books on this amazingly diverse group.

294. Linsenmaier, Walter. **Insects of the World**. Trans. from the German by Leigh E. Chadwick. New York: McGraw-Hill, 1972. 392p. LC 78-178047. ISBN 07-037953-X.

This unique work, translated from the German, is a major accomplishment. Walter Linsenmaier, a painter and self-trained entomologist, has created some of the most beautiful illustrations of insect life yet produced and added them to a well-written and informative text. He is also the photographer of the many color photos of mounted specimens. His approach is that of the artist that appreciates the beauty displayed in insects. Indeed, one chapter is titled "Living Works of Art." After the introductory chapters covering insect anatomy, mimicry and camouflage, and life stages of insects, Linsenmaier describes and beautifully illustrates the different groups of insects under the major headings of wingless insects, winged insects, social insects, and water insects. Each order within these groups is described, and two of the largest groups, the beetles and the butterflies and moths, are described in 44 and 95 pages respectively. The illustrations are included in the main index under the name of the insect with the page number and illustration number included. There is also a separate index of illustrations in numerical order. Linsenmaier's artistic talent and love of insects has produced a remarkable work that is a classic in entomology literature.

295. O'Toole, Christopher, ed. **The Encyclopaedia of Insects.** New York: Facts on File, 1986. 160p. $24.95. LC 85-29226. ISBN 0-8160-1358-6.

Beautiful color photographs and drawings combined with an authoritative yet easily readable text produce a good insect encyclopedia for the general reader and student alike. The book covers more than the insects and their 28 different orders. It also covers the millipedes and centipedes, as well as the scorpions, ticks, and spiders. After each class is discussed, the various orders within the class are covered in two to six pages. The headings for each order are by common name (e.g., crickets and grasshoppers), with the scientific name for the group (Orthoptera) and other taxonomic and anatomical features included in a fact box at the beginning of the text. Full-color photos and drawings accompany the text. A nice glossary, a one-page bibliography, and an index arranged mainly by common name with a few cross-references from the scientific name complete the book.

296. Stanek, V. J. **The Pictorial Encyclopedia of Insects**. London, New York: Hamlyn, 1969. 543p. LC 76-481367. ISBN 0-6000-3085-7.

This pictorial encyclopedia includes more than 1,000 color and black-and-white photos on 543 pages. The pictures are fascinating and the accompanying text informative with brief facts on each group. It is a translation of *Das Grosse Bilderlexikon der Insekten* (Prague: Artia, 1969). The arrangement is by major group beginning with insect relatives from the Arthropoda, spiders and scorpions and ticks, then primitive insects to social insects. There are no headings for each group, but rather the text and pictures flow from one group to the other. The index is to common and scientific name. It serves as a good introduction to the amazing diversity of insects with examples like the giant grasshopper from New Guinea that is pictured covering the length of a man's hand. Good to just pick up and browse through.

BIBLIOGRAPHIES

Chelicerata

ARACHNIDA (Spiders, Scorpions, Mites, and Ticks)

297. Bonnet, Pierre. **Bibliographia Araneorum**. Toulouse, France: Douladoure, 1945-1961. Reprint (of bibliography only). 3 vols. in 7. Lanham, Md.: Entomological Society of America, 1969. 832p. $40.00. LC 57-58745. ISBN 0-686-09299-6.

This set is reviewed in the checklist section (see entry 589). It is a catalog of the spider species described in the literature up to 1939. The text is in French and is in three parts. Part 1 contains a bibliography of the works, in all languages, used in compiling the list. It is very extensive, providing references from the older literature, and is currently in print from the Entomological Society of America.

Crustacea

298. Hart, C. W., Jr., and Janice Clark. **An Interdisciplinary Bibliography of Freshwater Crayfishes: Astacoidea and Parastacoidea from Aristotle Through 1985, Updated Through 1987**. Washington, D.C.: Smithsonian Institution Press, 1989. 498p. $35.00. LC 88-600328. ISBN 0-87474-464-4.

This is the second edition of one that was published in 1986. It contains the reprinted original bibliography and index with the additions appended in a separate section at the end with its own index. It contains references, arranged in alphabetical order by author, to books and journal articles published since the time of Aristotle on freshwater crayfishes. Each reference contains the author, date, title, and pagination. Also included are alphanumeric subject codes in brackets and the genera or species discussed in the work. Each citation is assigned a number that is referred to in the index. The index is arranged under broad subject headings such as "Distribution, Asia," "Distribution, Europe," and "Nervous System, Physiology, General." Lists of reference numbers that refer back to the bibliography are included under each heading. The references number up to 11,194 in the original bibliography plus 1,287 more listed in the appended references.

299. Kempf, Eugen Karl. **Index and Bibliography of Marine Ostracoda**. Koeln: Geologisches Institut der Universitaet zu Koeln, 1986. 5 vols. LC 88-213464.

See next reference.

300. Kempf, Eugen Karl. **Index and Bibliography of Nonmarine Ostracoda**. Koeln: Geologisches Institut der Universitaet zu Koeln, 1980. 5 vols. LC 81-162012.

These two sets cover the literature of marine and nonmarine ostracods from 1785 to 1980 or 1985. Each is similarly arranged. Parts 1 to 3 are the index in three versions, A, B, and C. The names of the genera, subgenera, species, and subspecies are arranged in different ways in each of the versions. In version A the names of the genera are arranged alphabetically, in version B the names of the species are arranged alphabetically, and in version C the arrangement is in chronological order by publishing date. The bibliographies themselves include citations to the taxonomic literature in bibliography A, and to all the nontaxonomic literature in bibliography B.

301. Sims, R. W., ed. **Animal Identification: A Reference Guide**. Vol. 2, **Land and Freshwater Animals (Not Insects).** London, British Museum (Natural History), New York: John Wiley, 1980. 120p. $59.95. LC 80-40006. ISBN 0-471-27766-5.

This three-volume set is reviewed under entry 29. A bibliography of the literature for crustaceans is covered from page 66 to 74. The references are listed under the major orders such as Brachiopoda, Ostracoda, and Copepoda. Within these orders the references are listed geographically by region, such as the Nearctic region, or the Neotropical region.

Uniramia

INSECTA (General)

302. **Bibliographies of the Entomological Society of America**. College Park, Md.: Entomological Society of America, 1983-1984. 3 vols. LC sc84-4010. ISSN 0749-2987.

This series of bibliographies on various entomological topics ceased in 1984 after only three issues. Each issue included four or five bibliographies on such topics as "Role of the Entomologist in Forensic Pathology, Including a Selected Bibliography," and "Genetics and Population Genetics of Grasshoppers and Locusts." Each bibliography was written by one or more research entomologists considered an authority in that subject.

303. Hollis, D. **Animal Identification: A Reference Guide**. Vol. 3, **Insects**. Chichester, N.Y.: John Wiley, 1980. 160p. $59.95. LC 80-40006. ISBN 0-471-27767-3.

This is the third volume of a three-volume set (see entry 29 for a complete description of the set). This third volume is devoted to the insects and it provides a list of the most important references for identification of the insect orders and families. It is arranged by order with further breakdown by family for the larger orders such as the Heteroptera. Within the orders and families, reference sources are grouped by major geographical region (i.e., Nearctic, Afrotropical, Neotropical, etc.). Both book and journal literature are included here. A very useful bibliography for the nonspecialist as well as the professional.

INSECTA (Systematic Section)

Isoptera (Termites)

304. Ernst, E., and R. L. Araujo, comps.; and Tropical Development and Research Institute Publications, Publicity and Public Relations Section. **A Bibliography of Termite Literature, 1966-1978.** Edited by P. Broughton and K. M. Fullarton. Chichester, N.Y.: John Wiley, 1986. 903p. LC 84-5173. ISBN 0-471-90466-X.

Created to supplement T. E. Snyder's *Annotated Subject-Heading Bibliography of Termites, 1350 B.C. to A.D. 1954* (see entry 305), this bibliography covers the world's periodical literature on termites from 1966 to 1978. It includes 3,165 entries arranged in alphabetical order by author. Each entry includes short (two sentences) to long (half-page paragraphs) annotations. There is a junior author index, and a 348-page index to subjects and scientific names.

305. Snyder, T. E. **Annotated Subject-Heading Bibliography of Termites, 1350 B.C. to A.D. 1954**. Washington, D.C.: Smithsonian Institution, 1956. 305p. (Smithsonian miscellaneous collections, vol. 130).

Updated by *A Bibliography of Termite Literature 1966-1978* (see entry 304), this bibliography includes 3,624 references to the termite literature from 1350 B.C. through A.D. 1954. The reference from 1350 is in Sanskrit and refers to an organism that destroyed wood, which was probably a termite. The work was done from 1909 to the author's retirement in 1951. The bibliography is done in two parts. The first part is a subject approach with references arranged under subject headings such as morphology, nutrition, parasites, flight, and predators. The second part lists authors of the references in alphabetical order, with articles listed under the author in chronological order. An index to subject helps find references by specific subject. Two supplements (1955-1960 and 1961-1965), were also completed by the author and published by the Smithsonian as part of the Smithsonian Miscellaneous Collections (vol. 143, no. 3 and vol. 152, no. 3).

Hemiptera, Homoptera (True Bugs and Aphids)

306. Metcalf, Zeno Payne. **A Bibliography of the Homoptera (Auchenorhyncha)**. Raleigh, N.C.: N.C. State College of Agriculture and Engineering of the University of North Carolina, 1943. 2 vols. LC a45-4197.

This bibliography of books and journal articles covering the literature up to 1942 is arranged into three parts, the bibliography itself, a list of periodicals cited, and a topical index. The bibliography is arranged alphabetically by author with articles listed in chronological order. A brief annotation of the contents of the article or book is often given. Approximately 8,000 titles are listed. Volume 2 includes the list of journal abbreviations in alphabetical order and the full form of the title. The subject and scientific-name index follows and is a very useful and complete index to the bibliography.

307. Oman, P. W., W. J. Knight, and M. W. Nielson. **Leafhoppers (Cicadellidae): A Bibliography, Generic Check-List, and Index to the World Literature 1956-1985.** Oxon, England: C.A.B. International, 1990. 368p. LC gb91-15157. ISBN 0-85198-690-0.

This supplements Metcalf's *General Catalog of the Homoptera. Fascicle VI. Cicadelloidea* (see entry 608). This supplement covers the 30 years since the publication of Metcalf's bibliography and includes 1,084 new genera and almost 8,000 new species. It is arranged in three sections: a bibliography with approximately 7,000 titles, an index to the bibliography, and a checklist of generic and family group names.

308. Sharma, M. L. **Bibliography of Aphidoidea**. Sherbrooke, Quebec, Canada: Editions Paulines, 1969-1971. 2 vols. (Publications de l'Universite de Sherbrooke). LC 73-855903.

Intended as a supplement to Z. P. Metcalf's *A Bibliography of the Homoptera (Auchenorhyncha)* (see entry 306), the first volume covers the literature on aphids up to 1966. Volume two goes to 1970. The arrangement is alphabetical by author. There is no subject index; rather, the author has provided codes in the margin next to each reference indicating what the article is about. For example, a *P* indicates "population studies," and *T* is "techniques with aphids."

Coleoptera (Beetles)

309. Arnett, Ross H., Jr., comp. and ed. **Bibliography of the Coleoptera of North America North of Mexico, 1758-1948, and Supplements Including Mexico, Central America, and the West Indies.** Gainesville, Fla.: Sandhill Crane, 1982. 180p. $23.00. LC 79-109222. ISBN 0-916846-07-5.

This is actually part of a loose-leaf work by the same author titled *Checklist of the Beetles of North and Central America and the West Indies* (see entry 610). The bibliography is arranged in alphabetical order by author and is useful in finding the full citation to the references mentioned in the checklist, as well as being a good source for older references to the Coleoptera literature.

Diptera (Flies, Mosquitoes, etc.)

310. Evenhuis, N. L. **An Indexed Bibliography of Bombyliidae (Insecta, Diptera).** Braunschweig, Germany: J. Cramer, 1983. 493p. LC 84-197942. ISBN 3-7682-1379-X.

This bibliography of books and journal articles includes references on the literature of bee-flies from 1758-1983. Most of the citations are on the systematics of bee-flies, but they cover all other areas as well. Although this is a specialized bibliography on a particular family, it is included here because of its comprehensiveness and potential use for anyone studying Diptera as a whole. In all, 3,089 citations are listed in alphabetical order by author. A handy list of journal abbreviations used in the citations is included after the bibliography. Taxonomic, subject, geographic, and species indexes provide a wide range of access to the citations. A useful source for anyone studying Diptera. A supplement to this work is N. L. Evenhuis' *An Indexed Bibliography of Bombyliidae (Insecta, Diptera). Supplement I* (Bishop Museum Press, 1992).

311. West, Luther S., and Oneita Beth Peters. **An Annotated Bibliography of Musca Domestica Linnaeus.** Folkestone, England; in conjunction with Northern Michigan University, 1973. 743p. LC 73-173626. ISBN 0-7129-0536-7.

This work includes around 8,000 entries, with useful, brief annotations for 5,800 entries. The references are grouped into three historical time periods according to date of publication: from early times 1949, from 1950 to 1959, and from 1960 to 1969. There is also a section on research techniques with articles on laboratory rearing of *Musca*, breeding, and so on. Arrangement is alphabetical by author. An index that is divided into 15 broad subject headings (ecology, genetics, physiology, insecticides, etc.) provides a small measure of subject access to this bibliography.

INDEXES AND ABSTRACTS

312. **Abstracts of Entomology**. Vols. 1- . Philadelphia: BIOSIS, 1970- . Monthly. $200.00/yr.; cum. index additional $95.00. ISSN 0001-3579.

Primarily a current awareness index, its contents come from the *BIOSIS Previews* database, which includes *Biological Abstracts* and *Biological Abstracts/RRM* (see entries 33 and 34). Citations and abstracts are specifically to applied research on insects and arachnids, and to their control. Indexes may be difficult to use; the cumulative index is purchased separately.

This is available online as part of the *BIOSIS Previews* database (see entry 34). It is also available on CD-ROM as part of *Biological Abstracts on Compact Disc* (see entry 34).

313. **Bibliography of Agriculture** . Vols. 1- . Phoenix, Ariz.: Oryx Press, 1942- . Monthly. $650.00/yr. ISSN 0006-1530.

This bibliography to the literature of agriculture and related sciences is prepared primarily by the National Agriculture Library. It is the most comprehensive source of U.S. agricultural information and includes citations to books, journal articles, theses, technical reports, software, audiovisual material, and patents. Entomology subjects can be accessed in this index under the sections for "Pests of Plants-Insects," "Pests of Animals-Insects," "Parasites of Human-Insects and Other Arthropods," or through the index under the various headings beginning with "Insect Attractants."

This database is also available online under the title *AGRICOLA*. Vendor is DIALOG (File 10, 1979 to present; File 110, 1970-1978). The database includes more than 3 million citations. Approximate cost to search the files is $45.00 per contact hour, plus a separate cost per citation viewed or printed.

This database is also available on CD-ROM as *AGRICOLA on Compact Disc*. SilverPlatter produces this product. The three archival disks cover 1975 to 1990. The current disk covers from 1991 to the present. Approximate minimum cost for a single user is $825.00 for the current disk per year and $950.00 for the archival set. The cost is higher for multiple users.

314. **Biological Abstracts** and **Biological Abstracts/RRM**. Philadelphia: BIOSIS. (See entries 33 and 34.)

These are indexes to the world's literature in the life sciences and therefore index a significant number of entomology journals. They should be consulted in doing a complete search of the entomology literature.

315. **Entomology Abstracts**. Vols. 1- . Bethesda, Md.: Cambridge Scientific Abstracts, 1969- . Monthly. $700.00/yr. ISSN 0013-8924.

This indexes and abstracts the world's literature on insects, arachnids, myriapods, onychophorans and terrestrial isopods. Monthly and annual indexes, easier to use than *Abstracts of Entomology* (see entry 312).

This database is available online from BRS and DIALOG as part of the *Life Sciences Collection* database. See entry 47 for a complete description. It is also available on CD-ROM as part of the *Life Sciences Collection* (see entry 47).

316. Hagen, Hermann August. **Bibliotheca Entomologica: die Litteratur uber das ganze Gebiet der Entomologie bis zum Jahre 1862**. Leipzig, Germany: Verlag von Wilhelm Engelmann, 1862. Reprint, 2 vols. in 1, Forestburgh, N.Y.: Lubrecht & Cramer, 1960. $140.00. LC ISBN 3-7682-0035-3.

This is an index to the historical literature in entomology. It is arranged by author and covers literature on entomology published up to 1863. It has an involved subject index that has its own index. Each author entry (usually) has short biographical information (e.g., "geb. 1730, gest. 1890. Professor in Helmstadt") followed by the citation.

317. Horn, Walther, and Sigm. Schenkling. **Index Litteraturae Entomologicae**. Berlin-Dahlem: Dr. Walther Horn, 1928-1975. 9Bd. LC 30-9140.

This historical index was published in two parts. The first covers the years up to 1863 (4 volumes), and the second from 1864 to 1900 (5 volumes). The second part was authored by Walter Derksen and Ursula Scheiding. The first part covers the same time period as Hagen (see entry 316), but with a few extra citations. A subject index (volume 5) is included with the second part. Entries are alphabetically arranged by author, with brief biographical information on author.

318. **Index of American Economic Entomology**. College Park, Md.: Entomological Society of America, 1905-1959. 18 vols. LC sn86-25116.

Another historical index, this 18-volume work provides a subject as well as a scientific-name and common-name index to the economic entomology literature from 1905 to 1959. Each subject entry lists the author, journal, page numbers, and year of publication where information on the subject can be found. There is no author index. The various compilers of this multivolume work gleaned the citations from the periodical literature primarily with some selected book literature indexed as well. The later indexes scanned the major journal indexes such as *Bibliography of Agriculture* (see entry 32) and *Zoological Record* (see entry 55) to find citations for this index. The index ended in 1959 due to lack of financial support.

319. **Review of Agricultural Entomology.** Vols. 1 -, nos. 1- . Oxon, U.K.: C.A.B. International, 1913- . Monthly. $431.00/yr. ISSN 0957-6762.

Formerly titled *Review of Applied Entomology, Series A: Agricultural*, this is an index to the world's journal literature on applied entomology. Abstracts are included with the citations, some up to one column in length. There is an author and subject index with each issue, and cumulated indexes for each year.

Available online and on CD-ROM as *CAB Abstracts*, a cumulation of all C.A.B.'s abstract journal databases. See entry 37 for details.

320. **Review of Medical and Veterinary Entomology**. Vols. 1- . Oxon, England: C.A.B. International, 1913- . Monthly. $229.00/yr. ISSN 0957-6770.

Previously titled, *Review of Applied Entomology. Series B: Medical and Veterinary*, this journal indexes the world's journal literature on insects and other arthropods injurious to man and animals. Includes long abstracts with each citation. Author and subject indexes provided for each monthly issue and cumulated for the year.

Available online and on CD-ROM as *CAB Abstracts*, a cumulation of all C.A.B.'s abstract journal databases. See entry 37 for details.

321. **Zoological Record**. Vols. 1 - . Philadelphia: BIOSIS, 1864- . Annual. $2,200.00/yr. ISSN 0084-5604.

This is the standard index to the world's journal literature on systematic works in all areas of zoology. It is reviewed in chapter 1 (see entry 55). It is published in 20 sections with 27 parts. Each section represents a particular class or phylum. The arthropods are covered under sections 12 and 13. Section 12 covers Arachnida together with Myriapoda, Merostomata, Pantopoda, Tardigrada, Symphylida, Pauropoda, Onychophora, Arthropleurida and Pentastomida. Section 13 covers Insecta, with six parts: 13A is general insecta and smaller orders, 13B is Coleoptera, 13C is Diptera, 13D is Lepidoptera, 13E is Hymenoptera, and 13F is Hemiptera. Each section or part has five indexes (i.e., author, subject, geographic, palaeontological, and systematic). Full bibliographic information is only found in the author index, to which the other indexes refer.

Available online and on CD-ROM (see entry 55 in chapter 1 for a full description).

DIRECTORIES

Most societies compile membership directories. A list of entomological societies is found in the associations section of this chapter. Write to the society for a copy of their membership directory. Entry 323 is an example of a membership directory.

322. Arnett, Ross H., Jr., G. Allan Samuelson, and Gordon M. Nishida. **The Insect and Spider Collections of the World**. 2d ed. Gainesville, Fla.: Sandhill Crane Press, 1993. 220p. (Flora and Fauna Handbook, no. 11). $30.00. LC 83-255583. ISBN 1-8777-4315-1.

Arranged in two parts, this work lists public and private insect and spider collections. The public collections are arranged alphabetically by country; the private collections by owner. There is an index to taxa as well (to lead the reader to collections on particular families, but it does not list orders). The collections are described as to address, the director and professional staff's names, the coverage of the collection, how many drawers it encompasses, and whether or not the material is available for study and if it can be loaned or exchanged.

323. Entomological Society of America. **Membership Directory.** Lanham, Md.: The Society, 1990. 149p.

This is an alphabetical listing of members of the Entomological Society of America. Each entry contains the name, mailing address, telephone number, membership classification, section affiliation, American Registry of Professional Entomologists membership, and a FAX, Bitnet, or e-mail number.

A description of the awards, honors, and grants sponsored by the Society as well as ESA committee reports for the year are included at the end of the directory.

HANDBOOKS

Arthropoda (General)

324. Clarke, Kenneth U. **The Biology of the Arthropoda.** London: Edward Arnold, 1973. 270p. LC 73-2309. ISBN 0-7131-23532; 0-7131-2354-0pa.

Includes basic information on this the largest group of animals. Chapters cover the cells, tissues, organs, and anatomical systems of arthropods. It also includes chapters on evolution, life patterns, and arthropod organization. Black-and-white line-drawings, electron micrographs, graphs, and diagrams are used liberally to illustrate the text. A systematic index and a subject index help find pertinent information in the text.

325. Manton, S. M. **The Arthropoda: Habits, Functional Morphology, and Evolution**. Oxford, England: Oxford University Press, 1977. 527p. $95.00. LC 77-5466. ISBN 0-19-857391-X.

This classic work reviews the Arthropoda, both living and extinct. The work covers the evolution and habits of the arthropods through discussion of functional morphology and comparative studies between the major groups. The first few chapters cover the crustaceans (crabs, lobsters, shrimp, etc.) and the evolution of arthropodan jaws and types of limbs as well as the physical requirements of land arthropods. It then moves on to a discussion of the subdivisions of the arthropoda and their locomotory mechanisms. The last three chapters cover the habits and evolution of the remaining major groups of arthropods: Myriapoda (centipedes and millipedes), Hexapoda (insects), and Chelicerata (spiders and mites). Many detailed and labeled black-and-white drawings, and eight pages of black-and-white photos, are included with the text.

326. Thorp, James H., and Alan P. Covich, eds. **Ecology and Classification of North American Freshwater Invertebrates**. San Diego, Calif.: Academic Press, 1991. 911p. $59.95. LC 90-46694. ISBN 0-12-690645-9.

This work is reviewed in the handbooks section of chapter 2, "The Invertebrates" (see entry 170), but is included here because it is a significant and up-to-date work for the freshwater arthropods. It includes chapters on water mites, the aquatic orders of insects, the Crustacea in general, and separate chapters on the Ostracoda, the Cladocera and other Brachiopoda, the Copepoda, and the Decapoda. It is particularly strong in taxonomic information and includes keys to identification of genera.

Chelicerata

ARACHNIDA (General) (Scorpions, Spiders, Mites, and Ticks)

327. Cloudsley-Thompson, J. L. **Spiders, Scorpions, Centipedes, and Mites**. New York: Pergamon Press, 1958. 228p. LC 57-1449.

This is an easy introduction to these groups. Although dated, it still contains valuable information, presented in a simple, readable style. Each chapter covers a separate group (i.e., the wood lice, millipedes, centipedes, scorpions, false scorpions, whip scorpions, harvest spiders, spiders, and mites and ticks. The chapters cover such information on each group as classification and distribution, behavior, food and feeding habits, enemies, and reproduction and life cycle; each ends with a bibliography. Black-and-white drawings, and 17 black-and-white photos, illustrate the text.

328. Comstock, John Henry. **The Spider Book: A Manual for the Study of the Spiders and Their Near Relatives, the Scorpions, Pseudoscorpiones, Whip-Scorpions, Harvestmen, and Other Members of the Class Arachnida, Found in America North of Mexico, with Analytical Keys for Their Classification and Popular Accounts of Their Habits.** Revised and edited by W. J. Gertsch. Ithaca, N.Y.: Comstock, a division of Cornell University Press, 1948. 729p. LC 49-4798.

Dated but interesting and easily readable accounts of the North American spiders and their relatives. It provides good reading on the life of spiders, the types of webs they weave, venom, food, development, silk, and the aeronautics of some spiders. It describes the common families, genera, and species in an easy manner, with many pictures and drawings to illustrate the text. Recommended for a good, entertaining introduction to spiders.

329. **Handbuch der Zoologie**: **Eine Naturgeschichte der Stämme des Tierreiches.** 3d Bd, 2 Hft., Erster Teil. **Chelicerata**. Gegrundet von Willy Kükenthal. Hrsg. von Thilo Krumbach. Berlin: De Gruyter, 1941. 315p.

This volume is part of a classic German work that is now over 15 volumes and is still in progress. See entry 68 for a description of the set. This third volume, second part, deals with the morphology, physiology, and taxonomy of the chelicerates (the spiders, scorpions, pseudoscorpions, mites, and ticks). Many black-and-white drawings and photos illustrate the German text.

330. Savory, Theodore. **Arachnida.** 2d ed. London, New York: Academic Press, 1977. 340p. LC 76-1099. ISBN 0-12-619660-5.

This handbook covers information on this entire class, which includes scorpions, spiders, mites, and ticks. It begins with a description of the phylum Arthropoda, then the class Arachnida. The class is described in 10 chapters covering morphology, physiology, embryology, ethnology, ecology, phylogeny, general habits, and taxonomy. Seventeen orders of Arachnida are then described in 5 to 20 pages each, with the order Araneae (spiders) being the largest. Line-drawings and range maps illustrate the text. The author also includes chapters on economic, practical, chemical, medical, historical, and even linguistic arachnology. An example of linguistic arachnology is the derivation of the word *spider*. The Old English word *spinnen*, for working with woven threads, went to the word *spinster*, of which the masculine form was *spinthron*, which eventually became *spider*. A long bibliography plus a section on bibliographies on Arachnida and separate indexes to subject and species completes this useful and interesting handbook.

331. **Traité de Zoologie**. **Tome VI. Onychophores, Tardigrades, Arthropodes (Généralités), Trilobitomorphes, Chélicérates**. Ed. by Pierre-Paul Grassé. Paris: Masson & Cie, 1949; Reprint, 1968. 979p. Fr 1,376.00. LC 49-2833. ISBN 2-225-58311-0.

This sixth volume is part of a multivolume, classic work, entirely in French, that summarizes the biology, anatomy, and systematics of the animal world. See entry 69 for a description of the set. In addition to information on other animal classes noted in the title, volume six includes information on the biology, anatomy, and systematics of the spiders, scorpions, pseudoscorpions, mites, and ticks. Many black-and-white drawings and some color plates illustrate the text.

332. Van der Hammen, L. **An Introduction to Comparative Arachnology.** The Hague, The Netherlands: SPB Academic Publishing, 1989. 576p. ISBN 90-5103-023-1.

The author says in the preface to this work that this is not a handbook, but a "general survey of personal insights, written after the conclusion of my investigations, i.e., after my retirement from the Rijksmuseum van Natuurlijke Historie, Leiden (where I had been curator of Arachnida for a long time)." It is a comprehensive work, drawing on many of the author's past publications, on comparative arachnology and systematics of this group. It is divided into two parts. The first part, covers comparative morphology, reproduction, development and the relationships, the evolution, and the natural classification of this group. The second part, the major part of the book, covers the systematics of 12 orders and superorders of Chelicerata. Each group is introduced with information on where it was first described and other references to the literature on the group. The morphology is very detailed, covering several pages and including many labeled line-drawings to illustrate the text. Additional remarks follow giving additional information such as distribution in the world, extinct species, and evolution. A 15-page bibliography is included, but there is no index at all. Readers should be experienced in arachnid studies to benefit from this comprehensive work.

ARACHNIDA (Systematic Section)

Scorpiones (Scorpions)

333. Leegan, Hugh L. **Scorpions of Medical Importance**. Jackson, Miss.: University Press of Mississippi, 1980. 140p. LC 80-16419. ISBN 0-87805-124-4.

This book includes information on scorpions whose stings may result in serious illness or death. It covers scorpion morphology and biology, geographic distribution, scorpion control, medical aspects of scorpion stings, and scorpion classification, with accounts of genera and species taking up most of the text. Each species is illustrated with a full-page black-and-white drawing and another page of specific anatomical features. There is one page or more of text on each species as well. Although this does not cover all scorpions, it does cover the more well known ones, and in great detail.

334. Polis, Gary A., ed. **The Biology of Scorpions**. Stanford, Calif.: Stanford University Press, 1990. $85.00. 587p. LC 84-40330. ISBN 0-8047-1249-2.

This comprehensive and up-to-date book covers a wide range of information on scorpion biology. Twelve chapters written by 10 different specialists cover such topics as anatomy and morphology, systematics (with keys to genera), behavior, ecology, life history, prey, predators, environmental physiology, venoms, neurobiology, collection and preservation of scorpions, and even a chapter on scorpions in mythology and history. The extensive bibliography, 62 pages, can be used in itself as a comprehensive resource for further information on scorpion biology. Black-and-white drawings, diagrams, and photos illustrate the text. A scholarly work of interest to researcher and amateur naturalist alike.

Araneae (Spiders, Tarantulas, Windscorpions, etc.)

335. Bristowe, W. S. **The World of Spiders**. Rev. ed. London: Collins, 1971. 304p. (New Naturalist; 38). LC 76-880889. ISBN 0-00-213256-7.

Although this is a study of British spiders, the book covers many of the spiders commonly found in the United States. Also, the general information it provides on anatomy, classification, and the life of the spider is universal. The natural history of the common spider families is described in separate chapters. The author draws upon his years of experience studying spiders to create a text, written in the first person, that draws the reader into interesting facts and observation on each spider family. Each chapter is illustrated first with a black-and-white drawing of the face of representative species in the family, and second with black-and-white full-page plates and line-drawings showing distinguishing anatomical details or specific behaviors for that family. An appendix lists spiders found in Britain up to 1958.

336. Emerton, James H. **The Common Spiders of the United States**. New York: Dover Publications, 1961. 227p. $6.95. LC 61-3981. ISBN 0-486-20223-2.

This short, inexpensive book is a classic text that identifies the more common spiders found in the United States south to Georgia and west to the Rocky Mountains. This includes a "New Key to Common Groups of Spiders," by S. W. Frost. It is easy reading and includes line-drawings and black-and-white photos to illustrate the text. Some of the nomenclature has changed since the original publication of this book, but the descriptions and the general common names such as "jumping spiders" are essentially the same. A more recent book such as *American Spiders* (see entry 338) may be helpful in finding the correct name. After general information on anatomy, colors, habits, cobweb descriptions, and information on catching and preserving spiders, the author discusses the common groups of spiders found in the United States. The most common species of each group is described in one paragraph or two with a line-drawing of the head, a top view, and other distinguishing anatomical details.

337. Foelix, Rainer F. **Biology of Spiders**. Cambridge, Mass.: Harvard University Press, 1982. 306p. $33.00; $15.95pa. LC 81-13269. ISBN 0-674-07431-9; 0-674-07432-7pa.

This is a translation and revision of the author's original German work, *Biologie der Spinnen* (George Thieme, 1979). Its purpose is to bring together all aspects of spider biology in one text. It emphasizes physiology and behavior but also covers the anatomy, metabolism, neurobiology, reproduction, development, ecology, phylogeny, and systematics of spiders in an informative, easy-to-read style. Black-and-white photos, electron micrographs, line-drawings, and diagrams illustrate almost every page of the text. An extensive, 28-page bibliography leads the reader to further reading.

338. Gertsch, Willis J. **American Spiders**. 2d ed. New York: Van Nostrand Reinhold, 1979. 274p. LC 78-6646. ISBN 0-442-22649-7.

This book provides excellent information on U.S. and Canadian spiders for the scholar and student as well as for the general reader. The author, then Curator Emeritus, Department of Insects and Spiders of the American Museum of Natural History, presents a readable text covering many aspects of spider studies. It covers spider life histories, habits, morphology, and the different kinds of spiders and their unique characteristics. It includes such interesting details as the lifespans of some tarantulas (30 years), that some amphibious spiders can catch small fish as prey, and that ships 200 miles from land have caught tiny "ballooning" (flying through the air on silk lines) spiders in their sails. Chapters on the silk-spinning and web-making process, courtship and mating, and evolution come before four chapters on the major groups of spiders. Thirty-two plates of color photographs and 32 plates of black-and-white photos illustrate some of the species of spiders and their behaviors. A four-page glossary and two-page bibliography for further reading precede a general index to subject and scientific or common name.

339. Preston-Mafham, Rod, and Ken Preston-Mafham. **Spiders of the World.** New York: Facts on File Publications, 1984. $24.95. LC 83-25435. ISBN 0-87196-996-3.

Intended for the nonspecialist, this book provides a good introduction to the world of spiders. Beautiful full-color and black-and-white photos by the authors, who are wildlife photographers, are found on almost every page. The chapters cover the anatomy, classification, behavior, and life history of spiders as well as the ingenious ways spiders capture their prey and how they defend themselves from enemies. An appendix covering the classification of spider families and a two-page glossary of terms used in the text are also helpful in learning more about spiders. A general index to the book by subject, as well as scientific and common names of spiders, is included at the end of the text.

Pseudoscorpionida (Pseudoscorpions)

340. Weygoldt, Peter. **The Biology of Pseudoscorpions**. Cambridge, Mass.: Harvard University Press, 1969. 145p. (Harvard Books in Biology; no. 6). $43.00. LC 70-82300. ISBN 0-7837-1737, 2057267.

Pseudoscorpions are very small (1 to 7 mm in length) and resemble true scorpions except they are without the tail and sting. This short book covers the biology of this little-known order of Arachnida. The 11 chapters cover morphology, anatomy and physiology, locomotion and behavior, reproduction, development, ecology, evolution, and systematics; as well as on the collection and preservation of pseudoscorpions. A species of interest to libraries is the common book scorpion (*Cheiridium museorum*), sometimes found in old libraries, which is a minuscule 1.1 mm long. Black-and-white drawings, diagrams, and photos illustrate almost every page. The book concludes with a six-page bibliography and separate species and subject indexes.

Acariformes (Mites and Ticks)

341. Balogh, J., and S. Mahunka, eds. **The Soil Mites of the World**. Amsterdam, New York: Elsevier, 1983. Vols. 1- . LC 90-12027. Each volume approximately $150.00. ISBN 0-444-98809-2 vol. 3.

This series is up to three volumes as of this writing. Volume 1 is *Primitive Oribatids of the Palaearctic Region*, and volumes 2 and 3 are *Oribatid Mites of the Neotropical Region I and II*. See entry 553 for a complete annotation.

342. Krantz, G. W. **A Manual of Acarology**. 2d ed. Corvallis, Oreg.: Oregon State University, 1978. 509p. $39.95. LC 78-56128. ISBN 0-88246-064-1.

Although this is primarily a systematic text, more than 100 pages of introductory material on morphology, reproduction, life stages, habits, and collection, rearing, and studying of ticks and mites are also included. The text provides detailed, illustrated keys to family except for the suborder Oribatida, which goes to superfamily. The ticks are keyed to genus. More than 800 black-and-white drawings and diagrams illustrate the keys and text. Each chapter concludes with a list of references for further reading. A detailed and lengthy index containing more than 2,000 entries makes for easy access to information in this work.

343. Sonenshine, Daniel E. **Biology of Ticks**. New York: Oxford University Press, 1991- . Vols. 1- . $90.00 vol. 1. LC 91-1954. ISBN 0-19-505910-7.

This is the most current and comprehensive text on tick biology and includes information on tick systematics, ecology, and biology as well as disease relationships and control of ticks. Only the first volume of this comprehensive work had been published when reviewed for this book, but it covered more in one volume on this group of Arthropoda than any other source found. The clearly written text discusses many aspects of tick biology (i.e., evolution, life cycles, internal and external anatomy, structure and function of organs and tissues such as salivary glands, circulatory system, respiratory system, nervous system, male and female reproductive systems, pheromones, genetics, embryogeny, and water balance. A chapter on systematics includes a key to families and genera. Volume two will cover ecology, behavior, and host-parasite interactions as well as diseases caused by ticks and the control of ticks. The book is illustrated with many electron micrographs of ticks and tick anatomy. The 25-page bibliography further enriches the usefulness of this work.

Crustacea
(Crabs, Lobster, Shrimp, etc.)

343a. Anderson, D. T. **Barnacles: Structure, Function, Development and Evolution**. London: Chapman & Hall, 1994. 357p. $116.00. ISBN 0-412-44420-8.

This work covers the functional morphology, physiology and development of barnacles with an eye to evolutionary aspects of barnacle studies. It provides both an introduction to the group and an up-to-date synthesis of current research. In 10 chapters it discusses the history and morphology of barnacles, the biology of muscles and circulation, digestion, respiration and excretion, control functions and environmental interactions, the reproductive system, gametes and fertilization as well as larval development, growth and phylogeny. Each chapter ends with a lengthy list of references. Appendix A contains a taxonomic synopsis of Cirripedes (barnacles). The text is illustrated with labeled black-and-white drawings to help explain the text. Primarily for postgraduates, researchers and professionals, but also useful to undergraduates as an introductory text to these remarkable animals.

344. Barnard, J. L., and C. M. Barnard. **Freshwater Amphipoda of the World**. Mt. Vernon, Va.: Hayfield Associates, 1983. 2 vols. LC 83-174719.

This is a two-volume set: evolutionary information is in the first volume, and a checklist, keys to the genera of gammaridan amphipods, a handbook of the world gammaridans and other freshwater amphipods, and a bibliography are contained in the second volume. This covers primarily information on the family Gammaridae because they are the main group of freshwater amphipods. In addition to evolutionary information in the first volume, distribution and descriptions of some groups are given. Black-and-white line-drawings and distribution maps illustrate volume one. The complete bibliography in volume two includes 1,735 citations.

345. Bliss, Dorothy E., ed. **The Biology of Crustacea**. New York: Academic Press, 1982-1985. 10 vols. $960.50/set; $113.00/vol. LC 82-4058. ISBN 0-12-106401-8 vol. 1; 0-12-106402-6 vol. 2; 0-12-106403-4 vol. 3; 0-12-106404-2 vol. 4; 0-12-106405-0 vol. 5; 0-12-106406-9 vol. 6; 0-12-106407-7 vol. 7; 0-12-106408-5 vol. 8; 0-12-106409-3 vol. 9; 0-12-106410-7 vol. 10.

This major work on crustacean studies, begun in 1982, is a cooperative work of 10 editors and more than a 100 contributors, all recognized experts in their field. Each volume covers an important aspect of crustacean biology: vol. 1, systematics, the fossil record, and biogeography; vol. 2, embryology, morphology, and genetics; vol. 3, neurobiology, structure and function; vol. 4, neural integration and behavior; vol. 5, internal anatomy and physiological regulation; vol. 6, pathobiology; vol. 7, behavior and ecology; vol. 8, environmental adaptations; vol. 9, integument, pigments, and hormonal processes; vol. 10, economic aspects, fisheries, and culture. Chapters in each volume contain lengthy references that may be consulted for further reading. Each volume ends with a systematic and subject index. This is an essential work for academic libraries.

346. Boxshall, Geoffrey A., and H. Kurt Schminke. **Biology of Copepods: Proceedings of the Third International Conference on Copepoda**. Dordrecht, Boston: Kluwer Academic, 1987. 639p. (Developments in Hydrobiology; 47). $274.50. LC 88-6416. ISBN 90-6193-654-3.

This work contains the proceedings of the Third International Conference on Copepoda held in London in 1987. It contains a wide variety of papers on all aspects of copepod studies. It is arranged into seven sections: "Rate Processes in Field Populations of Planktonic Copepods"; "Taxonomy and Biology of *Calanus*"; "Oceanic and Deep-Sea Copepods"; "Marine Plankton"; "Freshwater Copepods"; "Harpacticoid Copepods"; and "Parasitic and Associated Copepods." Each section has from five to fifteen papers written by experts on that topic. References listed with each paper. Black-and-white drawings and graphs illustrate the text and an index to subject and scientific name is provided. A handbook for the specialist.

347. Crane, Jocelyn. **Fiddler Crabs of the World (Ocyopodidae: Genus *Uca*)**. Princeton N.J.: Princeton University Press, 1975. 736p. $135.00. LC 70-166366. ISBN 0-691-08102-6.

This monumental work on a single genus is included here for its comprehensiveness and sheer effort. It includes 90 species and subspecies from around the world. Jocelyn Crane's work on this genus took her to all parts of the world in search of new *Uca* species. The book is divided into two sections. Part one covers the systematics of *Uca* based on morphology and behavior. Each species is thoroughly described morphologically (description of male and female, size, color), by social behavior, by relationships with other groups, and by the derivation of the name. Part two covers the evolutionary biology. A taxonomic key to subgenera and keys to *Uca* in major geographical areas are found in the appendix. The appendix also has other details on fiddler crabs in tabular form, a section on maintaining these crabs in captivity, and an extensive glossary of terms. A 23-page bibliography, an index to scientific names, and a general index follow the text.

348. **Handbuch der Zoologie: Eine Naturgeschichte der Stämme des Tierreiches**. Bd.3:1. **Tardigrada, Pentastomida, Myzostomida, Arthropoda: Allgemeines, Crustacea**. Gedr. von Willy Kükenthal, heraus. von Thilo Krumbach. Berlin: de Gruyter, 1926-1927. 1158p.

This volume of the Handbuch (see entry 68) covers primarily the Crustacea, but also includes other groups, as the title suggests. The morphology and phylogeny of the entire group of crustaceans is discussed first, followed by lengthy chapters on the Brachiopoda, Ostracoda, Copepoda, Branchiura, Cirripedia, Malacostraca, Isopoda, Amphipoda, Decapoda, and Stomatopoda. Entirely in German, with black-and-white drawings to illustrate the text.

349. Kaestner, Alfred. **Invertebrate Zoology**. Vol. 3, **Crustacea**. New York: Interscience, 1970. 523p. LC 67-13947. ISBN 471-45417-6.

This is part of a three-volume work by Alfred Kaestner, translated and adapted from the second German edition by Herbert W. Levi and Lorna R. Levi. This third volume provides thorough accounts on morphology and classification of the major groups of crustaceans. The first and second chapters deal with the entire crustacean class anatomy and reproduction as well as development and relationships. The following chapters cover the various subclasses, superorders, and orders. Each chapter covers the anatomy, reproduction, development, habits, locomotion, respiration, and classification of a particular crustacean group such as Ostracoda, Cirripedia, Malacostraca, and so on. Black-and-white drawings help to illustrate the descriptive text. A helpful list of references accompany each chapter.

350. Schram, Frederick R. **Crustacea**. New York: Oxford University Press, 1986. 620p. $65.00. LC 85-11526. ISBN 0-19-503742-1.

The author's aim in this book is to provide a basic text on crustacean evolutionary biology, excluding the physiology, histology, and ultrastructure. Each chapter is devoted to a particular order or subclass of Crustacea. The chapter gives a description of the group, the history, morphology, natural history, embryonic development, fossil record, taxonomy, phylogeny and evolution. A list of references is at the end of each chapter. Black-and-white drawings illustrating the anatomy, and maps showing geographical distribution are included. Two indexes, a subject and a taxonomic, are useful and complete.

351. Southward, Alan J., ed. **Barnacle Biology**. Rotterdam, The Netherlands: A.A. Balkema, 1987. 443p. (Crustacean Issues, 5). £55.00. ISBN 90-6191-628-3. ISSN 0168-6356.

This comprehensive work includes review articles on all aspects of barnacle biology. Barnacles make up an entire class, the Cirripedia. The book is arranged into four sections; "Evolution and Genetics," "Physiology and Function," "Larval Biology and Settlement," and "Pollution and Fouling." There are two to nine articles in each section, each authored by experts on that topic. Lengthy lists of references follow each chapter. The book was dedicated to Dennis J. Crisp, an authority on barnacles, and includes biographical information as well as a long list of references by Dr. Crisp. Black-and-white photos, drawings, and graphs are found throughout the text. Separate taxonomic and subject indexes complete the work.

352. Warner, G. F. **The Biology of Crabs**. New York: Van Nostrand Reinhold, 1977. 202p. LC 77-24512. ISBN 0-442-29205-8.

This short, easy-to-read work on crabs serves as a good introduction to the subject. Clear chapters are written here on the anatomy, movement, sense organs, rhythms, life styles, food and feedings, social behavior, life histories, temperature and environments where crabs are found, evolution and systematics, and the economic importance of crabs to man. Forty-three figures with black-and-white line-drawings illustrate the text. A good list of references, a short glossary, and a subject and scientific-name index are found at the end of the text.

353. Waterman, Talbot H., ed. **The Physiology of Crustacea**. New York: Academic Press, 1960-61. 2 vols. LC 59-7690.

This is a classic work in comparative physiology that has been updated and continued by *The Biology of Crustacea* (see entry 345). Many international scholars collaborated in this work, which was three years in the making. The two volumes cover metabolism and growth; and sense organs, integration, and behavior. Each chapter is a treatise in itself on topics such as light sensitivity and vision, molting, locomotion, the neuromuscular system, respiration, and biochemistry of pigments. Intended for advanced students and researchers, it is still a standard reference for any biologist.

354. Williams, Austin B. **Shrimps, Lobsters, and Crabs of the Eastern United States, Maine to Florida**. Washington, D.C.: Smithsonian Institution Press, 1984. 550p. $47.50. LC 83-6000095. ISBN 0-87474-960-3.

This is a thorough treatment, covering the history, classification, zoogeography, and species accounts in addition to the main text on the systematics of this important group. An illustrated four-page glossary precedes the systematics section. Dichotomous keys down to species are followed by full descriptions of each species. Information on the species includes recognition characters, citations to the literature where first described, habitat, type locality, range, and remarks on other useful information on the species. There is usually a nice black-and-white drawing of the species that accompanies the description. Specific, identifying anatomical details are also illustrated. In all there are 380 illustrated figures. There is also an extensive bibliography (59 pages), and an index to genus and species.

Uniramia

INSECTA (General)

355. Davies, R. G. **Outlines of Entomology**. 7th ed. London, New York: Chapman and Hall, 1988. 408p. $75.00; $37.50pa. LC 88-10865. ISBN 0-412-26670-9; 0-412-26680-6pa.

This is a descendant of A. D. Imms' *Outlines of Entomology,* first published in 1942. It is now in its seventh edition, with major changes from the last edition. These changes include a new chapter on insect population biology, a new section on injurious insects, and many changes and updates in the rest of the text. It includes more black-and-white drawings and diagrams, and a larger classified bibliography. It covers basic information on insects such as structure and function, development, classification and description of 29 orders, as well as evolution, modes of life, insect populations, and injurious insects. The classified bibliography is useful for finding more specialized information for further reading.

356. **Handbuch der Zoologie: Eine Naturgeschichte der Stämme des Tierreiches**. Bd.IV **Arthropoda: Insecta**. Gedründet von Willy Kükenthal, herausgegeben von Thilo Krumbach. Berlin, New York: de Gruyter, 1968- . 31 parts as of 1992. Approximately $157.00/vol.

Originally published in German, but recently issued in English, this is part of a classic set of works on the animal kingdom (see entry 68). Volume four, so far issued in 31 parts, treats the class Insecta. It covers the biology of insects including the physiology, phylogeny, geographical distribution, and morphology; the associated areas of entomology such as the medical aspects; and the individual orders. Each order is thoroughly described with historical and systematic accounts, keys to families, zoogeography, ecology, economic importance, morphology, anatomy and physiology, and a list of cited literature. An index to scientific name and a general index are included with each bound volume. Most parts of this section on the Insecta are still in print.

357. Hermann, Henry R. **Social Insects**. New York: Academic Press, 1979-1982. 4 vols $400.00 LC 78-4871. ISBN 0-12-342201-9.

A complete treatment of the social insects, this four-volume set covers the socia behaviors of insects such as ants, termites, and bees, and even spiders, not really insects, bu included here as a close relative. Systematics, keys to taxa, anatomy, biology, and chemistry are included as well. Line-drawings, electron micrographs, photos, diagrams, and maps illustrate the text. Volume one serves to introduce the social behavior of insects; volume 2 covers termites, presocial insects, Arachnida, and systematics of social Hymenoptera; volume 3, bumblebees, honey bees, stingless bees, and a chapter on symbionts, or other organism that live closely with social insects; and volume 4, social wasps, ants, army ants, and fungus ants. Extensive references follow each chapter. Each volume has its own species index and a subject index. There is no comprehensive index to the set.

358. **IIE Guides to Insects of Importance to Man**. Oxon, England; Tucson, Ariz.: C.A.B International for International Institute of Entomology, 1990- . Approximately £24.00 each

Originally titled *CIE Guides to Insects of Importance to Man*, this series is published irregularly on insects of importance, such as pests or beneficial insects in tropical agriculture and forestry. So far, three guides have been published: Lepidoptera, Thysanoptera, Coleoptera. Aimed at students, teachers, foresters, and others in applied entomology, this serie provides a good overview of each order. Introductory material covers the biology, economi importance, adult morphology, and collecting of the insects. Illustrated keys to identificatio of specimens to family, of both adult and larval stages, are provided. Each family is the described, as well as some economically important species. Some guides provide checklists most provide a glossary of terms and a list of references and index. Black-and-white drawing illustrate the text. Good information, especially for identification and description.

359. Imes, Rick. **The Practical Entomologist**. New York: Simon & Schuster, 1992. 160p (A Fireside Book). $15.00. LC 91-23905. ISBN 0-671-74695-2pa.

This introductory guide to insect study features easy-to-read text with 200 full-colc illustrations. It begins with the basics of "What is an insect?" and progresses through insec relatives, taxonomy, anatomy, life cycles, and insect senses, as well as discussing the tool of entomology, and how to capture and keep live insects for study. It also tells the amateu entomologist how to make an insect collection, how to photograph insects, and how to kee a field notebook. The major insect orders are then discussed in separate chapters, with a fina chapter on the minor insect orders. On almost every page the reader will find a strikin full-color photo or color illustration. This is a fine and inexpensive guide for the beginne and would be a good addition to public and school libraries.

360. Klausnitzer, Bernhard. **Insects: Their Biology and Cultural History**. New York Universe Books, 1987. 240p. $40.00. LC 87-5853. ISBN 0-87663-666-0.

This book, translated from the German, covers the major orders of insects in a new wa It discusses ancient references to insects, their uses and beliefs about them in ancient time and shows how this knowledge has developed into its present state. There is much interestir historical information, such as the story behind the name *termite* (from the Latin name *Terme* meaning "end," which Linnaeus gave to the true termite because it produces a knocking sou in wood, making some people believe this a warning that their end is near). Each order is discussed and illustrated with beautiful full-color and black-and-white photos that cover entire pages. Black-and-white drawings of representative species from each order are found at the beginning of each chapter and throughout the text. Highly readable, with interesting information not available in other insect texts.

360a. McGavin, George C. **Bugs of the World**. New York: Facts on File, 1993. 192p. $24.95. LC 93-9122. ISBN 0-13-948225-3.

The title of this book suggests that it covers bugs of the world, but this short book is actually a general introduction to the world of insects. It combines easy-to-read text with beautiful full-color photos. After introductory chapters on how to collect insects and descriptions of their basic anatomy, the author lists and briefly describes the major groups of insects. Further chapters cover the diseases and enemies of these insects as well as their defense mechanisms, food sources, reproduction and the interaction of insects and man. A one-page bibliography for further reading and an index complete the work. This book features excellent color photos by Ken Preston-Mafham.

361. Metcalf, Robert L., and Robert A. Metcalf. **Destructive and Useful Insects: Their Habits and Control**. 5th ed. New York: McGraw-Hill, 1993. 1 vol. (various pagings). $85.00. LC 92-18374. ISBN 0-07-041692-3.

This is the revised edition of *Destructive and Useful Insects* by C. L. Metcalf and W. P. Flint (published in 1962). It is a very thorough and reliable work that has served as a standard reference work in economic entomology for years, the first edition being published in 1928. The first half of the book includes chapters on insect biology, classification, and control. The second half of the book includes identification keys of insect pests, and descriptions of each species and its control. Chapters are arranged under the type of damage done (vegetable garden pests, fruit-tree insects, corn insects, etc.). It is similar to the fourth edition, but it has new information, tables, illustrations, and a section on integrated pest management. A classic source for all libraries.

362. Stehr, Frederick W. **Immature Insects**. Dubuque, Iowa: Kendall/Hunt, 1987-1991. 2 vols. $300.00/set. LC 85-81922. ISBN 0-8403-3702-7 vol. 1; 0-8403-4639-5 vol. 2.

This two-volume work is a comprehensive treatment of the immature insects. It is intended as a textbook, an introduction to the diverse literature on this subject, a means to identify larvae of all orders to family, and a way to identify the more common, commercially important species. It is primarily a guide to the larvae, with a smaller coverage of eggs and pupae. Volume one covers collecting, rearing, preserving, storing, and studying immature insects, and then covers a key to orders, followed by separate chapters on each order. Volume two finishes the chapters on orders. Each order is described as to anatomy and classification with a key to families. A description of each family is given with information on biology, ecology, relationships and diagnosis, anatomical description, and comments on important facts and geographical distribution. Lengthy bibliographies end each chapter. Detailed black-and-white drawings illustrate the text. Each volume contains a complete index and an independent glossary.

363. **Traité de Zoologie**. Tomes VIII-X, **Insectes**. Ed. by Pierre-Paul Grassé. Paris: Masson & Cie, 1949-1979. 9 vols. LC 49-2833.

Part of a multivolume classic work, entirely in French, that summarizes the biology, anatomy, and systematics of the animal world (see entry 69). This nine-volume section deals with the entire group of insects. Tome 8 covers the anatomy and morphology of the insects in six fascicles. Tomes 9 and 10 cover the palentology, classification, and description of the insect orders. Many black-and-white drawings and some color plates illustrate each volume. Each fascicle is between 500 and 1,100 pages in length. The following titles are translated from the French:

Tome 8, fascicle 1. *Head, Wing, and Flight*. Fr. 1,376.00. ISBN 2-225-37004-4.

Tome 8, fascicle 2. *Thorax, Abdomen*. Fr. 1,324.00. ISBN 2-225-62667-7.

Tome 8, fascicle 3. *Integument, Nervous System, and Sensory Organs*. Fr. 1,376.00. ISBN 2-225-41701-8.

Tome 8, fascicle 4. *Splanchnology, Phonation, Aquatic Life, and Plant/Insect Interaction*. Fr. 1,702.00. ISBN 2-225-42031-9.

Tome 8, fascicle 5-1. *Gametogenesis, Fecundation, Metamorphosis*. Fr. 1,376.00. ISBN 2-225-44980-5.

Tome 8, fascicle 5-2. *Embryology, Gall-forming, and Venomous Insects*. Fr. 1,133.00. ISBN 2-225-45584-8.

Tome 9. *Classification and Phylogeny of Fossil Insects and Living Orders to Coleoptera*. Fr. 1,587.00. ISBN 2-225-58322-6.

Tome 10, fascicle 1. *Neuroptera to Hymenoptera*. Fr. 1,482.00. ISBN 2-225-58344-7.

Tome 10, fascicle 2. *Hymenoptera to Thysanoptera*. Fr. 1,482.00. ISBN 2-225-58366-8.

364. Wilson, Edward O. **The Insect Societies**. Cambridge, Mass.: The Belknap Press of Harvard University Press, 1971. $20.00pa. LC 74-148941. ISBN 0-674-45490-1; 0-674-45495-2pa.

Edward O. Wilson, famous entomologist and two-time winner of the Pulitzer Prize for nonfiction, wrote this synthesis of the literature on insect sociology. It is a thorough and readable text that covers the main groups of social insects (i.e., ants, termites and social bees and wasps), as well as chapters on behavior communication, social evolution, symbioses, and population dynamics. Black-and-white drawings, photos, diagrams, and graphs illustrate the text. An 11-page glossary and 54-page bibliography complete the text.

365. Wooton, Anthony. **Insects of the World**. Poole: Blandford Press, 1984. 224p. $24.95. LC 83-25425. ISBN 0-7137-1363-1.

This is a short, basic introduction to insects, written on a popular level with beautiful full-color photos to add interest. The book covers the definition of what an insect is, its distribution, anatomy, communication, excretion, feeding, and digestion, as well as parasites, life histories, and reproduction. There are chapters on the social and aquatic insects, and on migration, coloration, and defenses. A short chapter on nomenclature and classification includes black-and-white drawings of representative species. A glossary and a list of books for further reading are also helpful for the amateur entomologist to find more information.

INSECTA (Geographic Section)

Australia

366. Naumann, I. D., ed. **The Insects of Australia: A Textbook for Students and Research Workers.** 2d ed. Ithaca, N.Y.: Cornell University Press, 1991. 2 vols. $215.00. LC 91-9407. ISBN 0-8014-2669-3/set.

This second edition of an acclaimed text was written primarily by entomologists with CSIRO (Commonwealth Scientific and Industrial Research Organisation). It is a comprehensive treatise covering all aspects of Australian insect studies. Skeletal anatomy, general biology, systematics, fossil history and phylogeny, biogeography, and biographies of famous Australian entomologists are covered in separate chapters of volume one. Separate chapters on the individual orders follow and are continued into volume two. Each chapter on orders is written by one or more experts in the field. The chapters cover anatomy of adult and immature stages, biology, natural enemies, economic significance, and special features about the order; then, keys to the families and descriptions of each family follow. Detailed black-and-white ne-drawings, photos, and electron micrographs, as well as beautiful color plates, illustrate ıe text. A lengthy 72-page list of references and index to the set are found at the end of olume two. An excellent work.

Canada

67. Canada. Agriculture Canada. **Insects and Arachnids of Canada**. Ottawa: Canada ept. of Agriculture, 1976- . Published irregularly. $25.00-$30.00 each. LC sf88-34077. SSN 0706-7313.

This series of monographs on the arthropods of Canada has now reached 20 parts, eginning with part 1 (actually published second in the series) on *Collecting, Preparing, and reserving Insects, Mites, and Spiders*; the other volumes are on bark beetles, Aradidae, nthocoridae, crab spiders, mosquitoes, plant bugs, spiders, spittlebugs, larval midges, halcidoid wasps, carrion beetles, grasshoppers, crickets and related insects, metallic woodoring beetles, horse flies and deer flies, wolf spiders, nurseryweb spiders and lynx spiders, ower flies, ground spiders, and sawflies. Each part is authored by a specialist in the field nd describes and provides identification keys for the group. For example, the flower flies of ıe subfamily Syrphinae are treated in part 18. In it the author provides introductory material n the distribution, economic importance, general biology, mimicry, anatomy, and collection nd preservation of this group. The main part is on the classification and taxonomy of the roup with illustrated keys to subfamilies, genera, subgenera, and species. Each species is ıen described in one or two pages, covering both male and female, their distribution ncluding distribution maps), specimens identified, their biology, and some miscellaneous otes on remarkable features. An excellent series; authoritative and useful.

Great Britain

68. **Handbooks for the Identification of British Insects**. London: Royal Entomological ociety of London, 1949- . Multivolume set. Issues come out irregularly. $6.00-$30.00 each. C 85-2109.

See entry 568 in the taxonomic keys section. This is a series of books that provide keys nd descriptions of British insects. These are primarily detailed, illustrated identification keys insects of Great Britain, but they do contain additional information about the insect roup.

Hawaii

69. Zimmerman, Elwood C., and D. Elmo Hardy. **Insects of Hawaii: A Manual of the nsects of the Hawaiian Islands, Including an Enumeration of the Species and Notes on heir Origin, Distribution, Hosts, Parasites, etc.** Honolulu: University of Hawaii Press, 948-{1981}. 15 vols. to date. LC 48-45482.

This comprehensive work is included here because it covers insects common to other acific islands. The volumes include checklists of insects in each volume, descriptions of ıperfamilies, families, genus, and a listing of species. Species include the author of the name, ıe citation of where it was named, and the type-species. Excellent, detailed black-and-white rawings show species or specific anatomical details. Each volume includes a bibliography f references cited and an index to scientific name. Volumes 1-9 by Zimmerman and Hardy; olumes 10-14 by D. Elmo Hardy and others; volume 15 by Kenneth Christianson and Peter ellinger.

Volume 1. *Introduction*. (Out-of-print.)
Volume 2. *Apterygota to Thysanoptera*. $10.00.
Volume 3. *Heteroptera*. $7.50.
Volume 4. *Homoptera: Auchenorhyncha*. $7.50.

Volume 5. *Homoptera: Sternorhyncha*. $10.00.
Volume 6. *Ephemeroptera-Neuroptera-Trichoptera and Supplement to Vols. 1-5*. $7.50.
Volume 7. *Macrolepidoptera*. $12.00.
Volume 8. *Lepidoptera: Pyraloidea*. $10.00.
Volume 9. *Microlepidoptera*. $60.00.
Volume 10. *Diptera: Nematocera-Brachycera (Except Dolichopodidae)*. $9.00.
Volume 11. *Diptera: Brachycera II-Cyclorrhapha I*. $10.00.
Volume 12. *Diptera: Cyclorrhapha II*. $20.00.
Volume 13. *Diptera: Cyclorrhapha III*. $30.00.
Volume 14. *Diptera: Cyclorrhapha IV*. $35.00.
Volume 15. *Collembola*. $48.00.

Micronesia

370. **Insects of Micronesia**. Honolulu, HI: Bernice P. Bishop Museum, 1954-1984. 15 vols. many parts.

Although titled *Insects of Micronesia*, this set also includes other arthropods such as scorpions, spiders, and pseudoscorpions. This publication ceased in 1984. Each part on a particular family was published separately by an individual author. The family was described as to zoogeography and systematics, with keys to genera and subgenera. Black-and-white drawings show representative species and anatomical detail for identification. Most issues in English, some in German and French.

North America

371. Arnett, Ross H. **American Insects: A Handbook of the Insects of America North of Mexico.** New York: Van Nostrand Reinhold, 1985. 850p. $105.00. LC 84-15320. ISBN 0-442-20866-9.

In this one remarkable volume are the keys needed to identify the 87,000 insect species native to America north of Mexico. Ten years in preparation, it lists and describes each order with a key to each family, listing each genus and describing the common species within the genus. Bibliographic references are included at the end of the section for each order to help in identifying those taxa not covered. Over 1,200 black-and-white photos and line-drawings in each section further aid in identification of the insect. It includes a good glossary to terms used in the descriptions and a complete index with both the Latin and common name as well as terms such as *phenetics*. The author readies the user of this book with general chapters on insect classification and systematics, insect collecting, and general considerations where he discusses how to use identification keys. Armed with this information, a hand lens, and in some instances a microscope, the amateur, student, and professional should be able to identify most of the common insects in America.

372. Essig, E. O. **Insects and Mites of Western North America: A Manual and Textbook for Students in Colleges and Universities, and a Handbook for County, State, and Federal Entomologists and Agriculturists As Well As for Foresters, Farmers, Gardeners, Travelers, and Students of Nature.** 2d ed. New York: Macmillan, 1958. 1050p. LC 58-5163.

Although this work can serve as textbook as well as handbook, it is included in the handbook section of this book because it is primarily descriptive. Twenty-five orders of insects of western North America are described; boundaries are: Alaska on the north, the Rocky Mountains to the east, Mexico and lower California on the south, and the Pacific Ocean on the west. Each order is contained in an individual chapter. A key to families in the order begins the chapter, followed by a description of each family, and then a description of a few common species. Each chapter ranges from 1 page to almost 200 pages in length.

Black-and-white drawings and photos illustrate the text. Footnotes are used on each page in place of a full bibliography.

373. Merritt, Richard W., and Kenneth W. Cummins, eds. **An Introduction to the Aquatic Insects of North America**. 2d ed. Dubuque, Iowa: Kendall/Hunt, 1984. 722p. $52.95. LC 83-83109. ISBN 0-8403-3180-0.

This work is a guide to the North American families of aquatic insects. This second edition has more coverage in general biology and morphology than the first edition, and two new chapters on respiration and life history have been added; plus, all aspects of the work have been updated from the recent literature. The book includes chapters on morphology, collecting and sampling, ecology and distribution, and phylogeny, and a chapter on general classification that has a key to orders. Thirteen orders and four families of aquatic insects are described in individual chapters. Keys to the families and genera of each order, adults and nymphs, are the primary focus in each chapter; detailed line-drawings illustrate the keys. Tables are provided that give species information on habitat, habit, trophic relationships, North American distribution, and references to entries in the bibliography at the end of the text. The comprehensive bibliography includes 2,800 references; the first edition included 1,712. A very thorough and useful work.

Panama and Mesoamerica

374. Quintero, Diomedes, and Annette Aiello, eds. **Insects of Panama and Mesoamerica: Selected Studies**. Oxford, New York: Oxford University Press, 1992. 692p. $195.00. LC 90-28613. ISBN 0-19-854018-3.

This monumental work was prepared by 52 biologists from 11 countries covering chapters on specific groups of insects. It covers a vital insect fauna that is presently threatened due to the destruction of tropical forests. This work attempts to describe and illustrate more than 20 insect orders plus the Collembola. Keys to identification of insect groups (to species in some cases) are provided with most chapters. Each chapter is individually authored by a specialist on that particular insect group. A lengthy list of literature cited is included at the end of each chapter. Many black-and-white drawings and photos illustrate the text. Separate taxonomic and subject indexes finish the work.

Puerto Rico and the Virgin Islands

375. **Insects of Porto (sic) Rico and the Virgin Islands**. New York: New York Academy of Sciences, 1932-1939. 3 pts. (Scientific Survey of Porto Rico and the Virgin Islands, vol. 14).

The individual parts cover the Odonata, or dragonflies, by Elsie Broughton Klots; Homoptera (excepting the Sternorhynchi) by Herbert Osborn; and the Hemiptera-Heteroptera (excepting the Miridae and Corixidae) by Harry Gardner Barber. Keys to family genus and species are here with numerous line-drawings and black-and-white photos to illustrate the keys and the text. Checklists are given as well as general information on the ecology, economic importance, taxonomy, and host plants of each group. Bibliographies of sources and indexes to genera and species are found at the end of each part.

376. Wolcott, George N. **The Insects of Puerto Rico**. Rio Piedras, Puerto Rico: University of Puerto Rico, Agricultural Experiment Station, 1948. 975p. (The Journal of Agriculture of the University of Puerto Rico, vol. 32, nos. 1-4).

Published in four parts, this work provides textual information on the insect groups and species that are found in Puerto Rico. Published in four parts, it covers the more primitive groups in the first part, from Thysanura to Hemiptera. The Coleoptera are described in the second part; the Diptera, Siphonaptera and Lepidoptera in the third part; and the Hymenoptera in the fourth part. Each family is described, then the genus and species are briefly described, with notes to where it was published in the literature, and in what location in Puerto Rico it is usually found. Black-and-white drawings are placed here and there to illustrate representative species. An index to all four parts is found in the last volume.

USSR

377. **Fauna SSSR**. Leiden: E. J. Brill, 19- .

This multivolume series, published by different publishers over the years, includes works that cover the entire fauna of the USSR, including many on insect orders. There are separate works on Coleoptera, Diptera, Hymenoptera, Lepidoptera, and Trichoptera so far. Each work is separately authored and concentrates primarily on the taxonomy of the group. Usually general information on the morphology, biology, geographic distribution, and classification of the group is given, then the taxonomy and a key to each genus and species in the group. Each genus is then described as well as the species in the genus. Black-and-white drawings of selected species as well as drawings of anatomical details to aid in identifying them are included in each text.

INSECTA (Systematic Section)

Not all orders are listed here, only those with significant book literature.

Collembola (Springtails)

378. Christiansen, Kenneth, and Peter Bellinger. **Collembola of North America North of the Rio Grande: A Taxonomic Analysis.** Grinnell, Iowa: Grinnell College, 1980-81. 4 vols. $35.00. LC 81-204229. ISBN 0-686-34383-2(1-vol. ed.).

Although titled a taxonomic analysis, this is included here under handbooks because it includes a wide variety of other information such as geographical distribution, characteristics of Collembola, habitats, biology and collecting, and preservation of this group. This four-volume work provides keys to families, genera, and species found in North America north of Mexico. Each taxa is then described in a paragraph covering color and other characteristics followed by a paragraph of remarks covering other information on the taxa. Type locality is given as well as other geographical localities where the taxa may be found. Black-and-white line-drawings illustrate specific anatomical details for identifying each species. The fourth volume contains a glossary, a 29-page bibliography, and a systematic index.

Ephemeroptera (Mayflies)

379. Edmunds, George F., Jr., Steven L. Jensen, and Lewis Berner. **The Mayflies of North and Central America**. Minneapolis, Minn.: University of Minnesota Press; UMI Books on Demand, 1976. 330p. $35.00. LC 75-39446. ISBN 0-8166-0759-1.

This book provides information on adult and nymph mayflies of North America, Mexico, and Central America. It includes brief information on the biology of mayflies, collecting methods, rearing and preservation of specimens, classification, and anatomy. There are two separate keys; one to nymphs and the other to adult mayflies. Each key is illustrated with black-and-white drawings of anatomical details. Following the keys are descriptions of the families of mayflies, which form the bulk of the book. Each family is thoroughly discussed. Nymphal and adult characteristics and habits are detailed as well as the life history, mating flights, taxonomy, and distribution. A full-page black-and-white illustration of representative species is included with each family. A 10-page bibliography for further reading completes the work.

380. Needham, James G., Jay R. Traver, and Yin-Chi Hsu. **The Biology of Mayflies with a Systematic Account of North American Species**. Ithaca, N.Y.: Comstock, 1935. 759p. LC 36-203.

This work, though dated, covers more biology than George Edmunds' more recent book on mayflies (see entry 379). It covers the structure of adult and nymph mayflies, the internal anatomy, embryology, eggs, and life cycle in 100 pages. Taxonomic characters; detailed information on wings, legs, and thorax, and mayfly phylogeny are also discussed and illustrated in the next 100 pages, as well as information on collecting mayflies and their economic importance. The second part of this work is a systematic account of species, both adult and nymphal, in North America. This second part is updated by Edmunds' book.

Odonata (Dragonflies and Damselflies)

381. Askew, R. R. **The Dragonflies of Europe**. Colchester, England: Harley Books, 1988. 291p. £56.00. LC gb89-11886. ISBN 0-946589-10-0.

The purpose of Dr. Askew's book was to simplify the identification of European dragonflies and encourage an interest in them. It is a comprehensive work, with separate chapters on the life history, information on the adult dragonfly, the distribution in Europe, and a description of 10 families of Odonata found in Europe. Each chapter is illustrated with black-and-white drawings, and two color plates illustrate the biology and habitats. Most impressive, however, are the 29 color plates of drawings, by the author, of 210 adult dragonfly species. The striking color and detail are enough to encourage an interest in this beautiful group. Keys to family and species are provided. Each species is described with paragraphs describing the adult, the biology, flight period, and distribution. Large distribution maps are placed next to the descriptions. Common names are given in English, French, and German. Another impressive feature is the complete list of references on this group. All in all, an excellent work that meets and exceeds the purposes of the author.

382. Hammond, Cyril O. **The Dragonflies of Great Britain and Ireland**. 2d ed., rev. by Robert Merritt. Colchester, England: Harley Books, 1983. 116p. £11.20. LC 84-198376. ISBN 0-946589-14-3pa.

Because of the success of the first edition, published in 1977, this second edition was brought out with additional identification details, the addition of a new species, updated information on distribution of species, more recent distribution maps, and nomenclatural changes. Primarily a taxonomic work, it includes keys to suborders, families, and genera of the adults, and a separate key, by A. E. Gardner, to the larvae. The adult key includes beautiful color drawings, 20 plates by C. O. Hammond of various species. The key to the larvae includes detailed black-and-white drawings of representative species and identifying characteristics. Forty-four distribution maps are found at the end of the text and also next to the individual species descriptions of the adults. A checklist of dragonflies of this area is also included. An excellent text for amateur as well as specialist.

383. McGeeney, Andrew. **A Complete Guide to British Dragonflies**. London: Jonathan Cape, 1986. 133p. LC 86-191178. ISBN 0-224-02307-1.

This guide is more geared to the amateur and less to the specialist than the guide by C. O. Hammond (see entry 382). Introductory chapters cover the structure, reproductive behavior, life cycle, habitat, conservation, observation and study, and the collection of larvae. The following chapters cover the British dragonflies. Each species is described in two pages with four or five full-color photos of male and female species at the top of the description. After a complete morphological description, the larvae are described, and then the behavior and habitat as well as distribution (but no distribution maps) and time of appearance. A glossary, illustrated keys to the adults and larvae, a checklist of Odonata of the British Isles, and an index to English and scientific names follow the descriptions of the individual species.

384. Miller, P. L. **Dragonflies**. Cambridge, N.Y.: Cambridge University Press, 1987. 84p (Naturalists' Handbooks, 7). $19.95; $12.00pa. LC 86-31719. ISBN 30162-9; 31765-7pa.

This is another in the Naturalists' Handbooks series of fine books primarily for amateu studies of British insects. It provides a good general introduction to the subject, coverin evolution and the kinds of dragonflies found in Britain, as well as information on the egg larvae, and adult stages. The keys to identify larvae and adults are illustrated. The key t adults goes down to species. There are 45 species in Britain. A checklist of British species i found in appendix 1. The book also has a good list of books and journal articles for furthe reading. Black-and-white drawings are found throughout the text, and four pages of colo plates illustrate common species. A fine text for amateur as well as specialist.

385. Needham, James G., and Minter J. Westfall, Jr. **A Manual of the Dragonflies of Nort America (Anisoptera) Including the Greater Antilles and the Provinces of the Mexica Border**. Berkeley: University of California Press, 1954; Reprint, 1975. 615p. $87.50. LC 54-6674. ISBN 0-520-02913-5.

Primarily a systematic classification of Dragonflies, this manual also includes 54 page of introductory material on the order itself, anatomy of the dragonfly, field studies, and a lis of genera and species treated in the work, with aids to pronouncing the names. Keys ar provided to family and subfamily. Each family and genus is thoroughly described in one pag or more with information on distribution and dates when discovered. Keys to species are als included for the larger genera. Black-and-white photos and drawings illustrate the text. A glossary and index to synonyms are included at the end along with a separate index to subjec and scientific name.

386. Walker, Edmund M. **The Odonata of Canada and Alaska**. Toronto, Canada: University of Toronto Press, 1953-73. 3 vols. LC a54-4344.

This work is an assemblage of 50 years of Edmund Walker's study of dragonflies. Dr Walker died before the completion of the third volume and Philip S. Corbet stepped in to see it to completion. Volume 1 covers general information on the entire group and the Zygopter or Damselflies. Volumes 2 and 3 cover the Anisoptera, or dragonflies. Primarily a taxonomi work, it includes keys to family, genus, and species. Species are fully described in one to three pages. Information provided includes descriptions of male and female species; measurements; description of nymph; habitat and range; distribution in Canada; and full field notes Black-and-white drawings of individual species, specific anatomical details for identification and wing venation pattern are included. Each volume has its own bibliography and index.

Blattaria (Cockroaches)

387. Bell, William J., and K. G. Adiyodi, eds. **The American Cockroach**. London, New York: Chapman and Hall, 1981. 529p. $96.00. LC 81-196113. ISBN 0-412-16140-0.

This work deals mainly with the American cockroach, *Periplaneta americana* (L.), bu includes comparative research with other species. Eighteen contributors wrote the 16 chapter on such topics as the circulatory system, nutrition, respiration, osmoregulation, metabolism nervous system, sense organs, rhythms, neurosecretions, muscles, reproduction, behavior and embryology and regeneration. A 76-page bibliography leads to further reading.

388. Cornwell, P. B. **The Cockroach**. London: Hutchinson, 1968-76. 2 vols. $26.00 vol. 1; $35.00 vol. 2. LC 78-354046. ISBN 09-088670-4 vol. 1; 0-85227-102-6 vol. 2.

This two-volume set covers information on the insect itself and on its role as a pest in factories and homes. The first volume covers classification, principal species, anatomy, reproduction and growth, diseases, and incidence of cockroaches in the British Isles. Volume two concentrates on the cockroach as a pest with chapters on control by repellents and various forms of insecticides. The bibliography lists 921 references.

389. Guthrie, D. M., and A. R. Tindall. **The Biology of the Cockroach**. New York: St. Martin's Press, 1968. 408p. LC 68-26564.

This work reviews the literature on cockroaches to 1967. It includes anatomy, reproduction, life cycle, the nervous system, behavior, metabolism, excretion, circulation, respiration, predators, and control. References are included at the end of each chapter.

Isoptera (Termites)

390. Krishna, Kumar, and Frances M. Weesner, eds. **Biology of Termites**. New York: Academic Press, 1969-70. 2 vols. LC 68-26643.

This comprehensive two-volume work proposes to assemble all the current information on termites up to that time. Each chapter is written by an authority in the field. The chapters cover all areas of termite research. Volume one covers the basic information such as anatomy, reproduction, digestion, behavior, flight, feeding, caste formation, and water relations. Volume two covers the taxonomy, with descriptions of family and genus groups from the major regions of the world. Chapter 4 includes a brief key to families. References are found at the end of each chapter, with separate author and subject indexes at the conclusion of each volume. Black-and-white photos, drawings, graphs, and diagrams illustrate the text.

Grylloblattaria or Notoptera (Ice or Rock Crawlers)

391. Ando, Hiroshi, ed. **Biology of the Notoptera**. Nagano, Japan: Kashiyo-Insatsu, 1982. 194p.

This very small order of insects lives in cold places such as at the edge of glaciers and in ice caves, and often at high elevations. This book is a compilation of papers given at the 19th International Congress of Entomology held in Kyoto, Japan, in 1980, and other papers important to this order. These were given as part of the results of a cooperative project under the Japan-U.S. Cooperative Science Program. The papers synthesize the important information on this group. They cover the systematics, distribution, habitat, behavior, life history, anatomy, and structure of the group. Line-drawings, black-and-white and color photos, and maps illustrate the papers. A nine-page bibliography, and author, taxonomic, and subject indexes complete the text.

Orthoptera (Grasshoppers, Crickets, and Katydids)

392. Blatchley, W. S. **Orthoptera of Northeastern America with Especial Reference to the Faunas of Indiana and Florida**. Indianapolis, Ind.: Nature, 1920. 784p. LC 20-10567.

Though dated, this work provides comprehensive coverage of all Orthoptera in the eastern United States. It was conceived as a manual for the beginner, but it contains information of interest to the specialist as well. Keys to families, genera, and 353 species are provided. Species are described in one to three pages with detailed morphological information, geographical range, and field notes. Black-and-white drawings illustrate some of the species. A comprehensive bibliography, five-page glossary, index to synonyms, and brief general index complete the text.

393. Brown, Valerie K. **Grasshoppers**. Slough, England: Richmond, 1983. 64p. (Naturalists' Handbooks, 2). $20.00; $12.00pa. LC 81-17091. ISBN 0-855-46278-7; 0-855-462277-pa.

This short text is a gcod introduction to British grasshoppers. It includes brief chapters on life history and reproduction, adaptations and habits, distribution, and techniques for capturing, preserving, and maintaining these insects. Identification keys for true grasshopper and bush-cricket adults are included. Four full-color plates illustrate the more common species. A useful list of references is provided for further reading.

394. Chapman, R. F., and A. Joern, eds. **Biology of Grasshoppers**. New York: John Wiley, 1990. 563p. $79.95. LC 89-22666. ISBN 0-471-60901-3.

Each of the 16 chapters is authored by an authority on the particular subject of the chapter. The chapters included "Chemoreceptors," "Food Selection," "Nutrition," "Water Regulation," "Hormonal Control of Flight Metabolism in Locusts," "Color Pattern Polymorphism," and more on the biology of grasshoppers. A list of references follows each chapter and a comprehensive subject and species index concludes the book. Black-and-white drawings, photos, graphs and diagrams, and two color plates illustrate the text. Good for a current work on grasshoppers, with references for further reading.

395. Marshall, Judith A., and E. C. M. Haes. **Grasshoppers and Allied Insects of Great Britain and Ireland**. Essex, England: Harley Books, 1988. 252p. LC 89-131131. ISBN 0-94658-913-5.

This work updates and expands Ragge's work on grasshoppers (see entry 398). It covers the grasshoppers, crickets, bush-crickets, cockroaches, stick-insects, and earwigs of Great Britain, including the Channel Islands. The book is divided into four sections: "Introduction," "Systematic Section," "Habitats," and "Atlas of British and Irish Orthoptera." Five appendices include information on offshore populations of orthopteroid insects, Vice-county records, outstanding sites in the British Isles for finding these insects, a list of localities mentioned in the text, and an appendix of Welsh, Scottish, and Irish Names for orthopteroids. The introductory section includes text on nomenclature and classification, pronunciation of names, common names, history, morphology, life history, song and courtship, predators, parasites and diseases, and a section on collecting, rearing, preserving, and photographing these insects. A nice six-page glossary of terms used in the descriptions and a select bibliography of classic and modern standard works are included. The systematic section includes a checklist and key to adults, followed by descriptions of 52 species arranged in taxonomic order. Each species description is one page or more in length and includes information on the physical description, life history and behavior, habitat, and distribution and status. A detailed distribution map is included with each description. The third section on habitats describes the types of habitats where British orthopteroids are found. The final section is a compilation of 10-kilometer-square dot-distribution maps, with all records recorded by the organizer of the Orthoptera Recording Scheme for Great Britain. Ten beautiful color plates at the end of the text illustrate all species described in the text. An excellent and comprehensive work for student, amateur, and specialist.

396. Otte, Daniel. **The North American Grasshoppers**. Cambridge, Mass.: Harvard University Press, 1981- . 2 vols. $60.00 vol. 1; $71.00 vol. 2. LC 81-6806. ISBN 0-674-62660-5 vol. 1; 0-674-62661-3 vol. 2.

This three-volume (projected) set is intended to help in identifying grasshopper species north of the Gulf of Panama, and those species in the West Indies. Volume one covers the subfamilies Gomphocerinae and Acridinae (slant-faced grasshoppers), and volume two covers the Oedipodinae (band-winged grasshoppers). The projected volume three will cover the smaller groups of grasshoppers. The introduction to volume one provides a key to the families and subfamilies of North American grasshoppers, then a key to the subfamily covered in that particular volume. The main part of the text of both volumes is the description of the various species within a particular genus. The description of the species includes distribution, recognition, habitat, behavior, life cycle, and a list of references where further information may be found. Distribution maps are also included with the description. Color plates illustrate the species, sometimes including both the male and the female. Several appendices are included in each volume covering references, a glossary, pronunciation of names, and lists of species within their families. Taxonomic indexes first by genus, then by species, are included at the end of each volume.

397. Preston-Mafham, Ken. **Grasshoppers and Mantids of the World**. New York: Facts on File, 1990. 192p. $24.95. LC 89-39999. ISBN 0-8160-2298-4.

Although the title of this work includes the grasshoppers and mantids, the book actually covers the cockroaches, katydids, crickets, mole-crickets, and stick insects as well. Each of these groups are described in chapters on classification and physiology, reproduction, egg-laying and development, food and feeding, their geographical distribution, their defenses and enemies, and their relationship with man. The beautiful color photographs enhancing the text were taken by the author in trips to 24 countries. This is another in the Of the World series authored mainly by Ken Preston-Mafham and his brother Ron. The text is easily read, short, and beautifully illustrated. There is a glossary and index, but no bibliography of sources is included in this text.

398. Ragge, David R. **Grasshoppers, Crickets, and Cockroaches of the British Isles**. London, New York: Frederick Warne, 1965. 299p.

Covering more than the order Orthoptera, the Blattodea, or cockroaches, and Phasmidae, or stick insects, are included as well in this respected work. This work is updated by Judith Marshall and E. C. M. Haes (see entry 395). The work is systematically arranged with each group described in separate chapters. Keys to families and to species are included. Each species is described in one to four pages with the following information: heading by common name, followed by scientific name, the synonym, a physical description, a section with measurements of males and females, the habitat, life history, habit, distribution, and distribution abroad. Full-page maps of the British Isles showing the recorded distribution of the species are included with the description of the individual species. Twenty-two color plates and 130 figures illustrate the text. Three appendices cover information on Latin names, county subdivisions in the British Isles, and a checklist of the Orthoptera found there. A glossary and index to subject, common name, and scientific name finish the volume. Nicely organized for informative reading for amateur as well as specialist.

399. Rehn, James A. G., and Harold J. Grant, Jr. **A Monograph of the Orthoptera of North America (North of Mexico).** Vol. 1. Philadelphia: Academy of Natural Sciences of Philadelphia, 1961. 255p. (Monographs of the Academy of Natural Sciences of Philadelphia, no. 12). LC 61-65643.

This volume covers the Acridoidea, or grasshoppers, of North America. It specifically covers the following families: Tetrigidae, Eumastacidae, Tanaoceridae, and Romaleinae of the Acrididae. Keys to the identification of taxa down to species are provided for these groups. Each species is described as to distinctive features, measurements, coloration, distribution, habitat, and general remarks. Eight pages of plates with black-and-white photos of representative species are found at the end of the text. Black-and-white drawings are also found throughout the text illustrating specific anatomical features to help in the identification of individual species. Distribution maps of the major families are also included. An index to the families, genera, and species in this volume is provided, but no bibliography of sources referenced in the text is included.

400. Uvarov, Boris, Sir. **Grasshoppers and Locusts: A Handbook of General Acridology**. Cambridge, England: Published for the Anti-Locust Research Centre, at the University Press, 1966-1970. 2 vols. Vol. 1, out-of-print; $25.00 vol. 2. LC 64-21575. ISBN 0-85135-072-0 vol. 2.

This is a revised edition of the author's 1928 book titled *Locusts and Grasshoppers*. The emphasis of these volumes is now on the grasshoppers. Volume one covers anatomy, physiology, development, phase polymorphism, and an introduction to taxonomy; volume two covers behavior, ecology, biogeography, and population dynamics. This work only briefly discusses taxonomy with an outline of the classification listing families, with genera and species in alphabetical order under the family. Black-and-white illustrations of the more distinctive species are included. The entire text is illustrated with black-and-white drawings, graphs, maps, and photos. Comprehensive bibliographies as well as subject and species indexes are found at the end of each volume.

Plecoptera (Stoneflies)

401. Claassen, Peter W. **Plecoptera Nymphs of America (North of Mexico)**. Springfield, Ill.: Published for the Thomas Say Foundation by Charles C. Thomas, 1931. 199p. LC 32-10183.

After the publication of *A Monograph of the Plecoptera or Stoneflies of America North of Mexico* (see entry 402), Claassen spent time studying the nymphs, or immature forms, of stoneflies. In this work he describes the nymphs of 68 species, representing almost all genera. After introductory information on distribution, habits, life cycle, and methods of collection, a key to the families and genera of nymphs is given. Keys to representative species are also provided. Each family, genus, and species is described in one page or more. Thirty-five plates of black-and-white photos and drawings of selected species are found at the end of the text. A distribution and emergence table is provided as well as an index to scientific name. Updated by K. W. Stewart and B. P. Stark in *Nymphs of North American Stonefly Genera (Plecoptera)* (see entry 403).

402. Needham, J. G., and Peter W. Claassen. **A Monograph of the Plecoptera or Stoneflies of America North of Mexico**. LaFayette, Ind.: Thomas Say Foundation, 1925. 397p. (Thomas Say Foundation Publication, vol. 2). LC 36-11631.

This handbook describes 207 species of stoneflies in 24 genera. Keys to families, genera, and species are all provided. Introductory material covers the biology, collecting and preservation, diagnostic characters, illustrations of wing venation patterns used in distinguishing specimens, genital characters, and a note on classification. The descriptions of each family, genus, and species covers one paragraph to one page or more. Each species is described as to appearance of male and female and general characteristics as well as the geographical distribution. Fifty plates of black-and-white drawings illustrate anatomical characteristics described in the text. A bibliography, checklist of species described, and an index to subject and name finish the work.

403. Stewart, Kenneth W., and Bill P. Stark. **Nymphs of North American Stonefly Genera (Plecoptera).** Lanham, Md.: Entomological Society of America, 1988. 460p. (Thomas Say Foundation Monograph, vol. 12). $108.00 (nonmember); $65.00 (member). LC 89-122234. ISBN 0-938-52233-7.

Updates Claassen's work on Plecoptera nymphs (see entry 401). It reviews the major literature, both systematic and ecological, on Plecoptera nymphs (larvae). It includes new family and generic keys, and detailed descriptions of type-species. It also includes a complete species list and references, plus illustrations.

Anoplura (Sucking Lice)

404. Kim, Ke Chung, Harry D. Pratt, and Chester J. Stojanovich. **The Sucking Lice of North America: An Illustrated Manual for Identification**. University Park, Pa.: State University Press, 1986. 256p. $39.50. LC 84-43060. ISBN 0-271-00395-2.

This book provides a summary of information on North American sucking lice, an order of insects of medical and veterinary importance. Information on collecting, morphology, biology, and public and veterinary importance is contained in the first five chapters. The remaining chapters include illustrated keys to the families, genera, and species, and descriptions of 76 species. Parasite-host and host-parasite lists follow the text. Nine pages of references and a subject and scientific-name index finish the work.

Thysanoptera (Thrips)

405. Lewis, Trevor. **Thrips: Their Biology, Ecology, and Economic Importance**. New York: Academic Press, 1973. 349p. LC 72-12273. ISBN 0-12-447160-9.

Complementing the descriptive works on thrips, this work attempts to cover all other information on this order. The work is divided into four sections: thrips biology, techniques (laboratory methods and sampling), ecology, and economic importance. In addition, six appendices cover checklists, preservation techniques, thrips parasites, predators and preys, insecticides used in controlling thrips, and a list of thrips species and common names. Written in nonspecialist terminology for the most part, with black-and-white drawings, photos, and graphs to illustrate the text, this work can be useful to many groups.

406. Palmer, J. M., L. A. Mound, and G. J. du Heaume. **CIE Guides to Insects of Importance to Man. No. 2: Thysanoptera**. Ed. by C.R. Betts. Oxon, England; Tucson, Ariz.: C.A.B. International, International Institute of Entomology, and The Natural History Museum, 1990. 80p. LC gb90-24528. ISBN 0-85198-634-X.

See entry 358, *IIE Guides to Insects of Importance to Man.*

407. Stannard, Lewis J. **The Phylogeny and Classification of the North American Genera of the Suborder Tubulifera (Thysanoptera)**. Urbana, Ill.: University of Illinois Press, 1957. 200p. (Illinois Biological Monographs, no. 25). LC 57-6958.

This book covers the morphology and classification of this group. It includes keys to genus, subgenus, and species. Each species account is thorough with a description of male and female and holotype. The head, mouthparts, thorax, and abdomen are characterized, and additional notes on related species are included. Fourteen pages of references to literature cited in the text come before 14 plates of black-and-white illustrations to help identify species.

Hemiptera-Homoptera (True Bugs, Cicadas, Hoppers, Whiteflies, Aphids, and Scale Insects)

408. Blackman, R. L., and V. F. Eastop. **Aphids on the World's Crops: An Identification Guide**. Chichester, N.Y.: John Wiley, 1984. 414p. $130.00. LC 83-25889. ISBN 0-471-90426-0.

Although primarily an identification guide, this work is included here because it includes a wealth of other information on this large group. The introductory section includes information on life cycles, host-plant interactions, geographical distribution, and morphology. A very useful section lists crops in alphabetical order with names of aphids known to colonize the crop and a key to identify each species. Another section reviews the various aphid species, describes them, and gives geographical ranges as well as notes on their biology and host plants. A very complete list of references comes before a photographic section of slide-mounted aphids illustrating many species described in the text. A worthwhile text for agricultural entomologists, but useful for all entomologists for its concise information on this group.

409. Blatchley, W. S. **Heteroptera, or True Bugs of Eastern North America, with Especial Reference to the Faunas of Indiana and Florida**. Indianapolis: Nature, 1926. 1116p. LC 26-19247.

This is a thorough work on this group. Arranged taxonomically, it provides detailed keys down to species with full descriptions including habitat, general distribution, food habits, measurements, and field notes. Black-and-white drawings, photos, and plates illustrate many species. A lengthy bibliography lists works cited in the text. W. S. Blatchley's many years of experience with this group makes this work useful to contemporary researchers, though the classification may be dated.

410. Dolling, W. R. **The Hemiptera**. Oxford, England: Oxford University Press, 1991. 274p. $90.00. LC 91-12973. ISBN 0-19-854016-7.

This up-to-date work provides basic information on this large group of around 80,000 species. The beginning chapters cover food, diseases, natural enemies, defense, biorhythms, symbiotic relationships, dispersal, distribution, and collecting and preserving specimens. The final chapters include keys and systematics of the British families of Hemiptera, of which there are approximately 1,700 species that occur in the British Isles. However, these keys and systematics can be used to identify Hemiptera of the world to family, especially if used with the chapter on "British Hemiptera as a Sample of the World Fauna," because the British families are representative of those families found worldwide. There are eight pages of color plates and numerous, detailed black-and-white drawings of representative species of the various families. A useful 13-page glossary of terms used in the text is found at the end of the text. An extensive 17-page bibliography provides recent literature citations for further reading.

411. Ferris, Gordon Floyd. **Atlas of the Scale Insects of North America**. Stanford, Calif.: Stanford University Press, 1937-1955. 7 vols. $275.00 set. LC 37-4591. ISBN 0-404-08500-8 set.

The purpose of this multivolume set is to make possible the identification of all species of scale insects known in North America at that time. It is primarily a collection of plates illustrating each species with some basic information on the species like synonyms, hosts, distribution, where it may be found (on bark, leaves, etc.), distinguishing features, and field notes included on the page facing the plate. The plates are large, covering an entire 11-by-7-inch page. The atlas covers nearly 700 species.

Volume 1-4. *Diaspididae*. $155.00.
Volume 5-6. *Pseudococcidae*. $80.00.
Volume 7. *Acleridae, Asterolecaniidae, Conchaspididae, Dactylopiidae, and Lacciferidae*. $40.00.

412. McKenzie, Howard L. **Mealybugs of California: With Taxonomy, Biology, and Control of North American Species (Homoptera: Coccoidea: Pseudococcidae)**. Berkeley, Calif.: University of California Press, 1967. 525p. $87.50. LC 67-10279. ISBN 0-520-00844-8.

Although the title indicates this is a work concerning mealybugs of California, much of this information treats this group as found in North America. Information on control, ecology, biology, field methods, and morphology as well as a keys to the genera and species of North America are all included here. It is nicely illustrated with full-page line-drawings of 205 species, 193 California distribution maps, and 24 surprisingly beautiful watercolor paintings of mealybugs in various habitats. One very nice drawing is of african violets with an enlargement of mealybugs living in the roots. A very interesting work, with excellent illustrations.

413. Miller, N. C. E. **The Biology of the Heteroptera**. 2d rev. ed. Hampton, England: E. W. Classey, 1971. 206p. $60.00. LC 72-186641. ISBN 0-9000848-45-6.

This suborder of the Hemiptera is composed of approximately 25,000 species. This text is divided into two parts: part 1, giving general information on the group, and part 2, the major portion of the text, containing information on the families of Heteroptera. Part one covers the family and subfamily names, development, legs, stridulation, enemies, and species affecting mammals and birds. Part two describes the families in a few paragraphs to a few pages, but does not include an identification key. Five plates of black-and-white photos show representative species. Black-and-white line-drawings of some species and identifying characteristics are provided throughout the text. An 18-page list of references and an index to name and subject complete the volume. A work for the specialist or the serious amateur.

414. Nault, L. R., and J. G. Rodriguez, eds. **The Leafhoppers and Planthoppers**. New York: John Wiley, 1985. 499p. $89.95. LC 85-5383. ISBN 0-471-80611-0.

This work intends to provide comprehensive information on these two groups to the student and specialist alike. Systematics and keys to family begin the chapters on leafhoppers and planthoppers. The remaining 19 chapters cover nutrition, sensory mechanisms, plant defensive mechanisms, migratory behavior, evolution, ecology, and insect pathogens and parasites. Long lists of references follow each chapter. The work is illustrated with line-drawings, electron micrographs, graphs, tables, and other figures. A good introduction to these groups, with plenty of information and further reading for the specialist.

Coleoptera (Beetles and Weevils)

415. Ball, George E., ed. **Taxonomy, Phylogeny, and Zoogeography of Beetles and Ants**. Dordrecht, Boston: W. Junk. Distributed, Hingham, Mass.: Kluwer Academic, 1985. 514p. $218.00. (Series Entomologica, vol. 33). $218.00. LC 84-21782. ISBN 90-6193-511-3.

This volume was conceived as a memorial to Philip J. Darlington, Jr., a naturalist, systemaist, and a specialist in the beetle family Carabidae. Each chapter is written by an expert in the field and most are primarily on beetles. Taxonomy is the emphasis, but it also includes related subjects such as zoogeography. Keys to identification are provided in many chapters. The book is illustrated with many drawings, maps, electron micrographs, and photos. Included here because of the currency of information on a wide range of topics on beetles.

416. Blatchley, W. S. **An Illustrated Descriptive Catalogue of the Coleoptera or Beetles (Exclusive of the Rhynchophora) Known to Occur in Indiana**. Indianapolis, Ind.: Nature, 1910. 1386p.

This work is included here because it is a comprehensive work on all beetles found in a particular area. It discusses the major groups found in similar climates, allowing its use in other areas as well. Keys down to species are included with short descriptions of each, and where and at what time of year they are found. Five hundred ninety-four black-and-white drawings of the various species and identifying parts are scattered throughout this lengthy text. Useful to the beginner and specialist alike. A classic work.

417. Blatchley, W. S., and C. W. Leng. **Rhynchophora or Weevils of North Eastern America**. Indianapolis, Ind.: Nature, 1916. 682p. LC 16-1247.

Intended as a supplement to the above work by Blatchley (see entry 416) to complete work on the Coleoptera of this region, it was expanded to cover a larger area when C. W. Leng joined his work on the Atlantic Coast species with the work of Blatchley. This book has the same purpose, to provide a simple manual for the amateur and specialist alike to determine the names of the weevils as well as their arrangement and classification. Keys down to species are provided here as well as short descriptions of each species. Illustrated with 155 black-and-white drawings.

418. Booth, R. G., M. L. Cox, and R. B. Madge. **IIE Guides to Insects of Importance to Man, 3. Coleoptera**. Oxon, England; Tucson, Ariz.: International Institute of Entomology, C.A.B. International, and The Natural History Museum, 1990. 384p. ISBN 0-85198-678-1.

See entry 358, *IIE Guides to Insects of Importance to Man.*

419. Cooter, Jonathan. **A Coleopterist's Handbook**. General editor, P. W. Cribb. Middlesex, England: Amateur Entomologists' Society, 1991. 294p. $21.00. LC gb91-37206. ISBN 0-900054-53-0.

Primarily covering British beetles, this handbook is useful to any amateur new to the study of beetles. It provides practical information on collecting equipment, starting and building a collection, how to identify beetles, their anatomy, and keeping diaries and notebooks. The main part of this handbook is the description of 31 beetle families. Each family is described in two to six pages with a black-and-white drawing of a representative species at the beginning of the description. The descriptions are written by various authors and are easy to read, ending with a list of references for further reading. There are also sections on beetle larvae morphology, identification, and rearing as well as beetle associations with plants, stored food, and ants. A glossary and an index to genera discussed in the book finish this informative book for the amateur.

420. Crowson, R. A. **The Biology of the Coleoptera**. London, New York: Academic Press, 1981. 802p. LC 77-71815. ISBN 0-12-196050-1.

This comprehensive review of the biology of Coleoptera was written by a world authority on beetle systematics. It covers a wide range of information on beetles: a history of beetle study, internal and external anatomy of adults, larvae and pupae, food, digestion, blood, osmoregulation, excretion, locomotion, respiration, the senses, cuticular properties, behavior, development, cytology and genetics, predation, ecology, water beetles and herbivorous beetles, and the evolution and geographical distribution of beetles. This work does not cover the taxonomy and systematics of beetles, but does include a classification of families as an appendix to the text. Many black-and-white drawings, photos, electron micrographs, and graphs illustrate the text. An extensive 45-page bibliography of references is useful for further study. Includes a taxonomic and subject index. A very scholarly and comprehensive work.

421. Dillon, Elizabeth S., and Lawrence S. Dillon. **A Manual of Common Beetles of Eastern North America**. Evanston, Ill.: Row, Peterson, 1961. Distributed by Peter Smith. 2 vols. $33.00 set. LC 60-8281. ISBN 0-8446-4538-9.

Covering more territory than Blatchley's book (see entry 416), this manual covers the beetles from the 100th meridian, east to the Atlantic Coast and south to Mexico. It covers the common species, not the entire fauna of beetles in this area. Nearly 1,200 species in 64 families are treated here. Following a key to families, each family is treated in separate chapters. Keys to genera and to species are found in each family chapter. Plates with black-and-white drawings of representative species are included in each family chapter. Four color plates of the more colorful beetles are found at the beginning of the text. A glossary of terms used in the text is found before a useful bibliography subdivided by topic (ecology) and family.

422. Erwin, Terry L., et al., eds. **Carabid Beetles: Their Evolution, Natural History, and Classification**. The Hague, Boston: W. Junk. Distributed, Hingham, Mass.: Kluwer Academic, 1979. 635p. $210.50. LC 80-462582. ISBN 90-6193-596-2.

This is actually the proceedings of the First International Symposium of Carabidology held at the Smithsonian Institute in 1976. Carabid or ground beetles include tiger beetles and bombardier beetles. Each chapter is contributed by an expert in the field and they cover many aspects of study for this group. Historical literature, systematics, speciation, classification and phylogeny, keys to tribes of North American larvae, zoogeography, paleontology, natural history, ecology, and data computerization are all covered in separate chapters. Each chapter has a complete list of references. The text is illustrated with black-and-white drawings, electron micrographs, maps, diagrams, and graphs. A good assemblage of information on this large group.

423. Forsythe, Trevor G. **Common Ground Beetles**. Richmond, England: Richmond, 1987. 74p. $20.00; $12.00pa. (Naturalists' Handbooks, 8). ISBN 0-85546-264-7; 0-85546-263-9pa.

As with others in the Naturalists' Handbooks series, this emphasizes British species, but also gives a good, basic introduction to the group as a whole. Introductory materials cover the natural history, anatomy, feeding, predation, modes of locomotion, sound production, and identification characteristics. A final chapter on collection and preservation techniques and a list of references for further reading are also provided. Illustrated keys to the more common British species of ground beetles are given, and six plates of color and black-and-white drawings of the common species are placed after the keys.

424. Morris, M. G. **Weevils**. Slough, England: Richmond, 1991. 76p. $19.00. (Naturalists' Handbooks, 16). ISBN 0-85546-282-5; 0-85546-281-7pa.

Although primarily a text on British weevils, the introductory material of this short text gives a good background to the weevils as a whole. It describes weevils and their host plants, their biology and life histories, feeding, communication, development, predatory defenses, locomotion (walking, flying, jumping, swimming, and skating), predators, and parasites, and there is an interesting section on questions about weevils such as "How do weevils respond when they fall into the water?" Keys to family and selected species of the 570 British species are included. Four color plates of 24 species are inserted into the middle of the text. A nice list of references for further reading and a good chapter on collecting and study techniques round out this good introductory text on weevils.

425. Papp, Charles S. **Introduction to North American Beetles.** Sacramento, Calif.: Entomography, 1984. $17.50pa. 335p. ISBN 9-608-8484-1-9.

This is a beginner's guide to beetles, that includes the basics of beetle study and briefly describes 1,100 species. It includes more than 1,000 black-and-white figures illustrating various species or providing detail drawings or photos for identification purposes. The species accounts are easy to read, nontechnical descriptions that include geographical regions where found, a physical description, information on eggs, larvae, foodplants, and hibernation as well as other interesting facts. Black-and-white drawings of each species described are found lined up at the top of each page. Unfortunately, no color illustrations are found here. A glossary of terms, a list (with portraits) of early entomologists who studied and described beetles, and an index to scientific and common names complete this interesting introduction to North American beetles.

Siphonaptera (Fleas)

426. Ewing, H. E., and Irving Fox. **The Fleas of North America: Classification, Identification, and Geographic Distribution of These Injurious and Disease-Spreading Insects**. Washington, D.C.: U.S. Dept. of Agriculture, 1943. 142p. (Miscellaneous Publication, no. 500). LC agr43-32.

This short pamphlet includes descriptions and keys to identification of families, subfamilies, and genera of North American fleas. Each of 209 species and 63 subspecies are also described as to type-host, type-locality, range, and other identifying information. References to original description, figures, biology, economic importance, or redescriptions are provided for each name. Illustrations of identifying characteristics are included in an appendix. Dated but useful.

427. Fox, Irving. **Fleas of Eastern United States**. Ames, Iowa: Iowa State College Press, 1940. 191p. LC 40-5367.

This is a guide to the fleas found east of the 100th meridian, excluding Texas. It describes five families, 33 genera, and 55 species. Brief keys to the suborders, families, genera, and species are provided. Each family is treated in separate chapters and each species is fully described as to male and female, hosts, records of where found, localities, and type material. References to the literature on each taxon are listed after the name as well as a reference to where the species is described in figures and plates in the book. Thirty-one plates of black-and-white line-drawings illustrate distinguishing anatomical characteristics of different species. A selected bibliography and index to host organisms, and an index to subject and names complete the volume.

428. Hopkins, G. H. E., and Miriam Rothschild. **An Illustrated Catalogue of the Rothschild Collection of Fleas (Siphonaptera) in the British Museum (Natural History).** Oxford, England: Oxford University Press, 1966-{1987}. 7 vols. $135.00/vol. LC 54-3740.

Volume 6 is authored by D. K. Mardon, and volume 7 by F. G. A. M. Smit. This is not only a catalog of the amazing collection of fleas that Charles Rothschild gave to the British

Museum in 1913, it is also a classified list and description of the fleas of the world. The classification system was devised by Dr. Karl Jordan, who studied this order for 60 years. Each volume contains many black-and-white drawings of selected species and anatomical details for identification. Each volume also includes around 50 black-and-white photographic plates of the individual species.

429. Hubbard, Clarence Andresen. **Fleas of Western North America: Their Relation to the Public Health**. Ames, Iowa: Iowa State College Press, 1947. 533p. LC 47-2320.

This work is a guide to fleas found west of the 100th meridian. It covers more than 230 species of fleas, compared to the 55 species found in the eastern U.S. (see entry 427). It is divided into three parts. The first part covers some interesting background information on pioneers of flea research, medical significance of fleas, and laboratory and field technique. The second part describes the various families in individual chapters with keys to families and genera. Species are thoroughly described with records of where the species are found in the western states (each state listed separately). Sex predominance, seasonal distribution, biology, economic importance, medical importance, and control are also provided in the descriptions of each species. A nice geographical index shows what species are found in the individual states. The third part is a thorough treatment of flea hosts in rodents, carnivores, insectivora, bats, and birds. The host is described with the species of flea that parasitizes it, and in what states the fleas are found. A very complete work.

430. Lewis, Robert E., Joanne H. Lewis, and Chris Maser. **The Fleas of the Pacific Northwest**. Corvallis, Oreg.: Oregon State University Press, 1988. 296p. $53.95. LC 88-1612. ISBN 0-87071-355-8.

Although this work describes the fleas of the Pacific Northwest, its introductory chapters provide current general information on fleas. It covers the life history, morphology, disease, a literature review, and a systematic review of fleas. Keys to species are provided for fleas of this area. Each species is described in one long paragraph with a distribution map placed next to the species account. Literature citations to descriptions of each species are provided with each account also. Detailed line-drawings of anatomical characteristics used to identify a particular taxon are provided throughout the text. A bibliography of literature cited and a host/flea index are also provided.

431. Traub, R., and H. Starcke, eds. **Fleas**. Rotterdam, The Netherlands: A. A. Balkema, 1980. 420p. ISBN 90-6191-018-8.

This volume records the proceedings of the 1977 International Conference on Fleas, a conference meant to lay a basic foundation for the study of the Siphonaptera. Chapters are written by experts on many aspects of flea studies and include those on taxonomy (a new genera and subgenera), evolution and zoogeography, medical and veterinary aspects of flea studies (especially plague), physiology and morphology, and ecology (especially flea fauna on hosts). Each chapter has a respectable bibliography. Black-and-white drawings, electron micrographs, maps, and diagrams illustrate the text. Unfortunately, there is no index.

Diptera (Flies and Mosquitoes)

432. Carpenter, Stanley J., and Walter J. LaCasse. **Mosquitoes of North America (North of Mexico)**. Berkeley, Calif.: University of California Press, 1955. Reprint, 1974. 360p. $77.50. LC 55-7555. ISBN 0-520-02638-1.

Written for all levels of interest in mosquitoes, this comprehensive work describes 11 genera and 143 species and subspecies. Keys to genera and species are provided for adults and larvae. The original reference and synonymy are listed with each species along with descriptions of the adult female, male, and male terminalia and larva. A feature of this book are the 127 elegant full-page drawings of mosquito species. Other line-drawings of anatomical detail are found throughout the text. In addition to the descriptions, introductory material on life histories, internal and external anatomy, and collecting and preservation are supplied.

433. Clements, A. N. **The Biology of Mosquitoes**. London, New York: Chapman & Hall, 1992- . Vols. 1- . $89.95 vol. 1. ISBN 0-412-40180-0.

Volume one of this set covers the development, nutrition, and reproduction of mosquitoes. The second volume will cover all aspects of mosquito behavior as well as reproductive capacity. This new set reviews and brings up to date information on mosquito biology. Volume one includes information on embryology, larval feeding, and nutrition as well as excretion and respiration, growth and development, metamorphosis, and the various systems such as the circulatory system and the endocrine system. It also includes information on adult feeding and digestion, and diuresis and excretion. Spermatogenesis and oogenesis are also discussed. A thorough work, primarily for the student and specialist.

434. Cole, Frank R. **The Flies of Western North America.** Berkeley, Calif.: University of California Press, 1969. 693p. LC 68-10687.

This is a comprehensive guide to the study of western U.S. flies. The western region includes the area west of the 100th meridian, including Baja California, and north along the coast to Alaska. Introductory information on distribution, life history, maps and life zones, geological history, and general anatomy lead into the main text on families of Diptera. Illustrated keys to family, tribe, and genus are provided. Information on family, genus, and species includes a physical description, type species, references where first named, and geographical areas where specimens were collected. Fine black-and-white illustrations of selected species are found throughout the text, some showing specific anatomical details used in identification of a particular species. A glossary, selective bibliography of sources, and an index to scientific names and synonyms complete this comprehensive guide.

435. Crosskey, Roger W. **The Natural History of Blackflies**. Chichester, N.Y.: John Wiley, 1990. 711p. $180.00. LC 90-12453. ISBN 0-471-92755-4.

This text brings together all the current information on the natural history of blackflies available up to June 1989. It includes 1,200 citations to the literature and is aimed at synthesizing this information published by specialists into a text that is readable and useful to the general reader with a need for this information. It covers the history of blackfly studies, taxonomy, geographical distribution, larval development and function, natural enemies, pupa and adult flies, mating and reproduction, flight, hosts, and egg laying, with a whole section on the influence of blackflies on man.

436. Gillett, J. D. **Mosquitos**. London: Weidenfeld and Nicolson, 1971. 358p. LC 72-76160. ISBN 0-297-00404-2.

Aimed at the general reader but thorough enough for the specialist, this work provides a fine overview and summary of information on mosquitoes as a whole, not specific information on families, genera, or species. The individual chapters cover the eggs, larvae, pupae, and adult descriptions; their flight, feeding, and ovarian cycle; there are chapters on diseases caused by mosquitos, and their parasites and predators. Nine appendices provide checklists of known species of mosquitoes in Australian Region, New Guinea, Ethiopian Region, Madagascar, Nearctic Region, Neotropical Region, Oriental Region, Palaearctic Region, and the United Kingdom and Ireland. A glossary of terms, a bibliography of sources arranged by chapter, and an index finish the volume. Interesting reading.

437. Griffiths, Graham C. D., ed. **Flies of the Nearctic Region**. Lubrecht & Cramer, 1980- . Vols. 1- .

Intended as a counterpart to the set *Die Fliegen der palaarktischen Region*, begun in 1924, this comprehensive work is intended to be a basic reference work for flies of the Nearctic Region. This region includes Canada, Alaska, and the contiguous United States, but not Hawaii. Volumes to this set are published irregularly. In the first volume, part one includes a history of Nearctic dipterology. Each part thereafter is on a separate family and is authored by an authority on that family. Each family is thoroughly described with keys to genera of adults, pupae, and larvae. Species accounts are detailed and are accompanied by black-and-white drawings of specific anatomical details useful in identification of the species. Range maps, a list of literature cited in the text, and an index are included in each family account. A work for the dipteran specialist.

Volume 1. *History of Nearctic Dipterology*. Part 1. $45.00. ISBN 3-510-7000-15.
Volume 2. *Archaeodiptera and Oligoneura*. Part 4. *Blephariceridae*. $120.00. ISBN 3-510-70013-9.
Volume 5. *Bombyliidae*. Part 13, no. 5. $173.00. ISBN 0-685-43745-0.
Volume 5. *Bombyliidae*. Part 13, no. 1. $56.50. ISBN 3-510-70002-3.
Volume 5. *Bombyliidae*. Part 13, no. 2. $56.50. ISBN 3-510-70003-1.
Volume 5. *Bombyliidae*. Part 13, no. 3. $60.50. ISBN 3-510-70005-8.
Volume 5. *Bombyliidae*. Part 13, no. 4. $28.25. ISBN 3-510-70008-2.
Volume 5. *Bombyliidae*. Part 13, no. 6. $50.50. ISBN 3-510-70015-5.
Volume 6. *Orthogenya*. Part 6. *Dolicopodidae: Hydrophorinae*. $63.30. ISBN 3-510-70010-4.
Volume 8. *Cyclorrhapha*. Many volumes for this group. Price per volume approximately $100.00.

438. Hull, Frank M. **Bee Flies of the World: The Genera of the Family Bombyliidae.** Washington, D.C.: Smithsonian Institution Press, 1973. 687p. $85.00. LC 73-1581. ISBN 0-87474-131-9.

This is comprehensive work on all aspects of bee flies. It gives full background information on the history of bee fly studies, host relationships, ethology, morphology of immature and adult bee flies, fossil bee flies, distribution, phylogeny, and zoogeography. Keys to subfamilies and genera are detailed. Each genus is fully described, but not each species. Species descriptions are included with the genus. The genus gives references to where first described and other references. The description covers one page or more and details the morphology, the immature stages, the ecology and behavior of adults, and the distribution. Some species are illustrated in the text with striking black-and-white drawings, and, in addition, 123 pages of drawings illustrate specific anatomical parts described in the text. A lengthy bibliography of sources and an index to genera, subgenera, and species complete the text.

439. Hull, Frank M. **Robber Flies of the World: The Genera of the Family Asilidae.** Washington, D.C.: Smithsonian Institution, 1962. 2 vols. (Bulletin/United States National Museum, 224, pts. 1-2). LC 62-61957.

Robber Flies are one of the largest families in the insect world. It is found in almost all parts of the world and its fossil record goes back to the Eocene era. This two-volume work provides comprehensive information on this large group. It gives keys to genus, and describes each genus. Black-and-white drawings of selected species are included throughout the text. A detailed set of drawings on 200 pages of volume 2 provide drawings of the antennae, wings, heads, legs, genitalia, and associated parts of all genera described in the text. A bibliography that covers 200 years of literature on robber flies lists 1,344 titles. A definitive work on one of the largest groups of insects.

440. Johannsen, O. A. **Aquatic Diptera.** Los Angeles, Calif.: Entomological Reprint Specialists, 1969. 370p. $17.50. LC 78-7782. ISBN 0-9118336-01-2.

This is a reprint edition of four separately published works in the Cornell University Agricultural Experiment Station *Memoirs*. Part one covers the "Nemocera, exclusive of Chironomidae and Ceratopogonidae"; part 2, "Orthorrhapha-Bachycera and Cyclorrhapha"; part 3, "Chironomidae, Subfamilies Tanypodinae, Diamesinae, and Orthocladiinae"; and part 4, "Chironomidae, Subfamily Chironominae" and part 5 (published with part 4), the "Ceratopogonidae." Keys to larvae and pupae of American aquatic diptera down to species are provided. Detailed anatomical descriptions are given for each taxon down to species. Black-and-white line-drawings of identifying anatomical details are found on separately paged plates at the end of the text. References are also provided, but no index. Primarily a taxonomic work.

441. **Manual of Nearctic Diptera**. Coordinated by J. F. McAlpine and others. Ottawa: Research Branch, Agriculture Canada, 1981-1989. 3 vols. LC 81-158553. ISBN 0-660-10731-7 vol. 1.

This is primarily a systematic work for identifying the families and genera of two-winged flies of America north of Mexico, but it also serves as a basic manual on Diptera of this area. It includes keys to families of adults and larvae, and within each chapter devoted to a particular family there are illustrated keys to the genera. Each chapter on a family includes a detailed black-and-white drawing of a representative species, descriptions of adult, egg, larva, and pupa as well as a section on biology, behavior, classification, and distribution. Each chapter ends with a key to the genera and a list of references for further reading. Volume 1 introduces the information in volume one and two, and covers the families of the Nematocera and the Brachycera. Volume 2 continues with the families of the infraorder Muscomorpha plus some additions and corrections to volume 1. Volume 3 contains three chapters on the phylogeny and classification of Diptera. A very complete reference for the serious amateur to the professional.

442. Oldroyd, H. **The Natural History of Flies.** London: Weidenfeld and Nicholson, 1964. 324p. LC 65-7926.

Anyone looking for general, easily readable information on flies and their relatives, such as mosquitoes and midges, will want to consult this text. It is divided into three parts. Part one gives an overview of flies as a general group. Part two, the main part of the book, includes 15 chapters on individual groups of flies such as black flies, crane flies, water midges, coffin flies, fruit flies, mosquitoes, horse flies, snipe flies, bee flies, hover flies, dung flies, house flies, blow flies, and land midges. Part three includes three chapters on flies and man, fly swarms, and the history and future of flies. The individual chapters on types of flies are about 20 pages each and are interesting reading with black-and-white drawings and photos to illustrate the text. A nice bibliography for further reading adds to the usefulness of this text.

443. Skidmore, Peter. **The Biology of the Muscidae of the World**. Dordrecht, Boston: W. Junk, 1985. 550p. (Series Entomologica, vol. 29). $176.00. LC 84-3871. ISBN 90-6193-139-8.

This work provides a summary of the world literature on the biology and immature stages of 440 species of Muscidae, or houseflies. This represents around 25 years of study by the author. The work includes keys to subfamilies, genera, and species. Species are described as to their larval and pupal stages as well as their biology and distribution. Line-drawings illustrate anatomical details in 160 figures found throughout the work. A lengthy list of references provides further reading.

444. Snow, Keith R. **Mosquitoes**. Slough, England: Richmond, 1990. 66p. (Naturalists' Handbooks, 14). $20.00; $12.00pa. ISBN 0-85546-276-0; 0-85546-275pa.

Another of the Naturalists' Handbooks series that provides a good, short introduction to the study of mosquitoes, though the emphasis is on British species. Introductory material deals with the life stages of mosquitoes, how to recognize the stage, and where to find mosquito eggs, larvae, and pupae. Feeding, dispersal, and flight range, and a chapter on collecting, examining, and rearing mosquitoes provide more information for the beginning student. The work includes keys to genera for all stages of development of British mosquitoes. Keys to species for adults are given as well as a checklist of British species of mosquitoes. The keys are illustrated with black-and-white drawings. A selective list of references provides for further reading.

Lepidoptera (Butterflies and Moths)

445. Brooks, Margaret, and Charles Knight. **A Complete Guide to British Butterflies; Their Entire Life Histories Described and Illustrated in Colour from Photographs Taken in Their Natural Surroundings.** London: Jonathan Cape, 1982. 159p. $10.50. LC 82-107099. ISBN 0-244-01958-9.

Most butterfly books treat the adult forms only, but this comprehensive source provides information on and illustrates every egg, caterpillar, chrysalis, and adult form of every British species. It is illustrated by photos of the living insects in their natural surroundings. All 60 different species that breed in the British Isles are superbly photographed by the author in each of the four life cycle stages (i.e., egg, caterpillar, chrysalis, and adult butterfly). Each species is described and illustrated in two pages with descriptions of each stage as well as information on distribution, habitat, life cycle, and larval foodplants. An excellent book that also shows that the delights of butterfly photography may be more rewarding than the capture and collection of the live species.

446. D'Abrera, Bernard. **Birdwing Butterflies of the World**. Melbourne: Lansdowne Press, 1975. 259p. ISBN 0-600-31380-8.

The birdwing butterflies are the largest and most spectacular of all butterflies. Because of this they are in danger of extinction from over-collection and lack of adequate conservation measures. This beautiful book preserves their beauty, although the color photos are not always of the best quality as in D'Abrera's other butterfly books. Full-color photos of live and set specimens fill the book. All set specimens are shown in their natural size, so even in this oversize book often only two butterflies will fill a page. Three genera are discussed and displayed: *Ornithoptera*, *Trogonoptera*, and *Troides*. These are found from North India to the Solomon Islands. Each genera is described in text, followed by descriptions of the species, both male and female. The color photos show male and female, and indicate if viewing the specimen from verso (below) or recto (above). Most set specimens are from the British Museum (Natural History).

447. D'Abrera, Bernard. **Butterflies of the Afro-Tropical Region**. London: E. W. Classey, State Mutual Book, 1980. 613p. $295.00. LC 82-141682. ISBN 0-317-07042-8.

This beautiful book, like most of D'Abrera's butterfly books, is actually an illustrated checklist of butterflies of this region. This work is based on R. H. Carcasson's *Synonymic Catalogue of the Butterflies of the Ethiopian Region*. It provides no more textual information than that of the checklist; no natural history, descriptions, or other helpful information is provided. It is, however, the only source on butterflies of this region, and the full-color photos along with the checklist information make it good for identifying butterflies. The checklist is arranged by family from Papilionidae to Riodinidae, and lists the name, the author of the name, and the date it was published. It also gives geographical range and some notes on the male and female descriptions. The outstanding feature of this and other D'Abrera books are the color photographs. They fill each page of this lengthy (613 page) work and show the amazing color and variety of butterflies in this region.

448. D'Abrera, Bernard. **Butterflies of the Australian Region**. 2d ed. Melbourne: Hill House, 1990. 416p. LC 90-195494. ISBN 0-9473-5202-3.

This beautiful second edition on butterflies of the Australian Region has additions and corrections from the first edition plus a few more color plates. It is intended as an illustrated guide for the identification of butterflies of this region. Brief introductory information includes butterfly life history, larvae, pupa, imago, diet, respiration, circulation, wing venation, the nervous system, and distribution. The geographical area covered includes New Guinea, Australia, the Moluccas, New Zealand, and the Pacific Islands. The family descriptions begin with the family Papilionidae and end with the family Riodinidae. Species accounts give the name, source of original description, range, references to illustrations, early stages, and food plants. Full-color, natural-size photos of set specimens are found next to the species account. Color photos of live species are also found throughout this beautiful book.

449. D'Abrera, Bernard. **Butterflies of the Holarctic Region**. Victoria, Australia: Hill House, 1990- . £129.35 vol. 1; £117.19 vol. 2. LC 91-157014. ISBN 0-646-01202-9 vol. 1; 0-646-06255-7 vol. 2.

This work is up to two volumes as of this writing. Volume one covers the Papilionidae, Pieridae, Danaidae, and Satyridae (partim); volume two covers the Satyridae (concl.) and Nymphalidae (partim). Another beautiful work by D'Abrera.

450. D'Abrera, Bernard. **Butterflies of the Neotropical Region**. Victoria, Australia: Hill House, 1981- . 5 vols. £87.50 vol. 3; £99.80 vol. 4; £98.80 vol. 5. LC 82-141689. ISBN 0-9593639-5-5 vol. 3; 0-9593639-6-3 vol. 4.

Now in its fifth volume, this guide to the identification of the butterflies of the Neotropical Region is as beautiful as the others by D'Abrera. The Neotropical Region, for this work, includes all of South America, Central America from Mexico to Panama, and the entire West Indies, including Cuba and Hispaniola. Volume 1 covers the Papilionidae and Pieridae; volume 2, the Danaidae, Ithomiidae, Heliconidae, and Morphidae; volume 3, Brassolidae, Acraeidae, and Nymphalidae (partim); volume 4, Nymphalidae (partim); volume 5, Nymphalidae (concl.), and Satyridae. It has the same arrangement and includes the same checklist information as his other sets.

451. D'Abrera, Bernard. **Butterflies of the Oriental Region.** Victoria, Australia: Hill House, in association with E. W. Classey, 1982-1986. 3 vols. $370.00. LC 82-217884. ISBN 0-9593639-0-4.

Another of D'Abrera's beautiful books on butterflies, this one to those found in the Oriental Region. The Oriental Region encompasses India (south of the Himalayas), west to China, and southeast Asia. The three volumes cover the major families found there. Volume 1 includes Papilionidae, Pieridae, and Danaidae; volume 2, Nymphalidae, Satyridae, and Amathusidae; and volume 3, Lycaenidae and Riodinidae. It is comprehensive but not easy for an amateur to use. Full-color plates fill every page facing the brief descriptions of each species. Red dots next to some species indicate a type specimen. A few color photos of live butterflies found in the field are included here and there in the text. A list of references and an index to species complete the volume.

452. D'Abrera, Bernard. **Sphingidae Mundi: Hawk Moths of the World, Based on a Checklist by Alan Hayes and the Collection He Curated in the British Museum (Natural History)**. Oxon: E. W. Classey, 1986. 226p. £97.50. ISBN 086096-02206.

This is a beautifully illustrated systematic list of the known species of hawk moths of the world. Its beauty is in the full-color plates of preserved specimens from the collection in the British Museum (Natural History). The plates face each page of the checklist, illustrating those species listed. The list itself is in synoptic form (i.e., each subfamily, genus, and species name includes the author who named it, the date of the publication, and the journal citation). Other information included is the geographic range, a description of the larva, pupa and host plants, and other distinguishing information of the species. A two-page bibliography, and generic and species indexes complete this beautiful work.

453. Dickens, Michael. **The World of Butterflies**. New York: Macmillan, 1972. 127p. LC 72-86311.

Arranged like his other publication, *The World of Moths* (see entry 454), this work describes and pictures 108 species of butterflies found throughout the world. Each page provides a species account, with a color photo of the species at the top of the page. Information provided includes the scientific name, family name, common name, wingspan, range, habits and habitat, food plants, sexual dimorphism, subspecies, and similar species.

454. Dickens, Michael. **The World of Moths.** New York: Macmillan, 1974. 128p. LC 74-32.

Pictured and described in this small book are 103 species of moths; only the most commonly found species are included. One hundred three species of moths are pictured and described in this small book, only the most commonly found species. Introductory information includes the life history of the moth, how to trap and rear moths, and the anatomy and classification of moths. Descriptions of each moth cover one page with a color photo at the top of the page. The scientific name, the family, the common name, wingspan, range, habits, and food plants as well as descriptions of male and female differences and relationship to other species are all covered in this succinct description. A common name index is found at the end.

455. Dominick, R. B., ed. **Moths of America North of Mexico, Including Greenland**. Washington, D.C.: Wedge Entomological Research Foundation. Distributed, Faringdon, England: E. W. Classey, 1971- . Multivolume. $125.00/vol. LC 78-149292.

This multivolume work has been published irregularly, as books on the various families are finished by individual authors. When complete, the series will provide identification manuals for all species of moths occurring in North America north of Mexico. It is planned to be published in about 30 fascicles and 150 parts. Each family is treated with complete descriptions of the family in general, with host plants, description and characters, immature stages, classification, and keys to the subfamilies, genera, and species. Each species account includes references to the literature where described and the type locality. Many black-and-white drawings of anatomical details help in the identification of species. Full-color plates of photos of species described in the text are found at the end of the text. An excellent and thorough series.

456. Ferris, Clifford D., and F. Martin Brown, eds. **Butterflies of the Rocky Mountain States.** Norman, Okla.: University of Oklahoma Press, 1981. 442p. $45.00; $22.95pa. LC 80-22274. ISBN 0-8061-1552-1; 0-8061-1733-8pa.

This work is included here for its comprehensiveness in thoroughly exploring this region in detail. The guide includes information on life histories, food plants, flight seasons, genetic relationships, and also detailed range maps where these butterflies may be found. The book has introductory chapters on the geology and biogeography of the region plus taxonomic and biological notes as well as collecting and preserving methods. The species are described in order of the four major butterfly families in North America. Each species is described as to appearance, range and habitat, life history, and mention of subspecies. A black-and-white photo (unfortunately not color) of each species is included with the description (but colors are given in the text). Only four full-color plates are included in the center of the text. Range maps are grouped at the end. A checklist of the butterflies of this region is included as well as an index to general topics and to the scientific and common butterfly names.

457. Holland, W. J. **The Butterfly Book: A Popular and Scientific Manual, Describing and Depicting all the Butterflies of the United States and Canada**. New and thoroughly rev. ed. Garden City, N.Y.: Doubleday, 1947. 424p.

This classic work includes 77 color plates illustrating over 2,000 North American and Canadian butterflies. Introductory chapters cover the anatomy and life cycles of butterflies, the methods for their capture and preservation, and classification of butterflies. The main text covers the families, subfamilies, and then each genus and species. The descriptions cover the adult and sometimes the egg, caterpillar, and chrysalis stages. Food plants are also mentioned as well as similar species. This is a charming book for what the author calls "digression and quotations," such as the first picture of an American butterfly (from the first Virginia settlement in 1587). Updated by *The Butterflies of North America* (see entry 461).

458. Holland, W. J. **The Moth Book: A Popular Guide to a Knowledge of the Moths of North America**. New York: Dover Publications, 1968. 479p. LC 68-22887. ISBN 486-21948-8.

This is an unabridged republication of the work originally published in 1903. It is similar in arrangement to Holland's *The Butterfly Book* (see entry 457). Although dated, with old nomenclature, the book provides an interesting and easy-reading introduction to moths of the United States and Canada. It is arranged by family with genus and species accounts illustrated with black-and-white drawings. Each account is written so the amateur can understand the descriptions. Digressions of quotes and prose passages are found throughout the text to add more leisure reading interest. It includes 48 color plates of approximately 1,400 species of moths.

459. Holloway, J. D., J. D. Bradley, and D. J. Carter. **CIE Guides to Insects of Importance to Man. No. 1: Lepidoptera**. Oxon, England; Tucson, Ariz.: C.A.B. International Institute of Entomology, British Museum Natural History, 1987. 262p. ISBN 0-85198-605-6; 0-85198-594-7 spiral bound.

See entry 358, *IIE Guides to Insects of Importance to Man.*

460. Howarth, T. G. **Colour Identification Guide to Butterflies of the British Isles**. Rev. ed. New York: Viking, 1984. 151p. LC gb84-23979. ISBN 0-670-80355-3.

Although this is primarily an identification guide, it is included here instead of in the field guide section because it is not pocket size; comes bound in white material more appropriate for inside work. This work is arranged so that identification of the common British butterflies can be determined first by checking the color plates, or using the identification key, and then referring to the tabular information on each butterfly arranged in alphabetical order by common name. Color plates for identification of egg, caterpillar, or pupa are also provided. The tabular information includes the common name, the scientific name, the subspecific name, the number of described aberrations, brief mention of variations, a calendar for showing general time of appearance of each life stage, foodplants, number of larval stages, and brief information on the general habitat, distribution, and abundance.

461. Howe, William H., ed. **The Butterflies of North America**. Garden City, N.Y.: Doubleday, 1975. 633p. LC 73-15276. ISBN 0-385-04926-9.

Another classic in the butterfly literature, this scholarly work features 97 full-color plates containing 2,093 hand-painted species, all carefully executed by William Howe. This book is an update of *The Butterfly Book* by W. J. Holland (see entry 457). It is a comprehensive work on the butterflies of Canada and the United States, including Alaska and Hawaii. William Howe served as editor of this book with 20 contributors, although Howe authored the greatest number of chapters on the various butterfly families and subfamilies. Introductory material covers the basic "What is a butterfly?" to information on how to collect butterflies, how to make a collection, as well as anatomy, life history, and classification. The following text describes the family, genus, and species with information on where it is found geographically, the early stages and host plants the larvae feed on, flight emigration, and other pertinent facts. The plates are placed in the center of the text with references from the plates to the descriptions and vice versa. A glossary, a bibliography, and a list of chief collections of butterflies in North America and Europe follow the text. An index to food plants, and a detailed index to the butterflies by common and scientific name, as well as by subject, complete the volume.

462. Lewis, H. L. **Butterflies of the World**. Chicago: Follett, 1973. 312p. LC 73-822201. ISBN 695-80434-0.

The beauty and worth of this volume on the world's butterflies is in its plates. Seven thousand color photographs of almost all genera and 90% of most species are here. The specimens are from the collection of the Department of Entomology, British Museum (Natural History). The plates are arranged first by large geographical areas: Europe, North America, South and Central America, Africa south of the Sahara, Asia south of the Himalayas, and Asia north of the Himalayas. Within these geographical groups, the butterflies are arranged by family, then alphabetically by genera and species. The text follows the plates. The "text" is simply a numbered listing of butterflies in the plates. The list gives the scientific name, and the abbreviation of the author of the name, then its geographical range, and may additionally give information on similar species, the different varieties, and subspecies. Range maps are included at the beginning of each major geographical area to show the area where the butterflies are chiefly found. The index is to the plate number and butterfly number on the plates, and is by scientific name only.

463. Opler, Paul A., and George O. Krizek. **Butterflies East of the Great Plains: An Illustrated Natural History**. Baltimore, Md.: Johns Hopkins University Press, 1984. 294p. $49.50. LC 83-6197. ISBN 0-8018-2938-0.

This is the definitive work on butterflies of this region. Paul Opler has also published a field guide for this region titled *A Field Guide to Eastern Butterflies* (see entry 500). This work is arranged first with introductory chapters on the study of butterflies, anatomy, size, distribution, habitats, behavior, and life history. The main part of the book gives species accounts for all butterflies known to occur in the United States east of the Great Plains. The species accounts are arranged by family and include (usually) one full page of information such as etymology, description of the adult butterfly, range, habitat, life history, early stages, adult nectar sources, and caterpillar host plants. A range map is provided for each species account. References to where the species was first described in the literature are also provided. Fifty-four color plates of live butterflies are inserted into the middle of the text. A glossary and complete bibliography, and an index to scientific and common butterfly names complete this impressive work.

464. Preston-Mafham, Rod, and Ken Preston-Mafham. **Butterflies of the World.** New York: Facts on File, 1988. 192p. $24.95. LC 87-33228. ISBN 0-8160-1601-1.

This short introduction to butterflies covers all the basics with an easy-to-read text. The authors, both of them wildlife photographers, have included their beautiful full-color and black-and-white photos to enhance the text. The text includes chapters on the classification of butterflies, structure and life-cycle, as well as adult behavior, variation, ecology, butterfly enemies, defense mechanisms, and a chapter on butterflies and man. It has a glossary, a guide to further reading, and an index to subjects as well as common and scientific names of butterflies mentioned in the text.

465. Scoble, Malcolm J. **The Lepidoptera: Form, Function, and Diversity**. London: Natural History Museum Publications, Oxford University Press, 1992. 404p. $75.00. LC 92-4297. ISBN 0-19-854031-0.

This work reviews the general biology and diversity of the world's Lepidoptera with descriptions of the major taxa. It is divided into three main sections: a section on morphology and anatomy, one on environmental considerations, and one on the major taxa. The first section includes chapters with detailed information on the adult head, thorax, and abdomen as well as descriptions of the immature stages and the hearing, sound, and scent of the Lepidoptera. The second section is brief and covers the environmental importance of Lepidoptera, their roll in pollination, and the production of silk among lepidopteran larvae, among others. The third section on the major taxa provides introductory information on the classification of this large group, plus descriptions of the superfamilies, families, and sometimes subfamilies. Each family is briefly described, followed by a summary of adult structure, immature stages, general biology, classification, and phylogenetic relationships. This work is not intended as an identification tool, so keys and illustrations are absent except for 17 pages of plates, at the end of the text, containing photos of selected species of the major families. This work provides a useful summary of Lepidoptera form, function, and classification for the student and specialist.

466. Scott, James A. **The Butterflies of North America: A Natural History and Field Guide.** Stanford, Calif.: Stanford University Press, 1986. 583p. $49.50. LC 82-60737. ISBN 0-8047-1205-0.

This text covers a wide range of information on the butterflies of North America. It is divided into three main sections, one on butterfly biology, another on keys to the identification of the eggs, larvae, pupae and adult butterflies, and a final section, the largest, on descriptions of family, subfamily, tribe, and so on down to species. The section on butterfly biology has a wealth of information on courtship and mating, evolution, anatomy, flight, metamorphosis, life cycles, habitats, and even a section on butterfly intelligence. The section on keys provides very useful and detailed guides to the identification of butterflies in all their life-cycle stages. The keys are illustrated by black-and-white line-drawings as well as electron micrographs to aid in the identification. The final section describes the butterflies in taxonomic order with information on identification, distribution, behavior, and ecology. A feature of this section is the inclusion of range maps in the margin next to the species description showing where these butterflies may be found. Sixty-four color plates with photos illustrating the species described in the text are inserted into the middle of the third section of the text. The use of these plates is difficult at times, that is in matching the plate to the description of the butterfly. A section on studying butterflies, a short bibliography, a hostplant catalogue, a glossary, and an index of scientific and common names as well as subjects covered in the book are found at the end of the text.

467. Skinner, Bernard. **Colour Identification Guide to Moths of the British Isles (Macrolepidoptera)**. Harmondsworth, England: Viking, 1984. 267p. ISBN 0-670-80354-5.

Similar to Howarth's book on butterflies (see entry 460), this work attempts to aid the reader in identifying almost all large adult species of moths found in the British Isles. Larval forms are not included in the plates. By first consulting the 42 colored plates of adult moths found at the end of the text, the reader may identify the moth there. From the plates, the reader is referred to a page number where the species is described. Each description is listed under the common name, followed by the scientific name. The brief description consists of information on size, conservation status, similar species, times of appearance of both adult and larva, type of habitat, distribution, and larval food plants. A list of books for more information is found at the end of the text, along with a list of scientific names of food plants and an index to scientific and English common names.

468. Watson, Allan, and Paul E. S. Whalley. **The Dictionary of Butterflies and Moths in Color.** New York: McGraw-Hill, 1975. 296p. LC 74-30433. ISBN 0-07-068490-1.

The purpose of this work, to cover the entire order of Lepidoptera, has been achieved in a beautiful book. The 405 color photos of museum and live specimens depicting almost every known family in this group fill the first section with amazing color. The dictionary that follows is an alphabetical arrangement by scientific name of family and genus with species names listed under the genus. Cross references are given from the common name to the scientific name. Each name is defined in a short paragraph with information such as the size, where it is found geographically, male and female descriptions, colors, and other factual information. Plate numbers for those species pictured in the plates (not all are pictured) are included at the end of the paragraph.

Hymenoptera (Ants, Bees, Wasps)

469. Bohart, R. M., and A. S. Menke. **Specid Wasps of the World: A Generic Revision**. Berkeley, Calif.: University of California Press, 1976. 695p. LC 72-87207. ISBN 0-520-02318-8.

This comprehensive work covers all but 5 of the 226 genera of Specid wasps, a diverse family of solitary hunting wasps. It includes introductory chapters on the behavior, morphology, and systematics of these wasps as well as a generic catalog and key to the subfamilies and tribes. The main part of the text describes the subfamilies, families, genera, and subgenera with checklists of species under each genus. The detailed descriptions give appearance, geographic range, systematics, and biology. Black-and-white line-drawings show full body as well as anatomical detail to aid in identification. The 38-page section on literature cited shows the effort made to fully cover the literature on this subject. A detailed index down to specific and subspecific names is provided to end this scholarly work.

470. Gauld, Ian, and Barry Bolton, eds. **The Hymenoptera.** Oxford, England: British Museum (Natural History), 1988. 332p. $80.00. LC 88-12526. ISBN 0-19-85821-7.

This is the most current treatment of the this large group of insects available today. Created primarily by the British Museum (Natural History) staff, its emphasis is on the British Hymenoptera, but the work has relevance to North America and western Europe as well. The chapters cover the biology, economic importance, collecting, structure, classification, evolution, and works on the individual families of Hymenoptera occurring in Britain. A key to adults of superfamilies is included as well as keys to families for the larger superfamilies. For those wanting to identify a particular genus or species, the authors have included references to identification keys with the description of each family group. It concludes with a comprehensive 47-page bibliography and an index. Many black-and-white drawings are included throughout the text as well as eight color plates. Especially noteworthy are the black-and-white whole-insect drawings by David Morgan. A valuable reference book for college and university libraries as well as large public libraries.

471. Holldobler, Bert, and Edward O. Wilson. **The Ants.** Cambridge, Mass.: Belknap Press of Harvard University Press, 1990. 732p. $65.00. LC 89-30653. ISBN 0-674-04075-9.

Winner of the 1991 Pulitzer Prize for general nonfiction, this monumental work covers information on all aspects of myrmecology. It is written by two recognized authorities on the subject and reflects their appreciation of the ants through many years of research. The 20 chapters cover such topics as classification, colony life, communication, caste studies, organization of communities, and symbioses among themselves and other groups. Then are separate chapters on the major groups of ants (e.g., army ants, fungus growers, harvesting ants, and weaver ants). The classification chapter includes keys to subfamilies and genera. The text is illustrated with excellent black-and-white drawings, photos, and electron micrographs as well as 24 full-color plates. A glossary is included as well as an extensive bibliography of 54 pages with approximately 2,000 entries. An outstanding work.

472. Kimsey, Lynn S., and Richard M. Bohart. **The Chrysidid Wasps of the World.** Oxford, New York: Oxford University Press, 1990. 652p. $140.00. LC 90-40114. ISBN 0-19-854010-8.

This is a very thorough work on the Chrysidid wasps, a family of parasitic wasps, with the common name of gold or cuckoo wasps that numbers 84 genera and four subfamilies. The work is primarily systematic with introductory chapters on the biology, biogeography, and morphology of the group. Dichotomous, illustrated keys to family, subfamily, then to genus and species, with detailed synonymic species lists included under genus, are the framework to the text. Each genus is described with diagnostic characters, hosts, and geographic distribution information. Detailed, side-view, black-and-white drawings of representative species illustrate the keys as well as line-drawings of specific anatomical details. A complete 25-page bibliography and an index to Chrysidid names finish this comprehensive work.

473. Mitchell, Theodore B. **Bees of the Eastern United States**. Raleigh, N.C.: North Carolina Agricultural Experiment Station, 1960. Reprinted by J. Johnson. 2 vols. (North Carolina Agricultural Experiment Station Technical Bulletin, 141 and 152). $20.00 set. ISBN 0-910914-05-2.

This comprehensive work is the result of the author's 38 years of research on the taxonomy of bees. It describes in two volumes the bee flora of the United States east of the Mississippi River. Volume one covers the families Colletidae, Andrenidae, Halictidae, and Mellitidae. Volume two covers the families Megachilidae, Anthophoridae, Xylocopidae, and Apidae. Each family is described and a key to genera is provided; and each genera is described and a key to species is provided. The individual species are described with detailed descriptions of the female and the male, the geographical distribution, and the flower or flowers that the bee is dependent on for pollen. Black-and-white line-drawings illustrate anatomical details useful in identification of species. Indexes to taxa are found at the end of each volume. A comprehensive and authoritative work.

474. O'Toole, Christopher, and Anthony Raw. **Bees of the World**. New York: Facts on File, 1992. 192p. $24.95. LC 90-49023. ISBN 0-8160-1992-4.

This is a good introduction to the study of bees. The text is easy to read and full-color pictures, line-drawings, and diagrams are found on almost every page. The book includes introductory chapters on the origin of bees, their anatomy and development, the 11 families of bees, where they live, and the solitary life of most bees. The solitary bees are treated in chapters on the miner, mason, leaf-cutting, carpenter, and other solitary bees. These are followed by a chapter on the more well-known social bees, the honeybee, and the stingless bee. Final chapters on male bees, the enemies of bees, and the pollination of flowers by bees round out the book. A short glossary, a bibliography, and a general index add to the worth of this short introduction to bees.

475. Richards, O. W. **Hymenoptera: Introduction and Key to Families**. 2d ed. London: Royal Entomological Society of London, 1977. 100p. (Handbooks for the Identification of British Insects, vol. VI, pt. 1). LC 78-305802.

More than half of this work is on general information on the entire group of Hymenoptera, and so the work is included here, even though the last half deals with the identification and description of British Hymenoptera. Good introductory chapters are provided on the anatomy of the Hymenoptera, especially the major parts of head, thorax, prothorax, mesothorax, metathorax, propodeum, wings, legs, and abdomen. Detailed black-and-white drawings of all anatomical parts described are inserted throughout the text. Fine drawings of representative species in the British Isles are found at the end of the text. Keys to families of British Hymenoptera are also useful to other geographical areas with similar fauna. A short but authoritative introduction to one of the largest insect orders.

476. Richards, O. W. **The Social Wasps of the Americas Excluding the Vespinae**. London: British Museum (Natural History), 1978. 580p. LC 80-485409. ISBN 0-565-00785-8.

This is a scholarly work on the taxonomy and biology of American social wasps except Vespinae. Keys to the genera and to American species and subspecies as well as to the larvae are provided. Species accounts are detailed and include references to the literature where first described, to plates and figures, and to holotype location. It also describes the species, with notes on nests, and other field notes, and gives the geographic distribution. A checklist is included at the end of the text as well as an index to scientific name, a long list of references, and four color plates of nests.

477. Ruttner, Friedrich. **Biogeography and Taxonomy of Honeybees**. Berlin, New York: Springer-Verlag, 1988. 284p. $121.00. LC 87-21468. ISBN 3-540-17781-7.

This book discusses the subfamily Apinae, or honeybee. It includes information on taxonomy, ecology, and evolution as well as behavior and adapability to various climates. It is divided into two sections: part one discusses honeybees of the world, and part two discusses the western honeybee, *Apis mellifera*. Part one covers information on the genus *Apis*, stingless bees, evolution, geographic variability, honeybee taxonomy, and three chapters on each of the main species of honeybee other than the western honeybee, which is dealt with in part two. In part two, general information on the western honeybee is presented along with its unusual range of adaptation. Following this are chapters on the various subspecies of honeybees found in the Near East, Tropical Africa, Western Mediterranean, and the Central Mediterranean and Southeastern Europe. A 20-page list of references and an index finish the volume. A work for the specialist.

478. Spoczynska, Joy O. I. **The World of the Wasp**. New York: Crane, Russak, 1975. 188p. LC 74-13621. ISBN 0-8448-0560-2.

Another introductory work on wasps, this one describes the most interesting representatives of each group of wasps from all over the world. It focuses on the description, life history, and habits of the wasp groups. It has chapters on the paper-making social wasps, the solitary wasps, and the Ichneumons, Chalcids, and Braconids as well as the gall-wasps. Detailed black-and-white drawings illustrate representative species and a few pages of black-and-white photos of wasps and galls are inserted into the text. Final chapters on how to study wasps and making a wasp collection are useful for the beginning hymenopterist.

479. Spradbery, J. Philip. **Wasps: An Account of the Biology and Natural History of Solitary and Social Wasps with Particular Reference to Those of the British Isles**. Seattle, Wash.: University of Washington Press, 1973. 424p. (Biology Series). $50.00. LC 73-7872. ISBN 0-295-95287-3.

This work serves as a good introduction to wasp studies for the amateur and is a useful handbook for the specialist. It covers adult wasp anatomy and morphology, behavior, parasites and predators, developmental stages, nests, feeding, wasp populations, social organization, males and females, classification, and evolution. Several appendices include the taxonomy of British wasps with keys to families and species, a checklist of British Vespoidea, a classification of wasp nests, distribution of British wasps species (with range maps), and information on the collection, preservation, and study of wasps. A 20-page bibliography leads other sources of information. A classic work.

Chilopoda and Diplopoda (Centipedes and Millipedes)

480. Blower, J. Gordon, ed. **Myriapoda**. London: Academic Press, 1974. 712p. (Symposia of the Zoological Society of London, no. 32). £99.00. LC 73-7035. ISBN 0-12-613332-8.

This volume includes French and English papers from the First International Congress of Myriapodology, held in Paris in 1968. It covers many areas of both fossil and living myriapod studies (e.g., taxonomy, morphology, classification, reproduction, courtship, food preference, ecology, life cycle, habitat and distribution, economic importance, and control). Each paper ends with a useful list of references and an interesting discussion that followed the presentation of the paper. A subject and systematic index, and a classification scheme for the Myriapoda are found at the conclusion of the text.

481. Camatini, Marina, ed. **Myriapod Biology.** New York: Academic Press, 1979. 456p. LC 79-41559. ISBN 0-12-155750-2.

This is actually proceedings from the International Congress of Myriapodology held in Milan, Italy, in 1978. This was the fourth International Congress. The papers are arranged in six sections covering myriapod cytogenetics, systematics, anatomy and embryology, ecology and biogeography, endocrinology, and evolution. Although not a handbook in actual form, this covers many aspects on myriapod biology not found in other works.

482. Hopkin, Stephen O., and Helen J. Read. **The Biology of Millipedes**. Oxford, New York: Oxford University Press, 1992. 233p. $90.00. LC 91-40375. ISBN 0-19-857699-4.

This is a complete and up-to-date work on the biology of millipedes. It includes chapters on taxonomy, anatomy, feeding and digestion, metabolism, excretion and water balance, the nervous system, fertilization, development, predators and parasites, and ecology. Each chapter is clearly written and illustrated with many black-and-white drawings, photos, and scanning electron micrographs. An impressive 38-page bibliography of references attests to the thoroughness of the work, and the enthusiasm of the authors. Clear enough for the amateur, and complete enough for the specialist.

483. Lewis, J. G. E. **The Biology of Centipedes**. London: Cambridge University Press, 1981. 476p. £50.00. LC 80-49958. ISBN 0-521-23413-1.

This is a comprehensive work on centipedes, bringing the diverse literature on this group into one book with 24 chapters. It progresses through chapters on the external morphology and internal anatomy, such as the musculature, nervous system, endocrinology, alimentary canal, poison glands, respiratory system, pigments, circulatory system, glands, excretory and reproductive system, and feeding and digestion. The last chapters cover the ecology and physiology, parasites, and taxonomy of the group as well as relationships with other orders, and a final chapter on the classification of centipedes. Each chapter is illustrated with line-drawings and diagrams. A very complete 26-page bibliography provides further reading on all aspects of centipede biology.

FIELD GUIDES

Arthropoda (General)

484. Gosner, Kenneth L. **A Field Guide to the Atlantic Seashore: Invertebrates and Seaweeds of the Atlantic Coast from the Bay of Fundy to Cape Hatteras**. Boston: Houghton Mifflin, 1979. 329p. (The Peterson Field Guide Series). $17.95; $12.95pa. LC 78-14784. ISBN 0-395-24379-3; 0-395-31828-9pa.

This field guide includes a 50-page section on the arthropods, and so is included here as a general, easy-to-use field guide to this vast group. See entry 196 in chapter 2, "The Invertebrates" for a full description. The 50 pages of text (pages199-252) describe the major classes found in, on, and around the Atlantic seashore. Each description refers to a plate illustrating a representative species. Thirteen black-and-white and color plates contain illustrations of more than 130 representative species, including the sea spiders, the biting flies, saltwater mosquitoes, fish lice, sea pill bug, and beach fleas, as well as many species of shrimp, crabs, and lobsters. Each species is described in the text as to identification, similar species, where they may be found, and what family they belong to. Since no field guide is available to this vast group, this section may serve in a small way as an introductory guide.

Chelicerata

ARACHNIDA (Scorpions, Spiders, Mites, and Ticks)

485. Jones, Dick. **The Larousse Guide to Spiders**. New York: Larousse, 1983. 320p. LC 83-81216. ISBN 0-88332-324-9.

Although this is primarily a guide to British and European spiders, many species found here are also found in the northern hemisphere. It describes and pictures 350 spiders and harvestmen. Introductory material is given on the study of spiders including anatomy and life cycle, as well as on finding and collecting spiders. Identification of a species begins by using the illustrated key to spider families. This key leads to a page number where species of that family are described in the book. Each species is described in one paragraph listing the size, description of carapace, abdomen and legs, the habitat, season when found, and its distribution in Britain and Europe. On the page opposite the written description is a full-color photo (taken by the author) of the species. These photos make identification easy. An index to scientific and common name is found at the end.

486. Levi, Herbert W., and Lorna R. Levi, under editorship of Herbert S. Zim. **Spiders and Their Kin**. Racine, Wis.: Golden Press, Western, 1969. 160p. (A Golden Guide). $4.50. LC 68-23522. ISBN 0-307-24021-5.

This is a very easy, yet authoritative little guide to the common spiders and their ne relatives, the whipscorpions, windscoprions, pseudoscopions, scorpions, harvestmen, mite centipedes, millipedes, wood lice, and land crabs. It includes species found worldwide. Lik others in the Golden Guide series, it has brief introductory information on classificatio anatomy, courtship, growth, enemies, and how to collect and preserve them. Many colo drawings of the most common species are included with short descriptions of the group (e.g a one-paragraph description of Lynx spiders with color illustrations of four of the mo common species). A short bibliography for further reading, and an index to common ar scientific name conclude this little, but informative work.

487. Milne, Lorus, and Margery Milne. **The Audubon Society Field Guide to Nort American Insects and Spiders**. New York: Alfred A. Knopf, 1980. 989p. $17.50. L 80-7620. ISBN 0-394-50763-0.

This guide describes and pictures 60 of the common spiders found in North Americ For complete review of this guide see entry 491.

Uniramia

INSECTA (General)

488. Arnett, Ross H., Jr., and Richard L. Jacques, Jr. **Simon & Schuster's Guide to Insect** New York: Simon & Schuster, 1981. 511p. LC 80-29485. ISBN 0-671-25013-2; 0-67 25014-0pa.

This popular field guide features full-color photos and descriptions of 350 insec commonly found in North America. Introductory text contained in the first 64 pages of th guide covers basic information on insects such as where do they live, which insects are pest the science of entomology, classification, what an insect is, the orders of insects, inse anatomy, coloration, behavior, and information on the social insects, the aquatic insect insect pests, and insects that are beneficial. It also tells how to collect insects, the equipme needed, and how to label and store the collection. To use the guide to identify an insect, th reader is directed to a table, beginning on page 65, of the common insect orders. They a grouped under very broad categories such as "without wings at all stages," "without wings certain stages," or "always with wings as adults." The reader finds the correct order and provided a page number where that order is briefly described. From there the reader is directe to the description and color photos of selected species from that order found in the main pa of the guide. Each species is described as to what family and order it belongs, the lengt recognition remarks, distribution, and field notes. This guide is not as technical nor does describe as many families as the guide by Borror and others (see entry 489), but it is easie for the beginner to use, covering the most commonly found insects and featuring strikin full-color photos for easy identification.

489. Borror, Donald J., and Richard E. White. **A Field Guide to the Insects of Americ North of Mexico.** Boston: Houghton Mifflin, 1970. 404p. (The Peterson Field Guide Series $13.95pa. LC 70-80420. ISBN 0-395-07436-3;0-395-18523-8pa.

This standard, classic field guide will help novice and specialist alike identify the majo insect groups in North America north of Mexico. It describes and pictures an amazing 57 insect families. Although it covers the largest group in the animal kingdom, it is a relativel easy-to-use guide. It begins the identification of an insect with a pictured key to the majo orders of insects found on the front and back endpapers of the guide. Once the order has bee identified there, a page number is provided to the description of the order in the guide. B reading the descriptions of the family groups and looking at the black-and-white drawings o representative species, the insect may then be identified as to which family it belongs an possibly to genus and species. Size lines on each illustration show the approximate size o each insect. A section of color plates is inserted into the middle of the guide that illustrate nd briefly describes selected species from each of the major orders. A glossary of terms used n the descriptions and a four-page list of references, under headings for the major insect rders, follows the text. An index to common and scientific name finishes this excellent guide.

90. Chinery, Michael. **Insects of Britain and Northern Europe**. 3d ed. London: Harper-ollins, 1993. 320p. $22.75. (Collins Field Guide). ISBN 0-00-219918-1.

This classic field guide is an easy-to-use and informative guide to the major orders of nsects in Europe and Britain. Now in its third edition, it has been completely rewritten. It egins with a section on how to use the book, instructing the reader to go to the keys to the rders of insects, where they can identify the insect to order. All the major families of that rder are illustrated by color plates inserted in the center of the book. The page facing each late lists the names of the species illustrated and provides brief information on that particular rder and family. The reader can then turn to the textual information on that order and family. he textual information includes several pages on recognition features as well as collecting nd preserving information. Keys to the families of each order are also included, except for he Lepidoptera, Hymenoptera and Coleoptera, which are keyed to super-family. Black-and-hite illustrations are found in the keys and the descriptive text. A twelve-page glossary and nine 8-page bibliography follow the text. An index to common and scientific name ompletes the volume.

91. Milne, Lorus, and Margery Milne. **The Audubon Society Field Guide to North merican Insects and Spiders**. New York: Alfred A. Knopf, 1980. 989p. $17.50. LC 0-7620. ISBN 0-394-50763-0.

This field guide to the 550 common insects and 60 common spiders in North America an easy and enjoyable way to get to know two of the largest groups in the animal kingdom. eautiful full-color photos of each common species described in this guide are grouped gether before the text. A thumb-tab guide with the silhouette of a typical member of the roup helps the reader find a particular group (e.g., beetle group) in the color plates. The text at follows is arranged by family with a short description of the family and then a more etailed description of a common species in that family. For example, under the family hiteflies (scientific name Aleyrodidae) there is one paragraph on their general appearance ith other information useful to identify this family, then the description of the greenhouse hitefly follows. Included in the description is a section on what habitat it prefers, its eographical range, the food it prefers, and its life cycle.

92. Swain, Ralph B. **The Insect Guide: Orders and Major Families of North American nsects**. Garden City, N.Y.: Doubleday, 1948.

This is an old standby source enjoyed by many amateur entomologists. It is a guide to he 26 orders and major families of insects in North America north of Mexico. It is arranged om the most primitive order to the most specialized. Each order is described in a 1 to 12 or nore pages, with each family being described as to adult and larval morphology, economic nportance, and an example of a representative species. Color plates grouped in the center of he book depict a single representative species of each principle family. The numbers on the lustration correspond to the numbered section in the book where the description of the family nd species may be found. The end pages of the book feature an illustrated table of insect rders with principle morphological characteristics listed and a silhouette of a representative pecies so the amateur can recognize a particular order "at a glance."

OLEOPTERA (Beetles)

93. White, Richard E. **A Field Guide to the Beetles of North America**. Boston: Houghton lifflin, 1983. 368p. (The Peterson Field Guide Series). $19.50; $13.50pa. LC 83-60. ISBN -395-31808-4; 0-395-33953-7pa.

One hundred eleven beetle families are described in this field guide, with at least one pecies described in each family. Black-and-white drawings and 12 color plates aid in identifying these species. The arrangement is by family with the heading listing the commo name then the scientific name (e.g., "Tiger Beetles: Family Cicindelidae"). Information o identification, similar families, range, and habits, as well as collecting methods and example of species in the family are all succinctly given for each family. Introductory chapters on hov to collect beetles, preparing and identifying them, basic beetle structure, developmental stage of beetles, and the classification scheme used in naming beetles all aid the student or amateu entomologist in beetle research. A six-page glossary of terms used in the descriptions, a annotated bibliography, and an index by common or scientific name or synonym are include at the end.

LEPIDOPTERA (Butterflies and Moths)

494. Carter, David J. **Butterflies and Moths**. 1st Am. ed. New York: Dorling Kindersley 1992. 304p. (Eyewitness Handbooks). $29.95; $17.95pa. LC 91-58221. ISBN 1-56458-034 2; 1-56458-062-8pa.

More than 500 species of butterflies and moths from around the world are described an illustrated in this informative and colorful guide. It is written by David Carter, senior scientifi officer in the Entomology Department of the Natural History Museum, London, who ha authored several other guides to butterflies. This guide includes all five butterfly families an 22 of the major moth families. The introductory material covers 26 pages and includes suc information as life cycle, early stages, conservation, survival, rearing, and geographica distribution. The arrangement of the main part of the guide is by family, butterfly familie first, then moth families. Each family is introduced with a short paragraph on the basi characteristics, followed by descriptions of selected species. Each species is described wit brief general information as well as notes on early stages and distribution, with a ma highlighting the general area where they may be found. A feature of this guide is the use o full-color photos next to each description, rather than separating the color plates from th text. The photos are annotated with brief notes pointing out the distinguishing features suc as "bright red eyespot." Each description is framed in a color-coded box with headers fo family name, species name, author name, time of flight, and habitat, plus wingspan dimen sions. A visually pleasing and easy-to-use guide.

495. Carter, David J. **A Field Guide to Caterpillars of Butterflies and Moths in Britai and Europe**. London: Collins, 1986. 296p. $17.00. ISBN 0-00-219080-X.

This work treats the identification of the caterpillar stage of butterflies' life cycle Caterpillars are seen in gardens, trees, shrubs, and carried inside on clothes and shoes. "Wha kind of caterpillar is it?" or "What does the butterfly stage look like?" are questions answere by this easy-to-use guide. Also, because the caterpillar stage is the main feeding stage of th life cycle, horticulturists want to know what kind of caterpillar is chewing on their plants s they can find a way to control the damage. This guide helps to identify the more commo caterpillars encountered in Britain and Europe. The guide is arranged systematically by famil with species accounts giving distribution, a physical description, habitat, food plants, an biology. Common names are given as well. Thirty-five full-color plates with over 50 paintings of caterpillars and their food plants, as well as 300 paintings of the correspondin butterflies, are inserted into the center of the text. A selective bibliography, a list of foo plants and the caterpillars that feed on them, a general index, and a food plants index ar found at the end of the text.

496. Chinery, Michael. **New Generation Guide to the Butterflies and Day-Flying Moth of Britain and Europe**. Austin, Tex.: University of Texas Press, 1989. 315p. (Corrie Herrin Hooks Series; no. 13). $22.95. LC 88-51396. ISBN 0-292-75539-2.

This is a colorful guide to all European species of butterfly eggs, caterpillars, pupa, an adults, as well as day-flying moths. It is arranged by family, with species accounts liste within each family. Each species is described by size, color, flight, foodplants, pupa, egg status (common, extinct, threatened), geographic range (small range-map for each species) and other facts about the species. A color drawing of adult, caterpillar, pupa, and egg is found on the page opposite the species account. The guide also includes information on butterflies in general, their conservation, life cycle, reproduction, hibernation, migration, populations, and varieties. A glossary of terms, and an index to scientific and common name and illustration finish the guide. A good, easy-to-use guide for the amateur.

497. Covell, Charles V., Jr. **A Field Guide to the Moths of Eastern North America**. Boston: Houghton Mifflin, 1984. 496p. $18.45; $15.45pa. (The Peterson Field Guide Series). LC 83-26523. ISBN 0-395-26056-6; 0-395-36100-1pa.

This is a guide to 59 families and more than 1,300 common species of moths found east of the 100th meridian, which runs through the middle of North Dakota down through Texas. Sixty-four color and black-and-white plates are inserted into the middle of the text. Page numbers where the textual descriptions of a particular species is found in the book are listed opposite the plates. The way to use this guide is to find the species in the plates section and then go to the text for a full description. The full description includes information on identification, the food the larvae eat, and the geographical range where the species may be commonly found. Introductory chapters on moth anatomy, its life cycle, and collecting and preparing moths are interesting and give a good introduction to the study of these night creatures. A glossary and a bibliography for further reading are found at the end of the text. An index to scientific and common names of families, subfamilies, genera, and species in the guide is easy to use and includes page numbers to the plates in bold print. The end papers have illustrations of moth anatomy and dorsal views of moths at rest so the reader can identify those moths resting on the undersides of outdoor lights.

498. Daccordi, Mauro, Paolo Triberti, and Adriano Zanetti. **Simon & Schuster's Guide to Butterflies & Moths**. New York: Simon & Schuster, 1988. 383p. $21.95; $11.95pa. LC 87-26440. ISBN 0-6716-6065-9; 0-06716-6065-7pa.

Two hundred eighty-nine of the most common species of butterflies and moths of the world, including 34 caterpillar forms, are described in this colorful guide. They are arranged geographically by Palearctic Region, Nearctic Region, Neotropical Region, Afrotropical Region, and Indo-Australian Region. Each species description includes information on the family, the geographic distribution within the region, one paragraph giving a physical description (including wingspan length), as well as information on related species, and, finally, the particular habits of the species. A full-color picture of each species is found on the page facing the description, with two descriptions per page. Symbols are used next to each description as a quick reference to such information as habitat (symbols for tropical forests, mountain regions, etc.), whether it is diurnal (active during the day time) or nocturnal, and what geographical area in the region the species may be found. A two-page glossary and bibliography are included as well as an index to the scientific name, but there is no common-name index. Published by Macdonald Orbis in 1988 under the title *The Macdonald Encyclopedia of Butterflies and Moths*.

499. Higgins, Lionel G., and Norman D. Riley. **A Field Guide to the Butterflies of Britain and Europe**. 4th ed. London: Collins, 1980. 384p. $13.60. (Collins Guides). ISBN 0-00-219241-1.

This fourth edition of a classic work includes descriptions and full-color drawings of all the butterflies known to occur in Europe, including both sexes and all major subspecies. The endpapers illustrate eight of the nine families of butterflies covered in the book. Members of the ninth family, Danaidae, are illustrated in the plates section. To identify a particular butterfly, these endpaper illustrations, where the family can be found with illustrations of representative species and a plate number given to see other species in the family, should be consulted first. Once the butterfly is identified in the plates, a page number is given that leads to a textual description of the species. The species is described as to range, physical description, flight dates, habitat, geographical distribution, and similar species. The distribution map number is provided for the maps inserted at the end of the text. A concise checklist of species listed in this guide follows the maps. Separate indexes to common and scientific name may be consulted to find information and plates in the text. An easy-to-use and complete work.

500. Opler, Paul A. **A Field Guide to Eastern Butterflies**. Boston: Houghton Mifflin, 1992. 396p. (Peterson Field Guide Series, 4). $24.95; $16.95pa. LC 91-444477. ISBN 0-395-36452-3; 0-395-63279-Xpa.

This new edition of a 40-year-old classic by Alexander Klots, *A Field Guide to the Butterflies of Eastern North America* (Houghton Mifflin, 1951), promises to be a classic itself. Paul Opler's excellent text and Vichai Malikul's beautiful color paintings make a winning combination not only for accurate identification of eastern butterflies but for a true appreciation of them. Introductory chapters on butterfly study, gardening, conservation, and habitats lead into the chapters on identification of true butterflies (swallowtails, whites and sulphurs, harvesters, coppers, hairstreaks, and blues, metalmarks, and brushfoots), and the skippers. In all, 524 known species found east of the 100th meridian are described. The description covers the size range (in inches and millimeters), color and outward appearance of both male and female, a description of the early stages (caterpillar), food plants, flight dates, geographic range, habitat, and remarks on other distinguishing features. A clear, easy-to-read range map is provided for most species and is usually found on the same page as the species account. Supplementary material is found after the species accounts and includes a lift list, a glossary of terms used in the descriptions, references for further reading, a directory of butterfly societies, and butterfly-collecting tips. Indexes to the host and nectar plants and to butterfly names finish this comprehensive guide. A feature of this guide are 541 fine, color paintings by Vichai Malikul. These and 104 color photos of live specimens are found on 48 plates inserted into the center of the text. This excellent guide will be used by amateur and professional lepidopterists for many years to come.

501. Pyle, Robert Michael. **The Audubon Society Field Guide to North American Butterflies.** New York: Alfred A. Knopf, 1981. 924p. $17.95. LC 80-84240. ISBN 0-394-51914-0.

Another excellent book in the Audubon Society series of field guides. Seven hundred fifty-one full-color photos of butterflies are included before the text with the thumb-tab guide to help the reader find a particular group (the "skippers," for example). Arrangement of plates is by shape and color, so white butterflies are together, swallowtails are together, and so on, for easy identification. Descriptions of the more common butterflies with information on life cycle, flight period, similar species, and range follow the color plates. Introductory material covers butterfly anatomy as well as how to observe butterflies and how to identify them. A good glossary of terms used in the text and chapters on butterfly watching (two essential tools are a field guide and a pocket notebook), and how to photograph butterflies (need a 35 mm single-lens reflex camera with an aperture of at least f/16) are found at the end of the text. There are two indexes, one for the common and scientific names of host plants eaten by caterpillars, and another for the scientific and common names of the butterflies.

502. Scott, James A. **The Butterflies of North America: A Natural History and Field Guide**. Stanford, Calif.: Stanford University Press, 1986. 583p. $49.50. LC 82-60737. ISBN 0-8047-1205-0.

This work is more than a field guide and so is found in the handbook section of this chapter at entry 466. It is the only complete guide to the North American butterflies.

503. Tilden, James W., and Arthur Clayton Smith. **A Field Guide to Western Butterflies**. Boston: Houghton Mifflin, 1986. 370p. (Peterson Field Guide Series). $19.00; $14.95pa. LC 85-30501. ISBN 0-395-35407-2; 0-395-41654-Xpa.

Similar in arrangement to *A Field Guide to Eastern Butterflies* (see entry 500), this guide covers species west of the 100th meridian, including Hawaii. These number 512. Introductory chapters are on learning about butterflies, butterfly conservation, life history, butterfly classification, and keys to families. Species accounts include a physical description, the early stages, range, similar species, and dates when adults can be found. Illustrations include 48 color and black-and-white plates inserted into the middle of the text, with 16 figures found scattered throughout the text. A glossary of terms, a life list, a list of Hawaiian and Alaskan butterflies, a bibliography, and, a directory of organizations and supply houses are all provided after the species accounts. A handy, useful, easy-to-use guide.

504. Walton, Richard K., and Paul A. Opler, consultant. **Familiar Butterflies**. New York: Knopf. Distributed by Random House, 1990. 191p. $5.95. LC 90-052502. ISBN 0-679-72981-X.

This beginner's guide to butterflies describes and pictures 80 of the most common and widespread butterflies in North America. Introductory material on the life cycle of butterflies and identifying them leads into the descriptions and full-color photos of the individual species. Species are grouped by family in this order: swallowtails, whites and sulfurs, gossamer wings, snouts, brushfoots, and skippers. The species are described on one page with a full-color photo of a live specimen on the page facing the description. Textual descriptions cover the physical appearance (wingspan, color, top, and bottom), habitat (open meadow, fields, etc.), geographical range, and life cycle, including host plants. Appendices cover the anatomy of the butterfly, the butterfly family descriptions (in short paragraphs), a glossary, and an index to scientific and common names of the butterfly. A very easy-to-use guide to the most common butterflies, with striking, clear, color photos to identify them.

TEXTBOOKS

Arthropoda (General)

505. Manton, S. M. **The Arthropoda: Habits, Functional Morphology, and Evolution.** Oxford, England: Oxford University Press, 1977. 527p. $95.00. LC 77-5466. ISBN 0-19-857391-X.

This book is reviewed in the handbook section (see entry 325). It would also make a fine textbook for a course on the arthropods.

506. Snodgrass, R. E. **A Textbook of Arthropod Anatomy**. Ithaca, N.Y.: Comstock, 1952. 363p. LC 52-14887.

Although dated, this is still one of the best texts on arthropod anatomy. Each chapter is devoted to a particular class, the Crustacea for example. Anatomy for each class is described using representative animals from the group. For the Crustacea, a crayfish, a small shrimplike crustacean called *Anaspides*, and an isopod, *Ligyda exotica*, are described. The description covers general biological information followed by a prose description of the anatomy starting with the head and mouth parts, to the thorax and legs, to the abdomen. Each description is accompanied by detailed line-drawings labeled with abbreviations of the anatomical parts. The abbreviations are explained at the end of each chapter. The 13-page list of references is useful for finding sources for more detailed information.

Crustacea

507. Schram, Frederick R. **Crustacea**. New York: Oxford University Press, 1986. 620p. $65.00. LC 85-11526. ISBN 0-19-503742-1.

The author's aim in this book is to provide a basic text on crustacean evolutionary biology, excluding the physiology, histology, and ultrastructure. It is the most up-to-date text on the crustaceans. For a complete annotation, see the handbooks section of this chapter under entry 350.

Insecta

508. Borror, Donald J., Charles A. Triplehorn, and Norman F. Johnson. **An Introduction to the Study of Insects**. 6th ed. Philadelphia: Saunders College, 1989. 875p. $40.00. LC 88-043541. ISBN 0-03-025397-7.

This is a well-respected college-level text now in its sixth edition. Provides a very clear and well-organized text that combines the general biology of the insect groups with good keys to their identification. Indeed, upper-level entomology students may use the identification keys here for systematics work. The text begins with chapters on general information on insects, including those beneficial and injurious to man, as well as insect anatomy and physiology, behavior and ecology, and classification and identification. The main part of the book covers systematics and provides keys to the phylum Arthropoda and the other main groups besides the insects, the crustaceans and arachnids. The 26 orders of insects from Ephemeroptera (mayflies) to Hymenoptera (ants, wasps, and bees) are then described. Each chapter covers a particular order and includes a general description of the group, then classification information, physical features that are used in the identification of the order, then a key to identification (usually to family and subfamily), followed by a description of each family. Black-and-white drawings, diagrams, and photos are inserted into the text and keys for further use in identification. A good 19-page glossary explains terms used in the text and supplements the terms used in the index.

509. Chapman, R. F. **The Insects: Structure and Function**. 3d ed. Cambridge, Mass.: Harvard University Press, 1982. 919p. $55.00. LC 81-85964. ISBN 0-674-87535-4.

This is a more advanced text with an emphasis on morphology and physiology and their effects on insect behavior. This does not include insect classification or keys to their identification. It is divided into five sections covering the main parts of the body and their individual functions (e.g., the head, ingestion and utilization of the food, the thorax and movement, etc.). It concludes with a section on blood, hormones, and pheromones including the circulatory system, endocrine organs, and exocrine glands. References are included at the end of each chapter. A taxonomic index provides page numbers where specific species are mentioned and a detailed subject index is included as well. Black-and-white drawings and diagrams illustrate the text.

510. Davies, R. G. **Outlines of Entomology**. 7th ed. London, New York: Chapman and Hall, 1988. 408p. $75.00; $37.50pa. LC 88-10865. ISBN 0-412-26670-9; 0-412-26680-6pa.

This is a descendant of A. D. Imms' *Outlines of Entomology*, first published in 1942. Now in its seventh edition, it is a well-respected work. See entry 355 in the Handbooks section for a complete description.

511. Elzinga, Richard J. **Fundamentals of Entomology**. 3d ed. Englewood Cliffs, N.J.: Prentice-Hall, 1987. 456p. $52.60. LC 86-17055. ISBN 0-13-338203-6.

This introductory text provides basic information on many aspects of insect studies. It includes a detailed chapter on classification, with keys to identification of common orders and families, but the main part of the text is concerned with the overall picture of insects and their influence on the ecosystem. It begins with a chapter on the "Arthropod Plan," where insects are generally described along with the other classes of the arthropods, the crustaceans, arachnids, and millipedes and centipedes. Keys to common adults of each classes are presented here as well as keys to common arachnid orders. The chapters following describe the anatomy and morphology of the insect internally and externally, insect development, ecology, behavior, parasitism, and interactions with plants and humans. Each chapter ends with a list of questions the student can use to review the information in the chapter. Many black-and-white photos and drawings are found throughout the text. A detailed chapter on classification describes the major insect orders and provides keys to their identification down to common families of each order. A final chapter on making an insect collection provides practical information on where and how to collect and display insects in a collection. Includes a glossary, list of selected references, and an index to scientific and common name as well as subject.

512. Evans, Howard E. **Insect Biology: A Textbook of Entomology**. Reading, Mass.: Addison-Wesley, 1984. 436p. $52.75. LC 83-15481. ISBN 0-201-11981-1.

This textbook explains the biology of insects for the student who has completed a basic course in biology. It concentrates on the biology of insects rather than providing descriptions and identification keys to the insect orders as seen in the textbooks of Borror (see entry 508) and Richards (see entry 514). It is arranged into the following seven parts: "The Structure and Diversity of Insects," "Function and Development," "Behavior," "The Relationships of Plants and Insects," "The Relationships of Insects with Animals," "Insects and Their Environments," and "The Natural and Artificial Regulation of Insect Populations." Within these sections are two to four chapters. For example, the section on function and development includes a chapter on "Major Life Systems" with information on the cuticle, respiration, thermoregulation, digestion, excretion, and so on; chapters on "Development and Reproduction"; "Pheromones and Allomones"; and the "Nervous System." Each chapter ends with a summary and a list of selected readings. The book is illustrated with many black-and-white photos, drawings, and diagrams. An eight-page glossary of terms, and an index complete the volume.

513. Gillott, Cedric. **Entomology**. New York: Plenum Press, 1980. 729p. $79.50. LC 79-21675. ISBN 0-306-40366-8.

The aim of this textbook is to provide a basic undergraduate general entomology course with a well-rounded treatment of the subject. It is therefore not primarily taxonomic, but is divided into four main sections: "Evolution and Diversity," "Anatomy and Physiology," "Reproduction and Development," and "Ecology." The taxonomy and information on the various insect orders is included in the first section. Black-and-white drawings and line-drawings as well as diagrams and graphs illustrate the text. Each chapter ends with a section on the literature and includes a bibliography. The text concludes with separate author and subject indexes. Scientific and common names of insects are included in the subject index.

514. Richards, O. W., and R. G. Davies. **Imms' General Textbook of Entomology**. 10th ed. London: Chapman and Hall; New York: John Wiley, 1977. 2 vols. $29.95 vol. 1; $49.95 vol. 2. LC 76-47011. ISBN 0-412-15200-2; 0-412-15210-Xpa.

Now in two volumes, this is the 10th edition of a standard text first published in 1925. The major change to this edition is the removal of keys to families; otherwise it is in the same arrangement. It covers the anatomy, physiology, and embryology in volume one, and information on 29 orders of insects in volume two. Each chapter describes a particular order in 5 to 100 pages or more. External and internal anatomy are extensively covered as well as the life cycle stages, with black-and-white drawings to illustrate the text. Classification information down to family is given with illustrations. The chapter is concluded with a bibliography of literature on that order.

JOURNALS

Chelicerata

ARACHNIDA (Spiders, Scorpions, Mites, and Ticks)

515. **Bulletin of the British Arachnological Society**. Vols. 1- , nos. 1- . London: The Society, 1969- . 3 vols./yr. $20.00/yr. ISSN 0524-4994.

This periodical publishes papers and short notes on all aspects of arachnology except the group Acari (mites and ticks). Each issue of approximately 25 pages is in English and contains photos, line-drawings, and diagrams to illustrate the text. The emphasis seems to be on systematics, but also includes ecology and behavior.

516. **Journal of Arachnology.** Vols. 1- , nos. 1- . American Arachnological Society, 1973- . 3 vols./yr. $80.00/yr. ISSN 0161-8202.

This journal includes papers in English, French, Portuguese, and Spanish on all aspects of arachnology. It is published by the American Arachnological Society, formed in 1972 to foster cooperation between amateur and professional arachnologists. The Society also publishes a biannual newsletter, *American Arachnology*. Each issue of the journal is about 125 pages long and includes line-drawings, electron micrographs, graphs, and tables to illustrate the text.

Crustacea

517. **Crustaceana: International Journal of Crustacean Research**. Nos. 1- . Leiden, Netherlands: E. J. Brill, 1960- . Bimonthly. $79.50/yr. ISSN 0011-216X.

Covering all aspects of crustacean biology except biochemistry, this journal is the major international periodical on this subject. Papers are in English, Spanish, or German. Two volumes a year are published in three parts each, with each volume containing approximately 320 pages. A "Notes and News" section at the end of each issue includes short papers (one or two pages) usually on a systematics issue. Short book reviews and announcements are found at the end.

Uniramia

INSECTA

518. **Advances in Insect Physiology**. Vols. 1- . London: Academic Press, 1963- . Annual. $75.00-$200.00/vol. ISSN 0065-2806.

This serial was ranked number two in impact factor in the 1989 *Science Citation Index, Journal Citation Report* on entomology serials. Number one in ranking is also a review publication (see entry 521). This publishes four or five review articles in this annual volume on any aspect of insect physiology. Articles are long, from 50 to more than 100 pages, and the bibliographies are comprehensive. Each volume includes a subject index. Excellent for in-depth coverage of current research in insect physiology.

519. **American Entomologist**. Vols. 36- , nos. 1- . Lanham, Md.: Entomological Society of America, 1990- . Quarterly. $55.00/yr. ISSN 1046-2821.

Formerly titled *Bulletin of the Entomological Society of America* from 1955 to 1989, this journal contains articles of general interest in entomology for students, amateurs, and specialists. It also relates news to members of the Entomological Society of America. It contains glossy, full-color photos and illustrations, with these photos also forming the front and back covers. The journal is arranged into six areas: (i.e., columns [letters to the editor, book reviews], presidential address, forum [student information, usually], commentary [viewpoints on entomology subjects], features [general interest, nonresearch; the Summer 1992 issue had an article on Victorian scientific art], and research [reports on research]). Preference is given to publish member papers. A section on research briefs lists citations with abstracts to journal articles on entomology that the journal editors believe may be useful to their members. Book reviews are of two types, long and brief. The long take up two or more columns and provide opinions of the reviewers on the worth of the book; the brief review is usually one column in length and is mainly descriptive. An interesting, easy-reading journal.

20. **Annals of the Entomological Society of America**. Vols. 1- . Lanham, Md.: Entomo-
ɔgical Society of America, 1908- . Bimonthly. $150.00/yr. ISSN 0013-8746.

Papers accepted for publication in this journal are on the basic biology of arthropods. apers are printed in each issue under the following headings: systematics, ecology and opulation biology, arthropod biology, physiology, biochemistry and toxicology, and behav-ɔr and genetics. Book reviews (four to six of them) are included at the end of the journal and an be one column to more than one page in length. This well-established journal has been ublished since 1908.

21. **Annual Review of Entomology**. Vols. 1- . Palo Alto, Calif.: Annual Reviews, 1956- . nnual. $40.00/vol. ISSN 0066-4170.

This serial was ranked number one for impact factor in the 1989 *Science Citation Index ɔurnal Citation Report* for entomology journals. It publishes around 20 review articles on ll aspects of entomology in the single volume it publishes each year. It is used in many ways y students, specialists, and amateurs as well. Primarily, it provides review articles of current ·search that are both in-depth and indicate future studies. The extensive bibliographies rovide for further reading. It is also useful to find out the activities of many areas of ntomological studies in one volume. For many entomologists, this one volume helps to keep em up-to-date on current research without their having to keep up with the masses of formation published in entomology today. Additionally, older issues provide valuable istorical perspectives to current subjects. All this, and at a very reasonable price. Contains ıbject, author, and chapter-title indexes. The author and chapter-title indexes cumulate the revious five volumes.

22. **Aquatic Insects: International Journal of Freshwater Entomology**. Vols. 1- , nos. - . Lisse, Netherlands: Swets Publishing Service, 1979- . Quarterly. $125.00/yr. ISSN)65-0424.

This journal publishes research on taxonomy and ecology of aquatic insects. It does not ıblish faunistic studies or other papers with regional interest. Papers are illustrated with ack-and-white maps, drawings, or photos. Each issue is approximately 64 pages and ɔntains up to eight papers.

23. **Archives of Insect Biochemistry and Physiology**. Vols. 1- . New York: Wiley-Liss,)83- . Monthly. $573.00/yr. ISSN 0739-4462.

This serial was ranked number three in impact factor in entomology in 1989 *Science itation Index Journal Citation Report*. It is now published in collaboration with the ntomological Society of America. It publishes original research articles, announcements of eetings, and review articles on current topics in the field of insect biochemistry and ıysiology.

24. **Bulletin of Entomological Research**. Vols. 1, nos. 1- . Oxon, UK: C.A.B. Interna-ɔnal Institute of Entomology, 1910- . Quarterly. $231.00/yr. ISSN 0007-4853.

This journal focuses on economic entomology, especially insect control. There are ually 16 or so original articles, and occasionally a review article, published in each issue. lack-and-white illustrations in the form of maps, photos, and drawings are found throughout e journal. Annual indexes of authors, genera, and species are included with the last issue of e year.

25. **Bulletin of the Entomological Society of Canada**. Vols. 1- . Ottawa: Entomological ɔciety of Canada, 1969- . Quarterly (comes with subscription to *Canadian Entomologist*). SN 0071-0741.

This bulletin includes business meetings of the Society, announcements of meetings, ɔtes on members, announcements of grants and prizes, book notices and reviews (usually e page in length), job announcements, and notifications of scholarships and grants.

526. **Canadian Entomologist**. Vols. 1- . Ottawa: Entomological Society of Canada, 1868- . Bimonthly. $180.00/yr. ISSN 0008-347X.

This journal, in existence since 1868, is the primary entomological journal of Canada and publishes papers on all aspects of original research in entomology. It does not usually publish checklists or bibliographies, however. Abstracts are provided in English and French, but most papers are in English. Usually around nine papers are published in each issue, with a section on "Scientific Notes" that includes very brief one- or two-page papers on topics such as "Axle grease as an alternative adhesive for use on sticky traps" or "Portable, solar-powered charging system for blacklight traps" (both found in the May/June 1992 issue).

527. **Ecological Entomology**. Vols. 1- . Oxford, England: Published for the Royal Entomological Society by Blackwell Scientific Publications, 1976- . Quarterly. $195.00/yr. ISSN 0307-6946.

This journal focuses on original research in insect ecology. The papers are on the natural history or field biology of aquatic and terrestrial insects, insect/host interactions, migration, dispersal, adaptation, population biology, rhythmic behavior, and ecological apparatus or methods. There are usually 10 or 12 papers in each issue, with each paper usually no more than 12 pages in length. Supersedes *Transactions of the Royal Entomological Society of London*, which was published from 1933-1975.

528. **Entomologia Experimentalis et Applicata**. Vols. 1- , nos. 1- . Dordrecht, The Netherlands: Published for the Nederlandse Entomologische Vereniging by Kluwer Academic, 1958- . Monthly. $139.50/yr. ISSN 0013-8703.

This journal includes papers on experimental biology and ecology of insects and other land arthropods. It leans toward agricultural aspects of entomology, including control and management of insects as pests. Usually 10 or so papers are published in each issue, but some issues are devoted to symposia proceedings. These papers can be in English, French, or German, but are primarily in English.

529. **Entomological News**. Vols. 36- . Philadelphia: American Entomological Society, 1925- . Bimonthly. $18.00/yr. ISSN 0013-872X.

This journal publishes news of the Society and short papers on all aspects of entomological studies. It contains usually no more than 100 pages, with usually six or so papers plus notices of meetings, awards, and book reviews. Longer papers on systematics or taxonomy are published in the Society's other publication, *Transactions of the American Entomological Society* (see entry 548).

530. **The Entomologist**. Vols. 10, nos. 164- . London: Royal Entomological Society, 1988- . Quarterly. $45.50/yr. ISSN 0013-8878.

This journal began in 1877 with various publishers, absorbed a few other British entomology journals, and was suspended from publication from 1973 to 1988, when the Royal Entomological Society began as its publisher. It is aimed at the amateur as well as the specialist and includes papers that are in the area of natural history; more sophisticated papers are not found here. Usually there are around five papers along with news on conferences, festivals, Entomological Club grants, and other news for members of the Royal Entomological Society (who may subscribe to this journal at a discount).

531. **Environmental Entomology**. Vols. 1- . Lanham, Md.: Entomological Society of America, 1972- . Bimonthly. $150.00/yr. (institutions). ISSN 0046-225X.

Papers published in this journal concern primarily the interactions of insects with their environment, especially the biological, chemical, and physical aspects. The 30 or so papers published in each issue of 200 to 250 pages are grouped under main headings such as forum, pest management and sampling, community and ecosystem ecology, population ecology, physiology and chemical ecology, and biological control. A section on book reviews includes reviews of one page or two on recently published books.

532. **Insect Biochemistry and Molecular Biology.** Vols. 1- , nos. 1- . Oxford, England: Pergamon Press, 1971- . Bimonthly. $700.00/yr. ISSN 0965-1748.

This journal was ranked number five in impact factor in the 1989 *Science Citation Index Journal Citation Report* on entomology serials. Originally titled *Insect Biochemistry*, it now reflects a more molecular approach. Usually 10 to 14 papers on original research in this area are published in each issue. Black-and-white plates as well as occasional color plates may be found. International journal, but papers all in English.

533. **Insectes Sociaux: International Journal for the Study of Social Arthropods**. Vols. 1- , nos. 1- . Basel, Switzerland: Birkhauser Verlag, 1954- . Quarterly. $184.00/yr. ISSN 0020-1812.

This is the official journal of the International Union for the Study of Social Insects. It publishes papers on all aspects of the biology and evolution of social insects and other arthropods. Around 8 to 10 research papers are published in each issue with some short papers called "communications," of two to three pages each, included at the end. Papers are usually in English, though occasionally in German or French.

534. **International Journal of Insect Morphology & Embryology**. Vols. 1- , nos. 1- . Oxford, New York: Pergamon Press, 1971- . Quarterly. $415.00/yr. ISSN 0020-7322.

This journal publishes papers on all aspects of morphology and embryology of insects including gross morphology to fine ultrastructure morphology. Usually 7 to 12 papers are published in each issue. Electron photomicrographs are plentiful in this publication. Papers are primarily in English, although occasional papers in French and German are accepted. Occasional biographies of famous entomologists with bibliographies of their works are also included; for example, one on Professor Sir Vincent B. Wigglesworth was printed in the April 1992 issue on the occasion of his receiving the 1992 Distinguished International Award in Morphology and Embryology.

535. **Journal of Economic Entomology**. Vols. 1- , nos. 1- . Lanham, Md.: Entomological Society of America, 1908- . Bimonthly. $190.00/yr. ISSN 0022-0493.

Focusing on all aspects of economic entomology, this journal publishes around 60 papers in each issue. The papers are arranged under the following topics: insecticide resistance and resistance management, physiology, toxicology and biochemistry, apiculture and social nsects, biological and microbial control, ecology and behavior, sampling, household and structural insects, veterinary entomology, extension, forest entomology, horticultural entomology, row crops and forage, stored-product entomology, and plant resistance. Book eviews are included at the end of the journal.

536. **Journal of Insect Behavior**. Vols. 1- , nos. 1- . New York: Plenum, 1988- . Bimonthly. $185.00/yr. ISSN 0892-7553.

This journal publishes papers on all behavioral aspects of insects and other terrestrial rthropods (spiders). Each issue contains 8 to 10 research papers and two to three short papers n a section for "Short Communications." Book reviews and software reviews are also ncluded.

37. **Journal of Insect Physiology**. Vols. 1- . Oxford, New York: Pergamon Press, 957- . Monthly. $980.00/yr. ISSN 0022-1910.

Although the title suggests that this journal treats insect physiology, it may also include ie physiology of other arthropods as well. It includes such topics as endocrinology, pharmaology, nutrition, excretion, reproduction, and behavior. Usually 8 to 10 papers are published each issue. Although this is an international journal, all papers are published in English.

538. **Journal of Medical Entomology**. Vols. 1- , nos. 1- . Lanham, Md.: Entomologica Society of America, 1964- . Bimonthly. $150.00/yr. ISSN 0022-2585.

This journal publishes research papers on medical, public-health, and veterinary aspect of insect and tick studies, including articles on biological control. It publishes 18 or s research papers per issue and usually seven brief articles under the heading "Short Commu nications." Contributors are primarily from medical research labs, but can be from an environmental, veterinary, immunological, or entomological lab as well.

539. **Journal of the Australian Entomological Society**. Vols. 1- . Indooroopilly, Queens land, Australia: Australian Entomological Society, 1962- . Quarterly. $80.00/yr. ISSN 0004 9050.

This journal contains papers dealing with all areas of entomological research wit emphasis on the Australian fauna. Previously titled *Journal of the Entomological Society c Queensland*; the present title began in 1967. The papers are illustrated with black-and-whi figures, maps, drawings, or photos. Each issue contains around 20 papers ranging from on page to several pages in length. The fourth issue of the year contains a taxonomic index, general index, and an author index.

540. **Journal of the Kansas Entomological Society**. Vols. 1- . Lawrence, Kans.: Kansa Entomological Society, 1928- . Quarterly. $50.00/yr. ISSN 0022-8567.

This journal publishes research in all areas of entomology. It includes not only 8 to 1 research papers, but also "Short Communications" of less than three pages on narrow aspect of entomology or for rapid communication of a finding. Book reviews, notices of meeting and comments are also found here.

541. **Journal of the New York Entomological Society**. Vols. 1- . Lawrence, Kans Published for the Society by Allen Press, 1893- . 4/year. $45.00/yr. ISSN 0028-7199.

This journal publishes research papers on classification, taxonomy, phylogeny, beha ior, natural history, biogeography, and other areas of entomological studies. Usually, 10 c so research papers and one or two notes or comments are published in each issue. Black-anc white drawings are frequently found in the illustrations, as well as occasional color plates.

542. **Memoirs of the American Entomological Institute**. Nos. 1- . Gainesville, Fla American Entomological Institute, 1961- . Published irregularly. ISSN 0065-8162.

As of 1992, 49 issues of this monographic series had been published. Each hardboun issue covers a separate topic on insect systematics. Some selected titles are: *Annotate Checklist of the Weevils* (Curculionidae sensu lato) *of North America, Central America, an the West Indies (Coleoptera: Curculionøidea)* (see entry 615); *Indo-Australian Ichneumon dae, Water Mites from Australia*; *A Synopsis of the Chrysididae in America North of Mexic* and *A Catalog of the World Bethylidae*. Each monograph ranges in size from 100 to ove 1,000 pages in length.

543. **Memoirs of the American Entomological Society**. Nos. 1- . Philadelphia: Publishe by the American Entomological Society at the Academy of Natural Sciences, 1916- Published irregularly. $7.00-$25.00/vol. ISSN 0065-8170.

This is a series of monographs primarily on insect systematics. For example, a 199 issue, *A Synopsis of the Sawflies (Hymenoptera: Symphyta) of America South of the Unite States: Argidae*, by David R. Smith, provides detailed descriptions and keys to species sawflies of this geographical area. This series has been published since 1916, with work published approximately once a year.

544. **Pan-Pacific Entomologist**. Vols. 1- , nos. 1- . San Francisco: Pacific Coast Entom logical Society, c/o California Academy of Sciences, 1924- . Quarterly. $30.00/yr. ISS 0031-0603.

Papers published here center on the natural history, taxonomy, and observation of insects primarily in the western coastal United States. Usually eight or so papers are included with a few short "notes," especially on new species introductions into this area.

545. **Physiological Entomology**. Vol. 1, nos. 1- . Oxford, England: Published for the Royal Entomological Society by Blackwell Scientific Publications, 1976- . Quarterly. $195.00/yr. ISSN 0307-6962.

This journal publishes papers that deal primarily with the physiological basis of behavior in insects and other arthropods, but it also includes general physiology as well. Neurobiology and sensory physiology, endocrinology, circadian rhythms, and photoperiodism are other topics covered here. Usually 10 to 14 papers are published in each issue.

546. **Proceedings of the Entomological Society of Washington**. Vols. 1- . Washington, D.C.: Entomological Society of Washington, c/o Dept. of Entomology, Smithsonian Institution, 1884- . Quarterly. $50.00 (nonmember). ISSN 0013-8797.

This journal publishes primarily taxonomic works in entomology, but includes research papers from all other areas of entomology as well. Usually around 10 papers per issue are printed, with lengthy book reviews of two or three books following the papers. Obituaries of well-known entomologists are also included. The book reviews may be up to three pages in length, and can provide a lot of good information. Another well-established journal, published continuously since 1884.

547. **Psyche: A Journal of Entomology**. Vols. 1- . Cambridge, Mass.: Cambridge Entomological Club, 1874- . Quarterly. $30.00/yr. ISSN 0033-2615.

This journal publishes articles on all aspects of insect and other arthropod studies, especially those on taxonomy. The 10 or so articles published each quarter may also include checklists. Another publication with a long history; this one is associated with Harvard University.

548. **Transactions of the American Entomological Society**. Vols. 1- , nos. 1- . Philadelphia: American Entomological Society, 1867- . Quarterly. $17.00/yr. ISSN 0002-8320.

This journal publishes papers on all aspects of insect taxonomy and systematics. Usually, three or four long papers of 50 to more than 100 pages are printed in each quarterly issue. Papers include phylogenetic systematic studies, descriptions of species groups, cladistic analysis of a taxon, status of a particular group, and keys to identification of a group or species. Shorter papers and those involved with other areas of entomology besides taxonomy and systematics are published by this society in its other publication, *Entomological News* (see entry 529).

TAXONOMIC KEYS

Many identification keys may be found in handbooks for a particular group of Arthropoda. See the Handbooks section for these.

Arthropoda (General)

549. Brues, Charles T., A. L. Melander, and Frank M. Carpenter. **Classification of Insects: Keys to the Living and Extinct Families of Insects, and to the Living Families of Other Terrestrial Arthropods**. Cambridge, Mass.: Printed for the Museum of Comparative Zoology, Harvard College, 1954. 917p. (Bulletin of the Museum of Comparative Zoology at Harvard College, vol. 108). LC a46-353.

Although the title suggest this is a classification for insects, this comprehensive work attempts to provide keys to the living and extinct insect families of the world as well as keys to the families of other terrestrial Arthropoda. It is divided into four parts. Part I covers the keys to recent Insecta families; part II has keys to families of other terrestrial Arthropoda such as the spiders, scorpions, isopods, and ticks; part III has keys to the extinct families of Insecta; and part IV includes the glossary of terms used in the work, and an index. For those wishing to further identify an insect beyond the family level provided here, a list of references to the literature for that group is provided. Keys for the identification of the immature stages, in many cases, are also included. Black-and-white drawings to illustrate the keys are found throughout this large volume.

550. Kevan, D., Keith McEwan, and G. G. E. Scudder. **Illustrated Keys to the Families of Terrestrial Arthropods of Canada**. Ottawa: Biological Survey of Canada (Terrestrial Arthropods), 1989- . $16.00 vol. 1. LC cn90-71102. ISBN 0-9692727-1-5 vol. 1.

The Biological Survey of Canada (Terrestrial Arthropods) decided there was a need for keys to the families of the terrestrial arthropods in Canada, and from this decision this publication was devised. Several separate keys to the various groups of arthropods will be written for this series. The first key is to the Myriapoda, or centipedes and millipeds, of Canada. There is a basic introduction to the Myriapoda followed by the general morphology of the group, pictured in detail with labeled line-drawings. The work is spiral-bound for easy use of the keys. There are keys to class, order, family of the Diplopoda, Chilopoda, Symphyla, and Pauropoda. A glossary of terms used in the key and an appendix of the "Classification of Canadian and Northern United States' Families of Myriapods, with Their Genera," follow the key. A list of references and an index complete this useful work.

551. **Pictorial Keys to Arthropods, Reptiles, Birds and Mammals of Public Health Significance**. Atlanta, Ga.: National Communicable Disease Center, 1967. For sale by the U. S. Gov. Print. Off., Superintendent of Documents. 192p. 74-603677.

Although the title indicates this work covers more than the arthropods, 174 out of 186 pages devoted to keys are on the arthropods. This was developed to aid public-health officials in the identification of animals that may pose a public-health problem. The keys are simple and easy to follow. Black-and-white drawings and brief descriptions fill each page, giving the reader a choice of a characteristic ("wings well developed" or "wings absent") until the organism is identified. Identification is also aided by grouping organisms under headings such as "Household and Stored-Food Pests: Key to Common Adults, or Ectoparasites of the Dog." The other keys are to main group such as ticks or scorpions. A useful work and reasonably easy to use.

552. Thorp, James H., and Alan P. Covich, ed. **Ecology and Classification of North American Freshwater Invertebrates**. San Diego, Calif.: Academic Press, 1991. 911p. $59.95. LC 90-46694.

This work is reviewed in the handbooks section of chapter 2, "The Invertebrates" (see entry 170). It is included here for its up-to-date keys to identification. It has chapters on water mites, aquatic insects, a general chapter on the Crustacea, and chapters on the Brachiopoda, Copepoda, and Decapoda. The keys provide identification down to genus, and to family for the insects. It has a medium level of difficulty, and so is useful for college students and the specialist.

Chelicerata

ARACHNIDA (Spiders, Scorpions, Mites, and Ticks)

553. Balogh, J., and S. Mahunka, eds. **The Soil Mites of the World**. Amsterdam, New York: Elsevier, 1983. LC 90-12027. Approximately $150.00/vol.

This series is up to three volumes as of this writing. Volume one is *Primitive Oribatids of the Palaearctic Region*, and volumes two and three are *Oribatid Mites of the Neotropical Region I and II*. These works are almost entirely detailed identification keys to species level. The keys also provide a type locality for each species. Plates inserted at the end of the text illustrate the anatomical details used for identification in the keys. Lengthy bibliographies and an index to taxa complete each volume.

554. Kaston, B. J. **How to Know the Spiders**. 3d ed. Dubuque, Iowa: Wm. C. Brown, 1978. 272p. (Pictured Key Nature Series). $10.00. LC 77-82892. ISBN 0-697-04899-3; 0-697-04898-5pa.

More than 600 black-and-white drawings illustrate the key to 223 genera and 519 species of the more common spiders in the United States. Part of the Pictured Key Nature Series. (See entry 566.) Supplemented by Vincent Roth's *Spider Genera of North America* (see entry 557).

555. Legg, Gerald. **Pseudoscorpions (Arthropoda; Arachnida), Keys and Notes for the Identification of the Species**. New York: Published for the Linnean Society of London and the Estuarine and Brackish-Water Sciences Association by E.J. Brill, 1988. 159p. (Synopses of the British Fauna [new series], no. 40). LC 88-4312. ISBN 90-04-08770-2.

Although this is primarily a key to the British species of psueodscorpions, it includes 44 pages of information on general structure, reproductive biology, ecology, distribution, collection, preservation, and classification of pseudoscorpions. This work lists 26 species and describes them in detail as to anatomy, color, and distribution. Each of the 26 species is illustrated in a black-and-white drawings that cover almost an entire page. Additional drawings of specific identifying anatomical details are also included as well as distribution maps showing where that species is located in Britain. A nice glossary, a 10-page bibliography, and a taxonomic index complete this work.

556. McDaniel, Burruss. **How to Know the Mites and Ticks**. Dubuque, Iowa: Wm. C. Brown, 1979. 335p. (Pictured Key Nature Series). $12.00. LC 78-55768. ISBN 0-679-04756-3; 0-697-04757-1pa.

Six hundred eighty-two pen-and-ink drawings illustrate this key to the common mites and ticks of the United States. Part of the Pictured Key Nature Series. (See entry 566.)

557. Roth, Vincent D. **Spider Genera of North America with Keys to Families and Genera and a Guide to Literature**. N.p., American Arachnological Society, 1985. Available from Jon Reiskind, Dept. of Zool., Univ. of Florida, Gainesville, FL 32611.

This is a title change and second edition of *Handbook for Spider Identification* published by the author in 1982. This new title better describes the geographic range of the book. The book provides keys to identify 54 families of spiders to genus with a list of references to consult to identify to species level. The spider families are arranged alphabetically under the two large infraorders Mygalomorphae and Araneomorphae. The keys are illustrated with black-and-white drawings of specific anatomical characteristics. There is only one drawing of a whole species. A glossary of terms not found in Kaston's *How to Know the Spiders* (see entry 554) is included. There is, however, no index. A thorough work for the specialist.

Crustacea

558. Athersuch, J., D. J. Horne, and J. E. Whittaker. **Marine and Brackish Water Ostracods (Superfamilies Cypridacea and Cytheracea): Keys and Notes for the Identification of the Species.** New York: Pub. for the Linnean Society of London and the Estuarine and Brackish-Water Sciences Association by E.J. Brill, 1989. 343p. $94.25. (Synopses of the British Fauna [new series], no. 43). LC 89-22206. ISBN 90-04-09079-7.

This is part of a series on British fauna; however, the species covered here are also found in other marine and brackish water areas. Ostracods are small crustaceans (0.3-30 mm long) that look like small clams; most live in aquatic environments. This book is not just a key to the identification of ostracod species, but it also includes a significant amount of information on the general structure of ostracods, reproduction and life cycle, ecology, predators, and collecting methods. The dichotomous keys to the genera and to species are illustrated with black-and-white line-drawings. Each genus and species are described, with accompanying black-and-white drawings. The description includes a paragraph on the distinguishing morphology, type species, literature cited to when species is first named, type locality, and geographical distribution. Eight plates of electron micrographs of ostracods are useful in their identification. A glossary of terms, a long list of references, and an index to scientific names down to species complete this useful guide.

559. Burukovskii, R. N. **Key to Shrimps and Lobsters**. New Delhi: Oxonian Press, 1982. 174p. $21.00. LC 83-902711. ISBN 0-8364-2557-X.

This is a translation of a Russian work published in 1974 titled *Opredelitel' Krevetok, Langustov i Omarov*. It provides a key to the identification of the commercially valuable shrimps and lobsters found in the waters around Russia. The keys go down to family, genus, and species. It includes a chapter on principal morphological features complete with black-and-white line-drawings so that the reader may understand the anatomical terminology in the key. The key also includes detailed line-drawings to help in identification. A bibliography of sources used in creating this key is inserted before the index, which was reproduced from the Russian original. Page numbers from the original are inserted in the left-hand margin of this English text.

560. Fitzpatrick, Joseph F. **How to Know the Freshwater Crustacea**. Dubuque, Iowa: Wm. C. Brown, 1983. 227p. (Pictured Key Nature Series). $12.00pa. LC 82-83542. ISBN 0-697-04783-0pa.

Two hundred sixteen black-and-white drawings illustrate this key to the more common freshwater crustaceans. Part of the Pictured Key Nature Series. (See entry 566.)

561. Holthuis, L. B. **Marine Lobsters of the World: An Annotated and Illustrated Catalogue of Species of Interest to Fisheries Known to Date**. Rome: Food and Agriculture Organization of the United Nations, 1991. Distributed, Lanham, Md.: UNIPUB. 292p. (FAO Species Catalogue, vol. 13; FAO Fisheries Synopsis, no. 125, vol. 13). $55.00. ISBN 92-5-103027-8.

This is the 13th volume of the *FAO Species Catalogue*, a series that focuses on commercially important marine animals. This volume covers 149 species of lobsters in three infraorders, 10 families and 33 genera. These are the species that are commercially valuable (i.e., they are used for food, sold for bait or other products, or have potential value). After introductory material on lobster anatomy and a very good illustrated glossary of terms, the systematic catalogue begins. It is arranged phylogenetically to family, with detailed species entries under family. Species information includes an FAO-designated common name (in English, French, and Spanish), type locality, geographical distribution (with range map), habitat and biology, size, interest to fisheries, and citations to the literature for further reading. Excellent, detailed black-and-white drawings of each species with some additional anatomical details are also featured. Illustrated dichotomous keys to family and subfamily are found throughout the text. Following the catalogue is a section, with map, showing what species are found in the various major fishing areas of the world. A 13-page bibliography and an index of scientific and common names finish the volume.

562. Kensley, Brian, and Marilyn Schotte. **Guide to the Marine Isopod Crustaceans of the Caribbean**. Washington, D.C.: Smithsonian Institution Press, 1989. 308p. $35.00. LC 88-38647. ISBN 0-8747474-724-4.

This guide provides keys to suborders, families, genera, and species of marine isopods found in the Caribbean. The guide is introduced by a short introduction, followed by a glossary of terms used in the text. The guide is arranged with taxa in alphabetical order, not by phylogenetic relationships. For example, the suborder Anthuridea is discussed first, with the family Anthuridae and the genus *Amakusanthura* being described first in that group. For each group, a diagnosis and an indication of where it is found geographically in the Caribbean are included. Next to each taxa name is the name of the author and the date when first named. References to these are provided in a section of cited literature, which is 16 pages in length. The guide is filled with fine black-and-white line-drawings that help in the identification of taxa. Some black-and-white photos and electron micrographs are also included to illustrate the text. A chapter on zoogeography and a table listing isopod species occuring in depths greater than 200 meters (on both sides of the Ismus of Panama, in the Gulf of Mexico, and occurring at Bermuda) are found at the end of the guide. A good up-to-date guide, primarily for the student and specialist.

563. Schultz, George A. **How to Know the Marine Isopod Crustaceans**. Dubuque, Iowa: Wm. C. Brown, 1969. 359p. (Pictured Key Nature Series). LC 76-89536. ISBN 697-04864-9; 697-04865-9pa.

This work provides keys to all the 174 genera and most of the 444 species found in the North American waters. The 572 black-and-white drawings illustrate the keys and glossary. Part of the Pictured Key Nature Series. (See entry 566.)

564. Williams, Austin B. **Shrimps, Lobsters, and Crabs of the Eastern United States, Maine to Florida**. Washington, D.C.: Smithsonian Institution Press, 1984. 550p. $47.50. LC 83-6000095. ISBN 0-87474-960-3.

This is a thorough treatment, covering the history, classification, zoogeography, and species accounts in addition to the main text on the systematics of this important group. An illustrated four-page glossary precedes the systematics section. Dichotomous keys down to species are followed by full descriptions of each species. See the handbook section entry 354 for a full annotation.

Uniramia

INSECTA (General)

565. Arnett, Ross H. **American Insects: A Handbook of the Insects of America North of Mexico**. New York: Van Nostrand Reinhold, 1985. 850p. $105.00. LC 84-15320. ISBN 0-442-20866-9.

For a full description of this work, see the handbook entry 371. Included in this remarkable handbook are keys to identify 87,000 species of insects native to America north of Mexico.

566. Bland, Roger G., and H. W. Jacques. **How to Know the Insects**. 3d ed. Dubuque, Iowa: Wm. C. Brown, 1978. 409p. (Pictured Key Nature Series). $12.00pa. LC 77-88344. ISBN 0-697-04783-0pa.

This is one of several excellent spiral-bound books in the Pictured Key Nature Series published by the William C. Brown Company. Individual titles in this series are listed in this bibliography under the major arthropod groups. They are relatively easy-to-use dichotomous (giving two choices) keys to the orders and common families of a taxonomic group in the United States. Each book is introduced by several good chapters on how to collect, preserve, organize and maintain a collection; observe and rear; and where to look for, for instance, insects. It also includes a chapter in the structure and development of the group and a good bibliography for further reading and for more specific identification guides.

The key itself is easy to follow with two choices given (e.g., "wings present" or "wings not present"). A number is listed after either choice that leads the reader on to the next set of choices to further identify the insect until the reader is lead to the family group it belongs to. At the family level, several common species are described. Black-and-white line-drawings of different species and specific anatomical features are included in the keys. The index is als a glossary that explains the terminology used in the keys. Indexes to other keys in this Picture Key Nature Series also have black-and-white line-drawings to further illustrate the glossar words. These *How to Know the ...* books are great for the student and beginning amateur t identify the more common species.

567. Chu, H. F., and Laurence K. Kutkomp. **How to Know the Immature Insects.** 2d ed Dubuque, Iowa: Wm. C. Brown, 1992. 346p. (Pictured Key Nature Series). $12.00. LC 91-73292. ISBN 0-697-05596-5.

This is a key to the various stages an insect goes through before it reaches its adult stage These include the egg, nymph, larvae, and pupae. Generally, insects spend more time in thei immature stages than in the adult, so this guide to these stages is an important part in insec study. Part of the Pictured Key Nature Series. (See entry 566.)

568. **Handbooks for the Identification of British Insects**. London: Royal Entomologica Society of London, Published by the British Museum (Natural History), 1949- . Multivolum set. Published irregularly. $6.00-$30.00/vol. LC 85-2109.

These are primarily detailed, illustrated identification keys to insects of Great Britain Each handbook covers a particular family or genus in great detail. For example, the handbool on orthocerous weevils is 108 pages long and includes 10 pages of information on the biology life history, economic importance, distribution, and methods of collecting, preserving, an examining these weevils. The rest of the text, after a checklist of species, is devoted to key to family, genus, and species. Twenty pages of plates of black-and-white drawings to illustrat the keys are found at the end of the text. A lengthy list of references for further reading i also included. Many volumes have been published so far, and they continue to be publishe irregularly.

569. Lehmkuhl, Dennis M. **How to Know the Aquatic Insects.** Dubuque, Iowa: Wm. C Brown, 1979. 168p. (Pictured Key Nature Series). $12.00pa. LC 78-55761. ISBN 0-697 04766-0; 0-697-04767-9pa.

Includes 600 black-and-white drawings to illustrate the descriptions in this identificatio key. Part of the Pictured Key Nature Series. (See entry 566.)

INSECTA (Systematic Section)

Blattaria (Cockroaches)

570. Rehn, John W. H. **Classification of the Blattaria as Indicated by Their Wing (Orthoptera)**. Philadelphia: American Entomological Society, 1951. 134p. (Memoirs of th American Entomological Society, no. 14).

This is a study of the wings to establish a modified classification of cockroaches. Key are formulated to families and subfamilies. Each family is described in two to several pages Thirteen plates of black-and-white drawings of the wings illustrate the various wing detail used to identify a species. An index to the subfamilies, tribes, genera, and species complete the volume.

Orthoptera (Grasshoppers, Crickets, Katydids).

571. Dirsh, V. M. **Classification of the Acridomorphoid Insects**. Faringdon, Oxon: E. W Clasey, 1975. 171p. $16.75. LC 76-357742. ISBN 900848-82-0.

This classification scheme is arranged taxonomically and provides keys to superfamilie and families. There are no keys to genus, but a type genus or sometimes all genera are liste under the family. The author describes six superfamilies and 20 families. Each family i described by diagnosis, including the overall appearance, phallic complex, spermatheca karyotype, and type genus. Geographic distribution is listed and a black-and-white line-drawing of a representative species is provided. A glossary of terms, a four-page bibliography, and an index to Latin name complete the text. A more advanced and detailed work than Jacques Helfer's work (see entry 572), but not as up-to-date.

572. Helfer, Jacques R. **How to Know the Grasshoppers, Crickets, Cockroaches, and Their Allies**. New York: Dover, 1987. 363p. $10.95. LC 86-29322. ISBN 0-486-25395-3.

Seven hundred sixty U.S. varieties are identified and, in most cases, illustrated in black-and-white drawings. The text is also illustrated with amusing drawings of crickets, grasshoppers, and so on sprinkled throughout the text. This is a corrected and slightly enlarged edition of the 1972 edition published by the Wm. C. Brown Company as part of its Pictured Key Nature Series. (See entry 566.)

Plecoptera (Stoneflies)

See the Handbooks section under Plecoptera, and pages 182-230 of *An Introduction to the Aquatic Insects of North America* (see entry 373).

Anoplura (Sucking Lice)

573. Kim, Ke Chung, Harry D. Pratt, and Chester J. Stojanovich. **The Sucking Lice of North America: An Illustrated Manual for Identification**. University Park, Pa.: State University Press, 1986. 256p. $39.50. LC 84-43060. ISBN 0-271-00395-2.

See entry 404 in the handbook section. Includes illustrated keys to the families, genera, and species, and descriptions of 76 species.

Thysanoptera (Thrips)

574. Stannard, Lewis J. **The Phylogeny and Classification of the North American Genera of the Suborder Tubulifera (Thysanoptera).** Urbana, Ill.: University of Illinois Press, 1957. 200p. (Illinois Biological Monographs, no. 25).

This descriptive work on thrips provides information on morphology, a key to the adults, and description, of genera found in North America. Each genus is described as to head, mouthparts, thorax, abdomen, and general information on the group. Keys to species (when necessary) are included with descriptions of each genera. Black-and-white drawings are included in 13 plates at the end of the text to illustrate the keys. A section on phylogeny and a 14-page list of references as well as an index to scientific name contribute to the usefulness of this work.

Hemiptera, Homoptera (True Bugs)

575. Slater, James A., and Richard M. Baranowski. **How to Know the True Bugs (Hemiptera—Heteroptera)**. Dubuque, Iowa: Wm. C. Brown, 1978. 256p. (Pictured Key Nature Series). $12.00. LC 76-24534. ISBN 0-697-04893-4; 0-697-04894-2pa.

There are 496 black-and-white drawings to illustrate this key to the most common and widespread species of true bugs. These numerous illustrations and the easy-to-use keys to the families of nymphs and adults and the more common genera in each family make this a good source for identifying the true bugs in North America. Part of the Pictured Key Nature Series. (See entry 566.)

Coleoptera (Beetles)

576. Arnett, Ross H., Jr. **The Beetles of the United States (A Manual for Identification).** Washington, D.C.: Catholic University of America Press, 1960-62. 110 fascicles.

This comprehensive work on the identification of adult beetles in the continental United States to family and genus was published in 110 fascicles, each fascicle being devoted to a particular family. The scientific name and the common name of the family are given, followed by a complete description of the features used to identify family. A paragraph on the family's ecology or where they are found and what kind of habitat they prefer is included. This is followed by a paragraph on the status of the classification for this group and another on its distribution in the world. A key to the genera comes next followed by a list of references used in that fascicle. A black-and-white drawing (6-8 cm) of a representative of the family is placed at the beginning. Additional drawings are included in some keys as an aid in identifying a particular anatomical feature mentioned in the key. Each fascicle is 2 to 40 pages in length and has its own index. A cumulated index at the end of the final fascicle covers the entire work.

577. Arnett, Ross H., Jr., N. M. Downie, and H. E. Jaques. **How to Know the Beetles**. 2d ed. Dubuque, Iowa: Wm. C. Brown, 1980. 416p. (Pictured Key Nature Series). $12.00. LC 79-55045. ISBN 0-697-04776-8.

Several hundred of the most common beetles are included in this pictured identification key. It includes 922 black-and-white illustrations to aid in the identification. Part of the Pictured Key Nature Series. (See entry 566.)

578. Boving, Adam G., and F. C. Craighead. **An Illustrated Synopsis of the Principal Larval Forms of the Order Coleoptera**. New York: New York Entomological Society, American Museum of Natural History, 1930. 351p. (Entomologica Americana, vol. 11, no. 1).

The purpose of this text is to provide basic keys and descriptions of the larvae of families and subfamilies of beetles. Detailed black-and-white drawings on 125 plates take up half of this text. The other half consists of the keys. No bibliography is provided, but references are included as footnotes. The index to name also lists the plate where an illustration may be found. More up-to-date keys to each family may be found either in the journal literature or in texts such as Paul O. Ritcher's *White Grubs and Their Allies: A Study of North American Scarabaeoid Larvae* (Oregon State University Press, 1966).

579. Crowson, R. A. **The Natural Classification of the Families of Coleoptera**. 1st ed., reprint with 6-page addenda and corrigenda. Middlesex, England: E. W. Classey, 1967. 187p. LC 75-405296.

Reprinted from a series of articles in the *Entomologist Monthly Magazine* from 1950-1954, this system of classification is devised as a natural or phylogenetic one, or one that is ordered by evolutionary descent. Keys to the families are illustrated with black-and-white line-drawings. Each family is described along with their phylogeny. An index to genera and higher taxa is provided. Because a significant knowledge of the Coleoptera is necessary to use this work, it is mainly for the use of the professional entomologist.

580. Trautner, Jurgen, and Katrin Geigenmuller. **Tiger Beetles, Ground Beetles: Illustrated Key to the Cicindelidae and Carabidae of Europe.** 1988. 488p. $75.00; $55.00pa. ISBN 3-924333-05-X; 3-924333-04-1pa.

In German and English, this work provides keys to family, subfamily, genus, and species in Europe. The keys are illustrated with line-drawings of distinguishing anatomical parts for identification. Each species has a line-drawing of the complete animal and a short paragraph describing it. Range maps are included for some taxons.

Siphonaptera (Fleas)

581. Benton, Allen H. **An Illustrated Key to the Fleas of the Eastern United States.** Fredonia, N.Y.: Marginal Media, 1983. 34p. (Bioguide, no. 3). $2.00. ISBN 0-942788-09-5.

This simple key to species found in this region was prepared for use by beginners and specialists alike. It is well illustrated with black-and-white drawings showing the detail referred to in the key. It is strictly a key and provides no other information. Easy enough for almost anyone to use.

582. Hopkins, G. H. E., and Miriam Rothschild. **An Illustrated Catalogue of the Rothschild Collection of Fleas (Siphonaptera) in the British Museum (Natural History).** Oxford, England: Oxford University Press, 1966-{1978}. 7vols. $135.00/vol. LC 54-3740.

See entry 428 in the Handbooks section. This set provides many detailed, illustrated keys to genera.

Diptera (Moquitoes and Flies)

583. Darsie, Richard F., Jr., and Ronald A. Ward. **Identification and Geographical Distribution of the Mosquitoes of North America, North of Mexico**. Fresno, Calif.: American Mosquito Control Association, 1981. 219p. (Mosquito Systematics. Supplement 1). LC 81-50441. ISBN 0-9606210-1-6.

This work contains keys to the identification of the adult females and fourth-stage larvae of 167 mosquito species and subspecies in North America, north of Mexico. The keys are heavily illustrated with 983 black-and-white drawings to make identification easier. Full-page range maps of North America follow a chapter on the geographical distribution of the mosquitoes, which includes tables listing each species and in what states it may be found. More than 500 references from 1955 to 1979 form a useful bibliography to mosquito systematics literature. A very complete and well-illustrated key to mosquitoes of this region.

Lepidoptera (Butterflies and Moths)

584. Ehrlich, Paul R., and Anne H. Ehrlich. **How to Know the Butterflies**. 1st ed. Dubuque, Iowa: Wm. C. Brown, 1961. 262p. (Pictured Key Nature Series). LC 61-2394.

This work provides keys to all butterfly species in North America north of Mexico. Species accounts are listed under each genus with information on distribution, habits, and larval food. Unfortunately, only black-and-white drawings of the individual species are provided, but they are in such detail as to help in the identification provided by the written keys. Part of the Pictured Key Nature Series. (See entry 566.)

Hymenoptera (Ants, Bees, and Wasps)

585. Richards, O. W. **Hymenoptera: Introduction and Key to Families**. 2d ed. London: Royal Entomological Society of London, 1977. 100p. (Handbooks for the Identification of British Insects, vol. VI, pt. 1). LC 78-305802.

The last half of this work deals with the identification and description of the British Hymentopera, but because many are the same as found in North America it may be useful for this area as well. See entry 475 for a complete annotation. Keys are provided to family.

Chilopoda and Diplopoda (Centipedes and Millipedes)

586. Blower, J. Gordon. **Millipedes: Keys and Notes for the Identification of the Species**. London: Pub. for the Linnean Society of London and The Estuarine and Brackish-Water Sciences Association, by E. J. Brill, 1985. 242p. (Synopses of the British Fauna, no. 35) ISBN 90-04-07698-0.

Another volume in the series to identify the fauna of the British Isles, this is actually a second edition of *British Millipedes* (Linnean Society, 1958) by the same author. The book covers only 52 species of an estimated 10,000 species in the world, but it is a good introduction to this diverse group. It covers those millipedes living in the British Isles. Introductory material covers the anatomy, reproduction, life history, life cycle, and the unusual movements of millipedes. Feeding, economic importance, chemical defence, predators, and habitats of millipedes is also covered as well as their collection and preservation and basic classification information. The systematic section includes a key to six orders of the British Diplopoda and then keys to individual species within the order. Each species is described with its distribution in the British Isles and Europe covered in detail. Detailed line-drawings illustrate the keys and the descriptions of species. A lengthy glossary, a 10-page bibliography, and a taxonomic index to genera and species finish this very complete volume on British millipedes.

587. Kevan, D., Keith McEwan, and G. G. E. Scudder. **Illustrated Keys to the Families of Terrestrial Arthropods of Canada**. Ottawa: Biological Survey of Canada (Terrestrial Arthropods), 1989- . $16.00 vol. 1. LC cn90-71102. ISBN 0-9692727-1-5 vol. 1.

Volume one of this set deals with the Myriapoda. (See entry 550.)

CHECKLISTS AND CLASSIFICATION SCHEMES

Arthropoda (General)

588. Pittaway, A. R. **Arthropods of Medical and Veterinary Importance: A Checklist of Preferred Names and Allied Terms**. Wallingford, Oxon, U.K.: C.A.B. International, 1991. 178p. $28.50. LC 91-93628. ISBN 0-85198-741-9.

This checklist provides a list of the scientific names and taxonomic positions of some of the most important species and genera of arthropods in medicine and veterinary studies. Also included in the list are those species that are used to control harmful arthropods, organisms that they carry, intermediate hosts, natural enemies, and all important synonyms. The names were gleaned from the *CAB Abstracts* database, which includes records from its 26 abstract journals. There are separate lists of names under the following headings: arthropods, microorganisms, viruses, fungi, helminths, fish, and other organisms. Under each heading the names are arranged in alphabetical order on the left-hand side of the page, followed by the name of the author of the species. To the right of the name is the relevant family, order, subclass, or class. Useful for keeping the usage of names consistent in the literature.

Chelicerata

ARACHNIDA (Spiders, Scorpions, Mites, and Ticks)

589. Bonnet, Pierre. **Bibliographia Araneorum**. Toulouse, France: Douladoure, 1945-1961. Reprint (of bibliography only), 3 vols. in 7. Lanham, Md.: Entomological Society of America, 1969. 832p. $40.00. LC 57-58745. ISBN 0-686-09299-6.

This is a catalog of the spider species described in the literature up to 1939. The text is in French and is in three parts. Part one is primarily an introduction and bibliography of works used in compiling the list. Part two is the catalog itself, arranged in alphabetical order by genera, comprising four volumes. Part three contains the index to the catalog with lists of genera and species in alphabetical order. A similar catalog, but arranged by family, is *Katalog der Araneae* by C. F. Roewer (see entry 592).

590. Brignoli, Paolo M. **A Catalogue of the Araneae Described Between 1940 and 1981.** Manchester: Manchester University Press, in association with the British Arachnological Society, 1983. 755p. $220.00. LC 83-7937. ISBN 0-7190-0856-5.

This is intended as a supplement to C. F. Roewer's *Katalog der Araneae* (see entry 592). It is a catalog to all the genera and species of spiders described in the literature after 1940 for most of the families, and after 1954 for the rest. It includes 96 families and about 7,000 species.

591. Platnick, Norman I. **Advances in Spider Taxonomy, 1981-1987.** New York: Manchester University Press, in association with the British Arachnological Society. Distributed by St. Martin's Press, 1989. 673p. $210.00. LC 88-39228. ISBN 0-7190-2782-9.

This is a supplement to Brignoli's *Catalogue of the Araneae Described Between 1940 and 1981* (see entry 590). It lists 105 families in the arrangement of Brignoli and Roewer and covers the literature from 1981 through 1987.

592. Roewer, C. F. **Katalog der Araneae von 1758 bis 1940; bzw. 1954**. Bremen, Germany: Kommissions-Verlag von "Natura," 1942-1954. 2 vols. in 3.

This catalog contains much of the same information as Bonnet's work (see entry 589), but is arranged by family, rather than Bonnet's alphabetical arrangement. It is this format that was continued in updates to this work by Brignoli (see entry 590) and Platnick (see entry 591). Genus and species are in alphabetical order under the family.

Crustacea

593. Holthuis, L. B. **Marine Lobsters of the World: An Annotated and Illustrated Catalogue of Species of Interest to Fisheries Known to Date**. Rome: Food and Agriculture Organization of the United Nations, 1991. Distributed, Lanham, Md.: UNIPUB. 292p. (FAO Species Catalogue, vol. 13; FAO Fisheries Synopsis, no. 125, vol. 13). $55.00. ISBN 92-5-103027-8.

This is the 13th volume of the *FAO Species Catalogue*, a series that focuses on commercially important marine animals. This volume covers 149 species of lobsters in three infraorders, 10 families, and 33 genera. These are the species that are commercially valuable (i.e., they are used for food, sold for bait or other products, or have potential value). See entry 561 for a complete annotation.

594. Sims, R. W., Ed. **Animal Identification: A Reference Guide.** Vol. 2, **Land and Freshwater Animals (Not Insects).** London: British Museum (Natural History); New York: John Wiley, 1980. 120p. LC 80-40006. ISBN 0-471-277665.

This guide is annotated under entry 29. Volume 2 of this set covers the crustaceans from page 66 to 74. Most of the checklists for the crustaceans are found in the journal literature. Citations to these checklists may be found in this source.

595. Williams, Austin B., and others. **Common and Scientific Names of Aquatic Invertebrates from the United States and Canada: Decapod Crustaceans**. Bethesda, Md.: American Fisheries Society, 1989. 77p. (American Fisheries Society Special Publication, 17). $17.00pa. LC 88-70618. ISBN 0-913235-62-8; 0-913235-49-0pa.

In an effort to standardize the common names of aquatic invertebrates in North America north of Mexico, the American Fisheries Society created the Committee on Names of Aquatic Invertebrates to study this problem. This committee developed guidelines for formation of common names and then handed over the actual task of developing these lists to specialized groups. The Crustacean Society was picked to develop this list and through the efforts of its Committee on the Names of Decapod Crustaceans, this first comprehensive list of North American decapod crustacean species was created. The list contains 1,614 species. This includes 509 shrimps; 364 lobsters, crayfishes, and like forms; 285 hermit crabs, squat lobsters, porcelain, mole, and sand crabs; and 456 brachyuran crabs. The list is arranged systematically down to species with the common name listed next to the species (no common names were assigned to rare or obscure species). A single index includes the common and the scientific names. Future editions of these lists are planned for every 10 years. Twelve color plates of representative decapod crustaceans are included at the end of the text.

Uniramia

INSECTA (General)

596. **Common Names of Insects and Related Organisms, 1989.** Manya B. Stoetzel, chairman, Committee on Common Names of Insects. Lanham, Md.: Entomological Society of America, 1989. 199p. $35.00. LC 90-214412. ISBN 0-938522-34-5.

This provides a list of common names approved by the Entomological Society of America of more than 2,000 insects and other invertebrates. These are insects that inhabit the United States, Canada, or the possessions and territories. It is arranged into four sections. The first section lists insects alphabetically by common name, the second by scientific name, the third by order and family, and the fourth by phylum, class, order, and family names. The first three sections are arranged in three columns with the alphabetized column on the left side of the page. For example, the common name section lists the common names in alphabetical order with the next two columns listing the order, family, and genus species names (e.g., aloe mite, Acari: Eriophyidae, *Eriophyes aloinis keifer*). The scientific name section lists these names in alphabetical order followed by two columns for the order and family and then common name.

INSECTA (Systematic Section)

Ephemeroptera (Mayflies)

597. Hubbard, Michael D. **Mayflies of the World: A Catalog of the Family and Genus Group Taxa (Insecta: Ephemeroptera)**. Gainesville, Fla.: Sandhill Crane Press, 1990. 119p. (Flora & Fauna Handbook, no. 8). $24.95. LC 90-48456. ISBN 1-87743-06-2.

This catalog of the world's recent and fossil mayflies is divided into four sections. The first section lists genus and subgenus names in hierarchical order from fossil to more advanced. Fossil names are shown with their geological age and living groups are shown with the geographical distribution. The second section is an alphabetical list of family-group names (family, subfamily, tribe, subtribe), and the third section (the largest section) is an alphabetical list of genus-group names. The genus name is followed by the author of the name, the date of publication of the name, and the beginning page number of the original description. The type-species is listed and synonyms are also given. The final section is a 27-page bibliography of references cited in the text. Authored by a well-known authority on Ephemeroptera.

Collembola (Springtails)

598. Mari Mutt, José A., and Peter F. Bellinger. **A Catalog of the Neotropical Collembola, Including Nearctic Areas of Mexico**. Gainesville, Fla.: Sandhill Crane Press, 1990. 237p. (Flora & Fauna Handbook, no. 5). $29.95. LC 89-70038. ISBN 1-877743-00-3.

Supplements and updates *An Index to the Collembola* (see entry 599).

599. Salmon, J. T. **An Index to the Collembola**. Wellington, New Zealand: Royal Society of New Zealand, 1964-1965. 3 vols. (Bulletin, Royal Society of New Zealand, no. 7). LC 65-84667.

This work is included here because the bulk of it is a checklist of the Collembola of the world. The checklist is found in volume two of this set and encompasses 498 pages. Volume one includes a bibliography of sources (with 2,603 entries) and a classification and key to the families and genera. Volume three is mainly an alphabetical index to genera and higher groups in the checklist. The checklist is arranged systematically with references arranged chronologically by year of publication followed by the specific name, the author's name, and the reference number in brackets. This reference number refers to the bibliography entry in volume one. Page numbers and references to illustrations, geographic distribution, and other information are also included. Repository of type material, when known, is given.

Odonata (Dragonflies)

600. Davies, D., Allen L. Davies, and Pamela Tobin. **The Dragonflies of the World: A Systematic List of the Extant Species of Odonata**. Utrecht: Societas Internationalis Odonatologica, 1984-1985. 2 vols. (Rapid Communications/supplements, Societas Internationalis Odonatologica), nos. 3 and 5.

Unable to get a copy of this work to review. Volume one includes the Zygoptera and the Anisozygoptera; volume two includes the Anisoptera.

Isoptera (Termites)

601. Snyder, Thomas E. **Catalog of the Termites (Isoptera) of the World**. Washington, D.C.: Smithsonian Institution, 1949. 490p. (Smithsonian Miscellaneous Collections, vol. 112). LC 49-47072.

Both fossil and living species are listed here. The catalog is arranged systematically from primitive to specialized, with lists for fossil and living species arranged separately. After the taxa name, the author of the name and the year published are given along with the page number. Type species are listed as well after the genus name information. Species names also include indications of where illustrations (figures or plates) may be found, distribution, and the location of the type species. An index to species names follows the checklist, as well as separate bibliographies on termite taxonomy for living and fossil termites.

Dermaptera (Earwigs)

602. Steinmann, H. **World Catalogue of Dermaptera**. Boston: Kluwer Academic, 1989. 933p. (Series entomologica, vol. 43). $299.00. LC 89-2704. ISBN 0-7923-0096-3.

This is a systematic revision of the Dermaptera. It is a comprehensive work covering more than 2,000 taxa. It is arranged in systematic order above genus, with genus and species arranged in alphabetical order. Each taxon is in boldface type, followed by the year of the original description, with the original name and literature citation where the description ca be found. The sex of the type specimen, where it may be found, and the type-locality follow Other literature citations are included that add to the history of the taxon. A 30-pag bibliography and an alphabetical index to listed taxa complete the text.

Thysanoptera (Thrips)

603. Jacot-Guillarmod, C. F. **Catalogue of the Thysanoptera of the World**. Grahamstowr South Africa: Pub. jointly by the Cape Provincial Museums at the Albany Museum, 1970 1975. 515p. (Annals of the Cape Provincial Museums [natural history], vol. 7, pts. 1 and 2

This world catalog of the Thysanoptera is arranged by family with genus and specie listed alphabetically. The names are followed by all literature references to that name. Th complete citations are given followed by type-species (for genus). For species, location o type, distribution, type locality, and habitat are listed.

Homoptera, Hemiptera, Heteroptera (True Bugs, Whiteflies, Aphids, Hoppers, Cicadas, Scale Insects)

604. China, W. E., and N. C. E. Miller. **Check-List and Keys to the Families an Subfamilies of the Hemiptera-Heteroptera**. London: British Museum (Natural History) 1959. 45p. (Bulletin of the British Museum [Natural History]. Entomology, vol. 8, no. 1).

This short work lists the families and subfamilies and their synonyms. Beside each nam is the author and reference to where the name was first published. The keys to families an subfamilies takes up most of this work. No illustrations are provided.

605. Distant, W. L. **A Synonymic Catalogue of Homoptera. Part I. Cicadidae**. London Trustees of the British Museum, 1906. Reprinted, New York: Johnson Reprint, 1966. 207p $19.00 ISBN 0-685-02765-1.

Still in print as of this writing, this catalog also includes key to genus and species. Eac name includes a reference to the literature where first named with dates and page number indicated. The country where the species is found is listed next to the name in the margin Type species are given with the genus name. An index to family, subfamily, genus, and specie is found at the end.

606. Eastop, V. F., and D. Hille Ris Lambers. **Survey of the World's Aphids**. The Hague The Netherlands: W. Junk, 1976. 573p. ISBN 90-6193-561-X.

This survey of the world's aphids is divided into two main parts. Part I is the "Aphi Genera of the World with their Species," part II is the "Index by Species-Group Names an Infra-Subspecific Names." The aphid genera are listed alphabetically. References when firs named are listed after the name. Type species are also given. Species names are listed unde the genus in alphabetical order. Synonyms are listed as well. The index is to specific subspecific, or varietal name, listed in alphabetical order. No bibliography of sources i included.

607. Henry, Thomas J., and Richard C. Froeschner, eds. **Catalog of the Heteroptera, o True Bugs, of Canada and the Continental United States.** Leiden, New York: E. J. Brill 1988. 958p. LC 87-38212. ISBN 0-916846-44-X.

Arranged by family, this catalog attempts to list each species in their respective families Each of the 45 families are briefly described, giving the general distribution, what habitats i prefers, and other notable information pertinent to this group; the major literature sources o the group are included at the end of the description. A representative bug from the group i illustrated in black and white on the first page of the description. Some illustrations take u a full page. The arrangement is then by subfamily, then genus, with species listed alphabeti cally under the genus. References to when the species was first named and subsequent using of different name combinations are listed next to the species name. An indication of geographical range by state or province is also included. The 131-page bibliography of the literature cited in the text is useful in itself as a resource for research on the Heteroptera.

608. **General Catalog of the Hemiptera (Later Homoptera).** G. Horvath, gen. ed., H. M. Parshley, managing ed., fasc. 1-4, pt. 2; W. E. China, gen. ed., and H. M. Parshley, managing ed., fasc. 4, pts. 3-10, and fasc. 5 ; W. E. China, gen. ed., and Z. P. Metcalf, managing ed., fasc. 4, pt. 11; remaining fasc. by Z. P. Metcalf, some by Virginia Wade. Began publication at Smith College, Northhampton, Mass.; final volumes published at North Carolina State University, Raleigh, N.C., 1927-1971. 8 fascicle, in many parts.

A comprehensive multivolume work written primarily by Z. P. Metcalf and completed by Virginia Wade with technical advice from David A. Young after Dr. Metcalf's death in 1956. Published irregularly in independent parts, each part contains usually one family. Fossil as well as living Homoptera are listed here. Bibliographies are also included, such as the *Bibliography of the Cicadelloidea (Homoptera: Auchenorhyncha)*, and are meant to be used with the corresponding fascicle or fascicles on a particular family or superfamily. They are also useful as an independent source. In the more recent publications, the family is divided into genera which are in phylogenetic order. Species are arranged under each genus in alphabetical order. References to when the name is first used are recorded. Other references are listed as well, with notations such as "ecology," "new species," and "described," to inform the researcher of what information on the species is found in the article. Geographical localities are also given.

609. Mound, L. A., and S. H. Halsey. **Whitefly of the World: A Systematic Catalogue of the Aleyrodidae (Homoptera) with Host Plant and Natural Enemy Data**. Chichester, N.Y.: British Museum (Natural History) and John Wiley, 1978. 340p. ISBN 0-471-99634-3.

This work includes the 1,156 species in 126 genera of the world whitefly. Species are listed alphabetically within each genus. Genera are listed alphabetically within two subfamilies, the Aleyrodinae and the Aleurodicinae. Genus names give the author of the name and the reference where named as well as the type species. Each species entry also gives the author and references along with host plants, parasites, predators, and geographical distribution. A 45-page systematic list of host plants of the family Aleyrodidae is found at the end of the checklist. A 14-page bibliography of references and an index to genera and species of the family Aleyrodidae finish this comprehensive list of the whitefly.

Coleoptera (Beetles and Weevils)

610. Arnett, Ross H., Jr., comp. and ed. **Checklist of the Beetles of North and Central America and the West Indies**. Gainesville, Fla.: Sandhill Crane, 1983. 10 vols. (loose-leaf). $250.00 set. $15.00-$40.00/vol. ISBN 0-916846-21-1.

This publication grew out of the North American Beetle Fauna Project that produced a "Red Version" and a "Yellow Version" of a *Checklist of the Beetles of Canada, United States, Mexico, Central America, and the West Indies*. Only two parts of the "Yellow Version" were completed. The "Red Version" was completed in 1976 but only a few copies were made and they quickly went out of print. This work took the old "Red Version," renumbered the old parts of the list to conform to updated classification information, revised parts, and added new parts that had been published elsewhere.

This 10-volume loose-leaf, which is an amazing 4½ inches thick and 2,100 pages, includes all the beetle families of the world, even if they are not found in the geographical region covered. Those not found in the region are simply listed in their phylogenetic position and only brief information is given. The various volumes cover different families and each volume is authored by an authority on the family: volume 1, *Ground Beetles, Water Beetles, and Related Groups*; volume 2, *Rove Beetles and Related Groups*; volume 3, *Scarab Beetles, Buprestid Beetles, and Related Groups*; volume 4, *Click Beetles, Fireflies, Checkered Beetles, and Related Groups*; volume 5, *Ladybird Beetles and Related Groups*; volume 6, *Darkling*

Beetles, Strepsiptera and Related Groups; volume 7, *Longhorned Beetles*; volume 8, *Leaf Beetles and Bean Weevils*; volume 9, *Fungus Weevils, Bark Beetles, Weevils, and Related Groups*. Volume 10 is titled *Bibliography of the Coleptera of North America, North of Mexico 1758-1948, and Supplements Including Mexico, Central America, and the West Indies*. The bibliography is arranged in alphabetical order by author and is useful in finding the full citation to the references mentioned in the catalog.

611. Blackwelder, Richard E., comp. **Checklist of the Coleopterous Insects of Mexico, Central America, The West Indies, and South America.** Washington, D.C.: U.S.G.P.O., 1944-1957. Reprint, Washington, D.C.: Smithsonian Institution, 1982. Pts. 1-6. (United States National Museum Bulletin 185, Pts. 1-6). $40.00. LC 81-607585. ISBN 0-87474-244-7.

This is the only checklist of beetles for Latin America and covers 5,000 species. It is written in three main sections: the checklist, a bibliography of references, and an index to generic and higher taxonomic names. The checklist is arranged by family and genera with species listed alphabetically under genus. References are provided to when the taxon was first named, and the geographical distribution by country is also given. The bibliography is found in part six and includes papers published before 1946 on all aspects of coleoptology. A list of the journals cited and their abbreviations makes finding these references a little easier.

612. Bousquet, Yves, ed. **Checklist of Beetles of Canada and Alaska.** Ottawa: Research Branch, Agriculture Canada, 1991. 430p. (Publication, 1861/E). $35.95; $29.95pa. ISBN 0-660-13767-4.

This is the first comprehensive checklist of the beetles occurring in Canada and Alaska. The checklist was no easy task since the beetles group consists of more than 250,000 species, of which approximately 24,000 occur in the United States and Canada. Beetles account for close to one-third of all known animals on earth and are found in almost all the earth's environments. This checklist provides the correct names and synonyms of the beetles and indicates in what area of Canada or Alaska where they may be found. It is arranged by suborder, then by family, and so on down to genus and species. Classification and basic information for each family group is written by one of the ten contributors to this checklist. This family information also includes references to key publications on the taxonomy of the group where further information on identification may be found. These references are included in a 29-page bibliography. An index to taxa, including synonyms, above species level is included at the end of the text.

This scholarly work was produced primarily by the Biosystematics Research Centre of Agriculture Canada with other contributors from universities in Canada and one from the University of Wisconsin, Madison. It is an indispensable list of the beetles for this area, and should be a standard source for years to come. Primarily for the professional entomologist and serious student, it is recommended for academic and large public libraries and those with strong collections in zoological sciences.

A comparable work is *A Catalog of the Coleoptera of America North of Mexico* (see entry 613), which covers the geographic ranges of Canada and Alaska. This catalog is published irregularly; there are 124 fascicles, each with information on a particular family.

613. **A Catalog of the Coleoptera of America North of Mexico**. Washington, D.C.: U.S. Dept. of Agriculture, 1979- . Many fascicles. (Agriculture Handbook, 529-).

This catalog, produced by the Agricultural Research Service of the Dept. of Agriculture, is meant to supersede the Leng catalog (see entry 614), published in 1920, and its two supplements. It is printed irregularly in fascicles, each fascicle representing a particular beetle family. Each fascicle is authored by a specialist in that family and includes an introductory text describing the family; a list of the genus and subgenus in that family, with references to other information on the genus such as taxonomy, ecology, and published keys; a bibliography of sources, and an index. The foreword in each fascicle says that this information is stored on computer tape, is updated periodically, and is available for computer searching in a database.

614. Leng, C. W. **Catalogue of the Coleoptera of America, North of Mexico.** Mount Vernon, N.Y.: John D. Sherman, Jr., 1920. 470p. LC 21-806.

This original catalogue of the Coleoptera is supplanted by one produced by the Agricultural Research Service, U.S. Dept. of Agriculture (see entry 613). Supplements for 1919-1924 and 1925-1932 were published with A. J. Mutchler; 1933-1938 and 1937-1947 by R. E. and R. M. Blackwelder.

615. O'Brien, Charles W., and Guillermo J. Wibmer. **Annotated Checklist of the Weevils (*Curculionidae sensu lato*) of North America, Central America, and the West Indies (Coleoptera: Curculionoidea)**. Ann Arbor, Mich.: American Entomological Institute, 1982. 382p. $45.00. (Memoirs of the American Entomological Institute, no. 34).

This is a checklist of 843 genera and 7,068 species of weevils north of South America. Arranged by family with species listed alphabetically under genus names. The author of the name is given followed by the year of publication and a page number where the description of the species begins. Geographical area where found is given next to the reference. Abbreviations of countries and states are provided. A lengthy bibliography containing references cited in the text, plus additional references for further reading, follows the checklist. A handy list of the abbreviated journals and the full form of the journal title is very useful, especially for older citations. Indexes to family and genus group as well as species group aid in access to these names in the text. A checklist of the weevils of South America is listed under entry 617. It is updated by the authors in *Additions and Corrections to Annotated Checklists of the Weevils of North America, Central America, and the West Indies, and of South America* (Southwestern Entomological Society, 1989).

616. Schenkling, S., ed. **Coleopterorum Catalogus**. The Hague, Netherlands: W. Junk, 1910-1940. 31 vols.

Each of the 30 parts of this set lists genus and species for an individual family and has its own author and index. It lists references where the name was first described and other citations to the literature on that taxa. References to the literature on the biology are also included as well as those to illustrations. The index for the entire set is in volume 31. Supplements to this set have been issued irregularly since 1950.

617. Wibmer, Guillermo J., and Charles W. O'Brien. **Annotated Checklist of the Weevils (*Curculionidae sensu lato*) of South America (Coleoptera: Curculionoidea)**. Gainesville, Fla.: American Entomological Institute, 1986. 563p. (Memoirs of the American Entomological Institute, no. 39). $50.00.

A checklist to the weevils of North America by the same authors is listed under entry 615. The arrangement is the same, with the inclusion of an expanded bibliography, a list of abbreviated periodical titles, and indexes to species, family, and genus names. This checklist lists the names of 1,010 genera and 9,046 species found in South America. It is updated by the authors in *Additions and Corrections to Annotated Checklists of the Weevils of North America, Central America, and the West Indies, and of South America* (Southwestern Entomological Society, 1989).

Siphonaptera (Fleas)

618. Hopkins, G. H. E., and Miriam Rothschild. **An Illustrated Catalogue of the Rothschild Collection of Fleas (Siphonaptera) in the British Museum (Natural History).** Oxford, England: Oxford University Press, 1987. 7 vols. $945.00. LC 54-3740.

See entry 428 in the Handbooks section. This work includes a checklist to the fleas of the world.

619. Lewis, Robert E., and Joanne H. Lewis. **A Catalogue of Invalid or Questionable Genus-Group and Species-Group Names in the Siphonaptera (Insecta)**. Koenigstein, Germany: Koeltz Scientific Books, 1989. 263p. (Theses Zoologicae, vol. 11). ISBN 3-87429-302-5.

This volume fills a gap in the nomenclature of fleas by dealing with the many invalid names. These are listed in alphabetical order with detailed explanations of where the name was first published, locality where found, host, collector, and dates of collection. Three appendices are found at the end of the text. One is for valid bird and mammal host names; another for abbreviations of scientific periodicals used in the catalog; and the last is a glossary of terms used in the work. A lengthy bibliography of references cited in the work is also helpful in reading more on where these invalid names were used.

Diptera (Flies and Mosquitoes)

620. **A Catalogue of the Diptera of the Americas South of the United States**. Sao Paulo, Brazil: Departamento de Zoologia, Secretaria da Agricutura, 1966- . Issued in fascicles. LC 72-169115.

Issued in fascicles with each one on a particular family, there are an intended 110 fascicles to complete the work. So far, 106 fascicles have been published. The last four will cover final remarks, a general bibliography, a general index, and a list of type-localities. This work covers the dipteran fauna of the area south of the U.S.-Mexican border, plus continental islands such as the Bahamas and the Galapagos Islands. Each fascicle treats an individual family and is authored by a specialist in that area. General information is given on the family followed by genus and species names. Each name includes the following information: the author, the date of first publication, and the page number, with figure and plate numbers of illustrations, the type locality, distribution, and references. If the reference is for a synopsis, a revision, or for morphological information, that is indicated. A bibliography of references cited and an index finish each fascicle.

621. Crosskey, R. W., ed. **Catalogue of the Diptera of the Afrotropical Region**. London: British Museum (Natural History), 1980. 1437p. LC gb80-31575. ISBN 0-565-00821-8.

This catalog lists 16,300 of the known species found in the Afrotropical Region, a region encompassing the continent of Africa south of the Sahara and including Madagascar and the surrounding islands. Forty specialists have contributed to this catalog. Not just a catalog of dipteran names, it also includes one page or more of introductory material for each individual family chapter with references to the most important works on the taxonomy and biology of the family. References to identification keys or other taxonomic information are found in these introductions. The catalog is arranged systematically with each family forming an individual chapter. Genera are alphabetical within the higher taxa, and species are alphabetical within genera. Each name is followed by the author, the year, and the page number where the name was first described. Type species information is also given as well as geographical distribution. A bibliography of the 4,700 cited works in the catalog follows the checklist, as well as a list of the full names of the authors of the cited works. An index to all the taxonomic names in the catalog is found at the end.

622. Delfinado, Mercedes D., and D. Elmo Hardy, comp. and ed. **A Catalog of the Diptera of the Oriental Region**. Honolulu: University Press of Hawaii, 1973-1977. 3 vols. $30.00 vol. 1; $30.00 vol. 2; $40.00 vol. 3. LC 74-174544. ISBN 0-8248-0205-5 vol. 1; 0-8248-0274-8 vol. 2; 0-8248-0346-9 vol. 3.

This three-volume set catalogs the dipteran fauna of the Oriental Region, which includes India and Asia south of the Himalayan-Tibetan mountain barrier, and the Australasian archipelago. Volume 1 includes the suborder Nematocera (moquitoes, black flies, midges, gnats, and related forms); volume 2, suborder Brachycera (horse flies, snipe flies, robber flies, and related forms), through division Aschiza, suborder Cyclorrhapha (big-eyed flies, bee- and hover flies); and volume 3, suborder Cyclorrhapha (two-winged flies). Together these volumes list 1,876 valid genera and 15,964 valid species in 101 families. Each family is authored by an authority in that field. General information on the family is given followed by subfamilies with genus and species listed alphabetically under family or subfamily. The author of the name, the geographic area where it is found, the type locality, and the citation where the taxon was first described are listed after the name. Each volume has its own bibliography and index, and has a map of the Oriental Region printed on the end papers.

623. Evenhuis, Neal L., ed. **Catalog of the Diptera of the Australasian and Oceanian Regions**. Honolulu: Bishop Museum Press and E. J. Brill, 1989. 1155p. (Bishop Museum Special Publication, 86). $100.00. LC 89-060913. ISBN 0-930897-37-4 (Bishop Museum Press); 90-04-08668-4.

This is the final work in the publications by various authors and editors to inventory the Diptera of the world. No other major insect order has been cataloged as completely as the Diptera. The other regions, Palaearctic, Neotropical, Nearctic, Oriental, and Afrotropical, have been published or are continuing to be published. All are included here in this section on Diptera. The region covered in this work includes the eastern Indonesian provinces of Maluku and Irian Jaya, Australia, New Zealand, and Oceanic islands. Antarctica and subantarctic islands are included in an appendix. Fifty-four contributors wrote the 115 chapters on the various families, which are arranged systematically. Each family is described in one page or more with genus and species names listed alphabetically within the family. Next to the name is a reference to where the name was originally proposed, then a type species (after genus name), and geographical distribution. A 181-page bibliography of literature cited and an index to all valid and invalid taxa listed in the catalog and appendices complete the work.

624. Knight, Kenneth L., and Alan Stone. **A Catalog of the Mosquitoes of the World (Diptera: Culicidae)**. 2d ed. College Park, Md.: Entomological Society of America, 1977. 611p. LC 77-82735.

This is an update to *A Synoptic Catalog of the Mosquitoes of the World* (see entry 627). It lists the subfamily, genus, and species names in alphabetical order. Each name is followed by the geographical area where it is found, the type location (depository where a specimen may be found), and then references to the literature for further information on its identification. A 146-page bibliography of references cited in the catalog is included as well as a gazetteer of the geographic terms used in describing where a species may be found. An index is provided for the scientific names of living as well as fossil (indicated by an asterisk) species. A supplement to this catalog was published in 1978.

625. Soos, A., ed.; and L. Papp, assist. ed. **Catalogue of Palaearctic Diptera**. Amsterdam, New York: Elsevier, 1986- . 14 vols. $150.00-$200.00/vol. LC 84-13534. ISBN 0-444-99600-1 set.

This is a comprehensive checklist of the diptera found in a region that includes Europe, Asia north of the Himalayan-Tibetan border, North Africa, and part of Arabia. Projected as a 14-volume set, these volumes have been published so far: volume 2, *Psychodidae and Chironomidae*; volume 3, *Ceratopogonidae and Mycetophilidae*; volume 4, *Sciaridae and Anisopodidae*; volume 5, *Athericidae and Asilidae*; volume 6, *Therevidae and Empididae*; volume 7, *Dolichopodidae and Platypezidae*; volume 8, *Syrphidae and Conopidae*; volume 9, *Micropezidae and Agromyzidae*; volume 10, *Clusiidae and Chloropidae*; volume 11, *Scathophagidae and Hypodermatidae*; volume 12, *Calliphoridae and Sarcophagidae*. The last volume will be an index to volumes 1-13. Each volume covers two families and is written by authorities on those families. Each catalog concludes with an extensive bibliography and index to name.

626. Stone, Alan, and others. **A Catalog of the Diptera of America, North of Mexico**. Washington, D.C.: U.S. Dept of Agriculture, 1965. 1696p. (Agriculture Handbook, no. 276). 1696p.

This catalog lists all the published names of Diptera in this geographic area wi references to the original publications of each name. The references next to the name in th catalog are abbreviated with only author's name, date, and page number. The full citation found in the lengthy bibliography at the end of the text. The names are arranged phylogenet cally, with species names listed alphabetically under the genus. Other information provide next to the name beside the reference are the type-species, nature of type fixation, and vali name of type-species. Geographical distribution is also provided to state, province, or countr

627. Stone, Alan, Kenneth L. Knight, and Helle Starcke. **A Synoptic Catalog of th Mosquitoes of the World (Diptera, Culicidae)**. Washington, D.C.: Entomological Socie of America, 1959. 358p. (Thomas Say Foundation, vol. 6). LC 59-15339.

Updated by *A Catalog of the Mosquitoes of the World (Diptera: Culicidae)* (see entr 624), this catalog attempted to be a guide to the literature of living mosquitoes (a few foss species are mentioned at the end of the text) and includes 110 valid genera and subgenera an 2,426 valid species. It provides a list of all generic, subgeneric, and trivial names mosquitoes prior to 1959. It is arranged systematically for subgeneric groups and above wi specific names listed alphabetically under generic names. The taxon name is followed by th author, the date, the original references, type species of genus, and selector of type. For speci names, the general distribution, type localities, present location of type species, and addition references are provided.

Trichoptera (Caddisflies)

628. Fischer, F. C. J. **Trichopterorum Catalogus**. Amsterdam: Nederlandsche Entomolo gische Vereeniging, 1960-1973. 16 vols. LC 68-136301.

Each volume of this comprehensive checklist includes lists for one or more familie With volume 11, all families in the order were covered. Four volumes (12-15) to suppleme the information in the original 11 were completed by 1973. Volume 16 is the index to all th names mentioned in the first 15 volumes, listed in alphabetical order. Each volume also ha its own index. The catalogue deals with the literature from 1758 to 1958. All species are liste in alphabetical order under the genus name. References to the literature where informatio and illustrations can be found on each name are arranged in chronological order after th name. Type species are listed and all accepted names of genera and species are printed in bo with synonyms in italics. The geographical distribution is listed in the right-hand margi Useful not only as a checklist but as a bibliography to the literature on caddisflies back 1758.

Lepidoptera (Butterflies and Moths)

629. Aurivillius, Chr., and others, eds. **Lepidopterorum Catalogus**. Berlin: W. Junk 1911-1939. 94 pts.

This catalogue is arranged by family name, with species names listed alphabeticall under genus. Next to the name are listed the author and reference (title, page, and date) where first named. Other references are listed, including those where illustrations may b found. In the right-hand margin next to the name are the geographical locations where th species may be found. This is an old work with outdated classifications; however, it is ver detailed and comprehensive, and the newer classifications may be derived from newer work

630. Bridges, Charles A. **Catalogue of Family-Group and Genus-Group Names (Lepi doptera: Rhopalocera)**. Urbana, Ill.: the author, 1988. 348p. $60.00.

This volume is published with ten parts. Part I contains an alphabetic list of th type-genera of the family-group names; part II contains a synonymic list of the family-grou names; part III contains an index to the authors of the family-group names; part IV is a alphabetic list of the genus-group names; part V contains a synonymic list of the genus-grou names; part VI contains an index to the authors of the genus-group names; part VII contain an index to the type-species of the genus-group names; part VIII contains the bibliography of sources; part IX indexes the bibliography by journal name; and part X indexes the bibliography by year of publication. The genus-group names are arranged in alphabetical order with references to the original description, type species, objective synonym, type method, subjective synonym, and other references provided next to the name. The author also published several catalogues of families within the order Lepidoptera.

631. D'Abrera, Bernard.

The books by Bernard D'Abrera listed in this paragraph are found in the Handbooks section of this chapter. They provide checklists for butterflies of these geographical areas. *Butterflies of the Afro-Tropical Region* (see entry 447), *Butterflies of the Neotropical Region* (see entry 450), *Butterflies of the Oriental Region* (see entry 451), *Butterflies of the Australian Region* (see entry 448), and *Butterflies of the Holarctic Region* (see entry 449). The following book also by D'Abrera provides a checklist of a type of moth: *Sphingidae Mundi: Hawk Moths of the World, Based on a Checklist by Alan Hayes and the Collection He Curated in the British Museum (Natural History)* (see entry 452).

632. Ferris, Clifford D., ed. **Supplement to A Catalogue/Checklist of the Butterflies of America, North of Mexico.** Los Angeles: Lepidopterists' Society, 1989. 103p. (Lepidopterists' Society Memoir, no. 3). LC 89-80153.

This supplements Miller and Brown's *A Catalogue/Checklist of the Butterflies of America, North of Mexico* (see entry 636). This work makes nomenclatural changes published since Miller and Brown's work was published, plus some corrections to errors in the text. It also aligns the catalogue with the 1985 revision of the *International Code of Zoological Nomenclature*.

633. Hemming, Francis. **The Generic Names of the Butterflies and Their Type-Species (Lepidoptera: Rhopalocera)**. London: British Museum (Natural History), 1967. 509p. (Bulletin of the British Museum (Natural History), Entomology, Supplement 9).

This work provides an alphabetical list of butterfly genera names used in the literature from Linnaeus in 1758 to December 1963. Each name is followed by a full bibliographical reference to the place and date where and when each name was first validly published. The type-species of the genus is also given with a reference to the place and date of publication of the name of that species. The names used in this work include established names, names with incorrect original spelling, names that are emendations of previously published names, names with incorrect subsequent spellings of earlier names, and rejected names.

634. Hodges, Ronald W., and others. **Check List of the Lepidoptera of America, North of Mexico.** London: E. W. Classey & the Wedge Entomological Research Foundation, 1983. 284p. $150.00. ISBN 0-86096-016-1.

This is strictly a list of Lepidoptera identified and in the literature as of 1978. The list is arranged taxonomically down to species and subspecies with a reference to where the taxon was first described, in abbreviated form (author abbreviated and date). An index to scientific name is included; no common names are referenced. A short bibliography of references is provided. For the specialist.

635. Miller, Jacqueline Y., ed. **The Common Names of North American Butterflies.** Washington, D.C.: Smithsonian Institution Press, 1992. 177p. $14.95pa. LC 91-21343. ISBN 1-56098-122-9.

Now that butterfly watching is fast becoming as popular as bird watching, a source on standardizing the common names of North American butterflies was needed so all would know and use the same common name and therefore avoid confusion in the literature and in communications between butterfly watchers. It is arranged in taxonomic order after Miller and Brown's work (see entry 636). For example, eastern swallowtail, common eastern swallowtail, American swallowtail, eastern black swallowtail, and black swallowtail are all common names for the same butterfly. This source lists all these common names beside the scientific name and puts the common name most frequently associated with that species in bold type. In this case, the black swallowtail name is bolded. Next to the bolded name are abbreviations to books that preferred that name over the others in their work. The index includes all names listed in the book, both common and scientific, so that the reader can easily find the preferred name. A useful and up-to-date work for the specialist and amateur alike.

636. Miller, Lee D., and F. Martin Brown. **A Catalogue/Checklist of the Butterflies of America, North of Mexico**. Los Angeles: Lepidopterists' Society, 1981. 280p. (Lepidopterists' Society Memoir, no. 2). LC 81-82185. ISBN 0-930282-02-7.

This catalog is arranged in taxonomic order beginning with superfamiy Hesperioidea. Each genus is consecutively numbered with roman numerals, and each species is numbered with arabic numbers; subspecies with letters are next to the species number. In all, 241 genera and more than 763 species are listed. The author of the name and the reference to the original description are provided as well as references to any subsequent modifications. Type species are indicated, as well as synonyms and geographical range. Numbered footnotes (699 of them) at the end of the catalog explain the author's decisions on taxonomy and nomenclature. Updated by Clifford D. Ferris in a supplement to this catalogue/checklist (see entry 632).

637. Nye, Ian W. B., ed. **The Generic Names of Moths of the World**. London: Trustees of the British Museum (Natural History), 1982-1991. 6 vols. LC 76-365435.

This six-volume work covers the genus-group names of the moths of the world. Each volume covers a specific superfamily. Each genus-group name is listed alphabetically within the family. Junior homonyms, junior objective synonyms, and other names are grouped together chronologically under the nomenclaturally available name. Type-species, type-locality, and depository are identified, and relevant references to first description and revisions are given. A black-and-white photo plate at the beginning of each volume shows representative species of the families covered in that volume.

Volume 1. *Noctuoidea*. $57.75. ISBN 0-565-00770-X.
Volume 2. *Noctuoidea*. $41.80. ISBN 0-565-00811-0.
Volume 3. *Geometroidea*. $41.80. ISBN 0-565-00812-9.
Volume 4. *Bombycoidea, Mimallonoidea, Spingoidea, Castnioidea, Cossoidea, Zygaenoidea, and Sesioidea*. $41.80. ISBN 0-565-00848-X.
Volume 5. *Pyraloidea*. $41.80. ISBN 0-565-00880-3.
Volume 6. *Microlepidoptera*. $76.00. ISBN 0-565-00991-5.

Hymenoptera (Ants, Bees, and Wasps)

638. Hedicke, Hans, ed. **Hymenopterorum Catalogus**. Gravenhage, The Netherlands: W. Junk, 1936- . LC 40-12043.

Written in 11 parts, this work attempts to list the Hymenoptera of the world. The 11 parts cover the following families: part 1, Tiphiidae; part 2, Cephidae and Syntexidae; part 3, Pamphiliidae; part 4, Xyelidae; part 5, Trigonalidae; part 6, Siricidae; part 7, Xiphydriidae; part 8, Sphecidae I (Astatinae-Nyssoninae); part 9, Evaniidae; part 10, Aulacidae; part 11, Gasteruptiidae. Each part has its own index. Each part is written by a separate author. A new edition, with parts issued at irregular intervals, continues this ongoing work.

639. Krombein, Karl V., and others. **Catalog of Hymenoptera in America North of Mexico**. Washington, D.C.: Smithsonian Institution Press, 1979. 3 vols. LC 78-606008.

This three-volume work is an update of the *Hymenoptera of America, North of Mexico, Synoptic Catalog* (see entry 640) and its two supplements. It lists each family, genus, and species name with references to where each species was first described in the literature. In addition, this catalog provides the basic biological and morphological information such as geographical distribution and preferred habitats. Volume one covers the suborders Symphyta, the sawflies and horntails, and Apocrita (Parasitica), mostly parasitic wasps; volume two covers Apocrita (Aculeata), the stinging ants, bees, and wasps; and volume three serves as an index to the set, with separate indexes for the taxa of Hymenoptera, hosts, parasites, prey, pollen and nectar sources, and predators.

640. Musebeck, C. F. W., Karl V. Krombein, and Henry K. Townes. **Hymenoptera of America, North of Mexico, Synoptic Catalog**. Washington, D.C.: U.S. Dept. of Agriculture, 1951. 1420p. (Agriculture Monograph, no. 2).

This catalog and its two updates, published in 1958 and 1967 and authored by Karl V. Krombein and others, were the first efforts to catalog the Hymenoptera of this area. Updated by *Catalog of Hymenoptera in America North of Mexico* (see entry 639).

Chilopoda and Diplopoda (Millipedes and Centipedes)

See the Handbooks section of this chapter under Chilopoda and Diplopoda. Checklists are included in these handbooks.

BIOGRAPHIES

641. Carpenter, Mathilde M. "Bibliography of Biographies of Entomologists." **American Midland Naturalist** 33 (1945): 1-116; 50 (1945): 257-348.

Although journal literature has not been cited in other areas of this guide, this article is an important source for biographies of entomologists. The list of biographies of entomologists from all countries is an update of one published in the *Annals of the Entomological Society of America* (21 [1928]: 489-520) by J. S. Wade titled "Bibliography of Entomologists." The bibliography is arranged in alphabetical order by the last name of the entomologist. Birth and death dates are listed after the name followed by references from journals and books to biographical information on the entomologist. The designation of *B* or *P* in italics in the list indicates a Bibliography or Portrait in the article. Updated by Pamela Gilbert's work (see entry 642).

642. Gilbert, Pamela. **A Compendium of the Biographical Literature on Deceased Entomologists**. London: British Museum (Natural History), 1977. 455p. LC 79-304978. ISBN 0-565-00786-6.

Based on Carpenter's work (see entry 641), this work contains biographical references to deceased entomologists and covers the literature up to 1975. Compiled by Pamela Gilbert, former Entomology Librarian at the British Museum, this work includes 7,500 entomologists listed in alphabetical order. Each entry includes the name followed by the birth and death dates. Citations to journal and book literature where biographical information may be found are listed in chronological order under the name. A list of periodical abbreviations used in the citations is provided at the end of the work, as well as 27 black-and-white portraits of the more prominent entomologists.

643. Mallis, Arnold. **American Entomologists**. New Brunswick, N.J.: Rutgers University Press, 1971. 549p. LC 78-152316. ISBN 0-8135-0686-7.

More than 200 short biographies are included in this interesting work on American entomologists. The author has tried not only to describe the accomplishments of these entomologists, but to include interesting information on the people themselves. The book begins with pioneer entomologists such as John Abbot (1751-1840?), who was one of the first naturalists in the Colonies to be interested in insects and who made more than 3,000 drawings of them. The biographies are then arranged into groups such as "early federal entomologists," "notable teachers in entomology," and "notable orthopterists." Each biography is one page to several pages long and features a black-and-white photo or drawing of the entomologist. This is not only a valuable reference work but an enjoyable book to read as well.

GUIDES TO THE LITERATURE

Only two guides are listed here, and they cover the entomology literature. Consult the Guides to the Literature section in chapter 1 for a general guide to life sciences literature that may include the other arthropod groups.

644. Chamberlin, W. J. **Entomological Nomenclature and Literature.** 3d ed., rev. and enlarged. Dubuque, Iowa: Wm. C. Brown, 1952. Reprint, Westport, Conn.: Greenwood, 1970. 141p. $38.50. LC 79-108387. ISBN 0-8371-3810-8.

Although dated, this work is still valuable for finding older materials on entomology, and it is still in print. It is divided into three sections: nomenclature, entomological literature, and scientific publications. The section on nomenclature covers the historical development of nomenclature and then explains the rules of nomenclature. It also discusses the *International Code of Zoological Nomenclature*, which are the old rules, but students may simply refer to the latest rules (1985 see entry 75a). The section on entomological literature has an especially complete section on U.S. government-agency publications.

645. Gilbert, Pamela, and Chris J. Hamilton. **Entomology: A Guide to Information Sources**. 2d ed. London: Mansell, 1990. 259p. $51.80. LC 89-13268. ISBN 0-7201-2052-7.

This is the only current guide to entomological literature and is very thorough, but it does have a leaning toward the British entomological literature. Pamela Gilbert was formerly the Entomology Librarian at the British Museum (Natural History). In all 1,854 citations, some with brief annotations, are included here. They are grouped into eight chapters covering the history and early literature, the naming and identification of insects, specimens and collections, the literature of entomology itself including journals, books and indexes (both print and computerized), searching and locating the literature, newsletters and conference literature, and entomologists and entomological organizations, as well as information on translation services. The larger chapters are broken down into sections. The chapter on naming and identification of insects is divided into taxonomy and nomenclature, then faunas, of particular countries. This new, second edition has a greatly increased section on journals and newsletters as well as the addition of three new sections (i.e., "Angling Entomology," "Conservation," and "Butterfly Gardening"). A very useful guide that can be used to supplement the reference sources on insects found in this chapter on the arthropods.

ASSOCIATIONS

Insecta (General)

646. **American Entomological Society**. Academy of Natural Sciences, 1900 Race Street Philadelphia, PA 19103. (215) 561-3978.

647. **American Registry of Professional Entomologists**. P.O. Box AJ, 4603 Calvert Rd. College Park, MD 20740. (301) 864-1336. (Absorbed by Entomological Society of America.

648. **Australian Entomological Society**. c/o Division of Entomology, CSIRO, P.O. Box 1700, Canberra, ACT 2601.

649. **British Entomological and Natural History Society**. The Alpine Club, 74 South Audley St., London W1Y 5FF, England.

650. **Council for International Congresses of Entomology**. British Museum (Natural History), Cromwell Rd., London SW7, England.

651. **Entomological Society of America**. 9301 Annapolis Rd., Lanham, MD 20706-3115. (301) 731-4535.

652. **Entomological Society of Canada**. 1320 Carling Ave., Ottawa K1Z 7K9, Canada.

653. **Entomological Society of Southern Africa**. (Entomologiese Vereniging van Suidelike Afrika). c/o D. Van Heerden, P.O. Box 103, Pretoria 001, South Africa. 12-2062623. FAX 12-3235275.

654. **Entomological Society of Washington**. c/o Dept. of Entomology, NHB 168, Smithsonian Institution, Washington, DC 20560.

655. **International Centre of Insect Physiology and Ecology**. P.O. Box 30772, Nairobi, Kenya. 2-802501. FAX 2-803360.

656. **International Union for the Study of Social Insects (Union Internationale pour l'Etude des Insectes Sociaux)**. c/o Dr. H.H.W. Velthuis. Laboratory of Comparative Physiology, Univ. of Utrecht, Postbus 80086, NL-3508 TB Utrecht, Netherlands. 30-535421.

657. **Kansas (Central States) Entomological Society**. University of Kansas, Snow Entomology Museum, Snow Hall, Lawrence, KA 66044.

658. **Korean Society of Applied Entomology**. c/o Dept. of Agricultural Biology, College of Agriculture, Seoul National University, Suwon, Kyonggi 441-744, Republic of Korea. 331-2913681.

659. **New York Entomological Society**. American Museum of Natural History, Central Park West at 79th St., New York, NY 10024-5192.

660. **Pacific Coast Entomological Society**. Dept. of Entomology, California Academy of Sciences, Golden Gate Park, San Francisco, CA 94118-4599.

661. **Royal Entomological Society of London**. 41 Queen's Gate, London, SW7 5HU, England. (01) 584-8361.

662. **Young Entomologists' Society**. 1915 Peggy Pl., Lansing, MI 48910-2553. (517) 887-0499.

663. **Xerces Society**. c/o Melody Allen, Ten Ash St., S.W., Portland, OR 97204. (503) 222-2788.

Insecta (Systematic Section)

ODONATA (Dragonflies)

664. **Dragonfly Society of America**. 469 Crailhope Rd., Center, KY 42214. (502) 565 3795.

COLEOPTERA (Beetles)

665. **Coleopterists' Society**. c/o Margaret K. Thayer, Field Museum of Natural History Div. of Insects, Chicago, IL 60605. (312)922-9410.

LEPIDOPTERA (Butterflies and Moths)

666. **Lepidoptera Research Foundation**. 1160 W. Orange Grove Ave., Arcadia, C 91006.

667. **Lepidopterist's Society**. 1041 New Hampshire St., Lawrence, KA 66044.

HYMENOPTERA (Ants, Bees, Wasps)

668. **International Bee Research Association**. Hill House, Gerrards Cross, Bucks SL9 0NR, England. (49) 85011.

ARACHNIDA (Spiders, Scorpions, Mites, and Ticks)

669. **American Arachnological Society**. Norman I. Platnick, American Museum of Natura History, Central Park West at 79th St., New York, NY 10024-5192.

670. **British Arachnological Society**. Hon. Sec: J. R. Parker, FZS, Stone Raise, 4 Lakeland Park, Keswick, Cumbria CA12 4AT. (0596) 73074.

Crustacea (Crabs, Lobsters, and Shrimps)

671. **Crustacean Society**. c/o Denton Belk, 840 E. Mulberry, San Antonio, TX 78212. (512 732-8809.

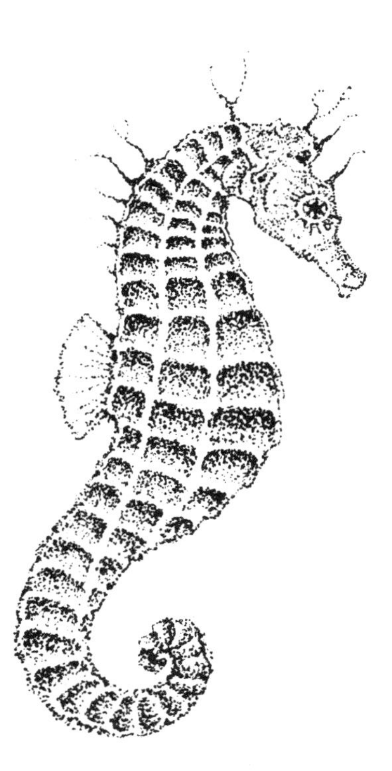

CHAPTER 4

THE VERTEBRATES

Unlike the invertebrates, all members of the vertebrates posses a dorsal vertebral column; that is, a backbone. In the middle of this column can be found the dorsal nerve cord, which emanates from the base of the brain. This vertebral column is either composed of cartilage as in the case of the sharks and rays, or bone, which can be found in the higher fishes, amphibians, reptiles, birds, and mammals. In addition, vertebrates show some degree of a developed protuberance known as the head. Such internal organs as a pancreas, kidney, liver, and a well-developed heart are also indicative of the vertebrate group in general.

Such chordates as the tongue worms, sea squirts, and amphioxus are considered transitional organisms between the invertebrates and vertebrate groups. They possess a notochord, which, it is believed, gave rise to the vertebral column. In the case of the tongue worm, it is half a notochord, hence the name of the group Hemichordata. There is no notochord in the adult sea squirt, however; it possesses one in the embryonic stage. The amphioxus is considered to be the highest form of chordate possessing a well-developed notochord as an adult.

The number of extant vertebrate species differs from source to source. As was stated in the introduction to chapter 2, Brusca considers approximately 47,000 extant species including the Hemichordates, Tunicates, and Cehpalochordates. However, Willson's work *Vertebrate Natural History* (see entry 696), suggests a total of approximately 42,460 species not including the hemichordates, tunicates, and cephalochordates. Since these protochordates number approximately 3,108 species, according to Brusca, there is a descrepancy of 1,432 species.

In any event, Willson's breakdown of extant vertebrate species (numbers are approximate) is listed as follows:

Agnatha (Lamprey Eels, Hagfishes):	60
Chondrichthyes (Sharks, Rays, Skates):	800
Osteichthyes (Bony Fish):	20,000

Amphibia (Frogs, Toads, Salamanders, etc.):	3,000
Reptilia (Snakes, Turtles, etc.):	6,000
Aves (Birds):	8,600
Mammalia (Mammals):	4,000

This chapter considers sources that treat the vertebrates as a whole. Suc sources that consider more than one vertebrate group have been included. Becaus of the massive amount of information on this subphylum, the number of source are few. However, for those institutions and individuals interested in generalize works, many of the volumes listed should provide ample information.

Man is a member of that series of living creature known as the vertebrates, or animals with a backbon a group including not only all the other warm-blooded hairy creature to which man is closely allied, but such varied forms as birds, reptile frogs and salamanders, and fishes. Their history is our histor and we cannot properly understand man, his body, his min or his activities, unless we understand his vertebrate ancestr
—Alfred S. Rome
The Vertebrate Story, *4th ed., 195*

DICTIONARIES

672. Jacobs, George J. **Dictionary of Vertebrate Zoology, English-Russian/Russian English: Emphasizing Anatomy, Amphibians, and Reptiles**. Washington, D.C.: Smith sonian Institution Press, 1978. 48p. $7.50pa. LC 78-16321. ISBN 0-87474-551-9.

The title of this work reveals its contents to a great extent. Scientific names, genera, an families of amphibians and reptiles are present along with a number of anatomical term Scientific names are translated into Russian common names and vice versa. The list wa compiled originally for personal reference.

ENCYCLOPEDIAS

673. Whitfield, Philip, ed. **Macmillan Illustrated Animal Encyclopedia**. New York Macmillan, 1984. 600p. $35.00. LC 84-3956. ISBN 0-02-627680-1.

Although the title of this volume would tend to indicate encyclopedic coverage of al animals, it does not. Only the vertebrate animals are included. Basically, it is broken dow into five sections that represent each of the major classes of vertebrates. The animals withi each section are arranged under their common group names (orders); the more primitive o the animals are first introduced within their respective sections. After a brief description o the order and family, the species are listed. For each vertebrate listed, the common name scientific name, geographical range, habitat, size in cm/m and in./ft., external anatomica features, breeding behavior, feeding behavior, and other interesting facts are given. The lengt of the description varies depending on the animal, although most are given at least a hal column. Superb colored paintings of these vertebrates can be found on the opposing page o the descriptive matter. It should be noted that not all vertebrate species are considered; rather select number was chosen that represent diversity for each group of vertebrates. The groups epresented, in the order in which they appear, are the mammals (180p.), birds (206p.), reptiles 56p.), amphibians (22p.), and the fish (94p.). Classifications are down to families, and in the ase of fish down to orders, of the five classes, and a separate common/scientific-name index ompletes the work. This is a first-class publication and represents the more familiar types f vertebrates. It is a good place to start for introductory information on these groups of nimals.

BIBLIOGRAPHIES

74. Baker, Sylva. **Endangered Vertebrates: A Selected, Annotated Bibliography, 1981-988**. New York: Garland, 1990. 197p. $32.00. LC 90-2863. ISBN 0-8240-4796-6.

Consisting of citations to monographic and journal literature, this compilation considers ertebrates that are endangered, threatened (vulnerable), rare, or extinct. References cited ontain biology, behavior, ecology, classification, reproduction, geographic range, and diet f the vertebrates. A total of 950 citations have been included. The references are listed under eir respective animal group, such as mammals, birds, amphibians and reptiles, and fishes. here is also a list of general references. In addition, there is an appendix listing names and ddresses of organizations that are concerned with endangered species. A separate common ame index, scientific-name index, author index, and geographic area index concludes the ibliography.

75. Blackwelder, Richard E. **Guide to the Taxonomic Literature of Vertebrates**. Ames, wa: Iowa State University Press, 1972. 259p. $10.95. LC 70-39613. ISBN 0-8138-1630-0.

Arranged phylogenetically, this work represents a list of books and journal articles that e relevant to the taxonomy of the vertebrates. Reference works such as bibliographies, ossaries, museum lists, checklists, regional works, and the like are included. The majority f the references are twentieth century with some late nineteenth century works included. The tations are presented under broad taxonomic groups, such as Pisces, with a large number of bheadings denoting geographical areas and families. This is a very comprehensive listing f the vertebrate taxonomic literature.

76. Wood, Casey A., comp. **An Introduction to the Literature of Vertebrate Zoology: ased Chiefly on the Titles in the Blacker Library of Zoology, the Emma Shearer Wood ibrary of Ornithology, the Bibliotheca Osleriana, and Other Libraries of McGill niversity, Montreal**. London: Oxford University Press, 1931. Reprint, **An Introduction the Literature of Vertebrate Zoology**. New York: Arno Press, 1974. 643p. (Natural ciences in America). $41.00. LC 73-17849. ISBN 0-405-05772-5.

Arranged under author, this catalogue of titles span works produced in the seventeenth, ghteenth, nineteenth, and twentieth centuries (up to 1930). As the title implies, it is a listing the holdings found in many of the libraries at McGill University dealing with all types of rtebrates. Each entry includes the author, title, date, and place of publication, pagination, mber of plates (if any), and notes. The majority of the entries are in English, German, ench, and Latin. In order to make this work more useful a subject/geographic index has en provided. This index lists entries (arranged chronologically) under broad animal groups, hich in turn are entered under a geographical area of the world. For example there is a ading "Africa—The Whole Continent" in which can be found subheadings such as General eatises, Birds, Mammals, and Fishes. The citations are arranged under the animal group ronologically (from earliest to latest). In addition to the catalog list a series of 19 chapters esent information on the history of the many vertebrate titles that have been produced down rough the ages. Important treatises published in the nineteenth and early twentieth centuries aling with the zoological literature are considered. This bibliography represents an

important contribution to zoological literature and would serve as a supplementary work t
guides in zoological literature.

INDEXES AND ABSTRACTS

677. **Zoological Record**. Vols. 1- , nos. 1- . Philadelphia: BIOSIS, Zoological Society o
London, 1865- . Annual. $2,200.00/yr. ISSN 0084-5604.

For a description of this index, see entry 55. The following section relates to this chapter
Section 14: Protochordata. (These organisms are not quite vertebrates, but are not inverte
brates; they belong to Chordata as do the vertebrates.)

HANDBOOKS

678. Davis, David E., ed. **CRC Handbook of Census Methods for Terrestrial Verte brates**. Boca Raton, Fla.: CRC Press, 1982. 397p. LC 81-18020. ISBN 0-8493-2970-1.

This work considers those procedures and techniques used in order to determine th
number of vertebrates within a specified area. The handbook is divided into two majo
sections. The first part, which is the larger of the two, presents specific types of vertebrate
and discusses the procedures necessary in order to record a population count of them. It als
considers, in many cases, the margin of error for a particular procedure. The length of text i
usually one page to two pages for each animal; a few citations to the literature accompan
the description. Examples of types of animals are snapping turtles, wood ducks, bluebirds
rice rats, wolves, and moose. The majority of the animals presented are either birds o
mammals, with a few amphibians and reptiles mentioned. Fish are not represented within thi
volume. The second part of the work consists of one- to two-page procedural descriptions c
certain animals within a habitat. The same type of information is provided as in the firs
portion of the book. Examples are birds in riparian woodlands, breeding birds in Texa
brush-grasslands, and rodents of sonoran desert habitats. A few references also accompan
the procedural descriptions in this portion. A combined index of subjects, common name
and scientific names completes the source.

679. Goodrich, Edwin S. **Studies on the Structure and Development of Vertebrate**
London: Macmillan, 1930. Reprint, Chicago: University of Chicago Press, 1986. 837
$32.50pa. LC 86-7002. ISBN 0-226-30354-3pa.

Based mainly on morphological considerations, this work examines the various extern
and internal structures found in the major vertebrate groups. Each chapter considers a specifi
anatomical area of vertebrates and illustrates the morphological changes through the verte
brate groups. Such areas as the skeleton, median fins, paired limbs, limb girdles, head regio
and skull are considered. In addition, the circulatory, respiratory, digestive, excretory, an
nervous systems are explored. A large number of labeled line-drawings are used to illustra
basic morphological principles. References to the literature are cited throughout the tex
Although some physiology is considered, it should be kept in mind that the basic theme c
the work is morphology. A total of 1,186 references can be found near the end of the work
A combined subject and scientific-name index completes the volume.

Although this work is outdated, it is still in print and is still useful for the anatomic
information presented. This is a very detailed book. It is useful for those wanting an exhaustiv
account of the morphological differences found among the many vertebrate groups.

680. **International Zoo Yearbook**. Vols. 1- . London: Zoological Society of London, 1959- . Annual. Price varies. ISSN UK 0074-9664.

The main purpose of these yearly publications is to discuss the activities carried out by the many zoos found throughout the world. Such information as breeding, nutrition, husbandry, exhibits, buildings, and the like are considered. In addition, sections on the care, handling, and introduction of specific vertebrate species to zoos are presented. Much of this information is written up in the form of research papers. Statistical information can also be found, such as vertebrates that are bred in captivity along with the number of births. Graphs, charts, tables, and black-and-white photographs provide the illustrative material. An author index, and a combined index of subjects, common names, and scientific names can be found at the end of the later volumes. Some of these indexes are cumulative for several volumes. For example, the subject index at the end of volume 6 is cumulative for volumes 1-6.

681. Jefferies, R. P. S. **The Ancestry of the Vertebrates**. Cambridge, New York: Cambridge University Press, 1986. 376p. $89.50. LC 87-126943. ISBN 0-521-34266-X.

This work, which is considered to be an interim report by the author, represents an evolutionary approach to the vertebrates. Its basic premise is that the vertebrates arose from fossils known as calcichordates. Detailed anatomical, morphological, and physiological materials are considered for the primitive vertebrates, embryological forms of the vertebrates, and internal adult structures of the vertebrates. Evolutionary relationships are shown as this factual material is being discussed. There are a large number of labeled diagrams to illustrate the many points brought out in the text. The diagrams are excellent. Chapters include metazoan relationships and the hemichordates and echinoderms, amphioxus and its relatives, the tunicates, adult anatomy and basic phylogeny of living vertebrates, embryology of vertebrates and the head problem, the cornutes, the mitrates, and the phylogeny of the deuterostomes. An 11-page bibliography and a combined index of subjects, common names, and scientific names concludes this evolutionary source. Citations in the bibliography have been cited in the text. Although this work was placed under handbooks, it could just as easily serve as a supplementary text to classes in vertebrate zoology.

682. King, Gillian M., and David R. N. Custance. **Colour Atlas of Vertebrate Anatomy: An Integrated Text and Dissection Guide**. Oxford, Boston: Blackwell Scientific Publications, 1982. 1 vol. (various paging). $37.00. ISBN 0-632-01007-X.

By means of extensive colored photographs and labeled line-drawings this work walks the reader through the dissections of a number of vertebrates representing major groups. Organisms represented are dogfish shark, skate, codling, frog, lizard, pigeon, and the rat. A section on the vertebrate brain, comparing the brains of the dogfish, frog, pigeon, and rat, is also present. The prochordates, which can be found in the introductory information, consists of general anatomy of the tunicates and amphioxus. The colored photographs depict the dissected animal. Internal organs and the skeletal system are included. Labeled line-drawings represent the dissection in order to bring clarity to the anatomical features. The text material considers the anatomical structures. Basic dissection procedures are also considered. Labeled line-drawings showing the circulatory system are colored in blue and red for arteries and veins (oxygenated and deoxygenated blood). A subject index completes the work. This source is useful for comparative anatomy and can serve as a supplementary text.

683. Owen, Richard. **On the Anatomy of Vertebrates**. London: Longmans, Green, 1866. Reprint, New York: AMS Press, 1973. 3 vols. $225.00 set. LC 72-1701. ISBN 0-404-08300-5.

This three-volume set considers in great detail the external and internal anatomy of the major vertebrate groups. After briefly presenting the basic characteristics of the vertebrates as a whole, the skeletal, muscular, nervous, digestive, circulatory, respiratory, urinary, integumentary, and reproductive systems of the vertebrate groups follow. Some attention is also given to the endocrine glands known at the time. Various vertebrate types are mentioned for specific anatomical principles. Labeled line-drawings enhance the detailed anatomical description given in the text. Bibliographies can be found at the end of each volume.

Volume 1. *Fishes and Reptiles.*
Volume 2. *Birds and Mammals.*
Volume 3. *Mammals.*

684. **Traité de Zoologie: Anatomie, Systématique, Biologie**. Tome XII, **Vertébrés Embryologie, Comparée, Caractéristiques Biochimiques**. Ed. by Pierre-Paul Grassé Paris: Masson, 1954. 1146p. Fr. 1,482. ISBN 2-225-58399-4.

A general description of the major set to which this volume belongs appears in chapter 1 (see entry 69). This volume deals with the vertebrates in general. In addition to general characteristics, this source considers embryology, comparative anatomy, and biochemical characteristics of the vertebrates. It is written in French. A four-page bibliography and combined subject and scientific-name index conclude the volume.

TEXTBOOKS

685. Alexander, R. McNeill. **The Chordates**. 2d ed. Cambridge, New York: Cambridge University Press, 1981. 510p. $79.50; $32.50pa. LC 80-41275. ISBN 0-521-23658-4; 0-521-28141-5pa.

Encompassing a large number of physiological principles and a number of mathematical models to support the principles, this textbook considers the vertebrates, including the primitive chordates. The emphasis of this text is on how individual vertebrate groups adapt to their environments in relation to the structures they possess, and on how those structures function. This is not one of the many texts whose empasis is on comparative anatomy. The chapters are arranged by large vertebrate groups such as fish, amphibians, reptiles, etc. In some cases these groups are broken down into a number of chapters. In the case of the fish there are sections devoted to the agnatha, chondrichthyes, teleosts, carps and perches, and lobe-finned fish. Certain types of activities are also included within the sections such as the physiology of swimming, locomotion, and flight. Various systems are considered as they relate to particular vertebrate types. Graphs, charts, and line-drawings provide the illustrative matter and are useful in clarifying principles. A list of selected references can be found at the end of each chapter. A subject, common, and scientific-name index concludes the volume.

686. Hildebrand, Milton. **Analysis of Vertebrate Structure**. 3d ed. New York: John Wiley, 1988. 701p. $52.95. LC 87-21545. ISBN 0-471-82568-9.

Not nearly as detailed as *Functional Vertebrate Morphology* (see entry 687), this textbook provides an excellent introduction to the various aspects pertaining to vertebrate zoology. Basically, divided into three sections, it consists of a survey of vertebrate animals, evolution of vertebrate structures, and structural adaptation of vertebrate structure in relation to their habitats and behavior. Following a phylogenetic approach, section 1 considers the various vertebrate classes and subclasses and presents some general history, evolution, and anatomy of these groups. Section 2 consists of the various vertebrate systems and, by means of certain vertebrate groups, shows the similarities and differences of the systems in the different types of vertebrates (comparative anatomy). Section 3 considers, using a physiological and anatomical approach, the various activities that vertebrates exhibit, such as climbing, flying, digging, running, etc. This section most closely resembles Hildebrand's *Functional Vertebrate Morphology* (see entry 687). Labeled and unlabeled line-drawings provide the illustrations and aid in the understanding of the text. A brief bibliography for each of the chapters and a combined index of subjects, common names, and scientific names complete the volume.

87. Hildebrand, Milton, and others, eds. **Functional Vertebrate Morphology**. Cambridge, Mass.: Belknap Press of Harvard University Press, 1985. 430p. $41.00. LC 84-19175. ISBN 0-674-32775-6.

Considered as a text for upper-division undergraduate as well as graduate students, this work considers the various biological activities in terms of the vertebrates. It is heavily involved in physiological principles as they pertain to these activities. Many examples from physics are considered. Each of the sections is devoted to a particular activity such as walking, flying, mastication, running, swimming, soaring, and swallowing. For each of these sections, specific vertebrates are used to illustrate the physiological principles. For example, in the section on flying, such principles as forces and velocities, power components, and scaling are presented. Birds are the obvious choice for this section. Graphs, charts, mathematical formulas, and an occasional black-and-white photograph aids in the explanation. This is not the basic comparative anatomy text. Over 1,100 references to the literature are listed near the end of the work. They are arranged under the names used for the chapters within the text. A combined index of subjects, common names, and scientific names concludes the volume.

88. Hyman, Libbie Henrietta. **Hyman's Comparative Vertebrate Anatomy**. 3d ed. Chicago: University of Chicago Press, 1979. 788p. $36.00. LC 79-731. ISBN 0-226-87011-1.

Following a comparative anatomical approach, this work considers the basic systems found within vertebrates and depicts the changes in the systems from one vertebrate group to another. The text begins with general information on classification, adaptive radiation, and characteristics of the vertebrates and the vertebrate groups. Preceeding the account of the various vertebrate systems, areas such as a description of the primitive chordates (balanoglossus, tunicates, amphioxus), and a section on general vertebrate embryology are explored. Evolutionary trends are considered. Vertebrate systems covered are integument, endoskeleton, muscles, digestive, respiratory, circulatory, urogenital, nervous, and the sense organs. Vertebrate groups are used to illustrate evolutionary trends within the systems. A large number of labeled line-drawings are used to illustrate basic anatomical principles. Summaries of each chapter and a list of selected references appear at the end of each chapter. A subject index can be found at the end of the textbook.

89. Kent, George C. **Comparative Anatomy of the Vertebrates**. 7th ed. St. Louis, Mo.: C.V. Mosby, 1991. 688p. LC 91-11704. ISBN 0-8016-6237-0.

Based on an evolutionary perspective, this source not only considers the vertebrate systems but gives some general information on the vertebrates themselves. The text opens with a discussion of the vertebrate body plan as well as the functional anatomy of the lower chordates (hemichordates, tunicates, and cephalochordates). Approximately 54 pages are devoted to the various animal groups making up the vertebrates. General external anatomy and evolutionary trends are considered. A section on vertebrate embryology is also present. The main portion of the work is devoted to the various vertebrate systems. For each of the systems covered, a description considers the anatomical and physiological differences for various vertebrate groups; evolutionary trends are shown among the groups. The systems discussed are skin, skeleton, skull, girdles, fins, limbs, locomotion, muscles, digestion, respiration, circulation, urogenital, nervous, sense organs, and the endocrines. A summary and list of selected references can be found at the end of each chapter. Labeled line-drawings aid in understanding the basic principles set forth in the text. The appendices consist of an abridged classification of the vertebrates (down to suborder), and a brief glossary of selected prefixes, suffixes, roots, and stems employed in the text. A combined index of subjects, common names, and scientific names concludes the volume.

90. Pough, F. Harvey, John B. Heiser, and William N. McFarland. **Vertebrate Life**. 3d ed. New York: Macmillan, 1989. 904p. $67.00. LC 88-22092. ISBN 0-02-396360-3.

Divided into four parts, this excellent text considers all the major biological aspects of the vertebrates. Line-drawings, many of which are labeled, as well as occasional black-and-white photographs, provide the illustrative material. Part 1 considers the diversity of vertebrates, their function, and evolution. Such areas considered are evolution, classification vertebrate body plan, evolution of organ systems, and homeostasis and energetics, as well a the geology and ecology found during the origin of the vertebrates. Part 2 deals with th earliest vertebrates, and goes on to discuss in subsequent chapters the agnathans, chondrich thyes, and osteichthyes, as well as the geology and ecology of the origin of tetrapods. Part includes the amphibians and reptiles and their evolution, while part 4 describes the birds an mammals. In each of these four sections, anatomy, physiology, evolution, ecology, an behavior are discussed. Some chapters within these sections deal exclusively with evolution while others consider physiological principles inherent to a particular vertebrate group. I any event, it is a well-balanced text. A summary and a list of references appear at the end o each chapter. The book concludes with a glossary, author index, and subject index.

691. Romer, Alfred Sherwood, and Thomas S. Parsons. **The Vertebrate Body**. 6th ed Philadelphia: Saunders College, 1986. 679p. $40.00. LC 85-8196. ISBN 0-03-058446-9 0-03-058443-4; 0-03-910754Xpa.

Following a comparative anatomical approach, this work has been considered a classi for some time. It is one of the classic texts in the field. A brief history of the vertebrate body a section on the anatomy and physiology of the lowerer chordates (balanoglossus, tunicates amphioxus), a general look at the classification scheme of the vertebrate groups, and a brie discussion on vertebrate embryology can be found in the opening sections. The rest of th work follows a systems approach. For each system included, the anatomy and some physiol ogy is considered for the various vertebrate groups. The sections follow the development o the systems from the agnatha to the mammals. Systems included are the skin, skeleton, skull muscles, body cavities, mouth and pharynx, respiratory, digestive, excretory, reproductive circulatory, sense organs, nervous, and endocrines. Labeled line-drawings, sketches of man of the vertebrates, and colored diagrams aid in the understanding of the principles foun within the text. The appendices consists of a synoptic classification of chordates, a brie glossary of anatomical terms, and a 21-page bibliography. A combined index of subjects common names, and scientific names concludes the volume.

692. Stahl, Barbara J. **Vertebrate History: Problems in Evolution**. Republication, rev. an enl. New York: Dover, 1985. 604p. $16.95pa. LC 84-21037. ISBN 0-486-64850-8.

This textbook is a revised and enlarged edition of the work published in 1974 b McGraw-Hill. A new preface, updated notes to supplement the original text, and supplemen tary references have been added.

From the most primitive of vertebrates to the mammals, this work considers th evolution of these animals as evidenced by the fossil record. Line-drawings of skulls, brains and skeletons are used to illustrate these relationships. Such topics as bone and cartilage i early vertebrates, the rise of jawless fishes to those bearing jaws, the rise of the modern fish water to land transition of vertebrates, and the relationships of the reptiles to the birds an mammals are considered in some depth. A closing chapter presents the most successful of th vertebrates and why they are considered so. The appendix consists of notes not found in th 1974 edition. A 24-page bibliography including the supplementary references and sources o illustrations, and a combined index of subjects, common names, and scientific names conclud the volume. It should be noted that there is not much material added from the 1974 edition.

693. Walker, Warren F. **Functional Anatomy of the Vertebrates: An Evolutionar Perspective**. Philadelphia: Saunders, 1987. 781p. $40.00. LC 50-12317. ISBN 0-03 064239-6.

Taking a comparative anatomical approach, this textbook presents vertebrate informa tion dealing with anatomy, physiology, physiological ecology, and evolution. The work i divided into 19 chapters. Chapter 1 provides an introduction dealing with homology, analogy phylogeny, surface-volume relationships, and so on. Chapter 2 is an introduction to th primitive chordates such as the hemichordates, tunicates, and cephalochordates. Chapter introduces the student to the various groups of vertebrate classes. Chapter 4 discusses th general embryology of vertebrates. The rest of the chapters represent different organ systems. For each of these systems, vertebrate groups are introduced showing anatomical, physiological, and evolutionary relationships. A summary is given at the end of each chapter. Labeled and unlabeled line-drawings provide the illustrative material. A 19-page bibliography consisting of references pertinent to each of the chapters, and a combined glossary/index conclude the textbook.

694. Waterman, Allyn J. **Chordate Structure and Function**. 2d ed. New York: Macmillan, 1977. 628p. LC 75-45173. ISBN 0-02-364800-7.

Arranged in three parts this text considers the anatomy, physiology, embryology, and evolutionary trends found among the vertebrate types. Preceeding the vertebrates, the work offers a detailed look at the chordate group and its relationship to vertebrates. Chordates discussed are the balanoglossus, tunicates, and amphioxus. In taking a comparative anatomy approach, chapters dealing with specific systems are presented. Differences in the systems are discussed as they apply to the various vertebrate groups. Systems include integument, skeletal, muscle, digestion, respiration, circulation, excretion, sensory, nervous, endocrine, and reproduction. Evolutionary trends of the different vertebrate types are considered as each system is presented. There are a large number of labeled diagrams, which greatly enhance the principles set forth in this text. A list of selected references appears at the end of each chapter. A subject index concludes the volume.

695. Weichert, Charles K. **Anatomy of the Chordates**. 4th ed. New York: McGraw-Hill, 1970. 814p. $50.12. LC 70-121668. ISBN 0-07-069007-3.

In addition to the classification and evolution of the chordates as well as some general embryology, this text explores the various systems found in vertebrates. For each of the systems included, the anatomy and some physiology of the different vertebrate groups are explored. This is accomplished through a comparative anatomy approach. The systems considered are integument, digestive, respiratory, excretory, reproductive, endocrine, skeletal, muscle, circulatory, nervous, and sense organs. A summary is given at the end of each chapter. A number of labeled line-drawings aid in the understanding of the textural material.

696. Willson, Mary F. **Vertebrate Natural History**. Philadelphia: Saunders College, 1984. 621p. $39.00. LC 83-10114. ISBN 0-03-061804-5.

Illustrated with graphs, charts, line-drawings, and geographical range maps, this textbook considers the ecology, behavior, taxonomy, and physiology of vertebrates. Such specific areas as the classification, characteristics, relationships, geographical distributions, perception, thermoregulatin, osmoregulation, respiration, locomotion, migration, food and feeding, problems with predators, social behavioral patterns, sex, courtship and mating, life histories, and parental care are presented. Specific vertebrate species and groups are used to illustrate these biological principles. A list of references appears at the end of each chapter. A combined index of subjects, common names, and scientific names complete the volume. This is a readable and informative text.

697. Young, J. Z. **The Life of Vertebrates**. 3d ed. New York: Oxford University Press, 1981. 645p. $32.50. LC 81-206006. ISBN 0-19-857172-0; 0-19-857173-9pa.

Arranged phylogenetically, this classic work considers all vertebrate groups and their relationships to each other. Each chapter either represents a class or an order of vertebrates. Each of these sections considers their anatomy, physiology, and evolution. Various anatomical structures and physiological features are followed and relationships are presented from one group of vertebrates to another. The taxonomy of the various vertebrate groups is also considered. Examples of topics included are adaptive radiation of bony fishes, evolution of the reptiles, bird behavior, evolution of placental mammals, origin of mammals, and evolutionary changes of the life of vertebrates. Line-drawings (many of them labeled) and graphs provide much of the illustrative matter. A list of references arranged by vertebrate group and combined index of subjects, common names, and scientific names complete the work.

JOURNALS

698. **Northwestern Naturalist: A Journal of Vertebrate Biology**. Vols. 1- , nos. 1- Olympia, Wash.: Society for Northwestern Vertebrate Biology, 1989- . 3/yr. $25.00/yr. ISSN 0027-3716 (**The Murrelet**).

Consisting of research papers dealing with all aspects of amphibian, reptilian, bird, and mammal biology, this journal is devoted to those species of vertebrates found in northwestern North America (northwestern Canada, northwestern United States, including Alaska). The number of research articles per issue varies from one to four or more. Lengths of these article are usually 10 or more pages. In addition, a number of shorter articles found under the genera notes section appear in each of the issues. These articles are approximately two to four page in length. Graphs, tables, charts, and photographs provide the majority of the illustrations. I should be noted that before 1989 this serial was titled *The Murrelet: A Journal of Northwes Ornithology and Mammalogy* and began publishing in 1920.

TAXONOMIC KEYS

699. Blair, W. Frank, and others. **Vertebrates of the United States**. 2d ed. New York McGraw-Hill, 1968. 616p. LC 67-18322.

Divided into six sections, this work provides basic taxonomic information on the classes orders, families, genera, and species of the vertebrate groups inhabiting the United States The groups under discussion are the fishes, amphibians, reptiles, birds, and mammals. Specie are arranged under their respective genus, which in turn are listed under their family, order and class. For most of the species included, basic anatomical characteristics, coloration, and geographical distribution are given. Information for the classes, orders, families, and gener consist of basic external anatomy, and other interesting facts such as eggs. Keys to orders suborders, families, genera, and species are included. An introductory part discussing genera information on the vertebrates is presented as well. Line-drawings of vertebrates in part or a a whole help in the identification process. A list of references can be found at the end of eac section. An adequate glossary, and a combined common/scientific-name index complete thi taxonomic work.

700. Whitaker, John O. **Keys to the Vertebrates of the Eastern United States, Excluding Birds**. Minneapolis, Minn.: Burgess, 1968. 256p. LC 68-5080.

Arranged phylogenetically, keys to families, genera, and species are presented tha reflect those vertebrates east of the Mississippi. As the title suggests, birds are not included Geographical maps depict the ranges of these species. A 17-page bibliography and scientific-name index conclude this compilation of keys.

CHECKLISTS AND CLASSIFICATION SCHEMES

701. Banks, Richard C., Roy W. McDiarmid, and Alfred L. Gardner, eds. **Checklist of Vertebrates of the United States, the U.S. Territories, and Canada**. Washington, D.C. U.S. Dept. of the Interior, Fish and Wildlife Service; Springfield, Va., distributed by the National Technical Information Service, 1987. 79p. (Resource Publication [U.S. Fish and Wildlife Service, no. 166]). LC 87-600208.

This work represents a listing of the scientific names of those vertebrates found throughout the United States, the territories of the United States, and Canada. The vertebrat groups considered are the amphibians, reptiles, birds, and mammals; the fish are excluded. Each species is arranged under its respective genus, which in turn is entered under family, order, and class. For each of the genera and species listed, the person who first named it, date of naming, and common name is given. Common names are also presented for the families. This checklist represents 226 amphibian species, 394 reptilian species, 1,100 bird species, and 467 species of mammals. Illustrations and index are lacking.

702. Committee of the Bermuda Audubon Society. **A Check List of the Birds, Mammals, Reptiles, and Amphibians of Bermuda**. [s.l.], The Society, 1959. 31p.

A total of three amphibians, three reptiles, six mammals, and 22 birds are included. The species are arranged under their respective classes. For each of the vertebrates mentioned, the common name, scientific name, and general information that varies from group to group are given. For example, breeding information is given for the birds. A reference list of the resident and migratory birds of Bermuda showing frequency of occurrence, seasonal distribution, and general abundance can be found at the end of the booklet. There is no index.

703. Philibosian, Richard, and John A. Yntema. **Annotated Checklist of the Birds, Mammals, Reptiles, and Amphibians of the Virgin Islands and Puerto Rico**. Frederiksted, St. Croix, U.S. Virgin Islands: Information Services, 1977. 48p. LC 77-670077.

The species are arranged under their respective families, which, in most cases, are entered under orders. The common and scientific name of the organism are given along with a chart indicating geographical distribution, seasonal information, status, and breeding information, if available. A total of 23 amphibians, 60 reptiles, 267 birds, and 42 mammals are presented. An index to English and Spanish common names concludes the checklist.

ASSOCIATIONS

704. **Society for Northwestern Vertebrate Biology**. Washington State Dept. of Wildlife, Wildlife Management Division, 600 Capitol Way N., Olympia, WA 98501-1091. (206) 753-2868.

705. **Society of Vertebrate Paleontology**. Natural History Museum, 900 Exposition Boulevard, Los Angeles, CA 90007. (213) 744-3310.

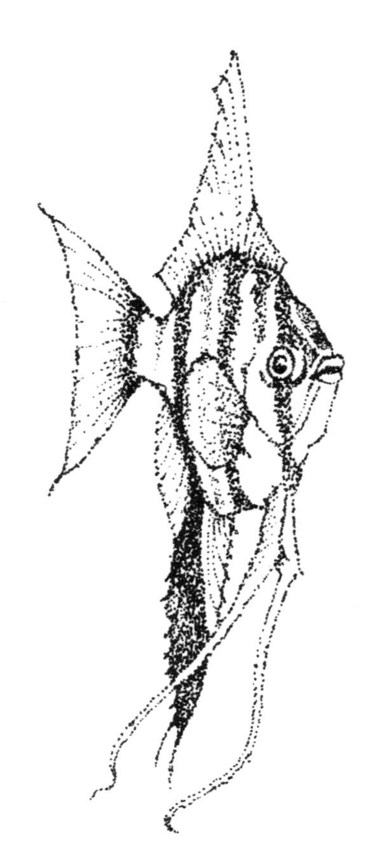

CHAPTER 5

THE FISHES

All fish are cold-blooded and carry ou their entire life cycle in either fresh, salt, o brackish waters, or combinations thereof. The ecological niches range from the depths of th ocean to the shallow waters of streams. They a possess scales, which come in assorted shape and forms. From an evolutionary point of view these scales became modified in the higher ve tebrate groups. The feathers on birds and the ha on mammals are examples of this modification

There are three basic groups of fish com prising approximately 20,860 species. Th Agnatha, known as the jawless vertebrates, con tain such organisms as the lamprey eels and th hagfishes. They are considered the most primi tive of the fish and contain approximately 6 species. The Chondrichthyes, which are employed frequently by Hollywood, con tain the sharks, rays, and skates. They possess a cartilaginous, rather than a bon skeleton, have a ventral jaw, and in some cases bear live young. Approximately 84 species are contained within this group. The most advanced of the fish belong t the group known as the Osteichthyes. They have a bony skeleton. Most of the fis likely to be encountered, eaten, or both are Osteichthyes. Approximately 20,00 species belong to this group.

We fish are upheld and supported to all sides
We lean confidently and harmoniously upon our element
We move in all dimensions and whatever course we take
the mighty waters out of reverenc
for our virtue change shape accordingly
—Isak Dinese
Seven Gothic Tales, *193*

DICTIONARIES

706. Kotlyar, A. N. **Dictionary of Names of Marine Fishes on the Six Languages.** Moscow: Russky Yazyk, 1984. 288p. $29.95. ISBN 0-8288-0326-9.

The main portion of the work presents the scientific names of the species arranged under their respective families (289 families are represented). The common names of these fishes, in Russian, French, German, Spanish, and English, are listed under the scientific names. Not all languages are present for many of the species. A scientific-name index, and an additional common-name index for the fishes can be found at the end of the dictionary. The common-name index is in English, Russian, French, German, and Spanish. There are no illustrations.

707. Krane, Willibald. **Five-Language Dictionary of Fish, Crustaceans and Molluscs.** New York: AVI, 1986. 476p. LC 89-22448. ISBN 0-442-30278-9.

Consisting of over 2,000 terms, this work considers species of fish, crustaceans, and molluscs that are of economic importance. Fish product names are also considered. The main portion of the dictionary includes the word in English (with synonyms) and then proceeds to list the equivalent of the word in Latin, German, French, Spanish, and Italian. The remainder of the volume represents a number of indices to indicate where the word can be found within the main portion of text. The indices, which are separate, are lists of names in English, Latin, German, French, Spanish, and Italian.

708. Lindberg, G. U., A. S. Heard, and T. S. Rass. **SlovarQ Nazvanifi Morskikh Promyslovykh Ryb Mirovofi Fauny Of. Multilingual Dictionary of Names of Marine Food-Fishes of World Fauna. Worterbuch der Namen der Marinen Nutzfische der Weltfauna. Dictionnaire des Noms des Poissons Marins Comestibles de la Faune Mondiale. Diccionario de Nombres de Peces Marinos Comunes de la Faune Mundial.** Leningrad: "Nauka," Leningradskoe otd-nie, 1980. 562p. $45.00. LC 81-143423. ISBN 0-8288-0327-7.

Consisting of 3,000 species of sea fishes having food value, this source provides nomenclature information in a variety of languages. Each species is entered under its respective genus, which in turn is entered under family. The person who first named it and date of naming accompanies each genus entry. For each of the species listed, the scientific name, person who first named it, date of naming, modern Russian names, and names of the fish in a variety of other languages are given. A small but adequate black-and-white drawing of the fish accompanies many of the species entries.

Introductory information is presented in Russian, English, French, German, Spanish, and Japanese. A six-page bibliography proceeds the introduction. A separate scientific-name index, Russian-name index, and a common-name index (many languages interfiled alphabetically) conclude the polyglot.

709. Organization for Economic Co-operation and Development. **Multilingual Dictionary of Fish and Fish Products. Dictionnaire Multilingue des Poissons et Produits de la Peche.** 2d ed. Farnham, Surrey, England: Fishing News Books, 1978. 430p. $95.00. ISBN 0-85238-086-0.

Comprising names of fish and fish products (1,117 entries), this dictionary lists equivalent terms, when known, in Danish, Dutch, English, French, German, Greek, Icelandic, Italian, Japanese, Norwegian, Portuguese, Serbo-Croat, Spanish, Swedish, and Turkish. Scientific names are also presented. The terms are alphabetically arranged in English, and defined in English and French. Other languages are listed below the definitions. The indexes, which are separate, consist of alphabetical arrangements of the terms in English, French, German, Danish, Spanish, Greek, Italian, Icelandic, Japanese, Norwegian, Dutch, Portuguese, Swedish, Turkish, and Serbo-Croat. In addition, an index for the scientific names of the fish is also present.

710. Rojo, Alfonso L. **Dictionary of Evolutionary Fish Osteology**. Boca Raton, Fla.: CRC Press, 1991. 273p. $69.95. LC 91-16508. ISBN 0-8493-4214-7.

The 375 terms dealing with fish osteology (mainly consisting of the names of the bones of fish), are arranged alphabetically in the English language. Under each of the entries the name is translated into the French, German, Latin, Russian, and Spanish languages. Definitions are also provided for the terms. The length of these definitions, which are written in English, vary from one paragraph to several pages. In some cases, references to the literature are presented at the end of the definition. Contained within the introduction is information on the skeleton plan of a fish along with techniques and chemicals that are useful in preparing fish skeletons. Preceding the list of terms there appears an analogical section that arranges the terms under broad subject categories. Therefore, if interested in terms dealing with the integumentary system, the reader will find a listing of these with corresponding page numbers in this section. Headings within the analogical section are basic terms, morphology, appendicular system, skeletal system, digestive system, reproductive system, organ of equilibrium, and integumentary system.

An 18-page bibliography proceeds the alphabetical list of terms. In addition, separate indexes comprising alphabetical lists of the terms in English, French, German, Latin, Russian, Spanish, and Latin appear near the end of the work. A large number of labeled line-drawings depicting the various skeletal parts, including the vertebrae of fish, conclude the dictionary.

ENCYCLOPEDIAS

711. Allyn, Rube. **A Dictionary of Fishes**. 3d ed. St. Petersburg, Fla.: Great Outdoors Association, 1953. 114p. $7.95pa. ISBN 0-8200-0101-5.

Consisting of over 500 species of freshwater and saltwater fishes found throughout the world, this source provides information in a brief encyclopedic-type format. The fishes are arranged under common name groups such as deep-ocean fishes, snake-like fishes, and snappers. For each of the species presented, the common name, scientific name, additional common names, weight and/or size, coloration, geographical range, diet, and habitat are given. A black-and-white drawing of the fish accompanies each descriptive account. The average length of the description is usually a half column. It should be noted that a number of vertebrates that are not fishes are included within the last few pages of the work. The non-fish vertebrates listed are dolphins, porpoises, whales, and turtles. Colored plates depicting a number of the fishes mentioned in the text can be found in the front and in the back of the book. A common-name index, found near the end of the volume, provides access to the fish species.

712. Grzimek, Bernhard, ed. **Grzimek's Animal Life Encyclopedia: Fishes I, Volume 4, 1974**. New York: Van Nostrand, 1974. 531p. LC 79-183178.

The discussion of the various fishes inhabiting the planet are contained within chapters designated by common-name groups. Examples are the tarpons, the eels, and the salmon. In addition to chapters on various fish groups, a discussion of larger groups such as the bony fishes and the cartilaginous fishes are also present. The fish groups are discussed in relation to their families and the species contained therein. Such information as coloration, geographical distribution, egg-laying, external anatomy, size, behavior, diet, habitat, and other interesting facts is included. A large number of colored photographs can be found interspersed among the pages of the book. A systematic classification of the fishes, a polygot dictionary listing the English, French, German, and Russian names of these creatures, a three-page bibliography, and a combined index of subjects, common names, and scientific names complete the volume.

713. Grzimek, Bernhard, ed. **Grzimek's Animal Life Encyclopedia: Fishes II and Amphibia, Volume 5, 1974**. New York: Van Nostrand, 1974. 555p. LC 79-183178.

Continuing where Grzimek's *Fishes I* stopped (see entry 712), this source continues the discussion on the fishes (267 pages) and considers the amphibian group for the remainder of the volume. Such fish types as the perch, mullets, barracudas, sticklebacks, flounders, puffers, coelacanths, and lungfishes, among others, are discussed. The volume follows the same format as *Fishes I*. The amphibian portion of the book will be discussed under the chapter on amphibians.

714. Herald, Earl Stannard. **Living Fishes of the World**. Garden City, N.Y.: Doubleday, 1972; Reprint. Garden City, N.Y.: Doubleday, 1961. 304p. (The World of Nature Series). LC 61-6384. ISBN 0-385-00988-7.

The families of fishes are arranged under their respective orders, which in turn are entered under their classes. Some brief descriptive information is given for the class and order. For each family listed, the number of genera and species, representative species, geographical locations, size, basic external anatomy, courtship, egg laying, incubation, and information regarding the hatchling are presented. The above descriptive information varies somewhat from family to family and the length varies from one column to several pages. The majority of the illustrations are colored photographs of the fishes in their natural habitats. Some black-and-white photographs are present as well. A one-page glossary, a two-page bibliography, and a combined common/scientific-name index concludes the source. This volume gives the reader an excellent introduction to the many families of fishes that inhabit the planet.

715. Le Danois, Edouard. **Fishes of the World**. Woodstock, Vt.: Countryman Press, 1957. 190p. LC 57-13951.

Arranged by geographical regions, this source provides general information on many of the species of fishes found throughout the world. The regions include the northern (20 pages), temperate (20 pages), tropical (52 pages), southern (13 pages), and those species found at great depths (5 pages). For each of these regions, behavioral patterns, size, names of a few of the species, water temperature, very basic external anatomy, coloration, weight, salinity, and other interesting facts are given. Colored as well as black-and-white photographs provide the illustrative matter. A two-page bibliography and a combined common/scientific-name index conclude the volume. Although interesting to read, *Living Fishes of the World* (see entry 714), gives the reader a more informative and organized account of the various fishes than does this work.

716. Migdalski, Edward C., and George S. Fichter. **The Fresh & Salt Water Fishes of the World**. New York: Greenwich House. Distributed by Crown, 1983. 316p. LC 83-8955. ISBN 0-517-41670-0.

Containing the better-known species of the world, this work considers the fishes found in both fresh and salt water. Basically, the volume is divided into three areas: the jawless fishes, the cartilaginous fishes, and the bony fishes. The families of fishes are described under their respective orders. Scientific and common names of the families are given. Family descriptions consist of general external anatomy, length, depth in which they are found, geographical distribution, brief remarks regarding representative species within the family, and other interesting facts about the group as a whole. Colored plates of those species mentioned in the family descriptions accompany the text. Introductory material includes a checklist of common and representative families of living fishes, a brief evolutionary history of fishes, general information on anatomy and physiology (15 pages), and a brief discussion on commercial and sport fishes. A one-page bibliography, a listing of the largest fishes taken in fresh and salt water, and a combined common/scientific-name index conclude the work.

This is a useful work for a general understanding of the various groups of fishes. It should be noted, however, that this volume does not represent detailed descriptive matter on specific species of fishes. It is a good place to start.

717. Wheeler, Alwyne. **The World Encyclopedia of Fishes**. London: Macdonald, 1985. 368p. LC 85-175105. ISBN 0-356-10714-9.

Divided into two sections, this work presents a large variety of the types of fishes inhabiting the waters of the earth. Section 1 is compiled of a group of 501 colored plates arranged by families. Section 2 represents the descriptive accounts of the species. The organisms are arranged under their families. The families are arranged alphabetically. For each species included, the scientific name, common name, size, brief external anatomy, geographical range, diet, and habitat are included. A small line-drawing of the fish accompanies the description. Each description is usually two paragraphs in length. In addition, some general information regarding the family is considered. Unfortunately, no index is provided.

BIBLIOGRAPHIES

718. Dean, Bashford. **A Bibliography of Fishes**. New York: Russell & Russell, 1962. 3 vols. LC 61-13774.

Based on Dean's personal reference research file, this three-volume set contains references to the literature relating to fishes, voyages and expeditions, and fish culture. The citations, which are arranged under the author's last name, consist of material dated from approximately the mid-eighteenth century to the early part of the twentieth century; references authored by Linnaeus are included. References depicting general bibliographies as well as those from the periodical literature are considered. Volume 3, which was extended and edited by Eugene Willis Gudger, consists of indices, general bibliographies, periodicals relating to fishes, early works, voyages and expeditions, addenda, and errata of volumes 1 and 2. The indices consist of a subject index, which is broken into a morphological section and a systematics section, as well as a finding index, which considers general subjects, common names, and scientific names. This work is a classic in the field.

It should be noted that the 1962 edition described was limited to a printing of only 350 sets.

719. Chock, Karl-Hermann, comp. **Bibliography of Antarctic Fish**. Hamburg, Germany: Bundesforschungsanstalt fur Fischerei, Informations und Dokumentationsstelle, 1989. 136p. (Literaturliste; no. 9). $8.60 (5¼-inch floppy disk).

Approximately 1,300 entries can be found in this volume consisting of citations to the literature dealing with Antarctic fish. Such fields as taxonomy, zoogeography, ecology, physiology, and exploitation are covered. The citations are arranged under author's last name and consist mainly of journal literature from the 1950s through the 1980s. Unfortunately, there are no indexes to the work, thereby lowering its usefulness. This work is available in print or on 5¼-inch floppy disk.

720. Ardizzone, G. D., comp. **A Bibliography of African Freshwater Fish, Supplement 1, 1968-1975. Bibliographie des poissons d'eau douce de l'Afrique, Supplement 1, 1968-1975**. Rome: Food and Agriculture Organization of the United States, 1976. 40p. (CIFA Occasional Paper; no. 5). $8.00pa. ISBN 92-5-000092-8.

721. Matthes, H., comp. **A Bibliography of African Freshwater Fish. Bibliographie des poissons d'eau douce de l'Afrique**. Rome: Food and Agriculture Organization of the United States, 1973. 299p. $25.00. LC 74-186158. ISBN 92-5-101595-3.

Together these two publications represent a list of citations that deal with freshwater fish as well as marine or brackish-water fish that can be found from time to time in freshwater throughout the African continent. Citations are arranged under the author. The majority of the citations refer to journal literature in the English, French, and German languages. A number of monographs are also present. There are no abstracts. The references compiled represent the last half of the nineteenth century and the twentieth century up to 1976. A subject, geographic, zoogeographic, and taxonomic index can be found at the end of the main volume as well as the supplement. The indexes are prepared under broad categories with no subheadings, thereby negating much of their usefulness. For example, one of the subjects in the subject index is swimbladder, followed by a list of numbers that represent the citation entries. If interested in swimbladders of a particular type of fish, the reader would be in for some rough waters.

INDEXES AND ABSTRACTS

722. **Fisheries Review**. Vols. 1- . Fort Collins, Colo.: U.S. Fish & Wildlife Service, 1955- . Quarterly. $16.00/yr. (individual subscription). ISSN 0038-786X.

Formerly known as *Sport Fishery Abstracts* (1955-1985), this source provides citations primarily to the journal literature. Such ichthyological areas as culture and propagation, limnology, oceanography, morphology, physiology, genetics, behavior, natural history, parasites and diseases, pollution, toxicology, and research and management are represented. Indexes to the citations include author, geographic, systematic, and subject. Abstracts were included in *Sport Fishery Abstracts*, but have been excluded in *Fisheries Review*. *Fisheries Review* is also available on compact disk.

It is available on two compact disk products. Both are available through the National Information Services Corporation, Suite #6, Wyman Towers, 3100 St. Paul Street, Baltimore, Maryland 21218. One CD is titled *Wildlife Review/Fisheries Review*. It includes the databases for both print indexes and begins coverage in 1971. The cost is $695.00 per year with semiannual updates.

The other CD is titled *Fish and Fisheries Worldwide*. In addition to *Fisheries Review* (1971-present), the CD also contains *FishLit* (1985-present), which is produced by the JLB Smith Institute of Ichthyology at Rhodes University in South Africa, *Aquaculture* (1970-1984), which is produced by the National Oceanic & Atmospheric Administration (NOAA), and *Fish Health News* (1978-1985), which is compiled by the U.S. Fish & Wildlife Service, Technical Information Services. The cost of the CD is $695.00 per year with semiannual updates.

FishLit. Contains about 30,000 records and covers tropical areas, Africa, and developing countries. Comprehensive coverage is given to African freshwater fish.

Aquaculture. Contains approximately 10,000 citations on marine, freshwater, and brackish organisms.

Fish Health News. Equivalent to the publication produced by the U.S. Fish and Wildlife Service, Technical Information Services. References published after 1985 are included in the Fisheries Review database.

723. **Zoological Record**. Vols. 1- , nos. 1- . Philadelphia: BIOSIS, Zoological Society of London, 1865- . Annual. $2,200.00/yr. ISSN 0084-5604.

For a description of this index, see entry 55. The section that relates to this chapter is Section 15 on Pisces.

DIRECTORIES

724. Becker, C. Dale, ed. **Fisheries Laboratories of North America**. Bethesda, Md.: American Fisheries Society, 1991. 180p. $16.00. LC 90-085282. ISBN 0-913235-67-9.

Consisting of 50 fisheries laboratories in North America, this source provides information on these institutes. For each of the laboratories included, the address, laboratory objectives, a list of personnel (directors and heads), laboratory history, laboratory features, areas of expertise, and cooperating agencies are considered. A map designating the area where the laboratory is located accompanies each description. The descriptions are usually two to three pages in length. Black-and-white photographs of the laboratories provide the illustrative material. An appendix listing the name and phone number of each laboratory concludes the reference source.

725. McAleer, Beth D., ed. **The Directory of North American Fisheries and Aquatic Scientists**. 2d ed. Bethesda, Md.: American Fisheries Society, 1987. 363p. $12.50. LC 86-72993. ISBN 0-913235-40-7.

Consisting of over 10,000 listings, this source provides the address, phone number specialties, and activities of each scientist included. The specialties and activities are entered in codes and can be deciphered in the specialties and activities key found near the beginning of the work. The specialty codes are arranged in a hierarchical format. Therefore physiology would be "1.16" (1.00 for biology and .16 for physiology). Major code headings are biology fish culture, economics, limnology, fisheries technology, oceanography, social sciences, and habitat and water quality. Some of the activity categories are education, research, communications, development, law enforcement, and habitat management. In addition to the alphabetical listing of the scientists, indices provide access by geography, activity, and specialty.

HANDBOOKS

Pisces (General)

726. Berra, Tim M. **An Atlas of Distribution of the Freshwater Fish Families of the World**. Lincoln, Nebr.: University of Nebraska Press, 1981. 197p. LC 80-24666. ISBN 0-8032-1411-1; 0-8032-6059-8pa.

Each page within this atlas is devoted to a particular family of fish. The upper third of each page consists of a world geographical range map showing distribution of the species within the family. Other information includes class, subclass, superorder, order, and family The scientific name, common name, and pronunciation of each family is given. In addition the descriptive account consists of geographical range, number of species, general external anatomy, and other interesting facts concerning the family. The length of these descriptions vary from several sentences to several paragraphs. This material is no where near as detailed as in Nelson's or Lindberg's *Fishes of the World* (see entries 817, 812). References to the literature can be found throughout the textural information. A 19-page bibliography and a combined common/scientific-name index complete the volume.

727. Grosvenor, Melville Bell, ed. **Wondrous World of Fishes**. Washington, D.C.: National Geographic Society, 1965. 368p. LC 65-11482.

Abounding with a large number of colored as well as some black-and-white photographs, this work includes over 300 fishes from Newfoundland to Hawaii. Its arrangement is nontechnical and should appeal to lay people who have an interest in this group of vertebrates. The work is divided into four sections, among which are "Fishing for Fun," "Coast and Deep-Sea Fishes," "Lake and Stream Dwellers," and "Warmwater Curiosities." Within these sections are a series of chapters that address the section focus. In addition to generalized chapters, there is one chapter per section that gives an account of species. These chapters are entitled "Gallery of Marine Fishes (Game Species)," "Gallery of Marine Fishes (Nongame Species)," "Gallery of Fresh-Water Fishes," and "Gallery of Sharks and Hawaiian Fishes." In each of these chapters the fishes are arranged under their common-name groups. For each species included, the common name, scientific name, basic external anatomy, geographical range, weight, length, and other interesting pieces of information are given. Colored photographs and paintings accompany the descriptive account. A combined index of subjects, common names, and scientific names can be found at the end of the work. This is a National Geographic publication and is formatted in the same style as *National Geographic*.

728. Gunther, Albert. **Catalogue of the Fishes in the British Museum**. London: British Museum (Natural History), Dept. of Zoology, 1859. Reprint, New Delhi, India: A.J. Reprints Agency, 1981. 8 vols. $295.00. ISBN 3-76827109-9.

Describing external anatomy and in some cases skeletal anatomy, this work considers all species of fishes found in the British Museum as well as specimens in other museums that the British Museum would be interesting in acquiring. The species are arranged under their respective families. In addition to a number of taxonomic references to the literature for each species, the description considers detailed information on the scales, bones, operculum, tail, teeth, and other external and skeletal features. Geographical distribution is mentioned in many cases. These descriptions vary greatly in length from one paragraph to several pages. Also considered is the condition of the specimen and from what collection the specimen was obtained.

A systematic index as well as an alphabetically arranged scientific-name index appear at the end of each section. Each order of fishes represents a section. This work is not for those with a general interest in the subject; rather, it is an extensive, detailed work for specialists.

729. **Handbuch der Zoologie: Eine Naturgeschichte der Stämme des Tierreiches**. Sechster Bd, Erste Hälfte, Erster Teil: **Acrania (Cephalochorda)—Cyclostoma—Pisces**. Gegrundet von Willy Kükenthal. Hrsg. von J. G. Helmcke and others. Berlin: W. de Gruyter, 1965- . An 11-part appendix is still being published.

For a description of this work, see entry 68 in chapter 1. Volume 6:1:1 contains voluminous information on the cephalochordates, cyclostomes, and pisces. As in many of the other volumes, labeled diagrams (many of them showing cross- and sagital-sections), black-and-white photographs, and photomicrographs provide the illustrative material. A great deal of morphology and internal anatomy is presented within this section. This volume is written in German. An 18-page bibliography and a combined subject/scientific-name index conclude the work.

730. Harder, Wilhelm. **Anatomy of Fishes**. 2d ed, rev. Stuttgart, Germany: E. Schweizerbart'sche Verlagsbuchhandlung, 1975. 2 vols. $165.00. ISBN 3-510-65067-0.

Comprising all the biological systems of a fish, this two-volume set considers the detailed anatomy of same. Volume 1 consists of the text, while volume 2 presents the illustrations. As a result, these two books will need to be used in unison. Volume 1, besides discussing the anatomical structures of the organ systems in great detail, contains an 87-page bibliography and a subject index. Volume 2 contains a large number of labeled line-drawings as well as a few photomicrographic plates. The labeled line-drawings represent detailed anatomy. Cross-references from the text (volume 1) to the illustrations (volume 2) and vice versa are provided.

731. Hoar, William S., and D. J. Randall. eds. **Fish Physiology**. New York: Academic Press, 1969- . 11 vols. to date. LC 76-84233.

Written in the form of review articles, this work considers all aspects of fish physiology. Areas such as embryology, respiration, circulation, and the like are covered. The review articles are written in a scholarly manner and a large number of references to the primary literature can be found throughout the text. Bibliographies appear at the end of each section. Graphs, charts, photographs, and photomicrographs comprise the majority of the illustrations. A separate author, systematic, and subject index can be found at the end of each volume.

Volume 1. *Excretion, Ionic Regulation, and Metabolism*. $125.00. ISBN 0-12-350401-5.

Volume 2. *The Endocrine System*. $125.00. ISBN 0-12-350402-3.

Volume 3. *Reproduction and Growth. Bioluminescence, Pigments, and Poisons.* $125.00. ISBN 0-12-350403-1.

Volume 4. *The Nervous System, Circulation, and Respiration.* $125.00. ISBN 0-12-350404-X.

Volume 5. *Sensory Systems and Electric Organs.* $135.00. ISBN 0-12-350405-8.

Volume 6. *Environmental Relations and Behavior.* $135.00. ISBN 0-12-350406-6.

Volume 7. *Locomotion.* $125.00. ISBN 0-12-350407-4.

Volume 8. *Bioenergetics and Growth.* $129.00. ISBN 0-12-350408-2.

Volume 9A. *Reproduction: Endocrine Tissues and Hormones.* $105.00. ISBN 0-12-350409-0.

Volume 9B. *Reproduction: Behavior & Fertility Control.* $105.00. ISBN 0-12-350429-5.

Volume 10A. *Gills: Anatomy, Gas Transfer, and Acid-Base Regulation.* $105.00. ISBN 0-12-350430-9.

Volume 10B. *Gills: Ion and Water Transfer.* $105.00. ISBN 0-12-350432-5.

Volume 11A. *The Physiology of Developing Fish: Eggs and Larvae.* $103.00. ISBN 0-12-350433-3.

Volume 11B. *The Physiology of Developing Fish: Viviparity and Posthatching Juveniles.* $103.00. ISBN 0-12-350434-1.

732. **Traité de Zoologie: Anatomie, Systématique, Biologie**. Tome XIII. **Agnathes et Poissons**. Ed. by Pierre-Paul Grassé. Paris: Masson, 1958. 3 vols. (various fascicules). LC 49-2833.

A general description of the set to which this part belongs appears in chapter 1 (see entry 69). Published out of sequence, this major, classic work devotes one volume (three parts) to the fish group. A large number of labeled line-drawings provides the majority of the illustrations. These three volumes provide comprehensive descriptions of the external and internal anatomy of fish, along with physiology and systematics. All three of these parts are in the French language. A combined subject/scientific-name index can be found at the end of part 3.

Tome XIII (Premier Fascicule). *Agnathes et Poissons: Anatomie, Ethologie, Systematique*. Fr. 1,376. ISBN 2-225-58410-9.

Tome XIII (Deuxieme Fascicule). *Agnathes et Poissons: Anatomie, Ethologie, Systematique*. Fr. 1,133. ISBN 2-225-58421-4.

Tome XIII (Troisieme Fascicule). *Agnathes et Poissons: Anatomie, Ethologie, Systematique*. Fr. 1,376. ISBN 2-225-58454-0.

Pisces (Geographic Section)

AFRICA

733. Boulenger, George Albert. **Catalogue of the Fresh-Water Fishes of Africa in the British Museum (Natural History)**. London: Printed by order of the Trustees [by Taylor and Francis], 1909-1916. Reprint, 4 vols. in 2, by permission of the Trustees of the British Museum [by Wheldon & Wesley, and Verlag J. Cramer], 1964. LC 10-6511.

Each species is arranged under its respective genus, which in turn is entered under its family, suborder, order, and subclass. Brief anatomical descriptions are presented for the genera, families, suborders, orders, and subclasses. In addition, biological keys are given for genera and species. For each of the fishes included, the scientific name, detailed measurements relating size and width of the whole fish and its parts, arrangement and number of scales, coloration, and geographical range are presented. A black-and-white drawing of the fish and a number of citations to the taxonomic literature accompany the descriptive information. A taxonomic index and a scientific-name index to all four volumes can be found in volume 4 of the set.

734. Jubb, R. A. **Freshwater Fishes of Southern Africa**. Cape Town: A. A. Balkema, 1967. 248p. LC 68-117283.

The majority of this source considers a descriptive account of the many freshwater fishes indigenous to Southern Africa. Species are entered under their respective genera, which in turn are entered under their families. For each species listed, the scientific name, person who first identified it, date of identification, common name, number of scales, shape of body, size, coloration, and other brief external anatomical characteristics are given. In addition, geographical distribution, history of the species as seen from the fossil record, habitat, behavior, and breeding information are presented. Length of the description varies as well as the parameters; not all the above are considered for every species. The length of the descriptive information usually runs a half column or more. Black-and-white photographs and colored plates of the fishes provide the illustrative material.

A checklist of indigenous fishes and a pictorial identification key can be found in the introductory chapter. The pictorial key cross-references to the page where the descriptive account can be found. A section on introduced and marine fishes, and fishing methods and farming is also present. Appendix 1 consists of 120 citations, while appendix 2 considers fishing regulations. Appendix 3 compiles freshwater angling records, and appendix 4 represents additional notes and species reported from the region. A separate scientific/common-name index completes the volume.

735. Smith, J. L. B. **The Sea Fishes of Southern Africa**. 5th ed. Johannesburg: Published for The Trustees of the Sea Fishes of Southern Africa Book Fund by Central News Agency, 1970. 580p. LC 56-28753.

Each species is arranged under its respective family, which in turn is entered under its order, suborder, order, subclass, and class. In addition, anatomical descriptions are discussed for each of these groups. The length of the description varies from one paragraph to several pages. For each of the species listed, the scientific name, person who first identified it, common names, shape of body, external anatomical description, habitat (including depth of water), geographical range, how it is caught, and its edibility are considered. A large number of line-drawings and colored plates provide the illustrative matter. Biological keys are also present that key down to families, genera, and species. A combined index of subjects, common names, and scientific names can be found at the end of the volume. This is a comprehensive listing of fishes found in this area of the world.

736. Smith, M. M., and P. C. Heemstra, eds. **Smith's Sea Fishes**. 6th ed. New York: Springer-Verlag, 1986. 1047p. $135.00. LC 87-100600. ISBN 0-86954-266-4.

Consisting of 144 pages of plates, many of which are in color, this work considers the many and varied species of fishes inhabiting the waters of the South African coast. It is understood that the species included occupy other oceanic areas of the world as well. The species are listed under their respective families, which in turn are entered under their orders. For each species included, the scientific name, person who first named it, size, teeth information, brief external anatomy, and where it was found are given. Brief descriptions of the families are also considered. A key to bony fishes, a bibliography of references cited within the text, and a separate common/scientific-name index complete the work. Although not a great deal of information is given for the fishes, the anatomical material included is detailed.

ATLANTIC OCEAN

737. Fish, Marie Poland, and William H. Mowbray. **Sounds of Western North Atlantic Fishes: A Reference File of Biological Underwater Sounds**. Baltimore, Md.: Johns Hopkins Press, 1970. 207p. LC 77-106135. ISBN 0-8018-1130-9.

Species are arranged under their respective families. For each fish included, the scientific name, common name, person who first named it, geographical distribution, habits, size, type of sounds, and mechanism by which the sound is produced are presented. Spectra and oscillograms of the sound can be found accompanying the description. One page is devoted to each of the species. A two-page bibliography and a combined common/scientific-name index can be found at the end of the volume.

738. Lythgoe, John, and Gillian Lythgoe. **Fishes of the Sea: The North Atlantic and Mediterranean**. Cambridge, Mass.: MIT Press, 1992. 256p. $35.00. LC 91-23354. ISBN 0-262-12162-X.

Not nearly as comprehensive as Whitehead's *Fishes of the North-Eastern Atlantic and the Mediterranean* (see entry 740), this work provides adequate information on many of the species found in these waters. Each species is arranged under its respective family, which in turn is entered under order and class. General information is given for the class, order, and family. For each of the species listed, the common name, scientific name, geographical range, basic external anatomy, coloration, size, habitat and water depths, behavior, reproduction, and other interesting facts are presented. Colored photographs and/or line-drawings accompany the descriptive accounts. The accounts average in length from a few paragraphs to one page. A separate scientific/common-name index can be found at the conclusion of the handbook.

The work represents an excellent publication and is a good place to start for information on Atlantic and Mediterranean species.

739. Tee-Van, John, and others., eds. **Fishes of the Western North Atlantic**. New Haven, Conn.: Sears Foundation for Marine Research, Yale University, 1948- . 8 vols. to date. LC 49-120.

Each species is arranged under its respective genus, which in turn is entered under family, suborder, order, subclass, and class. Descriptive information is given for each of the above taxonomic groups, and keys to each of these groups are also provided. Descriptive information for each of these groups includes fairly detailed external anatomy and in some cases behavior, size, breeding, development, intelligence, luminescence, diet, number of species, and commercial importance. For each of the species presented, the scientific name, person who first named it, date of naming, common names, and the number and size of the fish that were used for the descriptive study are given. In addition, detailed information on the size of the trunk at origin of pectorals, snout length, eye, mouth, nostril, gill openings, first dorsal fin, anal fin, caudal fin, pectoral fin, and other detailed size measurements are considered. Furthermore, the description also contains detailed external anatomy, coloration, size of eggs, embryos, hatchlings and adults, developmental stages, habits, diet, relation to man, geographical range, occurrence in the North Atlantic, occurrence in the western North Atlantic, and a large number of references to the literature. Line-drawings of the fish accompany the descriptive information for the species. A separate common/scientific-name index completes each of the volumes. It should be noted that this source provides extremely detailed information on fish species and should not be considered a general introduction to those fishes inhabiting the oceans of the western North Atlantic.

740. Whitehead, P. J. P. **Fishes of the North-Eastern Atlantic and the Mediterranean. Poissons de l'Atlantique du Nord-Est et de la Mediterranee**. Paris: UNESCO, 1984. 3 vols. $180.00 set. LC 85-239718. ISBN 92-3-002309-4 set.

This work is a direct result of *Check-List of the Fishes of the North-Eastern Atlantic and of the Mediterranean* (see entry 821). Each small section of this three-volume set represents a family of fishes found within the above geographical area. For each species listed under its respective family, the scientific name, person who first named it, date of naming, common synonyms (if any), black-and-white line-drawing, coloration, external anatomy, size, habitat, reproduction, geographical distribution, and a geographical range map are considered. In addition, the scientific name and common name of the family are presented along with general information on the family (anatomy, habitat, behavior). Keys to subgenera and species are listed within each family section. Within the introductory information a 40-page listing of keys to the families is incorporated along with line-drawings of representative fishes for each family.

Additional bibliographic references that compliment the bibliography found in *Check-List of the Fishes of the North-Eastern Atlantic and of the Mediterranean*; a separate common/scientific-name index completes the work. This three-volume set represents an advanced scholarly report on fish species within the northeast Atlantic and the Mediterranean.

AUSTRALIA

741. Allen, G. R. **Freshwater Fishes of Australia**. Neptune City, N.J.: T. F. H. Publications, 1989. 240p. $39.95. LC 90-135159. ISBN 0-86622-936-1.

Abounding with a large number of colored photographs depicting the fish species and the geographical locations, this source provides information on the fishes that inhabit the waters of Australia. The fishes are arranged under their respective families. Preceding the descriptive account of the species, general information on the family, along with biological keys down to species are presented. For each of the species listed, the common name, scientific name, person who first named it, alternative scientific names and dates, basic external anatomy providing detailed measurements, coloration, geographical distribution, habitat, and size are given. A geographical range map accompanies each of the descriptive accounts. Plate numbers, allowing access to the photographs, are entered as part of the account. Each of the fish photographs considers the scientific name, size, and geographic location. An illustrated biological key to the families is presented at the beginning of the text. Other sections within the source include a brief description of estuary fishes, a listing of Australia's threatened fishes, a six-page bibliography, and a gazetteer of Australian localities. A combined common/scientific-name index concludes the volume.

741a. Kuiter, Rudie H. **Coastal Fishes of South-Eastern Australia**. Honolulu: University of Hawaii Press, 1993. 437p. $85.00. LC 92-32504. ISBN 0-8248-1523-8.

Consisting of 855 species and 142 families, this work provides general information on those fish found in coastal waters from northern New South Wales to western Victoria. The species are arranged under their respective family, which in turn is entered under order. Orders are arranged phylogenetically. Scientific and common names are given for families and orders; general information for each of the families and orders is also included. Scientific keys down to families can also be found associated with the beginning of a new order.

For each of the species listed, the common name, scientific name, person who first named it, and date of naming is given. In addition, basic external anatomy including fin formulas, coloration, water depth, and size are considered within the description. The descriptions are usually one third of a column in length. Magnificient colored photographs representing each of the species listed can be found near the description of said species. A large map representing South-Eastern Australia, and a pictorial guide to families can be found near the beginning of the work. A brief glossary and a combined common/scientific-name index concludes the volume.

742. Marshall, Tom C. **Fishes of the Great Barrier Reef and Coastal Waters of Queensland**. Narberth, Pa.: Livingston, 1965. LC 65-3842.

Consisting of approximately 1,500 species of fishes, this work attempts to describe the majority of them that are found in and around the Great Barrier Reef and coastal waters of Queensland. Each fish is arranged under its respective family, which in turn is entered under suborder, order, and class. In addition to biological keys to the families and genera, som general information about these groups is presented. Much of this material is basic externa anatomy. For each species listed, the common name, scientific name, person who first name it, citations to the literature, other common names, coloration, geographical range, and siz are given. In some cases general information regarding egg laying is considered. Th descriptive matter ranges from several paragraphs to several pages depending upon th species. A large number of colored paintings as well as some black-and-white drawings c the fishes can be found interspersed among the pages. A brief glossary, and a separa common/scientific-name index conclude the volume. Although older, this work would mak a fine companion to the newer source, *Fishes of the Great Barrier Reef and Coral Sea* (se entry 743).

743. Randall, John E., Gerald R. Allen, and Roger C. Steene. **Fishes of the Great Barrie Reef and Coral Sea**. Honolulu: University of Hawaii Press, 1990. 507p. $60.00. L 90-38987. ISBN 0-8248-1346-4.

Consisting of 1,111 species, this work considers those fishes found throughout the Grea Barrier Reef and Coral Sea. The species are arranged under their respective families. Commo name and scientific name of the family are presented along with a brief anatomical descriptio of the family as a whole. For each of the species mentioned, the common name, scientifi name, person who first identified it, date of identification, brief external anatomy, size coloration, and geographical distribution are given. Each description is approximately one quarter of a column long. Excellent colored photographs of the fish accompany the descriptiv information. A two-page map of the area and a pictorial guide to the families can be found i the introductory information. A brief glossary and a combined common/scientific-name inde conclude the source. This is an excellent work. Many of the colored photographs are quit large and well presented.

BAHAMAS

744. Bohlke, James E., and Charles C. G. Chaplin. **Fishes of the Bahamas and Adjacen Tropical Waters**. 2d ed. Austin, Tex.: University of Texas Press, 1993. 771p. $100.00. L 92-20389. ISBN 0-292-70792-4.

This work represents nomenclatural changes and additions from the 1968 edition. Thes changes and additions were prepared by Eugenia B. Bohlke and William F. Smith-Vaniz. Th fishes (507 species from the 1968 edition as well as an additional 58 species) are arrange under their respective families. Scientific and common names of the families are given alon with a brief description that includes the number of species contained within the family group For each of the fishes included, the scientific name, person who first named it, common name size, basic external anatomy, coloration, habits and habitats, and geographical distribution i presented. Each entry is one page long. A black-and-white line-drawing of the fish accompa nies the descriptive matter. Keys to the genera and species are also present. A number o colored plates of selective fish can be found near the beginning of the volume. Genera information on the Bahama Islands, information describing the nomenclatural changes an additions, along with a four-page bibliography, and basic line-drawings depicting the shap of the fishes within families (Pictorial Key) can be found in the introductory material. A brie glossary, a bibliography containing 349 entries, and a common/scientific-name index can b found concluding the source.

CANADA

745. Scott, W. B. **Freshwater Fishes of Eastern Canada**. 2d ed. Toronto: University o Toronto Press, 1967. 137p. $15.95pa. LC 67-101255. ISBN 0-8020-6074-9.

Beginning with the lampreys and ending with the flatfishes, this work considers thos 154 species found in the provinces of Newfoundland, Prince Edward Island, Nova Scotia New Brunswick, Quebec, and Ontario. The species are arranged under their family names

Common and scientific names of the families are presented. For each of the species included, the common name, scientific name, person who first named it, other common names, coloration, basic external anatomy, size, geographical range, life history and habits, and food are given. A black-and-white photograph of the species accompanies the description. The descriptive matter for each fish is approximately one page in length. A key to the families, a one-page bibliography, and a combined common/scientific-name index conclude the volume.

EUROPE

746. **The Freshwater Fishes of Europe**. Wiesbaden, Germany: AULA-Verlag, 1986- . 9 vols. ISBN 3-89104-040-4 vol. 1, pt. 1.

Covering all fish species contained within the European continent, this work provides detailed biological information on all aspects of individual species. The fishes are arranged under their respective families. The scientific name, person who named it, date of naming, descriptive information, keys to the genera, and selected references are presented for each family included. The length of the descriptive account of each species varies from several pages to 45 pages. For each descriptive account, the scientific name, person who named it, date of naming, the common name (in English, French, German, and Russian), and other scientific names are given. In addition the holotype, etymology, size, weight, detailed morphology, karyotype, protein specificity, sexual dimorphism, variations, age and size variability, subspecies, hybrids, geographical distribution, habitat, migration patterns, hardiness, feeding habits, longevity, growth information, population dynamics, very detailed reproductive biological information, parasites and diseases, and economic importance are considered. A lengthy list of references can be found at the end of each species account. Graphs, charts, tables, and line-drawings enhance the descriptive information. Keys to the families and species can be found throughout the accounts. A separate scientific/common-name index concludes each of the volumes. Although no photographs or line-drawings of the fishes can be found throughout this set of books, this work, nevertheless, represents a scholarly and readable approach to all the biological parameters for each individual fish species in a detailed manner. It is very complete. The titles of the individual volumes follows:

Volume 1. *Petromyzontiformes*. Part 1.
Volume 1. *General Introduction to Fishes: Acipenseridae*. Part 2.
Volume 2. *Clupeidae, Anguillidae.*
Volume 3. *Salmonidae, Coregonidae, Thymallidae, Osmeridae, Umbridae, Esocidae.*
Volume 4. *Cyprinidae.*
Volume 5. *Cyprinidae.*
Volume 6. *Cobitidae, Siluridae, Ictalurideae.*
Volume 7. *Gadidae, Gasterosteidae, Syngnathidae, Cyprinodontidae, Poeciliidae, Atherinidae, Cottidae, Centrarchidae, Percidae.*
Volume 8. *Gobiidae, Blenniidae, Pleuronectidae.*
Volume 9. *Threatened Fishes of Europe.*

747. Wheeler, Alwyne. **The Fishes of the British Isles and North-West Europe**. London: Macmillan, 1969. 613p. $25.00. LC 69-19148. ISBN 0-87013-134-6.

Beginning with the lampreys and hagfishes, and ending with the angler fishes, this work describes in some detail the various fishes found off the coast of the British Isles and northwestern Europe. The fishes are arranged under their respective families. Common name, scientific name, and general description of the family are presented. For each of the species included, the common name, scientific name, person who named it, date of naming, and French, Dutch, German, Danish, and Norwegian names are listed. In addition, basic external anatomy, coloration, size, weight, breeding patterns, nesting and egg information, and geographical range are given. A geographical range map and a line-drawing of the fish species accompany the descriptive account. Descriptions usually range one to two pages in length. Keys (to family, genera, and species) can be found at the beginning of a suborder. A few colored and black-and-white plates can be found interspersed among the text. Within the introductory information a picture key is present. This key provides the reader with a black outline of the type of fish, page on which it can be found, marine or freshwater species, distinguishing anatomical features, and size. A 13-page bibliography, and a combined common/scientific-name index conclude the volume.

GREAT BRITAIN

748. Maitland, Peter S., and R. Niall Campbell. **Freshwater Fishes of the British Isles**. London: HarperCollins, 1992. 368p. $52.50. LC gb92-42912. ISBN 0-00-219383-3; 0-00-219380-9pa.

Each of the fishes included is arranged under its common-name group. Each of the common-name groups constitute chapters. For each of the species listed, the common name, scientific name, date of naming, person who named it, additional common names, size, weight, basic identifying external anatomy, number of rays and gill rakers, geographical distribution, habitat, hatching time, size of eggs, diet, and coloration are given. Stomach contents, in a tabular format, are presented for many species. The lengths of the accounts vary from 1 page to 12 or more pages. A black-and-white drawing and, in many cases, a black-and-white photograph of the fish accompany the descriptive account. In addition to the individualized species account within each chapter, biological keys down to species, and general information on the group as a whole are presented. A number of colored plates depicting a few of the species can be found within the text.

Introductory information consists of general discussions on fish form, function, distribution, habitat, conservation, and identification. A series of distribution maps, the names of British freshwater fishes, and growth curves appear in the appendices. A 14-page bibliography and a combined index of subjects, common names, and scientific names conclude the volume.

GULF OF CALIFORNIA

749. Thomson, Donald A., Lloyd T. Findley, and Alex N. Kerstitch. **Reef Fishes of the Sea of Cortez: The Rocky-Shore Fishes of the Gulf of California**. Tucson, Ariz.: University of Arizona Press, 1987. 302p. LC 86-24996. ISBN 0-8165-0985-0.

The species (271 of them) are arranged under their respective families. Common and scientific names are given to the families along with brief descriptions, which include the number of genera and species. For each of the fishes included, the common name, scientific name, person who first named it, date of naming, size, basic external anatomy and coloration, geographical distribution, ecology, and related species are presented. A line-drawing and in some cases a black-and-white photograph accompanies the description. The account is usually no more than one page or two pages in length. A series of colored photographs of the fishes can be found interspersed throughout the text. In addition, a section on nonresident reef fishes can be found near the end of the volume. Introductory information consists of the geography and the reef-fish communities of the Sea of Cortez.

Appendix 1 consists of a checklist of the species found in the Sea of Cortez, while appendix 2 lists the localities of the reef fishes that are illustrated (cross-references to colored plates). A brief glossary, a 14-page bibliography, and a combined common/scientific-name index conclude the source.

GULF OF MEXICO

750. Hoese, H. Dickson, and Richard H. Moore. **Fishes of the Gulf of Mexico, Texas, Louisiana, and Adjacent Waters**. College Station, Tex.: Texas A & M University Press, 1977. 327p. (W. L. Moody, Jr. Natural History Series; no. 1). $24.50; $14.95pa. LC 76-51654. ISBN 0-89096-027-5; 0-89096-028-3pa.

Arranged under families, this work considers those species endemic to the waters in and around the Gulf of Mexico. For each of the species mentioned, the common name, scientific name, person who named it, brief external anatomy, diet, geographical range, and citations to the literature are given. A generalized line-drawing of a representative fish can be found under the family name. Keys to the family and species can be found interspersed throughout the text. A series of 297 small colored plates can be found at the beginning of the work. Most of these are pictures of fishes not found in their natural habitats. The appendices consist of a listing of fish species along with their geographical location; a list of rarely caught species not covered in the keys or species accounts; a list of freshwater fishes likely to be found in marine waters; a listing of marine fishes that regularly invade northern gulf coastal freshwater; a listing of common fish species with protracted spawning seasons; some French (Cajun) common names for marine fishes; some Spanish (Mexican) common names for important species of marine fishes; and a listing of families that contain deeper-dwelling species that are occasionally found over the continental shelf. A brief illustrated glossary, a 20-page bibliography, and a combined common/scientific-name index conclude the source.

HAWAII

751. Tinker, Spencer Wilkie. **Fishes of Hawaii: A Handbook of the Marine Fishes of Hawaii and the Central Pacific Ocean**. Honolulu: Hawaiian Service, 1978. 532p. $29.95; $22.50pa. LC 77-93337. ISBN 0-930492-02-1; 0-930492-14-5pa.

Consisting of those marine fishes found in and around the Hawaiian Islands as well as the warm waters of the surrounding tropical, central Pacific Ocean, this work provides general information on these vertebrates. The species are arranged under their respective families. Such information as general external anatomy, coloration, habitat, reproduction, and number of species is given for each of the families presented. For each of the species listed, the common name, scientific name, person who first named it, date of naming, general external anatomy, size, coloration, and geographical distribution is given. Each account is usually a half page in length. A black-and-white drawing or black-and-white photograph accompanies each of the descriptive accounts. In addition, a number of colored photographs of some of the species can be found interspersed throughout the text. A total of 172 families are considered within this source.

The table of contents is arranged phylogenetically, thereby listing class, subclass, superorder, order, and family. A combined common/scientific-name index concludes the handbook.

JAPAN

752. Masuda, H., and others, eds. **The Fishes of the Japanese Archipelago**. Tokyo: Tokai University Press, 1984. 3 vols. $325.00 set. ISBN 4-486-05054-1.

The more than 3,200 species compiled within this book represent those fishes found in fresh, salt, shallow, and deep waters in and around Japan. The fishes are arranged under their respective families, which in turn are entered under orders and classes. A brief anatomical description is presented for class, order, and family. Scientific and common names of the families are considered. For each of the species listed, the scientific name, common name (in Japanese), person who first identified it, brief anatomical description, size, and geographical range are given. A typical description is one-quarter of a column long. In addition, references to the colored plates of the fishes are presented; a few are in black and white. The entire set of plates can be found in volume 3. In many cases, line-drawings representing the anterior portion of species can be found accompanying the description. A 34-page bibliography and a separate common-name (English) and scientific-name index complete volume 1. Volume is identical to volume 1 except it is written in Japanese; a common-name index in Japanese as well as the common-name index in English, and the scientific-name index can also be found at the end of the volume. Volume 3 contains 370 pages of plates; most of the plates are in color. This is an excellent set.

753. Okada, Yaichiro. **Fishes of Japan: Illustrations and Descriptions of Fishes of Japan**. Tokyo: Maruzen, 1955. 462p. LC 56-2741.

Considering the more typical type of fish found in coastal water and freshwater, this source consists of general descriptive information regarding these species. The fishes are arranged under their respective scientific family names. For each species mentioned, the scientific name, person who first identified it, common name in English and in Japanese, basic external anatomy (number of finlets, gill rakers, etc.), coloration, size, geographical distribution, water temperature habitat, and, in some cases, egg and breeding information. A black-and-white drawing of the fish accompanies the descriptive matter. The descriptive information usually runs one page to one and one-half pages in length. A separate common-name, and scientific-name index can be found at the end of the volume. The common-name index includes English and Japanese names.

NEW ZEALAND

754. McDowall, R. M. **New Zealand Freshwater Fishes: A Guide and Natural History**. Auckland: Heinemann Educational, 1978. 230p. LC 79-307520. ISBN 0-86863-130-2.

The species are arranged under their respective families. Each of the families constitutes one chapter. The scientific and common names of the families are presented along with a brief description consisting of size, habitat, coloration, external anatomy, geographical distribution, and number of species found within New Zealand. In addition, a key to the species within the family is given. For each of the species listed within the family chapter; the scientific name; common name; person who named it; date of naming; size; number of rays, gill rakers, and vertebrae; coloration; identifying features; taxonomic notes; and breeding and hatching information are considered. Line-drawings and colored photographs of the fishes along with geographical range maps are presented in each of the family chapters. In addition to the chapters dealing with the species accounts, a section on the study and structure of fishes, fishes and fisheries in New Zealand, diseases and parasites, and zoogeography are represented. An eight-page bibliography and a combined common/scientific-name index conclude the source.

NORTH AMERICA

755. Calabi, Silvio. **Game Fish of North America**. Secaucus, N.J.: Wellfleet Press, 1988. 171p. $17.98. ISBN 1-55521-275-1.

Although this work considers more species of fish than *McClane's Game Fish of North America* (see entry 764), it provides a lost less information on them. The volume is divided into freshwater and saltwater fish. The material entered for each of the fish is slanted toward the fisherman. It considers size, weight, the various lures, and bait needed for catching these beasts, and their behavioral patterns in the capturing process. The discussions are usually less than one page. A large number of colored photographs and paintings depicting these denizens of the deep provide the illustrations; they are excellent. A combined common-name and geographical index, and a list of fish caught that were considered records complete the book.

756. Carlander, Kenneth. **Handbook of Freshwater Fishery Biology: Life History Data on Freshwater Fishes of the United States and Canada, Exclusive of the Perciformes. Volume 1**. Ames, Iowa: Iowa State University Press, 1969. 752p. $42.95. LC 69-18736. ISBN 0-8138-2335-8.

This work follows the same format as *Handbook of Freshwater Fishery Biology: Life History Data on Centrarchid Fishes of the United States and Canada, Volume 2* (see entry 757).

757. Carlander, Kenneth. **Handbook of Freshwater Fishery Biology: Life History Data on Centrarchid Fishes of the United States and Canada, Volume 2**. Ames, Iowa: Iowa State University Press, 1977. 431p. $29.95. LC 77-4176. ISBN 0-8138-0670-4.

Not only does this two-volume set provide detailed data on the age, growth, and length-weight relationship of fish, it also serves as an index to the literature. The tables contain numerical and textural data on length-weight relationships, observed lengths and weights at various ages, calculated growth, age at maturity and reproduction data, discussion of growth data, and food habits. Each species is represented under its respective family. Common name, scientific name, and person who first named it are given along with the numerical and textural data mentioned above. There are no illustrations in either of these two volumes. Lengthy bibliographies can be found near the end of each volume. The works conclude with author indexes as well as subject indexes. The subject indexes represent subjects, common names, and scientific names.

758. Herald, Earl S. **Fishes of North America**. New York: Doubleday, 1972. 254p. LC 79-147353.

The fishes are considered with their respective families. Families are entered under their superorders and classes. For each family mentioned, the common and scientific names are presented. In addition, general external anatomy, weight, size, geographical ranges, specific species, breeding information, and other interesting facts are given. Black-and-white and colored photographs of many fishes can be found interspersed among the pages. It should be noted that this work provides a general account of the families and not the fishes themselves. A brief glossary and a combined common/scientific-name index complete the volume.

759. Johnson, James E. **Protected Fishes of the United States and Canada**. Bethesda, Md.: American Fisheries Society, 1987. 42p. $10.00pa. LC 87-71027. ISBN 0-913235-43-1.

Considering the native freshwater fishes of North America, this compendium lists those species that are receiving legal protection (special concern due to low numbers, limited distributions, or recent declines). The species are arranged under their respective families. Common and scientific names are given for the families and species. In addition, the states, provinces, and/or countries where the species are found are presented. Those areas (states, provinces, and/or countries) where the species is protected are outlined in bold type. Cross-references are also given to the colored plates of the fishes, which can be found in the middle of the book. A second table lists the fishes by common names under their geographical areas (state, province, country). A one-page bibliography completes the volume. There is no index.

760. Jordan, David Starr, and Barton Warren Evermann. **American Food and Game Fishes: A Popular Account of All the Species Found in America North of the Equator, with Keys for Ready Identification, Life Histories and Methods of Capture**. Garden City, N.Y.: Doubleday, Page, 1920. Reprint, New York: Dover, 1969. 574p. LC 71-84702. ISBN 0-486-22196-2.

This source considers all species of fishes north of Panama that are considered usable for either human consumption or sport. The species are arranged under their respective genera, which are entered under families. Common and scientific names for families and genera are given, along with biological keys to these groups. For each of the fishes presented, the common name, scientific name, person who first identified it, size, weight, geographical range, breeding, external anatomy, and, in some cases, information regarding its use as food or as sport are provided. Black-and-white drawings and photographs depict the species. An eight-page glossary, an artificial key to the families of American food and game fishes, and a combined common/scientific-name index complete the work. Although somewhat outdated, this volume provides a number of facts about these fishes.

761. Jordan, David Starr, and Barton Warren Evermann. **The Fishes of North and Middl America: A Descriptive Catalogue of the Species of Fish-Like Vertebrates Found in th Waters of North America, North of the Isthmus of Panama**. Washington, D.C.: Govern ment Printing Office, 1896-1900. Reprint, Jersey City, N.J.: published for the Smithsonia Institution by T.F.H. Publications, 1963. 4 vols. (Bulletin of the United States Nationa Museum, no. 47). LC 64-6498.

More comprehensive than Jordan and Evermann's *American Food and Game Fishe* (see entry 760), this four-volume source presents a descriptive account of those species o fishes known to inhabit North and Central America, the West Indies, the Caribbean Sea, an the Galapagos Archipelago. In addition, species found in the offshore banks and continenta slopes of the Atlantic and Pacific Oceans as well as the Gulf Stream waters are include Marine fishes found north of the equator, and all freshwater fishes found north of the Isthmu of Panama are also considered.

Each species is arranged under its respective genus, which in turn is entered under it family, order, and class. Descriptive anatomical features are presented for the class, orde family, and genus, along with biological keys and citations to the literature. For each of th species listed, the scientific name, person who first named it, common name, external anatom and anatomical measurements, coloration, geographical distribution, and habitat are giver The length of the descriptive account of the species varies from several sentences to one page

Volume 4, in addition to the completion of the species accounts, contains a number o indexes, keys, and plates. The sections included consist of artificial keys to the families o teleosts, an eight-page glossary of technical terms, a combined common/scientific-nam index, a systematic arrangement of the fishes of North and Middle America, an index to th genera and species illustrated, an explanation of the illustrations, and 392 plates. These plate contain black-and-white line-drawings of many of the species presented within the se Scientific names and cross-references to the descriptive accounts are provided. This work i a classic; it represents a major research and publication endeavor.

762. Lee, David S., and others. **Atlas of North American Freshwater Fishes**. [Raleigh N.C.]: North Carolina State Museum of Natural History, 1980. 867p. (Publication of the Nort Carolina Biological Survey, no. 1980-12). $25.00. LC 80-620039. ISBN 0-917134-03-6

Consisting of approximately 775 species, this work considers those freshwater fishe endemic to Canada and the United States. For each of the species listed, the scientific name person who first named it, common name, order, and family are given. In addition, citation to the literature, geographical distribution, habitat, systematics, size, diet, and general repro ductive information are provided. A small black-and-white drawing of the fish, along with large geographical range map, accompanies the descriptive account. Each account an accompanying illustrations are one page in length. A brief glossary, a list of books cited i the descriptive accounts, a list of regional publications (arranged by state and province), list of compilers, and a combined common/scientific-name index conclude the volume. Thi is a comprehensive work exhibiting a great deal of research.

763. Lee, David S., and others. **Atlas of North American Freshwater Fishes. Supplement** [Raleigh, N.C.]: North Carolina State Museum of Natural History, 1983. 67p. (North Carolin Biological Survey. Contributions, no. 1983-6).

Follows the same format and provides the same type of information as in *Atlas of Nort American Freshwater Fishes* (see entry 762). However, it concentrates its efforts on the fis species of the West Indies.

764. McClane, A. J., and Keith Gardner. **McClane's Game Fish of North America**. Nev York: Times Books, 1984. Reprint, New York: Bonanza Books, 1989. 376p. $22.99. LC 89-1012. ISBN 0-517-68852-2.

Divided into 30 chapters, this work considers a specific game fish for each of th chapters. From the atlantic salmon to the rainbow trout to the northern pike, this work present a large amount of information for each of the fish included. For each species mentioned, th common name (also represents the name of the chapter), the scientific name, geographical distribution, weight, size, breeding information, spawning, lures and bait for catching the fish, bodies of water where the fish are plentiful, coloration, and a number of other interesting facts are given. This information is not presented in a table; rather, it is written in a textural format that is very informative. A geographical range map accompanies each of the descriptions. The length of these accounts can run an average of 10 or more pages. Colored paintings of the fish and colored photographs of the geographical areas provide the illustrative material. A two-page bibliography, and a combined subject and common-name index conclude the work. This is a fine publication; although it emphasizes the catching of these fish, it nevertheless provides an adequate amount of biological information.

765. Ono, R. Dana, James D. Williams, and Anne Wagner. **Vanishing Fishes of North America**. Washington, D.C.: Stone Wall Press, 1983. 257p. $29.95. LC 82-62896. ISBN 0-913276-43-X.

The species of fishes that are considered to be endangered are arranged under their geographical habitats. Sections in the volume such as "Texas Spring Fishes," "Pyramid Lake Fishes," "Great Lakes Fishes," "Colorado River System Fishes" are some examples. For each of the fishes included, the common name, scientific name, breeding habits, threats from other fishes, environmental threats, documentation to prove fishes are in the decline, and other biological aspects of the species are considered. All fish descriptions will not necessarily cover all the above parameters. These minimonographs usually run an average of two pages in length. A series of colored plates of the fishes and a number of black-and-white photographs showing the geographical areas provide the illustrative matter. In addition, brief sections on extinct fishes, the biogeography of endangered and threatened fishes, acid rain, and the outlook for today's fishes can be found near the rear of the book. A list of endangered, threatened, and extinct fishes of the United States (common name, scientific name, distribution, and threats), a map of the United States depicting endangered and threatened fishes of North America, a brief glossary, a selected list of references, and a list of conservation organizations complete the volume. There is no index. The table of contents, however, can serve the reader in accessing specific species.

766. Schaffner, Herbert A. **Freshwater Game Fish of North America**. New York: Gallery Books, 1989. 144p. $7.98. LC 91-185721. ISBN 0- 8317-6426-0.

Follows exactly the same format as *Saltwater Game Fish of North America* (see entry 767). Twenty-one species are represented that are considered game fish found in the freshwaters of North America.

767. Schaffner, Herbert A. **Saltwater Game Fish of North America**. New York: Gallery Books, 1990. 144p. $14.98. ISBN 0-8317-6427-9.

Written in a popular style, this source discusses approximately 30 species that the author considers to be game fish. For each of the species included, the geographical range, coloration, identifying characteristics, size, weight, spawning period, behavior, and lures, bait, and tackle useful in catching these creatures are given. Colored photographs and pictures depicting these fish provide the majority of the illustrations. In addition, some general information on saltwater fish can be found in the introductory pages. A brief glossary and a common-name index conclude the volume. It should be emphasized that this work represents a popular treatment.

PACIFIC OCEAN

768. Burgess, Warren, and Herbert R. Axelrod, eds. **Pacific Marine Fishes**. 2d ed. Hong Kong, Neptune City, N.J.: T. F. H. Publications, 1973- . 10 vols. to date. $34.95/vol. LC 76-159802.

With an emphasis on colored photographs, this work considers those species found on or near Pacific coral reefs. Approximately one-fifth of the world's fish species can be located in this area.

Each of the volumes presents descriptive information on the families and includes such characteristics as external anatomy, coloration, diet, size, habitat, behavior, and brief statements on many of the species contained within the families. The majority of each volume consists of magnificently colored photographs of these sea creatures in their natural environments. The captions that accompany these photographs list the scientific name, size, where the photograph was taken, and, in some cases, the developmental stage of the fish (adult female, etc.). A combined common/scientific-name index can be found at the end of each volume. This book set can be used for general information on the families as well as an identification guide to the many and various forms of fishes inhabiting this part of the globe. The later volumes have specific titles and are outlined as follows:

Book 5. *Fishes of Taiwan and Adjacent Waters.*
Book 6. *Fishes of Melanesia (from New Guinea and the Solomons to Fiji).*
Book 7. *Fishes of the Great Barrier Reef.*
Book 8. *Fishes of California and Western Mexico.*
Book 9. *Fishes of Western Australia.*
Book 10. *Reef Fishes of the Indian Ocean.*

769. Fowler, Henry W. **The Fishes of Oceania**. Honolulu: Bernice P. Bishop Museum, 1928. Reprint, Lehre, Cramer; New York: Johnson Reprint, 1967. 540p. plus 3 Supplements (71p.; 84p.; 152p.). (Memoirs of the Bernice P. Bishop Museum, vol. X; vol. XI, no. 5/suppl. 1; vol. XI, no. 6/suppl. 2; vol. XII, no. 2/suppl. 3). $195.00. LC 68-86836. ISBN 3-7682-0444-8.

Considering the eastern region of the Indo-Pacific area, this work consists of descriptions of fishes taken from these boundaries. Such island groups as Melanesia, New Guinea, Micronesia, Polynesia, and Hawaii are included. The fishes are arranged under their respective families. For each of the species listed, the scientific name, person who first named it, bibliographical citations to the literature, external anatomical identifying characteristics (many in great detail), anatomical measurements, coloration, and geographical range are given. Black-and-white line-drawings of many of the species provide the illustrations. A series of 47 black-and-white plates depicting many of the species appears near the end of the main volume. A scientific-name index is provided for the main volume and each of the supplements. This work represents a detailed look at the many species of fishes found throughout the region.

770. Leis, J. M., and D. S. Rennis. **The Larvae of Indo-Pacific Coral Reef Fishes**. Honolulu: Published in association with the Australian Museum by University of Hawaii Press, 1989. 375p. $32.00. ISBN 0-8248-1265-4.

The larvae of families of fishes are considered in this source. Scientific names and common names of the families are presented. Descriptive accounts of the larvae consist of general information on the adults, spawning mode (size of eggs), development at hatching (size, brief external anatomy), and a lengthy morphological description of the larvae. Included is body shape, size, external anatomy, fin size, scale development, pigment, and similar families. In addition, a number of measurements are given dealing with pre-anal length, pre-dorsal fin length, head length, snout length, eye diameter, and body depth. Line-drawings of the larvae accompany the descriptive matter. A brief glossary, collecting and identification methods, and points on how to use this volume in identifying fish larvae can be found in the introductory information. An 11-page bibliography and a combined common/scientific-name index conclude the source.

'1. Walford, Lionel A. **Marine Game Fishes of the Pacific Coast from Alaska to the juator**. Berkeley, Calif.: University of California Press, 1937. Reprint, Washington, D.C.: nithsonian, 1974. 205p. $16.00. LC 74-80976. ISBN 0-87474-153-X.

The species are entered under their respective common-name families or tribe groups, ch as the sea-bass family, the mackerel tribe, and so on. General information on the tribe d/or family as well as biological keys to the species precedes the descriptive accounts. For ch of the species presented, the common name, scientific name, person who first named it, sic external anatomy, coloration, size, geographical distribution, diet, spawning, growth, pulations, and angling notes are given, in most cases. The length of the description ranges om one paragraph to several pages, as does the amount of information. A series of 69 plates picting each species in either a black-and-white or a colored drawing appears at the end of e volume. Other sections include a brief anatomical glossary, keys to the families of fishes und on the Pacific coast of the Americas, a two-page bibliography, and a separate mmon/scientific-name index.

HILIPPINES

'2. Fowler, Henry W. **Contributions to the Biology of the Philippine Archipelago and djacent Regions: The Fishes of the Groups Elasmobranchii, Holocephali, Isospondyli, d Ostarophysi Obtained by the United States Bureau of Fisheries Steamer "Alba-oss" in 1907 to 1910, Chiefly in the Philippine Islands and Adjacent Seas**. Washington, C.: U.S. Government Printing Office, 1941. 878p. (Smithsonian Institution, United States ational Museum, Bulletin 100, vol. 13).

Each species is arranged under its respective genus, which in turn is entered under mily, order, subclass, and class. Some general anatomical information is given for the class, bclass, order, family, and genus. Keys to the subclasses, orders, families, and genera are so presented. For each of the species listed, the scientific name, alternative scientific names, rson who named it, bibliographic citations to the literature, detailed anatomical measure-ents, coloration, and geographical distribution are given. Descriptive accounts are usually e page in length. Black-and-white line-drawings depicting the gross anatomy of a few of e fish species can be found interspersed throughout the volume. A scientific-name index mpletes the work.

Although this source has been placed under handbooks, it could just as easily serve as axonomic work of the fish species found throughout the Philippines.

NITED STATES

3. Hubbs, Carl L., and Karl F. Lagler. **Fishes of the Great Lakes Region**. Bloomfield lls, Mich.: Cranbrook Institute of Science, 1958. Reprint, Ann Arbor, Mich.: University of ichigan Press, 1964. 213p. $22.95. LC 58-7693. ISBN 0-472-08465-8.

Considering the many species found throughout the Great Lakes, this source contains formation on the family, the species within the family, and keys to the family. Descriptive formation on the family itself contains basic external anatomy, geographical distribution, bitat, behavior, diet, types of species, and other interesting facts. The keys can be used to entify down to species. In addition, cross-references present within the keys lead the reader black-and-white photographs of the fishes. For each of the species mentioned, the common me, scientific name, person who first named it, geographical distribution, geographical nge within the Great Lakes, and other facts pertinent to the species are considered. A number line-drawings, the black-and-white photographs, and 44 colored plates depicting the ecies provide the illustrative matter. The colored plates can be found at the front of the ok. Introductory information consists of a description of the waters of the Great Lakes, ld study and the collecting of fishes, preservation and identification of fishes, and keys to e families of Great Lake fishes. A 14-page bibliography can be found near the end of the lume. A combined common/scientific-name index concludes the source.

774. Sigler, William F., and John W. Sigler. **Fishes of the Great Basin: A Natural History**. Reno, Nev.: University of Nevada Press, 1987. 425p. (Max C. Fleischmann Series in Great Basin Natural History). $32.50. LC 86-7082. ISBN 0-87417-116-4.

The Great Basin is that area in the United States that drains internally into the states of Utah, Wyoming, Idaho, Nevada, California, and Oregon. The Salton Sea is excluded from the area. The species are arranged under their respective families. Families of fishes constitute chapters. After some general information on the family, the species are presented. For each of the species covered, the common name, scientific name, persons who are attributed in giving them their common and scientific names, economic importance, geographical range, coloration, basic external anatomy, size, longevity, limiting factors regarding their growth and development, food, feeding, breeding habits, habitat, and preservation procedures are given. Length of a description is anywhere from two to four pages. Approximately 88 species are described. A black-and-white drawing of the fish accompanies the description. Colored plates depicting a few of the species covered can be found in the center of the book. In addition, there are sections on the Great Basin dealing with drainage, fishing, endangered species act and desert fishes, and a biological key to the native and introduced fishes of the Great Basin. A small portion of the work is also devoted to the evolution and classification of fishes as well as the behavior of fishes and why they are studied. An annotated checklist of the fishes of the Great Basin, a list of the established fishes of the Great Basin, a 9-page glossary, a 14-page bibliography, and a combined index of subjects, common names, and scientific names can be found near the completion of the volume. A good deal of information on these species is provided by this source.

775. Tomelleri, Joseph R., and Mark E. Eberle. **Fishes of the Central United States**. Lawrence, Kans.: University Press of Kansas, 1990. 226p. $35.00; $17.95pa. LC 89-38658. ISBN 0-7006-0457-X; 0-7006-0458-8pa.

Beginning with the lampreys and ending with the sculpins, this work provides information on approximately 120 fish species occurring in the Ozarks as well as the large prairie region between the foothills of the Rocky Mountains and the Mississippi River; all or part of 15 states are represented. The inside front and back covers illustrate the geographical area. The species are arranged under their families. Common and scientific names of the families are given. For each fish included, the common name, scientific name, colloquial name(s), derivation of the scientific name, geographical distribution, size, status, and other interesting facts are presented. Each page represents descriptive material on a species. The colored plates of the fishes can be found at the center of the book; they are excellent. A two-page bibliography and a combined common/scientific-name index conclude the work.

Pisces (Systematic Section)

CHONDRICHTHYES
(Cartilagenous Fish; Sharks and Rays)

776. Compagno, L. J. V. **Sharks of the Order Carcharhiniformes**. Princeton, N.J. Princeton University Press, 1988. 486p. $99.50. LC 87-2557. ISBN 0-691-08453-X.

Based on Compagno's *Sharks of the World* (see entry 777), the majority of this work i devoted to a descriptive account of the families, genera, and species that comprise the orde Carcharhiniformes. Each of the chapters comprising the descriptive matter is devoted to family of sharks. Extremely detailed external anatomy, familial nomenclature, taxonomi problems, and keys to the genera are given for the family. In addition, the same type o information, as well as geographical distributions are also presented for the genera and specie contained within the family chapters. Keys to the species are presented as well. Citations t the literature can be found interspersed within the descriptive matter. In addition to the specie accounts, there are sections dealing with the carcharhinoid in general, their external morphology ion, pectoral fin skeleton, clasper morphology, detailed jaw information, head muscles, lar intestine, mode of reproduction, and phylogeny.

There are a series of tables near the end of the work that consider the number of families, a, and species contained within all the orders of sharks, the total tooth counts for arhinoids, the pectoral radial counts for all orders and families of sharks, the total oral counts, spiral valve counts, and modes of reproduction for those species within the arhinoids. A last table considers various size comparisons of Atlantic and Pacific lotriakis. A 22-page bibliography consisting of the cited literature, and a combined index ojects, scientific names, and common names complete the major part of the book. A large er of figures can be found at the end of the work comprising line-drawings of the sharks etailed, labeled diagrams of the vertebrae and teeth. Scanning electron micrographs of eth are also present.

This is an extremely scholarly work and presents a great deal of detailed information on rder as it relates to external anatomy and taxonomy. It is not for the novice.

Compagno, Leonard J. V. **Sharks of the World: An Annotated and Illustrated ogue of Shark Species Known to Date**. Rome: Food and Agricultural Organization of nited Nations, 1984. 655p. (in 2 parts). (FAO Species Catalogue, vol. 4, parts 1 and 2; Fisheries Synopsis, no. 125, vol. 4).

Presenting a broader variety of information than Compagno's *Sharks of the Order arhiniformes* (see entry 776), this two-part work discusses 342 species of sharks ging to eight orders and 31 families. The species are arranged under their respective a, which in turn are entered under families and orders. Information for each order les scientific name, person who named it, date of naming, citations to the literature, ymy, basic external anatomy, and an illustrated key to the families. Family information lers scientific name, person who named it, date of naming, citations to the literature, ymy, FAO names, names and number of genera, basic external anatomy needed for fication, and illustrated keys to the genera and species. For each genus presented, the ific name, person who named it, date of naming, citations to the literature, type species, ynonymy are given. The information contained within the species accounts consists of ific name, person who named it, date of naming, holotype, synonymy, basic external my, geographical distribution, habitat, reproduction, size, interest to fisheries, and graphic references. A black-and-white drawing of the fish and its teeth, and a geographi- nge map accompany each of the descriptive accounts. The accounts are usually one page gth. A table of contents, preceding the accounts, lists the shark groups systematically. ppendix, which is found near the end of part 2, consists of a discussion of shark attacks y relate to the various families of sharks. A list of species by major fishing areas, a ge bibliography, and a combined scientific/common-name index can be found at the f part 2.

EICHTHYES (Bony Fish)

Lophiiformes (Anglerfishes)

Pietsch, Theodore W., and David B. Grobecker. **Frogfishes of the World: Systemat- oogeography, and Behavioral Ecology**. Stanford, Calif.: Stanford University Press, 420p. $69.50. LC 84-51302. ISBN 0-8047-1263-8.

Of the 420 pages that make up this source, 227 of these pages are devoted to the species nt of the frogfish. Each species is entered under its respective genus, which in turn is d under its family. For each of the species listed, the scientific name, common name, n who first named it, citations to the literature, and taxonomic problems are presented. lition, detailed external anatomy, coloration, and geographical distribution are given. to the genera and species, line-drawings, and colored plates of the fishes can also be throughout the pages.

Other sections are devoted to external anatomy for the frogfishes as a whole, phylogenetic relationships, zoogeography, and behavioral ecology. A 33-page bibliography consising of references cited within the text, and a combined common/scientific-name inde conclude the volume.

This work is not for the novice. It concentrates primarily on detailed external anatom taxonomic problems, and phylogenetic relationships found in these fishes. It is a scholar work.

Perciformes

779. Axelrod, Herbert R. **The Most Complete Colored Lexicon of Cichlids: Ever Known Cichlid Illustrated in Color**. Neptune, N.J.: T. F. H. Publications, 1993. 864 $100.00. ISBN 0-86622-422-X.

Abounding with a large number of colored photographs, many consisting of an enti page, this work provides general information on all species of fish in the family Cichlid that are found throughout the world. Approximately 1,200 species are included. The gene and species are presented under geographical sections. These geographical sections incluc the cichlids of Lake Tanganyika, Lake Malawi, Lake Victoria, Southern and Eastern Afric Western Africa, Tilapias, Middle East and Asia, Madagascar, Central America, and Sou America.

Information provided for the genera include person who first described the genera an year of description, reproduction, behavior, number of species, external anatomical chara teristics, and coloration. The length and amount of information varies from genus to genu In like manner, the descriptive account of each species varies as does the information. Suc parameters as person who first described the species and year of description, size, nestin behavioral patterns, coloration, and taxonomic variations are usually included. The length these accounts ranges from one paragraph to several paragraphs. General information c cichlids is provided in the introductory section. A scientific-name index completes th work.

Although this work is much more encompassing than *Cichlids of North & Centr America* (see entry 780), there are a number of differences. Among these differences a genera and species taxonomy as well as the amount of information provided in the descripti accounts of genera and species. For example, *Cichlids of North & Central America* provi proper pH and temperature range for the species; this work does not. The length of th descriptive accounts also varies. In this work, one paragraph is devoted to the pastel cichli while in *Cichlids of North & Central America*, one and one-half columns are provided. addition, this work considers the scientific name of a certain species as *Herichthys lyons* while *Cichlids of North & Central America* considers it *Cichlasoma lyonsi*. It should also b noted that some of the photographs found in this work can also be found in *Cichlids of Nor & Central America*.

780. Conkel, Donald. **Cichlids of North & Central America**. Neptune, N.J.: T. F. H Publications, 1993. 191p. $49.95. ISBN 0-86622-444-0.

Abounding with large colored photographs, this work provides general information the species of fish found throughout North and Central America, in the family Cichlidae. O of the 1,200 or so species found within this group, 126 are covered in this source. Each fis is arranged under its subgenus; a general description of the subgenus is included. For each the species listed, the scientific name, person who first named it, date of naming, alternati scientific names, common names, geographical distribution, habitat (including pH and tem perature range), diet, size, coloration, and basic external anatomical features are presente Colored photographs of the species accompany the descriptive material. Large geographic range maps along with colored photographs of the geographical area can be found in th introductory section. Other areas in the introduction consider, in general terms, classificatio ecology, anatomy, feeding habits, reproduction, aquarium management, and holdin shipping, and acclimatization techniques. A one-page bibliography and a combined common/scientific-name index conclude the work.

Differences between this work and that of *The Most Complete Colored Lexicon of Cichlids* (see entry 779), can be found in that annotation. In addition, some of the photographs in this work can also be found in the other.

781. Gery, Jacques. **Characoids of the World**. Neptune City, N.J.: T. F. H., 1977. 672p. $49.95. LC gb78-28312. ISBN 0-87666-458-3.

Considering the American and African freshwater characoid fishes, this work represents a comprehensive treatment of the group. Fishes such as pike-like characoids, tetras, citharinids, trahiras, darters, pencilfishes, pyrrhulinins, headstanders, leporins, hemiodids, curimatas, hatchetfishes, pacus, silver dollars, and piranhas are all included within these pages. Many of these fishes are popular aquarium-type favorites. By the use of keys, textural descriptions, black-and-white photographs, and a large number of colored photographs, all the various families, subfamilies, and genera are described that belong to the suborder of characoids. The textural descriptions consist of the number of species found within a genus, coloration, size, and other useful anatomical facts. The photographs depict representative species within the genus. The keys allow identification down to genus. A glossary of technical terms, a 25-page bibliography, and a scientific-name index conclude the work.

782. Steene, Roger C., and Gerald R. Allen. **Butterfly and Angelfishes of the World**. New York: John Wiley, 1978. 2 vols. LC 78-17351. ISBN 0-471-04737-6 vol. 1; 0-471-05618-9 vol. 2.

Consisting of magnificently colored photographs, this two-volume set includes all species of butterflies and angelfishes from around the world. Volume 1 considers the geographical areas of Australia and New Guinea, while volume 2 concentrates on the Atlantic Ocean, Caribbean Sea, Red Sea, and the Indo-Pacific region. The species of these beautiful fishes are arranged under their respective genera. Each page represents a species. For each fish included, the scientific name, common name, person who first named it, date of naming, size, geographical distribution, dwelling depth, habitats, behavior, and a colored photograph(s) are presented. The opening pages of volume 2 consist of a taxonomic listing of the species contained within volumes 1 and 2. Volume 2 also contains, at the close of the work, a three-page bibliography, an index of synonyms, and a separate index of common and scientific names. For those interested in these type of fish, this two-volume set is a must.

Pleuronectiformes (Flatfishes)

783. Norman, J. R. **A Systematic Monograph of the Flatfishes (Heterostomata)**. London: Printed by Order of the Trustees of the British Museum, 1934. Reprint, New York: Johnson Reprint, 1966. 459p. ISBN 0-384-41950-X.

This work represents volume 1 (Psettodidae, Bothidae, Pleuronectidae) of a projected two-volume (Soleidae, Cynoglossidae) set. Research on this book would lead one to wonder whether volume 2 was ever published.

Each species is arranged under its respective genus, which in turn is entered under the family. After some brief anatomical descriptions of the families and genera, the species accounts ensue. For each of the fish listed, the scientific name, person who first identified it, citations to the literature, detailed measurements such as depth of body and length of head, external anatomy, coloration, and geographical distribution are given. A line-drawing of the flatfish accompanies the descriptive account. Keys to the subfamilies, genera, and species are also present. The introductory chapter consists of general information of the flatfish as a whole and considers evolution, development, sexual differences, geographical distribution, and so on. A combined scientific/common-name index concludes the volume. This monograph represents a scholarly work and focuses mainly on external anatomy and taxonomy.

FIELD GUIDES

Pisces (General)

784. La Monte, Francesca. **Marine Game Fishes of the World**. Garden City, N.Y Doubleday, 1952. 190p. LC 52-13793.

Although somewhat outdated, this is a useful guide for the identification of mari fishes. Colored as well as black-and-white plates provide the illustrations with no more th a few fishes on any page. These pictures accompany the text. Common names of the fish accompany the pictures. The species are arranged under their broad common names. Bri descriptions of the groups are presented. For each of the fishes entered, the common nam scientific name, person who first identified it, other common names, geographical distrib tion, distinguishing characters (scales and fins), size, diet, habitat, type of angling necessa in order to catch fish, and its commercial importance are given. A combined common/scientifi name index concludes the field guide.

Pisces (Geographic Section)

ALASKA

785. Kessler, Doyne W. **Alaska's Saltwater Fishes and Other Sea Life: A Field Guid** Anchorage, Alaska: Alaska Northwest, 1985. 358p. $19.95pa. LC 85-13483. ISBN 0-8824 302-8pa.

Of the 358 pages found in this work, 124 of them are devoted to the fishes (constituti 127 species); the rest deal with the invertebrates. For each fish included, an outline of t body of the type of fish, the common name, scientific name, coloration, size, and geographic distribution are given. The outline of the fish corresponds to the outlines found at the end the work in a section that is a picture index. This is an excellent way to begin the identificati process. Colored pictures of the fishes accompany the brief descriptions. This is a true gui in the sense that its sole purpose is to aid in the identification of a particular fish. Many the other descriptive parameters found in other field guides are missing. The invertebrat contained within this book are given the same treatment as the fishes. A brief glossar selected list of references, a separate common/scientific-name index, and the picture ind conclude the volume.

ARABIAN GULF

786. Relyea, Kenneth. **Inshore Fishes of the Arabian Gulf**. London, Boston: George All & Unwin, 1981. 149p. LC 80-41341. ISBN 0-04-597003-3.

Species are arranged under their respective families, which in turn are entered und their orders. Brief descriptions of the family and order are given. For each of the speci presented, the scientific name, person who named it, common name, geographical range, siz coloration, and brief external anatomy are presented. Line-drawings and colored plates of t fishes aid in the identification process. Keys to the orders of the Arabian Gulf fishes can found near the beginning of the guide. In addition, keys to the families can be found und each of the orders as they are introduced throughout the text. A two-page bibliography and separate scientific/common-name index can be found at the end of the field guide.

ATLANTIC OCEAN

787. Coad, Brian W. **Guide to the Marine Sport Fishes of Atlantic Canada and New England**. Toronto; Buffalo, N.Y.: University of Toronto Press, 1992. 307p. $50.00; $19.95pa. ISBN 0-8020-5875-2; 0-8020-6798-0pa.

Consisting of 70 species of fishes that have been classified as either fishes used for bait, fishes caught for the sport of it, or fishes that make good eating, this source provides general information on the identification of these vertebrates. The species of fishes, arranged phylogenetically, are presented under their respective families. The common name of the family is given along with the number of species contained within the group and general information on the external anatomy of the group. For each of the species presented, the common name, scientific name, person who first named it, date of naming, and other common names are considered. In addition, basic external anatomical features, coloration, geographical distribution, size and weight, behavioral patterns, diet, and commercial value are given. A black-and-white drawing showing gross morphology, and citations to the literature accompany the descriptive account. In some cases, geographical range maps are also present. Besides the black-and-white drawings, a number of black-and-white photographs of fishes can be found interspersed among the accounts of the species.

The introductory material consists of 55 pages devoted to the various aspects of fish biology including, but not limited to, such topics as how to identify fishes, preserving fishes for identification, commercial fisheries, fish farming, and the like. Bibliographic citations are also present in the introductory section.

There are a number of appendices preceding the descriptive accounts of the species. Such information as taxidermy, fish prints, knot tying, recipes, a bibliography, and tides, charts, and weather reports can be found in these sections. A 12-page illustrated glossary, an English-to-French lexicon containing fish terms, metric conversions, and a combined index of subjects, common names, and scientific names complete the volume.

788. Robins, C. Richard, and G. Carleton Ray. **A Field Guide to Atlantic Coast Fishes of North America**. Boston: Houghton Mifflin, 1986. 354p. (The Peterson Field Guide Series; 32). $20.95; $14.95pa. LC 85-18144. ISBN 0-395-31852-1; 0-395-39198-9pa.

Arranged under their respective families this guide is an excellent aid in the identification of 1,100 fish species that occur between the Canadian Arctic and the Gulf of Mexico. The guide is divided into jawless fishes, cartilaginous fishes, and bony fishes. Preceding an account of the species, a brief description is given for each family. For all fishes included, the common name, scientific name, coloration, general external anatomy, size, geographical range, habitat, similar species (when appropriate), and general remarks are presented. The black-and-white and colored plates are positioned in the center of the work. Each of the plates represents a two-page spread. Common name, scientific name, and brief identification characteristics can be found on one of the pages, while the pictures of the species are on the opposing page. Cross-references from plate to text and vice versa are given. Introductory information is brief and consists of distribution and habitat, dangerous and poisonous fishes, tips on using the guide, identification, size, range, habitat, and similar and related species of fish. A brief glossary, a selected list of references, and a combined common/scientific-name index concludes the guide.

789. Ursin, Michael J. **A Guide to Fishes of the Temperate Atlantic Coast**. New York: E. P. Dutton, 1977. 262p. $10.95. LC 76-30854. ISBN 0-87690-242-5; 0-87690-243-3pa.

A general pictorial key to the fishes will aid in the beginning of the identification process. This key shows an outline of different groups of fishes and cross-references to the pages within the text. Each fish is arranged under its respective family, which in turn is entered under order and class. Scientific and common names are given for class and family. A brief description of the class and family is considered. For each of the species listed, the common name, scientific name, person who first named it, size, coloration, geographical range, basic external anatomy identification features, water temperature of habitat, and other interesting

facts are presented. In some cases, this descriptive account is for a number of related specie
within a family. In addition to the pictorial key, line-drawings of the species accompany th
descriptive information. A chart describing possible hazards in handling fishes, and a separat
common/scientific-name index conclude the guide. There are no colored plates in this work

BERMUDA

790. Mowbray, Louis S. **A Guide to the Reef, Shore and Game Fish of Bermuda**. rev
ed. [s.l., s.n.], 1976. ([Bermuda], Island Press). 71p. LC 81-465139.

Fish are entered under common group names such as sharks, snappers, and so on. Fo
each fish listed, the common name, scientific name, coloration, brief anatomical identificatio
features, size, and geographical distribution are given. Each description is approximately on
paragraph long and is accompanied by either a black-and-white line- or colored drawing o
the whole fish. A few invertebrates and a number of turtles can be found at the rear of th
guide. A two-page glossary and a common-name index complete the source. The pictures o
the fish are quite large and, in many cases, take up more room on the pages than th
descriptions. This is useful for quick identification.

CARIBBEAN

791. Randall, John E. **Caribbean Reef Fishes**. 2d ed., rev. Neptune City, N.J.: T. F. H
Publications, 1983. 352p. $29.95. LC 83-180609. ISBN 0-87666-498-2.

Beginning with the sharks, this work lists the species under their respective families. A
brief description of the family is presented along with the common and scientific name o
same. For each species mentioned, the common name, scientific name, person who firs
identified it, date of identification, brief external anatomy, size, geographical range, habitat
and other interesting facts are given. Black-and-white and colored plates provide the illustra
tive matter and make identification easier. The plates are associated with the descriptiv
accounts of the species. A brief glossary and a combined common/scientific-name index ca
be found at the end of the guide. Although this work is entered under the field guides, it coul
just as easily serve as a miniencyclopedia.

792. Stokes, F. Joseph. **Handguide to the Coral Reef Fishes of the Caribbean, an
Adjacent Tropical Waters Including Florida, Bermuda and the Bahamas**. New York
Lippincott and Crowell, 1980. 160p. LC 79-27224. ISBN 0-690-01919-Xpa.

Abounding with large numbers of colored pictures, this work attempts to aid in th
identification of those species found in the waters that are mentioned in the title. Each fish i
arranged under the common name of the family; scientific name is also given. For each specie
considered, the common name, scientific name, and identification markings are given. Thes
descriptions are no more than a few sentences long. The majority of the pages are taken u
by the colored pictures. Some general information on some families of fishes, general fis
physiology, a brief glossary, and a separate common/scientific-name index complete the fiel
guide. A quick-reference, arranging fish by color, can be found at the beginning of the source
This helps in the initial stages of identification.

EUROPE (see GREAT BRITAIN)

GREAT BRITAIN

793. Muus, Bent J. **Collins Guide to the Sea Fishes of Britain and North-Wester
Europe**. London: Collins, 1974. 244p. $29.95. ISBN 0-00-212058-5.

Encompassing a large number of colored paintings, this guide provides information o
the fishes (161 species) of the North East Atlantic. The fishes are arranged under thei
common group names, such as the cod group, the pipefishes, and so on. For each of the specie
presented, the common name, scientific name, size, geographical range, egg-laying an

hatching, life span, behavior, diet, and other interesting facts are included. A colored painting of each fish accompanies the description. In addition, taxonomic keys can be found at the beginning of each of the fish groups. These keys provide information on the identification of the fishes presented within each section. Keys to the sharks and rays as well as the bony fishes can be found at the beginning of the field guide. These are also used for the identification of the fishes contained within the guide. Other sections include some general information on the biology of fishes, history of marine fisheries, capture techniques, the fishing industry, and fisheries biology. A separate common/scientific-name index concludes the work.

HAWAII

794. Randall, John E. **Guide to Hawaiian Reef Fishes**. Newtown Square, Pa.: Harrowood Books, 1985. 79p. $18.95; $14.95pa. LC 85-24551. ISBN 0-915180-07-3; 0-915180-29-4pa.

Consisting of 177 species of fishes this guide consists of those fishes most likely to be encountered when snorkeling throughout the reefs of Hawaii. The species are arranged under their respective families. The common name and scientific name of the family are given along with a brief description of the group. For each of the fishes listed, the scientific name, person who first named it, date of naming, common name, brief anatomical identifying characteristics, size, and coloration are presented. A series of colored plates depicting all the species can be found in the center of the guide. An eight-page glossary and a combined common/scientific-name index conclude the volume.

NEW ZEALAND

795. Armitage, R. O., and others, eds. **Guide Book to New Zealand Commercial Fish Species**. Wellington, New Zealand: New Zealand Fishing Industry Board, 1981. Reprint, Huntington, N.Y.: Osprey Books, 1983. 216p. $55.00. LC 83-15132. ISBN 0-943738-10-5.

Consisting of a variety of fish species, this guide aids in the identification of those New Zealand fish having commercial value. The fish are arranged alphabetically by common name under their family groupings. For each of the species presented, the common name, scientific name, family, other common names, and related species are given. In addition, the size, weight, coloration, brief anatomical description, habitat, geographical range, and its value as a food are presented. Each species entered is given a two-page spread. A colored and a black-and-white drawing of each fish accompanies the description along with a picture of its meat (musculature). Near the end of the work, a number of mollusks, crustaceans, and echinoderms are considered. A brief glossary, a combined common/scientific-name index, and a list of selected references complete the volume.

796. Heath, Eric, and John M. Moreland. **Marine Fishes of New Zealand**. Wellington, New Zealand: A. H. & A. W. Reed, 1967. 56p. LC 68-85835.

Composed of the more common species, this guide enables the enthusiast to identify fishes found off the coast of New Zealand down to a depth of approximately 600 feet. The species are entered under their common names and grouped according to broad categories. For each fish listed, the common name, habitat (including water depth), identifying anatomical features, and common names are given. Colored paintings of the fish described can be found on the opposing page with scientific names and size. Approximately five species are described on a page. A combined common/scientific-name index can be found on the last page of the guide. The colored paintings are excellent; an easy-to-use guide for the common species.

NORTH AMERICA

797. Filisky, Michael. **Peterson First Guide to Fishes of North America**. Boston: Houghton Mifflin, 1989. 128p. $4.95. LC 88-32887. ISBN 0-395-50219-5.

Comprising the common types of fishes, this guide has been published with the beginner in mind. Species are arranged under their common group names. For each fish included, the common name, size, coloration, and geographical range are considered. Colored pictures c the species can be found on the opposing page of text. Four to five species can be found o a page. Arrows are used with the colored pictures in order to point out those anatomical part that are necessary for proper identification. A common-name index completes the volum This guide is a good place to start for the novice.

798. La Monte, Francesca. **North American Game Fishes**. Garden City, N.Y.: Doubleday 1958. 206p. LC 59-96.

Divided into saltwater and freshwater, this field guide aids the fish enthusiast in makin proper identification of the many game fishes inhabiting both types of waters. The fishes ar entered under their common group names. For each species mentioned, the common nam scientific name, person who named it, other common names, geographical distributio coloration, basic external anatomy, size, diet, and habits are presented. Approximately tw to three entries can be found on a page. Black-and-white and colored plates of the fish hel in the identification process. A combined common/scientific-name index concludes th guide.

799. McClane, A. J., ed. **McClane's Field Guide to Freshwater Fishes of North Americ** New York: Holt, Rinehart and Winston, 1978. 212p. LC 77-11967. ISBN 0-03-021116-6.

Species are introduced under their family names. Both common names and scientifi names are given for the families and species. For each fish included, the geographic distribution, coloration, number of softrays in the dorsal and anal fins, number of scales, eg laying, weight, size, and economic considerations are presented. This material is usuall one-half of a page long. In addition, a colored picture of each fish described accompanies th information. A brief glossary, and information on the measurements and identification fishes can be found in the introductory section. A common-name index concludes the fiel guide.

800. McClane, A. J. **McClane's Field Guide to Saltwater Fishes of North America**. Ne York: Holt, Rinehart and Winston, 1978. 283p. LC 77-14417. ISBN 0-03-021121-2.

Follows the same format as *McClane's Field Guide to Freshwater Fishes of Nort America* (see entry 799).

801. Page, Lawrence M., and Brooks M. Burr. **A Field Guide to Freshwater Fishes: Nort America North of Mexico**. Boston: Houghton Mifflin, 1991. 432p. (The Peterson Fiel Guide Series, 42). $24.95; $16.95pa. LC 90-42049. ISBN 0-395-35307-6; 0-395-53933-1p

Sponsored by the National Audubon Society, the National Wildlife Federation, and th Roger Tory Peterson Institute, this work considers all freshwater fishes in North Americ excluding Mexico. Approximately 790 species are included. The fishes are arranged und their respective families. The common and scientific names of the family are present. general anatomical descriptive account of the family is also considered. For each of the specie considered, the common name, scientific name, basic external anatomy, geographical distr bution, habitat, similar species, and other interesting facts are given. The black-and-white an colored plates can be found in the middle of the source; cross-references are given from th descriptive accounts to the plates and vice versa. Each plate represents a two-page spread. O one page, a list of species considering the common name, scientific name, coloration, siz and very basic external anatomy can be found. On the opposing page, pictures of the specie are presented. A series of 377 geographical range maps representing the geographic distribution of the fishes, and a combined common/scientific-name index conclude the fiel guide.

802. Schrenkeisen, Ray. **Field Book of Fresh-Water Fishes of North America: North Mexico**. New York: G. P. Putnam, 1938; reissued, 1963.

Each fish is arranged under the common name of the family. The scientific name of th family is also present. For each species included, the common names, scientific name, basi xternal anatomy, coloration, size, behavioral patterns, and geographical distribution are iven. Line-drawings of the fish accompany the description. In order to make better use of is guide, it would be useful to know the type of fish to be identified (i.e., the common name f the family to which it belongs). There are no colored plates. A nine-page glossary, a few elected references, and a combined common/scientific-name index conclude the field guide.

ACIFIC OCEAN

03. Eschmeyer, William N., and Earl S. Herald. **A Field Guide to Pacific Coast Fishes f North America: From the Gulf of Alaska to Baja California**. Boston: Houghton Mifflin, 983. 336p. (The Peterson Field Guide Series). $19.95; $13.95pa. LC 82-11989. ISBN -395-26873-7; 0-395-33188-9pa.

Comprising over 500 species, this field guide, for the most part, considers those fishes ound off the Pacific coast from the Gulf of Alaska to Baja, at depths of 650 feet or less; a ew are described that are endemic to deeper waters. The species are arranged under their espective families, which in turn are entered under their orders. The common and scientific ames of the family as well as some general information on the family and order are given. or each of the species listed, the common name, scientific name, coloration, size, brief natomical identifying characteristics, geographical range, habitat, similar species, and, in ome cases, breeding information are considered. Descriptions usually run one-third of a olumn long. Cross-references to the sets of plates are also present. The plates (many of them olored) can be found in the center of the guide. Each set of plates represents a two-page pread. The common name, scientific name, and identifying characteristics can be found on ne page and the plates of those species on the opposing page. A brief glossary, a list of elected references, and a combined common/scientific-name index complete the field guide.

04. Goodson, Gar. **Fishes of the Pacific Coast: Alaska to Peru, Including the Gulf of alifornia and the Galapagos Islands**. Stanford, Calif.: Stanford University Press, 1988. 98p. $7.95pa. LC 87-60691. ISBN 0-8047-1385-5pa.

Greater in geographical coverage than *A Field Guide to Pacific Coast Fishes of North merica* (see entry 803), this work considers over 550 species. The fishes are arranged under eir common-name group, such as damselfishes, surfperches, and snappers. For each species sted, the common name, scientific name, Spanish name, size, coloration, identifying ana- omical characteristics, dwelling depth, mating information, geographical range, diet, and uality of edibility are given. A large colored picture of each fish accompanies the description. n addition, some general information on the family group as a whole is presented. This work s easier to use than *A Field Guide to Pacific Coast Fishes*, due to the size of the colored ictures of fishes. Geographical maps, a list of selected references, and a combined ommon/scientific-name index complete the field guide.

05. Gotshall, Daniel W. **Pacific Coast Inshore Fishes**. Ventura, Calif.: Western Marine nterprises, 1989. 96p. $18.95pa. LC 80-5327. ISBN 0-930030-31-1.

This work represents a revised edition of *Fishwatchers' Guide to the Inshore Fishes of he Pacific Coast*. Species (most commonly observed) are listed under their respective amilies. For each fish included, the common name, scientific name, coloration, egg laying, ize, geographical range, and habitat are given. In most cases, a colored photograph of the pecies can be found on the opposing page. A line-drawing is used in place of the photograph n a few instances. Each description is usually one-fourth of a page long; four species to a age. The fishes included in this work are those found from the northern tip of Washington o the southernmost point of Baja. An illustrated glossary, and a written key as well as a ictorial key to the families can be found in the introductory material. The keys are useful in he initial identification process. A few references to the literature, a separate common/ cientific-name index, and a listing of fishes arranged under their geographic areas conclude he guide.

TEXTBOOKS

806. Lagler, Karl F., and others. **Ichthyology**. 2d ed. New York: John Wiley, 1977. 506p. $59.95. LC 76-50114. ISBN 0-471-51166-8.

This text is concerned with the various biological principles that relate to fish. The majority of the work considers anatomical and physiological parameters. Such areas as basic fish anatomy, skin, food getting, nutrition, digestion, growth, the skeleton, blood and circulation, respiration, excretion, reproduction, nervous system, endocrine system, and ecology are presented. Some general material on genetics, evolution, and systematics rounds off the text. A taxonomic listing of classes, orders, suborders, and families are also considered within a chapter. A large number of labeled line-drawings showing external and internal anatomy, graphs, charts, and a few geographical range maps provide the majority of the illustrations. A list of selected references appears at the end of each chapter. A combined scientific/common-name index as well as a separate subject index conclude the work.

807. Moyle, Peter B., and Joseph J. Cech, Jr. **Fishes: An Introduction to Ichthyology**. 2d ed. Englewood Cliffs, N.J.: Prentice-Hall, 1988. 560p. $67.00. LC 87-34465. ISBN 0-13-319211-3.

In addition to chapters discussing the various fishes via a phylogenetic approach, this text considers the various physiological parameters and ecology of fishes as well as their geographical ranges. The chapters on the fish present anatomical descriptions of the families with an emphasis on evolutionary trends. Physiological topics consider buoyancy, thermal regulation, respiration, blood, circulation, digestion, growth, reproduction, sensory organs, behavior, and communication. In addition, a chapter on form and movement is also present. The geographical chapters discuss the various waters found throughout the world and the types of fishes and fish families that are present in these waters. The ecology section considers the various habitats and the means by which fishes survive in these various ecological niches. Labeled and unlabeled line-drawings, charts, and graphs provide the illustrative matter. A 42-page bibliography and a combined index of subjects, common names, and scientific names complete the text. The material in this book is well presented and well balanced.

JOURNALS

808. **Environmental Biology of Fishes**. Vols. 1- , nos. 1- . Dordrecht, The Netherlands: Kluwer Academic, 1976- . 4 issues/yr. $616.50/yr. ISSN 0378-1909.

From the upstream migratory behavior in landlocked Arctic char to the growth and reproduction of the Nile perch, this journal considers research articles dealing with all behavioral aspects of saltwater and freshwater fishes. Such areas as physiology, ecology, morphology, systematics, and evolution are considered. In addition to a number of research articles ranging in length from approximately 10 to 15 or more pages, this serial also provides shorter articles under a section of brief communications. Book reviews are also present. In a number of instances, an entire volume is devoted to a particular topic. For example, volume 32(1-4) deals with articles relating to the biology of *Latimeria chalumnae* and the evolution of coelacanths. Charts, graphs, tables, labeled line-drawings, and photographs, as well as electron and light microscopy photomicrographs, provide the majority of the illustrations.

809. **Journal of Fish Biology**. Vols. 1- , nos. 1- . London, San Diego: Published for the Fisheries Society of the British Isles by Academic Press, 1969- . Monthly. $623.00/yr. ISSN 0022-1112.

From the ultrastructure of the myotomal muscles of the golomianka to the infection of Atlantic salmon, this journal provides research articles in all branches of fish biology and fisheries science. Approximately 10 to 15 research articles ranging in length from 10 to 20 pages each can be found in each issue. In addition, other features include brief communication papers and book reviews. Book reviews, however, are not found in every issue. Graphs, charts, tables, and electron and light microscopy photomicrographs provide the majority of the illustrations.

810. **Journal of Ichthyology**. Vols. 1- , nos. 1- . New York: Scripta Technica, a subsidiary of John Wiley, 1953- . 9 times/yr. $750.00/yr. ISSN 0032-9452.

Consisting of translated papers from the Russian-language journal *Voprosy ikhtiologii*, this journal contains research articles and brief communications in all major areas of freshwater and saltwater fish biology. Such areas as physiology, biochemistry, systematics, zoogeography, and the like are covered. Approximately 10 to 15 articles ranging in length from 10 to 20 pages each can be found in each issue. The brief, communication articles are not found in every issue. Graphs, charts, line-drawings, and tables provide the illustrations.

TAXONOMIC KEYS

811. Eddy, Samuel, and James C. Underhill. **How to Know the Freshwater Fishes**. 3d ed. Dubuque, Iowa: Wm. C. Brown, 1978. 215p. (The Pictured Key Nature Series). $21.75pa. LC 77-82895. ISBN 0-697-04750-4pa.

Following in the same traditional format as the other books in this series, this volume presents keys to the families of fishes found in the fresh waters of the United States for purposes of identifying the species. The keys contain specific anatomical descriptions dealing with the type of scales, teeth, snout, type of body, fin position and fin size, and so on. Black-and-white line-drawings depicting the outline of the fishes accompany the keys. In addition, brief descriptions of the families are given. The introductory material consists of general information regarding external fish anatomy. A three-page bibliography of fish identification books, and a combined common/scientific-name index conclude this identification guide.

812. Lindberg, G. U. **Fishes of the World: A Key to Families and a Checklist**. New York: John Wiley, 1974. 545p. LC 73-16382. ISBN 0-470-53565-2.

Translated from the Russian by Hilary Hardin, half of this work represents a list of keys enabling the researcher to identify down to the family level. In addition, there are line-drawings of representative species within the families. The keys consist of brief, detailed anatomical descriptions. Other areas of the volume provide a checklist of the families, complete with bibliographic citations to the literature, and, in many cases, geographical distribution of the species found within each family. A 148-page bibliography (citations are arranged under subject headings and arranged by last name of author under each heading), a separate scientific-name index (Latin names), an index of names in various western languages (common names), an index of Russian names, and an index of Chinese and Japanese names in Latin transcription conclude this scholarly work.

It should be noted that this book lists the keys, whereas Nelson's *Fishes of the World* (see entry 817), does not. Both, however, are quite detailed in anatomical descriptions of families of fishes.

813. Maitland, Peter S. **A Key to the Freshwater Fishes of the British Isles: With Notes on Their Distribution and Ecology**. Ambleside, Westmorland: Freshwater Biological Association, 1972. 139p. (Freshwater Biological Association, Scientific Publication, no. 27). $29.00. ISBN 0-900386-18-5.

This work represents a series of keys to freshwater fishes of the British Isles. Keys to the families and species as well as a preliminary key to fish larvae and fish scales by family are included. In addition, a series of 55 geographical distribution maps can be found near the remainder of the source. The species are listed underneath each of the maps and provide scientific name, common name, geographical distribution, size of fish and eggs, and diet Cross-references to this information can be found in the biological keys. A number o black-and-white line-drawings of the fishes can be found interspersed throughout the keys these aid in the identification process. Some general information on the collection an preservation of fishes as well as a checklist of British fish species can be found in th introductory section. A one-page bibliography of selected readings closes the volume.

CHECKLISTS AND CLASSIFICATION SCHEMES

814. Bohlke, Eugenia B. **Catalog of Type Specimens in the Ichthyological Collection o the Academy of Natural Sciences of Philadelphia**. Philadelphia: Academy of Natura Sciences of Philadelphia, 1984. 246p. (Academy of Natural Sciences of Philadelphia, Specia Publication, no. 14). $15.00. ISBN 0-910006-41-5.

Consisting of 2,198 species worldwide from 12,326 specimens that can be found in th Ichthyological Collection of the Academy of Natural Sciences of Philadelphia, this sourc provides a checklist of same. The species are arranged under their respective families. Fo each of the species listed, the scientific name, person who named it, date of naming, citatio to the literature, and holotypes, paratypes, and syntypes, when appropriate, are given. A 31-page bibliography consisting of those references cited within the checklist, and a scientific-name index conclude the list. Access to the families can be found in the table o contents.

815. Eschmeyer, William N. **Catalog of the Genera of Recent Fishes**. San Francisco California Academy of Sciences, 1990. 697p. $55.00. LC 90-082789. ISBN 0-940228-23-8

Divided into three sections, this scholarly work considers all genera of fishes that ar known up to this time. Part 1 represents an alphabetical listing of the genera. For each of th genera considered, the person who first named it, date of naming, citations to the taxonomic literature, a history of the namings and persons involved (name changes, etc.), status of th genera, and the family are given. Part 2 consists of the names of the genera under thei respective subfamilies, families, orders, and classes. Name of the person who first named the genus and date of naming are also presented within this classification scheme. An index of the higher taxa (groups larger than genus) proceeds part 2. Part 3 is a bibliographic listing (136 pages) of those citations found throughout the text. Other sections appearing at the end of the volume consist of a brief glossary, information on the genera of recent fishes and the International Code of Zoological Nomenclature with an appropriate index to the above, and information regarding the opinions and other actions of the International Commission involving fishes. This is a definitive work on the genera of fishes.

816. Jordan, David Starr. **The Genera of Fishes and A Classification of Fishes**. Stanford, Calif.: Stanford University Press, 1963. 4 pts. in 1. $75.00. LC 63-22441. ISBN 0-8047-0201-2.

This source is a reprint of Jordan's *Genera of Fishes* (four parts, 1917-1920) and *A Classification of Fishes* (1923). It contains a list of genera of fishes arranged by the biologists who compiled the lists and who named many of the genus groups. The person(s) responsible for the naming of each genus is included within the list. The genus listings are divided into four parts.

Part I. "From Linnaeus to Cuvier, 1758-1833."
Part II. "From Agassiz to Bleeker, 1833-1858."
Part III. "From Guenther to Gill, 1859-1880."
Part IV. "From 1881 to 1920."

In addition, a classification of fishes appears near the remaining third of the work. Genus is entered under family, which in turn is entered under suborder, order, and class. A separate index to the genera, as well as the names of families and higher groups, can be found at the end of the checklist.

817. Nelson, Joseph S. **Fishes of the World**. 2d ed. New York: John Wiley, 1984. 523p. $59.95. LC 83-19684. ISBN 0-471-86474-7.

Although the title of this volume would indicate it is an encyclopedia with dozens of colored plates and a variety of facts regarding fishes inhabiting this planet, it is not. Rather, it consists of brief anatomical descriptions of classes, subclasses, superorders, orders, suborders, and families of fishes. The majority of the descriptive matter considers the families. This descriptive information considers external anatomy, such as type of scales, teeth, and fins. In addition, the length of the adults and larvae, names of genera within the family, references to the taxonomical literature, and line-drawings of many of the fish are included. Introductory information consists of a number of fish species, biological and morphological diversity, habitat, classification and systematics, and distribution and zoogeography.

Appendix 1 represents a checklist of the extant classes, orders, suborders, and families of fish, while appendix 2 contains 45 fish distribution maps. A 50-page bibliography and a combined common/scientific-name index conclude the work.

Pisces (Geographic Section)

AFRICA

818. **Check-List of the Freshwater Fishes of Africa. Catalogue des Poissons d'eau Douce d'Afrique**. Edited by J. Daget, J. P. Gosse, and D. F. E. Thys van den Audenaerde. Paris: ORSTOM; Tervuren: MRAC, 1984- . 3 vols. to date. LC 85-184297.

Covering the African continent and adjacent islands, including Madagascar, this work presents a list of freshwater and brackish-water fish species. The species are arranged under their respective genera, which in turn are entered under families, orders, and classes. For each of the species listed, the scientific name, person who first named it, date of naming, alternative scientific names, citations to the taxonomic literature, geographical distribution and habitat, and size are given. Volume 1 considers 30 families, 129 genera, and 993 species, while volume 2 concentrates on 43 families, 211 genera, and 1,045 species. Volume 3 consists of a list of bibliographic references that were cited in volumes 1 and 2. This bibliography is arranged alphabetically by author. Volumes 1 and 2 each contain a scientific-name index for their respective compilations. There are no illustrations in any of the three volumes.

819. Jackson, P. B. N. **Check List of the Fishes of Nyasaland**. [Salisbury]: National Museums of Southern Rhodesia, 1961. 536-621p. (National Museum of Southern Rhodesia. Occasional Papers, vol. 3, no. 25B). LC 80-481271.

The species (244) are arranged under their respective genera, which in turn are entered under families, suborders, orders, subclasses, and classes. For each genus listed, the person who first named it and date of naming are given. For each species presented, the scientific name, person who first named it, citations to the literature, geographical distribution, habitat, and other facts are considered. Some geographical information on Nyasaland (Malawi) is given in the introductory information. A biological key to the families, genera, and species of the fish of Nyasaland, and a three-page bibliography complete the checklist. There is no index.

ATLANTIC OCEAN

820. International Committee for the Check-List of the Fishes of the Eastern Tropica Atlantic. **Check-list of the Fishes of the Eastern Tropical Atlantic. Catalogue des Poisson de l'Atlantique Oriental Tropical**. Portugal: Junta Nacional de Investigacao Cientifica Tecnologica (JNICT), 1990. 3 vols. $110.25/3-vol. set. ISBN 92-3-002620-4.

This source can be considered a companion to *Check-List of the Fishes of the North Eastern Atlantic and of the Mediterranean* (see entry 821). Its coverage is similar an represents a geographical area south of the above title: waters off the eastern portion of mos of the African coast. Freshwater species found in estuaries, lagoons, mangroves, and so o are not included; they are listed in *Check-List of the Freshwater Fishes of Africa* (see entr 818).

The species are listed under their respective genera, which in turn are listed under famil names. For each species of fish included, the scientific name, person who first named it, dat of naming, and citations to the literature are given. In some cases, the habitat and geographica distribution are present. Volume 3 consists of a 249-page bibliography, a full listing o citations found within volumes 1 and 2. The bibliography is arranged by last name of author In addition, a scientific-name index is also present and completes volume 3. In conclusion, total of 236 families, 757 genera, and 1,572 species (including subspecies) can be found i this work.

821. International Committee for the Check-List of the Fishes of the North-Eastern Atlanti and Mediterranean. **Check-List of the Fishes of the North-Eastern Atlantic and of th Mediterranean. Catalogue des Poissons de l'Atlantique du Nord-Est et de la Mediterra nee**. Paris: UNESCO, 1979. 2 vols. LC 80-146210. ISBN 92-3-001762-0.

Considered a first step in compiling a more comprehensive treatment of the Europea fishes, this checklist includes those species found throughout the northeastern Atlantic an Mediterranean areas. The species are arranged under their respective families. For eac species mentioned, the scientific name, person who first identified it as such, date o identification, bibliographic references to the taxonomic literature, and, in some cases habitat, distribution, and abundance are given. Occasionally a reference will be presente dealing with the eggs, larvae, and young stages of the species. All the above material can b found in volume 1. Volume 2 consists of a massive bibliography (165 pages) listing thos citations found in volume 1. The citations in the bibliography are in a more complete form In addition, the references are arranged under last name of the author. An index of commor names, one of scientific names, and a supplement to volume 1 complete volume 2. N illustrations appear in either volume. The three-volume set entitled *Fishes of the North-Eastern Atlantic and the Mediterranean* (see entry 740) is a follow up to this work.

822. Scott, W. B., and M. G. Scott. **A Checklist of Canadian Atlantic Fishes With Key for Identification**. Toronto: Royal Ontario Museum, 1966. 106p. (Life Sciences Contributions, no. 66).

Consisting of 300 species of fishes contained within 115 families, this work provides a list of these vertebrates found in the Atlantic off the coast of Canada. The region extends from Cape Chidley in northern Laborador to Georges Bank. Each species is arranged under its respective family, which in turn is entered under order and class. For each of the orders and families mentioned, the scientific and common names are given. For each of the species presented, the scientific name, person who first named it, date of naming, and common name are supplied. In addition, keys to the classes, orders, families, genera, and species are provided in the second half of the work. A four-page bibliography and a scientific-name index complete the checklist.

AUSTRALIA

823. McCulloch, Allan R. **A Check-List of the Fishes Recorded from Australia**. Sydney: The Australasian Medical Publishing Company, 1929-1930. 534p. (Australian Museum, Sydney, memoir V). LC a33-794.

Issued in four parts, this source provides a listing of all fish species found in Australian waters. Each species is arranged under its respective genus, which in turn is entered under family, order, and class. For each genera, the person who first named it, and date of naming are listed. For each of the species included, the scientific name, person who named it, citation to the taxonomic literature, geographical distribution, and, in many cases, genotype (orthotype, haplotype, logotype, etc.) are given. A scientific-name index to the genera and species completes the checklist.

CANADA

824. Halkett, Andrew. **Check List of the Fishes of the Dominion of Canada and Newfoundland**. Ottawa: Printed by C. H. Parmelee, 1913. 138p. LC f14-43.

The species (566 of them) are arranged phylogenetically under their scientific names. For each of the species presented, the scientific name, person who named it, common name, habitat, and geographical distribution are given. A classification scheme is presented in the opening pages listing genera, families, orders, subclasses, and classes. This scheme follows the same order as the species listing. A series of 14 black-and-white plates depicting many of the fishes listed can be found scattered throughout the text. A separate scientific/common-name index concludes the checklist.

825. Scott, W. B. **A Checklist of the Freshwater Fishes of Canada and Alaska**. [Toronto]: Royal Ontario Museum, Division of Zoology and Palaeontology, 1958. 30p. LC 58-30834.

The species are arranged under their respective families. Scientific and common names of the families are given. For each of the species listed, the scientific name, person who named it, common name, and the general geographical distribution (province and/or state) are given. In some instances, a citation to the literature accompanies the species listing. A five-page bibliography concludes the checklist. There is no index.

CANARY ISLANDS

826. Dooley, James K., James Van Tassel, and Alberto Brito. **An Annotated Checklist of the Shorefishes of the Canary Islands**. New York: American Museum of Natural History, 1985. 49p. (American Museum Noviatates, no. 2824). $4.00.

The species are arranged under their respective families. For each of the species listed, the scientific name, person who first named it, and geographical distribution are considered. The species of fishes identified in this checklist are mainly those osteichthyes found at depths from 25 to approximately 200 meters. A number of maps depicting the Canary Islands can be found in the introductory section. A five-page bibliography completes the work.

EUROPE

827. **European Inland Water Fish: A Multilingual Catalogue. Poissons des eaux Continentales d'Europe: Catalogue Multilingue. Peces de Aguas Continentales de Europa: Catalogo Multilingue. Binnengewasser-Fische Europas Mehrsprachiger Katalog**. London: Published for the Food and Agriculture Organization of the United Nations by Fishing News, 1971. 1 vol. (various paging). LC 75-324593. ISBN 0-85238-056-9.

This source represents a multilingual list of names of inland-water fish of Europe. The fish are arranged under their respective scientific family names. The number of common names of the fish presented varies from species to species, from 1 to 20 or more. Russian French, German, Spanish, Italian, and Danish are some of the many languages presented. In addition to the list of multilingual names, the scientific name, person who first named it, date of naming, a geographical range map, and a line-drawing of the fish accompanies each entry A list of the species under their families can be found at the beginning of the source. The list of species is arranged in the same order as the species appear in the text. In addition, several colored plates of a few of the species can also be found in the opening pages. A scientific-name index concludes the source.

HIMALAYANS

828. Menon, A. G. K. **A Check-List of Fishes of the Himalayan and the Indo-Gangetic Plains**. Barrackpore, India: Inland Fisheries Society of India, 1974. 136p. (Special Publication, Inland Fisheries Society of India, no. 1). LC 75-905189.

Consisting of freshwater fishes, this work provides the reader with a checklist of species inhabiting the Himalayan and Indo-Gangetic plains. Each species is entered under its respective genus, which in turn is entered under family, suborder, order, superorder, subclass, and class. Person responsible for naming of the genus along with a taxonomic reference are given. For each of the species listed, the person who first named it, bibliographic citations, and geographical distribution are given. A total of 207 species can be found throughout the volume. Some general information on the geography of the region can be found in the introduction. A scientific-name index completes the checklist.

INDIAN OCEAN

829. Winterbottom, Richard, Alan R. Emery, and Erling Holm. **An Annotated Checklist of the Fishes of the Chagos Archipelago, Central Indian Ocean**. Toronto: Royal Ontario Museum, 1989. 226p. (Life Sciences Contributions, no. 145). $48.50. LC cn86-95096. ISBN 0-88854-329-8.

The species (over 585 of them) are arranged under their respective families. For each species listed, the scientific name, person who named it, citations to the literature, and notes denoting where specimens were captured are given. In some cases, general information on the family is considered. A 14-page bibliography, a map of the Chago Archipelago, and 454 black-and-white photographs of the fish can be found at the end of the volume. References to the photographs accompany the species listings. In addition, a series of eight pages of plates consisting of small, colored photographs of a number of the fishes can be found within the text. The scientific name and size of the fish accompanies the colored photograph. Eight photographs are depicted on each plate. In some cases, the photographs depict areas of the ocean floor. The table of contents lists the families along with the appropriate page where each family can be found in the source. There is no index.

KOREA

830. Mori, Tamezo. **Check List of the Fishes of Korea**. Sasayama, Japan: Hyogo University of Agriculture, 1952. 228p. (Memoirs of the Hyogo University of Agriculture, vol. 1, no 3 [Biological Series no. 1]).

The species (824 of them) are arranged under their respective genera, which in turn are entered under families, orders, classes, and superclasses. For each of the genera listed, the scientific name, person who first named it, and citations to the taxonomic literature are present. For each of the species listed, the scientific name, the person who first named it, and citations to the taxonomic literature are given. A historical sketch concerning the investigations of fishes in Korea, along with pertinent citations to the literature, can be found in the introductory account. A scientific-name index and an index to the common Japanese names can be found at the end of the checklist.

NORTH AMERICA

831. American Fisheries Society. Committee on Names of Fishes. **A List of Common and Scientific Names of Fishes from the United States and Canada**. 5th ed. Bethesda, Md.: American Fisheries Society, 1991. 183p. (American Fisheries Society, Special Publication No. 20). $32.00; $24.00pa. LC 90-86052. ISBN 0-913235-70-9; 0-913235-69-5pa.

This checklist represents 2,428 species of fishes that inhabit the fresh waters of the continental United States and Canada. Marine species found in the contiguous shore waters on or above the continental shelf to a depth of 200 meters are also listed. In addition, species found along the shores of Greenland, eastern Canada and the United States, and the northern Gulf of Mexico to the mouth of the Rio Grande are also considered. Species from the continental shelf of the Mexican-United States boundary to the Bering Strait as well as from the Arctic shore waters of Alaska and Canada are also presented.

The species are listed under their scientific names and arranged under their respective families, which in turn are listed under their respective orders and classes. The scientific and common names of the families are given. The list begins with the hagfishes and ends with the molas. For each species considered, the scientific name, occurrence (Atlantic, Pacific, freshwater), person who first identified the species, and the common name are presented.

Appendix 1 is a compilation of annotations to those species not annotated in the fourth edition. Annotations consist of the name of family and/or species along with a bibliographic citation to the literature. Appendix 2 is a list of established exotic fishes, while appendix 3 considers names that have been applied to hybrid fishes. Appendix 4 lists important references for determining publication dates of certain works on fishes. A combined common/scientific-name index completes the volume.

832. Jordan, David Starr, and Barton Warren Evermann. **A Check-List of the Fishes and Fish-Like Vertebrates of North and Middle America**. Washington, D.C.: U.S. Government Printing Office, 1896. 1p. X., pp. 207-584. LC f11-199.

At the time, this source was considered to be the most complete work on North American fishes. It contains 3,100 species contained within 1,053 genera. It was subsequently revised under the title *A Check List of the Fishes and Fishlike Vertebrates of North and Middle America North of Venezuela and Colombia* (see entry 833). Each species is arranged under its respective genus, which in turn is entered under family, order, and class. The scientific and common names are given for the class, order, and family. For each of the genera mentioned, the scientific name, person who first named it, and a bibliographic citation to the literature are presented. For each of the species listed, the scientific name, person who first named it, common name, geographical distribution, and citations to the literature are given. A separate common/scientific-name index concludes the checklist.

833. Jordan, David Starr, Barton Warren Evermann, and Howard Walton Clark. **A Check List of the Fishes and Fishlike Vertebrates of North and Middle America North of Venezuela and Colombia**. Washington, D.C.: U.S. Government Printing Office, [1928] 1963. 670p. (U.S. Bureau of Fisheries, doc 1055).

The species/subspecies (4,137 of them) are arranged under their respective genera, which in turn are entered under families, orders, and classes. In addition to the scientific names of the class, order, and family, the common names are presented as well. For each of the genera listed, the scientific name, person who named it, and citations to the literature are given. For each of the organisms included, the scientific name, person who first named it, common name, geographical distribution, and citations to the literature are given. A separate common/scientific-name index concludes the checklist. It should be noted that the majority of species included in this work are described in Jordan and Evermann's four-volume publication entitled *The Fishes of North and Middle America* (see entry 761). Furthermore, this work is to be considered a revision of Jordan and Evermann's 1896 work entitled *A Check-List of the Fishes and Fish-Like Vertebrates of North and Middle America* (see entry 832).

834. Robins, C. Richard, and others, eds. **World Fishes Important to North Americans: Exclusive of Species from the Continental Waters of the United States and Canada.** Bethesda, Md.: American Fisheries Society, 1991. 243p. (American Fisheries Society Special Publication, no. 21). $38.00; $30.00pa. LC 91-71562. ISBN 0-913235-54-7; 0-913235-53-9pa.

Each species is arranged under its respective family, which in turn is entered under order and class. Brief information on the order and class is given. For each of the fish included, the scientific name, person who first named it, date of naming, common name, alternative scientific names (if any), importance codes, and geographical distribution are given. The importance codes correspond to aquarium, bait, aquaculture, endangered or threatened, food, industrial, poisonous, sport, textbook, and experimental research. An index to the families can be found at the beginning of the work. A combined common/scientific-name index completes the list.

The species listed in this volume have no geographic or bathymetric limitation as do the fishes listed in *A List of Common and Scientific Names of Fishes from the United States and Canada*, 5th ed. (all North American freshwater and marine species north of Mexico that occur in waters shallower than 200 meters) (see entry 831).

PACIFIC OCEAN

835. Fowler, Henry W. **Fishes of Guam, Hawaii, Samoa, and Tahiti**. Honolulu: Published by The Museum, 1925. Reprint, New York: Kraus Reprint, 1971. 38p. (Bernice P. Bishop Museum. Bulletin No. 22). $10.00.

The species (158 of them) are arranged under their respective families, which in turn are entered under the island groups. The amount of information given for the species varies. The information ranges from scientific name and person who named it to size and basic external anatomy. There is no index.

836. Fowler, Henry W. **Fishes of the Tropical Central Pacific**. Honolulu: Published by The Museum, 1927. Reprint, New York: Kraus Reprint, 1971. 32p. (Whippoorwill Expedition. Publication No. 1; Bernice P. Bishop Museum. Bulletin No. 38). $10.00. ISBN 0-527-02141-5.

Consisting of species (171 of them) from the waters off the islands of Palmyra, Washington, Fanning, Christmas, Jarvis, and Tongareva (Penrhyn), this source provides an annotated checklist of same. The species are arranged under their respective families. The amount of information given varies from species to species. In some cases, such information as scientific name, person who named it, name of island, and size are presented. In addition to the above, other descriptive accounts include detailed anatomical measurements, coloration, and the like. A few black-and-white line-drawings depicting the fishes represent the illustrative material. There is no index.

837. Fowler, Henry W., and Stanley C. Ball. **Fishes of Hawaii, Johnston Island, and Wake Island**. Honolulu: Published by The Museum, 1925. Reprint, New York: Kraus Reprint, 1971. 31p. (Tanager Expedition. Publication No. 2; Bernice P. Bishop Museum. Bulletin No. 26). $10.00. LC 27-172. ISBN 0-527-02129-6.

The species (234 of them) are arranged under their respective families. The amount of information given for each species varies from scientific name, person who named it, and size, to detailed anatomical measurements. In many cases, the geographical region where the fish was taken is presented. There is no index.

838. Herre, Albert W. **Check List of Philippine Fishes**. Washington, D.C.: U.S. Government Printing Office, 1953. 977p. (U.S. Fish and Wildlife Service, Research Report, vol. 20). LC 53-63395.

The East Indies, of which the Phillipines form an integral part, are considered to be the greatest center of fish life in the world. Therefore, this work provides a checklist of many of the fish species found throughout the Pacific. The species are arranged under their respective genera, which in turn are entered under families and orders. Brief anatomical information is presented for the family and genus. For each of the species listed, the scientific name, person who first named it, geographical distribution, and citations to the taxonomic literature are given. A scientific-name index concludes the checklist.

839. Kami, Harry T., Isaac I. Ikehara, and Francisco P. DeLeon. **Check-List of Guam Fishes**. [s.l., s.n., 1968-1975], (Tokyo, Japan: Kokusai Bunken Insatsusha). 1 vol. (various paging).

This list represents a compilation of reprints from Micronesica: "Check-list of Guam fishes" (vol. 4, 1968), "Check-list of Guam fishes, supplement I" (vol. 7, 1971), and "Check-list of Guam fishes, supplement II" (vol. 11, 1975).

The species (465 of them) are arranged under their respective families. The scientific and common names of the families are given. For each of the species listed, the scientific name, person who first named it, where the fish was taken, size, and, in some cases, citations to the literature are given. The fishes considered are freshwater and saltwater varieties. A two-page bibliography and an index to families complete the main list. Supplements I and II proceed the main list, providing additional species and references. There are no indices for the supplements.

840. Wass, Richard C. **An Annotated Checklist of the Fishes of Samoa**. Seattle, Wash.: National Oceanic and Atmospheric Administration, National Marine Fisheries Service, 1984. 43p. (NOAA Technical Report SSRF-781).

The species (991 of them) are arranged under their respective families. Some general information on each family is considered. For each of the species listed, the scientific name, person who first named it, and date of naming are given. In addition, museum catalog numbers are provided for many of the unidentified and uncommon species. A two-page bibliography, a combined common/scientific-name index, and an index to Samoan fish names complete the checklist.

PERU

841. Ortega, Hernan, and Richard P. Vari. **Annotated Checklist of the Freshwater Fishes of Peru**. Washington, D.C.: Smithsonian Institution Press, 1986. 25p. (Smithsonian Contributions to Zoology; no. 437). LC 86-600237.

The species, found throughout the Peruvian countryside, are entered under their respective families, which in turn are entered under orders and classes. For each of the species considered, the scientific name, person who first named it, and date of naming are given. In addition, references to the taxonomic literature accompany the entry; full information to the reference can be found at the end of the booklet.

The introductory information contains, among other pieces of information, a table listing the names of the families, and the number of genera and species contained therein. Additional information in the introduction consists of the geography of Peru, and methodology in creating the list.

UNITED STATES

842. Association of Systematics Collections. **Checklist of Fishes of the United States and the U.S. Territories**. Lawrence, Kans.: Association of Systematics Collections, 198- . 72 leaves.

Prepared by the Association of Systematics Collections under cooperative agreement number 14-16-0009-1038 for Eastern Energy and Land Use Team, Fish and Wildlife Service, Office of Biological Services, this working draft lists those species of fishes found in the United States and its territories as well as those found in Canada. Each species is arranged under its respective genus, which in turn is entered under family, order, and class. Each of the genera listed presents to the reader the scientific name, person who named it, and date of naming. For each of the species listed, the scientific name, person who named it, date of naming, and common name are given. There is no index.

VIETNAM

843. Kuronuma, Katsuzo. **A Check List of Fishes of Vietnam**. [Saigon]: U.S. Operations Mission to Vietnam, 1961. 66p.

Supported by the Division of Agriculture and Natural Resources, U.S. Operations Mission to Vietnam, this source provides a list of 807 species of saltwater and freshwater fishes contained within 411 genera and 139 families. The species are arranged under their respective families. For each species listed, the scientific name, person who named it, and the common Vietnamese name, Japanese name, and English name are given. A separate scientific-name index, Vietnamese common-name index, and English common-name index can be found at the end of the checklist.

ASSOCIATIONS

844. **American Fisheries Society**. 5410 Grosvenor Lane, Ste. 110, Bethesda, MD 20814. (301) 897-8616. FAX (301) 897-8096.

845. **American Institute of Fishery Research Biologists**. Lehman Coll. of CUNY, Dept. of Biological Sciences, Bedford Park Blvd., W., Bronx, NY 10468. (212) 960-8239.

846. **American Killifish Association**. P.O. Box 1, Midway City, CA 92655. (219) 255-6356.

847. **American Medical Fly Fishing Association**. c/o Veryl Frye, M.D., P.O. Box 768, Lock Haven, PA 17745. (717) 769-7375.

848. **American Society of Icthyologists and Herpetologists**. Dept. of Zoology, Southern Illinois Univ., Carbondale, IL 62901-6501. (618) 453-4113.

849. **Aquatic Research Institute**. 2242 Davis Court, Hayward, CA 94545. (415) 785-2216. FAX (415) 784-0945.

850. **Association of Midwest Fish and Wildlife Agencies**. Michigan Dept. of Natural Resources, Box 30028, Lansing, MI 48909. (517) 373-1263.

851. **Association of University Fisheries and Wildlife Program Administrators**. Texas A&M Univ., Dept. of Wildlife and Fisheries Sciences, College Station, TX 77843-2258. (409) 845-1261. FAX (409) 845-3786.

852. **Atlantic Offshore Fishermen's Association**. P.O. Box 3001, Newport, RI 02840. (401) 849-3232. FAX (401) 847-9966.

853. **Atlantic Salmon Federation**. P.O. Box 429, St. Andrews, NB, Canada EOG 2X0. (506) 529-4581. FAX (506) 529-4438.

854. **Atlantic States Marine Fisheries Commission**. 1400 16th St. NW, Ste. 310, Washington, DC 20036. (202) 387-5330. FAX (202) 387-3830.

855. **Bass Research Foundation**. 1001 Market Street, Chattanooga, TN 37402. (615) 267-1680.

856. **Committee for the Development and Management of Fisheries in the South China Sea**. FAO Regional Office for Asia and the Pacific, Phra Atit Rd., Bangkok 10200, Thailand. 2 2817844. FAX 2 2800445. TELEX 82815 FOODAG TH.

857. **Desert Fishes Council**. P.O. Box 337, Bishop, CA 93514. (619) 872-8751.

858. **Fish Culture Section**. c/o Amer. Fisheries Society, 5410 Grosvenor Ln., Ste. 110, Bethesda, MD 20814. (301) 897-8616. FAX (301) 897-8096.

859. **FishAmerica Foundation**. 1010 Massachusetts Ave. NW, Ste. 320, Washington, DC 20001. (202) 898-0869. FAX (202) 371-2085.

860. **Fisheries Committee of the International Co-Operative Alliance**. ZENGYOREN, Co-op Bldg., 7th Fl., 1-1-12 Uchikanda, Chiyoda-ku, Tokyo 101, Japan. 3 32949617. FAX 3 2949602. TELEX 222 6234 ZENYO J.

861. **Fisheries Society of the British Isles**. Fisheries Laboratory, Ministry of Agriculture Fish and Food, Pakefield Rd., Lowestoft, Suffolk, NR33 OHT, England. 502 562244.

862. **Fishery Committee for Eastern Central Atlantic (Comite des Peches pour l'Atlantique Centre-Est)**. 56, avenue Georges Pompidou, Boite Postale 154, Dakar, Senegal. 220177. TELEX 3138.

863. **Florida Tropical Fish Farms Association**. P.O. Drawer 1519, Winter Haven, FL 33880. (813) 293-5710. FAX (813) 299-5154.

864. **General Fisheries Council for the Mediterranean (Conseil General des Peches pour la Mediterranee)**. Food and Agriculture Organization of the U.N., Via delle Terme di Caracalla, I-00100 Rome, Italy. 6 57976435. CABLE: FOODAGRIROME. FAX 6 57976500. TELEX 610181 FAO I.

865. **Goldfish Society of America**. P.O. Box 1367, South Gate, CA 90280. (213) 633-6016.

866. **Gulf and Caribbean Fisheries Institute**. Sea Grant Consortium, 287 Meeting St., Charleston, SC 29401. (803) 727-2078. FAX (803) 727-2080.

867. **Hawaiian International Billfish Association**. P.O. Box 30547, Honolulu, HI 96820-0547. (808) 836-0974. FAX (808) 836-0976.

868. **Indo-Pacific Fishery Commission (Commission Indo-Pacific des Peches)**. FAO Regional Office for Asia and the Pacific, Maliwan Mansion, Phra Atit Rd., Bangkok 10200, Thailand. 2 2817844. FAX 2 280045. TELEX 82815 FOODAG TH.

869. **Inland Commercial Fisheries Association**. c/o Green Island Fishing Co., Inc., 11 Ogden St., Marinette, WI 54143.

870. **Inter-American Tropical Tuna Commission**. c/o Scripps Institution of Oceanography, La Jolla, CA 92093. (619) 546-7100. FAX (619) 546-7133. TELENEX 69 7 115 TUNACOM.

871. **International Association of Fish and Wildlife Agencies**. 444 N. Capitol St., NW, Ste. 534, Washington, DC 20001. (202) 624-7890. FAX (202) 624-7891.

872. **International Baltic Sea Fishery Commission (Commission Internationale de Peches de la Baltique)**. ulica Hoza 20, PL-00-528 Warsaw 1, Poland. 22 288647. TELEX 813407 GOMO PL.

873. **International Betta Congress**. 923 Wadsworth St., Syracuse, NY 13208. (315 454-4792.

874. **International Center for Living Aquatic Resources Management**. MC P.O. Bo 1501, Makati, Metro Manila 1299, Philippines. 2 8180466. FAX 8163183. TELEX 67494.

875. **International Commission for the Conservation of Atlantic Tunas (Commission Internationale pour la Conservation des Thonides de l'Atlantique)**. Principe de Vergar 17, E-28001 Madrid, Spain. 1 4310329. TELEX 46330.

876. **International Commission for the Southeast Atlantic Fisheries (Commission Internationale des Peches de l'Atlantique Sud-Est)**. Paseo de la Habana 65, E-28036 Madrid Spain. 1 4588766. FAX 1 5710637. TELEX 45533 SEAF E.

877. **International Game Fish Association**. 3000 E. Las Olas Blvd., Ft. Lauderdale, Fl 33316. (305) 467-0161. FAX (305) 467-0331.

878. **International Institute of Fisheries Economics and Trade**. Oregon State Univ Office of Intl. Research and Development, Snell Hall 400, Corvallis, OR 97331. (503 737-2228. FAX (503) 737-3447.

879. **International North Pacific Fisheries Commission**. 6640 NW Marine Dr., Vancouver, BC, Canada V6T 1X2. (604) 228-1128. FAX (604) 228-1135.

880. **International Pacific Halibut Commission**. P.O. Box 95009, Seattle, WA 98145 2009. (206) 634-1838. FAX (604) 228-1135.

881. **National Fisheries Contaminant Research Center**. Fish and Wildlife Service, U.S Dept. of the Interior, Rt. 2, Columbia, MO 65201. (314) 875-5399.

882. **National Fisheries Education and Research Foundation**. 1525 Wilson Blvd., Ste 500, Arlington, VA 22209. (703) 524-9216.

883. **National Military Fish and Wildlife Association**. P.O. Box 128, Encinitas, CA 23509. (619) 725-4540. FAX (619) 725-3528.

884. **National Ornamental Goldfish Growers Association**. 6916 Black's Mill Rd., Thurmont, MD 21788. (301) 271-7475.

885. **North American Native Fishes Association**. 123 W. Mt. Airy Ave., Philadelphia, PA 19119. (215) 247-0384.

886. **North Atlantic Salmon Conservation Organization**. 11 Rutland Sq., Edinburgh EH 2AS, Scotland. 31 2282551. FAX 31 2284384.

887. **North-East Atlantic Fisheries Commission**. Nobel House, Rm. 425, 17 Smith Sq. London SW1P 3HX, England. 712385919. FAX 71 2385889. TELEX 21271 MAFWSL G.

888. **Northwest Atlantic Fisheries Organization (Organisation des Peches de l'Atlantique Nord-Ouest)**. P.O. Box 638, Dartmouth, NS, Canada B2y 3y9. (902) 469-9105. FAX (902) 469-5729. TELEX 01931475.

889. **Ornamental Fish International**. U.S. Branch Office, 102 Charlton St., New York, NY 10014. (212) 741-1023. FAX (212) 627-8608.

890. **Pacific Coast Federation of Fishermen's Associations**. Ft. Cronkhite Bldg. 1064, P.O. Box 989, Sausalito, CA 94966. (415) 332-5080. FAX (415) 331-CRAB. TOLL FREE NUMBER (800) 235-1444.

891. **Pacific Fisheries Devlopment Foundation**. P.O. Box 4526, Honolulu, HI 96812. (808) 548-3469.

892. **Pacific Fishery Management Council**. 2000 SW 1st Ave., Rm. 420, Portland, OR 97201. (503) 326-6352.

893. **Pacific Ocean Research Foundation**. 74-425 Kealakehe Pkwy., No. 15, Kilua Dona, HI 96740. (808) 329-6105. TELEX: RCA 8280.

894. **Pacific Salmon Commission**. 1155 Robson St., Ste. 600, Vancouver, BC, Canada V6E 1B5. (604) 684-8081. FAX (604) 666-8707.

895. **Pacific States Marine Fisheries Commission**. 2501 SW 1st Ave., Ste. 200, Portland, OR 97201. (503) 326-7025. FAX (503) 326-7033.

896. **Society for the Protection of Old Fishes**. Univ. of Washington, WH-10, School of Fisheries, Seattle, WA 98195. (206) 778-7397.

897. **South East Asian Fisheries Development Center**. P.O. Box 13-4, Phrapradeng, Samutprakarn 10130, Thailand. FAX 2352070. TELEXES 82156 and 87032 COMSERV TH.

898. **South Pacific Forum Fisheries Agency**. P.O. Box 629, Honiara, Solomon Islands. 21124. CABLE: FORFISH. FAX 23995. TELEX 66 336 FORFISH.

899. **Southeastern Association of Fish and Wildlife Agencies**. 1021 Rodney Dr., Baton Rouge, LA 70808. (504) 766-0519.

900. **Sport Fishery Research Program**. 1010 Massachusetts Ave., NW, Ste. 320, Washington, DC 20001. (202) 898-0770. (207) 371-2085.

901. **The Steamboaters**. 233 Howard Ave., Eugene, OR 97404. (503) 688-4980.

902. **Stripers Unlimited**. 880 Washington St., P.O. Box 3045, South Attleboro, MA 02703. (508) 761-7983.

903. **Trout Unlimited**. 501 Church St. NE, Vienna, VA 22180. (703) 281-1100.

904. **United States Tuna Foundation**. 1101 17th St. NW, Ste. 609, Washington, DC 20036. (202) 857-0610. FAX (202) 331-9686.

905. **Western Association of Fish and Wildlife Agencies**. 1416 9th St., 12th Fl., Sacramento, CA 95814. (916) 323-7319.

906. **Western Central Atlantic Fisheries Commission**. Via delle Terme di Caracalla, I-00100 Rome, Italy. 6 57976616. FAX 6 57976500. TELEGRAM: FOODAGRI ROME. TELEX 610181 FPOI ROME.

CHAPTER 6

THE AMPHIBIANS AND REPTILES

Approximately 3,000 species of amphib ans are in existence today. These vertebrate illustrate the initial attempt to go from a wat environment to a land environment. Like th moss, they are somewhat successful; however, water environment is still needed for reprodu tive purposes. In addition, amphibians, unlik the reptiles, lack claws and scales. Most of th familiar amphibians are the frogs, toads, and salamanders. However, there a amphibians known as caecilians that are footless and resemble worms; their habit is the tropics.

Unlike the amphibians, the reptiles made a complete transition to land. Th was accomplished by evolving a skin and an egg cover that are impermeable t water. The total number of reptilian species living today is approximately 6,00 They can be recognized by their dry scaly skin (snakes being slimy is a myth). The are cold-blooded, as are the amphibians, although there is some bone evidence t suggest that a number of dinosaurs were warm-blooded. It is believed that ancie reptilian stock divided into two branches of the evolutionary tree. One branch gav rise to the birds and the other to the mammals. The scales evolved into feathers f the bird group and into hair for the mammals. The types of organisms belonging t this vertebrate group are turtles, tortoises, crocodiles, alligators, snakes, tuatara and lizards.

Although there are many differences between the amphibian and reptilia groups, they are brought together under the study known as Herpetology.

These foul and loathsome animals ... their Creator has n
exerted his powers (to make) many of them
—Carolus Linnae
1700

Amphibians and reptiles are not degenerate
or inferior in comparison to birds and mammals;
they simply go about things in different ways and are,
in many respects, just as successful.
—Tim Halliday
The Encyclopedia of Reptiles and Amphibians, *1986.*

DICTIONARIES

907. Gotch, A. F. **Reptiles: Their Latin Names Explained: A Guide to Animal Classification**. Poole, N.Y.: Blandford Press, 1986. 176p. LC 86-197196. ISBN 0-7137-1704-1.

Basically, the terms contained within this source are arranged under their common group names (orders); each group constitutes a chapter. The terminology, which is arranged taxonomically within these chapters, consists of names of orders, suborders, families, and species. The species are presented under their common names. The words are dissected showing derivations (Latin or Greek) of the parts of the words as well as a literal meaning of each part. The major chapters include turtles, tortoises, and terrapins; alligators, crocodiles, and the gavial; the tuatara; lizards; and snakes. In addition, some general information regarding each order is presented. A one-page bibliography, a brief glossary, and a separate subject, common-name, and scientific-name index concludes the dictionary.

908. Peters, James A. **Dictionary of Herpetology: A Brief and Meaningful Definition of Words and Terms Used in Herpetology**. New York: Hafner, 1964. 392p. LC 64-23131.

All the terms contained within this source deal to some extent with the science of herpetology. The majority of them are morphological, although physiology, behavior, and collecting technique terms are present. Definitions range from a few sentences to a few paragraphs. A number of labeled line-drawings of these creatures can be found at the close of the book. Drawings consist mainly of external and internal anatomy of vertebrates. A variety of structural formulas of toad poisons are also present. Although somewhat outdated, this dictionary is still useful due to the large number of anatomical terms entered.

ENCYCLOPEDIAS

909. Breen, John F. **Encyclopedia of Reptiles and Amphibians**. Hong Kong: T. F. H., 1974. 576p. $34.95. ISBN 0-87662-203-X.

Arranged by broad animal groups, this work abounds with colored and black-and-white photographs of the many types of amphibians and reptiles inhabiting the planet. Various species are discussed within their respective chapters, accompanied by at least one photograph of each animal. The sections (chapters) consist of turtles, alligators and their relatives, native American lizards, exotic lizards and the tuatara, harmless North American snakes, nonpoisonous exotic snakes, poisonous snakes, newts and their relatives, and frogs and toads, as well as a few chapters dealing with collecting herptiles, housing herptiles, feeding the herptiles, and illnesses of these creatures. Descriptive information on the species consists of common name, geographical distribution, habitat, size, egg and laying information, and food. General descriptive information concerning the families is also presented. The length of the descriptive information varies from species to species, from a few sentences to a few pages. Scientific names of the amphibians and reptiles included are presented with their photographs. A classification of the living reptiles and amphibians, down to family, is given in the introductory material. A two-page bibliography and a combined common/scientific-name index conclude the work. The volume represents a good place to begin for general, nondetailed information concerning these vertebrates.

910. Grzimek, Bernhard, ed. **Grzimek's Animal Life Encyclopedia: Fishes II and Amphibia, Volume 5, 1974**. New York: Van Nostrand, 1974. 555p. LC 79-183178.

The variety of amphibians are arranged under their broad common-name groups, such as the urodeles, frogs and toads, lower anurans, and higher anurans. Factual information on the families and the species is contained within these sections. Such information as the number of genera and species, habitats, geographical distributions, basic external anatomy, behavior, ecology, reproduction, egg-laying, coloration, and the like are given. A large number of colored photographs of the animals provide the illustrations. In addition, a colored picture of the internal anatomy of a frog along with a labeled line-drawing of that anatomy appears in the frogs and toads section. A systematic classification of the remainder of the fishes as well as the amphibians; a polyglot animal-name dictionary in English, German, French, and Russian; a three-page bibliography, and a combined subject, common-name, amd scientific-name index complete the volume. A discussion dealing with the fish portion of the book can be found in chapter 5.

911. Grzimek, Bernhard, ed. **Grzimek's Animal Life Encyclopedia: Reptiles, Volume 6, 1974**. New York: Van Nostrand, 1974. 589p. LC 79-183178.

Following in the footsteps of the other volumes within this set, this source provides a wealth of information regarding the vertebrate group known as the reptiles. These animals are discussed under broad common-name groups, each of which constitutes a chapter. Such groups, among others, are the turtles, the iguanids, the skinks, the crocodiles and alligators, and the lizards. For each of these sections, the families and species within the family are considered. Such information as size, basic external anatomy, number of genra and species within a family, habitats, geographical distributions, behavior, reproduction, and a variety of other interesting facts are presented. As with the other volumes, a large number of colored photographs can be found interspersed among the pages of the book. A systematic classification of the reptiles; a polyglot animal-name dictionary listing the names of the animals in English, German, French, and Russian; a three-page bibliography, and a combined subject, common-name, and scientific-name index conclude the work.

912. Halliday, Tim R., and Draig Adler. **The Encyclopedia of Reptiles and Amphibians**. New York: Facts on File, 1986. 152p. $24.95. LC 85-29249. ISBN 0-8160-1359-4.

Abounding with large numbers of colored photographs, many taking up one page or more, as well as colored paintings, this work provides an introduction to some of the more interesting amphibians and reptiles found throughout the world. The chapters are arranged by broad animal groups such as the salamanders and newts, the frogs, the turtles and tortoises, and so on. For each chapter, information in tabular format is presented, such as order, number of genera and families in the group, geographical distribution, geographical range map, size, coloration, reproduction, and longevity. In addition, a large number of biological facts about the groups can be found in the text portions. Such biological considerations as body plan, temperature, water regulation, reproduction and development, diet, and information on specific groups within the group are presented. Each of the chapters conclude with a listing of the families, the number of species within the genera, geographical range maps, geographical distribution, coloration, brief external anatomy, the common and scientific name of a number of the species, and the number of species threatened. This above information is presented in a chart form. A two-page bibliography, a brief glossary, and a combined subject, common-name, and scientific-name index conclude the work. This source is a great place to start for general information on these two groups of vertebrates. If not initially interested in these animals, the reader will be after looking at some of the incredible photographs and browsing through the text.

13. Obst, Fritz Jurgen, Klaus Richter, and Udo Jacob. **The Completely Illustrated Atlas f Reptiles and Amphibians for the Terrarium**. Neptune City, N.J.: T. F. H. Publications, 988. 830p. $100.00. LC 92-110201. ISBN 0-86622-958-2.

Originally published by Edition Leipzig under the title *Lexikon der Terraristik und Herpetologie*, this work defines and/or discusses a large number of amphibian and reptilian erms. The types of entries included consider classes, orders, families, genera, species, ehavior, internal anatomy, ecology, physiology, diseases, reproduction, and other related erms dealing with those reptiles and amphibians that are kept in captivity. The length of the ntries vary from a few sentences to a few columns. A large number of colored photographs rovide the illustrative material. The colored photographs of these animals are excellent. ther illustrations consist of black-and-white photographs, line-drawings (some labeled), and eographical range maps. Coloration, external anatomical characteristics, geographical anges, terrarium environments, and the like are discussed for many of the taxonomic groups nentioned. In addition, the person who first named the group or gave it its scientific name, s well as the date of naming are provided. Information on how to treat diseases in these nimals is given under the disease entries.

Guidelines for the use of live amphibians and reptiles in field research, and a common-name index conclude this dictionary/encyclopedic work.

Do not be deceived by the title. This work is fairly comprehensive and can easily serve s a dictionary or encyclopedia for those interested in these two groups of vertebrates.

BIBLIOGRAPHIES

14. Chiszar, David, and Rozella B. Smith. **Fifty Years of Herpetology: Publications of Hobart M. Smith**. North Bennington, Vt.: John Johnson, 1982. 78p. $4.00pa. ISBN 0-10914-17-6.

Along with a brief biographical sketch of Hobart M. Smith, this source provides the eader with a list of his publications from 1931 through 1980. An index to the type of ublication, a coauthor index, an index of scientific names in the title of the works, and a new axa index can be found completing the bibliography.

15. Smith, Hobart Muir, and Rozella B. Smith. **Synopsis of the Herpetofauna of Mexico**. ugusta, W. Va.: Eric Lundberg, 1971- . 6 vols. $10.00 vol. 1; $12.50 v. 2; $25.00 vol. 3; 12.50 vol. 4; $12.50 vol. 5; $40.00 vol. 6. ISBN 0-910914-06-0 vol. 1; 0-910914-07-9 vol. ; 0-910914-08-7 vol. 3; 0-910914-09-5 vol. 4; 0-910914-10-9 vol. 5; 0-910914-11-7 vol. 6.

Volumes 1 and 2 represent a listing of citations, arranged alphabetically under the uthors' last names, to the literature dealing with the amphibians and reptiles of Mexico. olume 1 considers the literature of the axolotl while volume 2 considers the literature xclusive of the axolotl. The citations (mainly journal literature) are from the later half of the ineteenth century and the twentieth century up to the beginning of the 1970s. English, rench, Spanish, and German are the predominant languages. Volume 1 contains over 2,850 tations and provides a subject and coauthor index. Volume 2 contains over 4,800 citations nd provides an index of scientific names in the titles, an index of place names in the titles, n index of disciplines, of general terms, and of coauthors. Volume 3 contains a listing of eferences that did not appear in volumes 1 and 2 either because they were inadvertently mitted or had not yet been published. The bulk of volume 3, however, consists of a series f indexes to volume 2. The indexes are a number of primary indexes (alphabetical arrange-ment of all species by scientific names under large amphibian and reptilian groups), a nonym list representing all valid names occurring in the primary index (alphabetically rranged by scientific names), a species combination list (alphabetical order of all cited pecific names, and under each one all combinations in which it occurs, arranged alphabeti-ally), and a state list (names cited in the primary index that are accompanied by any indication f occurrence in any state, listed alphabetically). Volume 4 completes the index to volume 2, while volume 5 considers a bibliography and index to the suborder Amphisbaenia of the ord Squamata, and the order Crocodylia. Volume 6 delves into the turtles of Mexico. In additio to the list of references for individual species, this volume also presents a descriptive accou of many of the species. Such information as taxonomic status and history, ecologic considerations, coloration, basic external anatomy, geographical distribution, and conserv tional status are considered. A geographical index (gazetteer) giving locations of the specie a series of geographical range maps, and a series of black-and-white drawings depicting turt species can also be found, near the last third of the volume. A bibliographic addendum th lists 767 references not occurring in the other volumes proceeds the drawings. This biblio raphy takes into account the years 1976-1978 as well as citations that were previous overlooked in preceding volumes. A separate coauthor and scientific-name index conclud volume 6. It should be noted that this work is difficult to use. The publications are uneven i content and one must read the prefaces and introductory sections in order to use the effectively.

916. Villa, Jaime, Larry David Wilson, and Jerry D. Johnson. **Middle American Herpeto ogy: A Bibliographic Checklist**. Columbia, Mo.: University of Missouri Press, 1988. 131 $35.00. LC 87-19115. ISBN 0-8262-0665-4.

Partly a checklist and partly a bibliography, this source considers 782 species of reptile and amphibians found in Mexico and Central America. The species are listed under the respective families, which in turn are listed under orders and classes. For each specie considered, the scientific name, person who first named it, and one or more bibliographi citations can be found. The bibliographic citations are associated with categories such as reference representing key literature, one representing illustrations, one or more dealing wi geographical distribution, and so on. These references found under the species contain th author and year of publication. The complete citations can be found near the end of the wor and total 21 pages.

Ninety-one colored photographs of representative amphibians and reptiles are prese at the beginning of the source. In addition, a few black-and-white photographs of thes animals are scattered throughout the bibliography. A scientific-name index concludes th source.

INDEXES AND ABSTRACTS

917. **Zoological Record**. Vols. 1- , nos. 1- . Philadelphia: BIOSIS, Zoological Society c London, 1865- . Annual. $2,200.00/yr. ISSN 0084-5604.

For a description of this index, see entry 55. Relating to this chapter are section 1 Amphibia, and section 17, Reptilia.

HANDBOOKS

Amphibia and Reptilia (General)

918. Boulenger, G. A. **Catalogue of the Batrachia Gradientias. Caudata and Batrachi Apoda in the Collection of the British Museum**. 2d ed. London: Printed by Order of th Trustees, 1882. 127p. LC 08-21674.

Follows the same format and gives the same information on the species as *Catalogu of the Batrachia Salientias. Ecaudata in the Collection of the British Museum* (see entr 919).

)19. Boulenger, G. A. **Catalogue of the Batrachia Salientias. Ecaudata in the Collection of the British Museum**. 2d ed. London: Printed by Order of the Trustees, 1882; Reprint, ondon, Lubrecht & Cramer, 1966. 495p. $81.60. ISBN 3-7682-0291-7.

Originally written by Dr. Albert Ghunther and published by the Order of the Trustees n 1858, this work, which is called a second edition, is considered an original work. This is lue to the inclusion of additional species and a rework of the taxonomy. The species are rranged under their respective genera, which in turn are entered under families, suborders, nd orders. There are 800 species considered and they represent all parts of the world. 'receding the descriptive account, taxonomic keys to the families, genera, and species are resent. In addition, anatomical information for the family and genus along with taxonomic eferences are considered. For each of the organisms listed, the scientific name, taxonomic eferences to the literature, and detailed external anatomical descriptions for species identi- ication are given. Geographical distribution is also presented. A few line-drawings depicting asic anatomical parts of the frogs and toads along with line-drawings of the animals hemselves provide the illustrative material. A scientific-name index can be found preceding he plates of line-drawings.

'20. Boulenger, G. A. **Contributions to American Herpetology**. Ann Arbor, Mich.: Ohio Ierpetological Society, Society for the Study of Amphibians and Reptiles, 1971. 523p.

Being a facsimile reprint of collected papers authored by G. A. Boulenger, this volume onsiders in great detail the taxonomic qualities that identify one species from another.)escriptions of species from different parts of the world are covered. In addition to detailed natomical features, the accounts also consider coloration, and, in some cases, geographical listribution. Biological keys are also present in many instances. References to the taxonomic iterature can be found throughout the papers. A number of black-and-white line-drawings re present. Although the majority of the work is in English, a few papers are presented in 'rench as well as German. The journals from which these articles were taken are *Annals and Magazine of Natural History (London), Annali del Museo Civico di Storia Naturale di Genova Genoa), Bulletin de la Societe Zoologique de France (Paris), Jahresberichte und Abhandlungen us dem Naturwissenschaftlichen Verein in Magdeburg (Magdeburg), Journal of the Linnean ociety of London, Zoological Series (London), Nature (London)*, and the *Proceedings of the Zoological Society (London)*. Unfortunately, there is no index.

'21. **Handbuch der Zoologie: Eine Naturgeschichte der Stämme des Tierreiches**. Sech- ter Band, Erste Hälfte: **Reptilia**. Sechster Band, Zweite Hälfte: **Amphibia**. Gegrundet von Villy Kükenthal. Hrsg. von J. G. Helmcke and others. Berlin: W. de Gruyter, 1930-1964. ome volumes still in print.

A general description of the set to which these volumes belong appears in chapter 1 (see ntry 68). One and one-half volumes are devoted to the amphibians and reptiles, plus dditional parts, which were added until 1964. The subject areas in these volumes concentrate n morphology, physiology, and taxonomy. Line-drawings (many of them labeled) provide he majority of the illustrations. A bibliography appears at the end of each volume. These vorks are written in German.

'22. International Union for Conservation of Nature and Natural Resources. **The IUCN Amphibia-Reptilia Red Data Book: Part 1, Testudines, Crocodylia, Rhynchocephalia**. Gland, Switzerland: IUCN, 1982. 426p. ISBN 2-88032-601-X.

For each of the species entered, the common name, scientific name, person who first amed it, date of naming, synonyms, order, and family are listed. In addition, parameters such s geographical distribution, population, habitat, ecology, threats to survival, conservation neasures taken, conservation measures proposed, number of individuals bred in captivity, tatus, and a list of references are given. The status categories include extinct, endangered, ulnerable, rare, indeterminate, out of danger, and insufficiently known. A list of the scientific ames of the amphibians and reptiles found within the text can be found preceding the escriptive information. These organisms are arranged under their families; common name and status of the species are also considered. A combined common/scientific-name inde completes the work.

923. **Traité de Zoologie: Anatomie, Systématique, Biologie**. Tome XIV. **Amphibiens e Reptiles**. Ed. by Pierre-Paul Grassé. Paris: Masson, 1970-1986. 3 vols. (various fascicules) LC 49-2833.

A general description of the set to which these volumes belong appears in chapter 1 (se entry 69). Published out of sequence, this major, classic work devotes one of its volumes (fou parts) to the amphibians and reptiles (two parts to amphibians and two parts to reptiles). Suc areas as anatomy, taxonomy, embryology, physiology, and other biological principles ar considered. Photomicrographs, black-and-white photographs, and line-drawings (many o them labeled) provide the majority of illustrations. Bibliographies appear at the end of eac chapter. A combined subject and scientific-name index appears at the end of each part. Thi work is written in French.

Tome XIV (Fascicules IA and B). *Amphibiens: Reproduction, Formes Larvaires Éthoécologie*. 1986. Fr. 1,477. ISBN 2-225-66858-2.

Tome XIV (Fascicule II). *Reptiles: Caractéres Généraux et Anatomie*. 1970. Fr 1,255. ISBN 2-225-61512-8.

Tome XIV (Fascicule III). *Reptiles: Glandes Endocrines, Embryologie, Systématique Paléontologie*. 1970. Fr. 1,255. ISBN 2-225-61700-7.

Amphibia (General)

924. Cochran, Doris M. **Living Amphibians of the World**. Garden City, N.Y.: Doubleday 1961. 199p. LC 61-9491.

Encompassing a large number of life-size black-and-white and colored photographs this source provides general information on the various families of amphibians inhabiting th planet. The descriptive account of the family (common and scientific names are provided includes breeding information, coloration, molting, behavioral patterns, life history, geo graphical distribution, size, life span, and other interesting facts about the amphibians Emphasis seems to be on courtship, mating, and reproductive information of these animals A number of species are mentioned within the description in order to illustrate biologica principles. The length of this information runs from 1 page to 10 or more pages. A one-pag list of selected references and a combined common/scientific-name index conclude th volume. This source provides general information on these organisms. It offers a goo introduction to the amphibians.

Reptilia (General)

925. Cooper, John E., and Oliphant F. Jackson, eds. **Diseases of the Reptilia**. London, New York: Academic Press, 1981. 2 vols. $276.00 set. LC 81-66390. ISBN 0-12-187901 vol. 1 0-12-187902-X vol. 2.

Contained within two volumes, this work considers those agents responsible for reptilia diseases, types of reptilian diseases, and the clinical aspects of diagnosis and treatment. I addition to sections of anatomy and physiology, pathology and histopathological techniques and microbiology and laboratory techniques, volume 1 considers the viruses, bacteria, fung and actinomycetes, and protozoans, as well as a number of endo- and ectoparasites. Volum 2 concentrates its efforts on types of diseases such as traumatic and physical diseases nutritional diseases, and neoplastic diseases. It also considers diagnosis, treatment, and drug and dosages for these afflictions. Tables, charts, black-and-white photographs, labele line-drawings, and microscopy plates provide the illustrations. A list of references appears a

e end of each section. A combined subject, common-name, and scientific-name index for
th volumes 1 and 2 is presented at the ends of volumes 1 and 2.

6. Gans, Carl, ed. **Biology of the Reptilia**. London, New York: Academic Press, 1969-
92. 17 vols. (1-16, 18). LC 68-9113.

Written as review articles, this set of books considers all the various biological aspects
ributed to reptiles. The articles are written in a scholarly manner with a large number of
ferences to the primary literature given throughout the text. Graphs, charts, photomicro-
aphs, line-drawings, and photographs provide the illustrative material. Bibliographies
pear at the end of each section. An author index as well as a combined subject, scientific-
me, and common-name index can be found at the end of each volume.

It should be noted that a number of the later volumes were published by publishers other
an Academic Press and have been noted as such.

Volume 1. *Morphology A.*
Volume 2. *Morphology B.* ISBN 0-12-274602-3. $161.00.
Volume 3. *Morphology C.*
Volume 4. *Morphology D.* ISBN 0-12-274604-X. $177.00.
Volume 5. *Physiology A.* ISBN 0-12-274605-8.
Volume 6. *Morphology E.* ISBN 0-12-274606-6.
Volume 7. *Ecology and Behaviour A.* ISBN 0-12-274607-4. $141.00.
Volume 8. *Physiology B.* ISBN 0-12-274608-2. $182.00.
Volume 9. *Neurology A.* ISBN 0-12-274609-0. $177.00.
Volume 10. *Neurology B.* ISBN 0-12-274610-4. $161.00.
Volume 11. *Morphology F.* ISBN 0-12-274611-2. $161.00.
Volume 12. *Physiology C: Physiological Ecology.* ISBN 0-12-274612-0. $182.00.
Volume 13. *Physiology D: Physiological Ecology.* ISBN 0-12-274613-9. $161.00.
Volume 14. *Development A.* New York: John Wiley, 1985. ISBN 0-471-81358-3.
Volume 15. *Development B.* New York: John Wiley, 1985. ISBN 0-471-81204-8.
Volume 16. *Ecology B: Defense and Life History.* New York: A. R. Liss. ISBN 0-8451-4402-2.
Volume 18. *Physiology E: Hormones, Brain, and Behavior.* Chicago: University of Chicago Press, 1992. ISBN 0-226-28122; 0-226-28124-8pa.

7. Schmidt, Karl P., and Robert F. Inger. **Living Reptiles of the World**. Garden City,
Y.: Hanover House, 1957. 287p. LC 57-9783.

Encompassing a large number of life-size black-and-white and colored photographs,
s source relates general information on the various families of reptiles found throughout
planet. Descriptive information on the family (common and scientific name given)
nsists of very basic external anatomy, breeding and egg-laying facts, geographical distri-
tion, number of species in the family, coloration, size, diet, courtship displays, behavioral
tterns, habitat, and the like. Descriptive accounts range from 1 page to 10 or more pages in
ngth. A combined common/scientific-name index concludes the volume. It should be noted
at this source follows the same basic pattern and arrangement as *Living Amphibians of the
orld* (see entry 924). It provides good general information on these creatures and is a great
ace to start in order to become knowledgeable about the variety of reptiles.

8. Welch, K. R. G. **Handbook on the Maintenance of Reptiles in Captivity**. Malabar,
a.: Robert E. Krieger, 1987. 156p. LC 86-15325. ISBN 0-89874-830-5.

This source provides information on how to best handle snakes and lizards when kept
captivity. Basically, it is divided into three sections. Section 1 provides general information
ch as diet, medical problems, mite infestations, record keeping, and the like. Section 2
nsiders certain species of snakes and lizards and discusses such parameters as geographical

distribution, size, vivarium conditions, and breeding. Section 3 includes a list of species wit appropriate references, a list of venomous colubrids with references, a table listing specie and giving the temperature and length of incubation, and a list of journals and societies. 46-page bibliography, and a scientific-name index conclude the handbook.

Amphibia and Reptilia (Geographic Section)

AUSTRALIA

929. Cogger, Harold G. **Reptiles & Amphibians of Australia**. 5th ed. Chatswood, Austra lia: Reed Books; and Ithaca, N.Y.: Cornell University Press, 1992. 775p. $97.00. L 91-34103. ISBN 0-8014-2739-8.

Consisting of approximately 664 species, this volume represents a first-class publicatio on the amphibians and reptiles found throughout the Australian continent. It is broader i scope and provides more information than Hoser's *Australian Reptiles & Frogs* (see entr 930). Each species is arranged under its respective genus, which in turn is entered und family, order, and class. General information is given for the class, order, family, and genu For each of the genera listed, the person who first named it and date of naming are als presented. Biological keys to the families, genera, and species precede the descriptiv accounts. For each of the species considered, the scientific name, person who first named i date of naming, common name, coloration, geographical distribution, and habitats are give The descriptive accounts are usually one-quarter to one-half of a column in length. geographical range map for each of the species and a colored photograph for most of th species accompany the descriptions. The colored photographs are excellent. In additio line-drawings enhance the biological keys.

Information such as conservation and protection, collecting methods, vegetation an climatic zones, snakebite and treatment, and so on can be found in the introductory sectio The introductory chapter is well illustrated with a large locality map and a vegetation an climatic zone map of Australia. An appendix listing name changes of Australian reptiles an frogs can be found near the end of the source. A 5-page glossary, a 16-page bibliograph whose citations are arranged under families, and a combined common/scientific-name inde complete the volume.

930. Hoser, Raymond T. **Australian Reptiles & Frogs**. Sydney, Australia: Pierson. Dis tributed by Gary Allen, 1989. 238p. $49.95. ISBN 0-947068-08-2.

Abounding with magnificently colored photographs, this work considers those reptile and frogs endemic to Australia. It should be noted that, save for 19 pages (section on frogs the major portion of the work is devoted to the reptiles. The source is divided into three part Part 1 considers the orders and suborders of these beasts. The descriptive information deal with anatomy, habitat, reproductive information, feeding, coloration, mating, fighting, an the like. Part 2 consists of a descriptive account of the frog and reptile species. The organism are arranged under their respective families, which in turn are entered under orders. Th scientific and common name of the families are presented. Preceding the accounts, som general information is also presented on the family. For each species included, the commo name, scientific name, person who first named it, date of naming, population, coloration, siz diet, reproductive information including nests and eggs, and habitat are given. Geographic range maps and colored photographs of the species accompany the descriptive accounts. Th descriptions range from one-quarter of a column to one full column in length. The abov descriptive parameters may not all be present in every account. Part 3 deals with captivit and conservation. Such information as capturing, handling, housing, feeding, and disease are discussed. In addition, conservation measures are considered along with the variou habitats of these creatures. Colored photographs of the habitats are also presented. A five-pag bibliography and a combined subject, common-name, and scientific-name index can be found at the end of the volume. This is an excellent work.

931. Waite, Edgar R. **The Reptiles and Amphibians of South Australia**. Adelaide, Australia: Printed by H. Weir, Govt. Printer, 1929. 270p. LC 48-35784.

Containing a large number of lined drawings as well as black-and-white photographs, this source provides information on the many species inhabiting the southern portion of the Australian continent. The species are arranged under their respective genera, which in turn are entered under their families and orders. Biological information, such as anatomy, behavior, diet, coloration, habits, dismemberment, mimicry, enemies, parasites, and the like are given for the orders, while some general anatomy, behavioral patterns, geographical distribution, and keys to the genera are presented for the families. Very brief anatomy and geographical distribution are considered for the genera. For each of the species listed, the common name, scientific name, person who first named it, references to the literature, external anatomical characteristics, size, coloration, and geographical distribution are presented. A typical descriptive account runs from one page to several pages in length. A combined common/scientific-name index concludes the volume.

CANADA

932. Cook, Francis R. **Introduction to Canadian Amphibians and Reptiles**. Ottawa: National Museum of Natural Sciences, 1984. 200p. $12.95pa. LC 84-209062. ISBN 0-660-10755-4pa.

Species are arranged under their respective families. Approximately 41 amphibian and 47 reptilian species are included. Scientific and common names of the families are present. For each organism mentioned, the common name, French name, and scientific name are given. In addition, basic external anatomy, size, geographical range, habitat, courtship and mating, egg and larva information, and diet are considered. A line-drawing of the vertebrate along with a geographical range map accompanies the descriptive matter. Descriptions vary from several pages to one-half of a page. Interesting tips on carrying out field observations, a three-page bibliography, and a section on the proper care of pet amphibians and reptiles conclude the volume. The table of contents, which is a taxonomic listing of common and scientific names, can be used as the index.

EGYPT

933. Anderson, John. **Reptilia and Batrachia**. London: Bernard Quaritch, 1898. Reprint, Weinheim, Germany: Wheldon & Wesley; Verlag J. Cramer, 1965. 371p. (Zoology of Egypt Series, no. 1). $180.00. ISBN 3-7682-0240-2.

The Egyptian species (78 reptiles, 7 amphibians) are arranged under their respective genera, which in turn are entered under families, orders, and classes. Anatomical information on the genera is presented and precedes each descriptive account of a species. For each of the species listed, the scientific name, person who first identified it, citations to the literature, detailed external anatomy, geographical range, coloration, habitat, and a number of other facts are presented. The amount of the above information varies from one species to another. The average length of the description is several pages. Biological keys to the species are presented throughout the text. A series of black-and-white plates depicting mainly the reptiles can be found at the end of the volume. One plate is devoted to the amphibians and contains four species. A section discussing the physical features of Egypt can be found at the beginning of the source. A table illustrating the distribution of the species found within this work, and a scientific-name index can be found preceding the plates.

EUROPE

934. Bohme, Wolfgang, ed. **Handbuch der Reptilien und Amphibien Europas**. Wie baden, Germany: AULA-Verlag, 1981. 7 vols. $157.20/vol. ISBN 3-400-00463-4 Bd Echsen I; 3-89104-000-8 Bd II, Echsen I and II; 3-89104-001-6 Bd II, Echsen II and III.

This work, written in the German language, represents an in-depth treatment of the man amphibians and reptiles found throughout Europe. In some cases, over 50 pages are devote to a species account. These accounts are written by different authors. For each of th organisms listed, the scientific name, person who first named it, and date of naming are give In addition, very detailed information on size, external and internal anatomy, geographic distribution throughout Europe, variations within the species, subspecies, ecology, populatio dynamics, detailed information on the young, behavior, and a list of references that were cite in the account are given. Geographical range maps can be found throughout the text. scientific-name index can be found at the end of each volume.

Band I. Echsen I	*Gekkonidae, Agamidae, Chamaeleonidae, Anguidae, Amphisbaenidae, Scincidae, Lacertidae I.*
Band 2/I. Echsen II	*Lacertidae II (Lacerta).*
Band 2/II. Echsen III	*Lacertidae III (Podarcis).*
Band 3/I.	*Schlangen.*
Band 3/II.	*Schildkroten.*
Band 4.	*Schwanzlurche.*
Band 5.	*Froschlurche.*
Band 6.	*Glossar und Verzeichnis der europaischen Trivialnamen.*
Band 7.	*Die Reptilien der Kanarischen Inseln.*

935. Hellmich, Walter. **Reptiles and Amphibians of Europe**. London: Blandford Pres 1962. 160p.

Species are presented under their common name groups, such as newts and salamander For each of the organisms listed, the scientific name, common name, person who first name it, size, basic external anatomy, coloration, habitat, behavior, and geographical distributio are given. The length of the descriptive account varies from one-half of a page to sever pages. A series of 68 colored plates (each species taking up one-half of a page) of many c the species listed provide the illustrative material. Some general information on amphibiar and reptiles as well as their care and conservation can be found in the introductory sectio A separate scientific/common-name index concludes the volume.

936. Honegger, Rene E. **Threatened Amphibians and Reptiles in Europe**. Wiesbade Germany: Akademische Verlagsgesellschaft, 1981. 158p. (Handbuch der Reptilien un Amphibien Europas, Suppl.). $52.50. ISBN 3-400-00437-5.

Considering most of the western Palearctic region, this work lists those species that ar threatened at the present time. The species are listed under either amphibians or reptiles. F each of the organisms presented, the scientific name, person who first named it, date c naming, common names in a number of European languages including English, status, orde and family are given. In addition, worldwide and European geographical distribution, Eur pean populations (country-by-country account), habitat, reasons for their decline, and co servation measures taken and proposed are considered. There are no illustration Introductory material consists of reasons for the decline of these animals, protection measure classification, and a comparative list of threatened amphibians and reptiles as well as thos considered to be under a threat. A 17-page bibliography concludes the volume. Althoug there is no index, the table of contents can serve the same purpose.

GREAT BRITAIN

937. Frazer, Deryk. **Reptiles and Amphibians in Britain**. London: Collins, 1983. 254p. (The New Naturalist, no. 69). LC 83-186971. ISBN 0-00-219706-5.

The species are arranged under broad common names such as frogs and toads, newts, snakes, and so on. Each chapter represents one of these groups. Preceding the accounts are adequate discussions of the group as a whole. General anatomy, physiology, egg-laying, metamorphosis, and so on are considered. For each species entered, the common name, scientific name, anatomical identification characteristics, coloration, geographical distribution, habitats, breeding, behavior, feeding habits, and mortality are given. Geographical range maps accompany the descriptions. A great deal of information is present in each of these descriptive accounts. These accounts can run anywhere from 10 to 20 pages in length. Black-and-white photographs of many of the species can be found in the center of the work. A bibliography consisting of 424 entries, and a combined subject, common-name, and scientific-name index conclude the volume.

INDIA

938. Smith, Malcolm A. **The Fauna of British India, Including Ceylon and Burma: Reptilia and Amphibia**. London: Taylor and Francis, 1931- . 4 vols. Reprint, **Reptilia & Amphibia: Loricata, Testudines, Turtles & Crocodilians, Volume 1**. Sanibel, Fla.: Ralph Curtis Books, 1981. 217p. (Fauna of British India). $20.00pa. ISBN 0-685-11906-8. Reprint, **Reptilia & Amphibia: Sauria (Lizards), Volume 2**. Sanibel, Fla.: Ralph Curtis Books, 1985. 455p. (Fauna of British India). $12.50pa. ISBN 0-88359-014-X.

Each species is arranged under its respective genus, which in turn is entered under family. Information on the family consists of external anatomical characteristics, citations to the literature, and keys to the genera, while genus information contains citations to the literature, external anatomical information, and keys to the species. For each of the species listed, the scientific name, list of references, external anatomy, size, coloration, and geographical distribution are given. Length of the species accounts run from one page to several pages. Line-drawings of the species and their anatomical parts can be found interspersed throughout the text. Bibliographic references and scientific-name indexes are contained within each of the volumes. It should be noted that the information contained within *A Vertebrate Fauna of the Malay Peninsula from the Isthmus of Kra to Singapore, Including the Adjacent Islands: Reptilia and Batrachia* (see entry 939), can be found within these four volumes:

Volume I.	*Loricata, Testudines.*
Volume II.	*Sauria.*
Volume III.	*Serpentes.*
Volume IV.	*Amphibia.*

MALAY PENINSULA

939. Boulenger, George A. **A Vertebrate Fauna of the Malay Peninsula from the Isthmus of Kra to Singapore, Including the Adjacent Islands: Reptilia and Batrachia**. London: Taylor and Francis, 1912. 294p.

Published under the authority of the government of the Federated Malay States, this work provides general information on species of reptiles, frogs, toads, and caecilians inhabiting the Malay Peninsula. Each species is arranged under its respective genus, which in turn is entered under subfamily and family. Keys to the subfamilies, genera, and species are present. In addition, brief anatomical accounts of the family and genera are considered. For each of the species listed, the scientific name, Malay name, citations to the literature, brief identifying anatomical characteristics, geographical distribution, and, in many cases, coloration, size, habitat, and behavior are given. The species account usually runs one-half of a page or less. Line-drawings of a few of the organisms provide the illustrations. A map of the Mala Peninsula and a scientific-name index conclude the volume.

NEW ZEALAND

940. Sharell, Richard. **The Tuatara, Lizards and Frogs of New Zealand**. London: Collin 1975. 95p.

Species information can be found throughout the chapters that in each instance discusse a group of reptiles as a family. Such family information as coloration, behavior, basic extern anatomy, diet, egg-laying, mating patterns, and the like are presented. In addition, species ar mentioned within the above discussion. Such information as common name, scientific nam coloration, identifying anatomical characteristics, size, and behavior are given. Colored an black-and-white photographs accompany the family and species accounts. In addition, ger eral information on the history and evolution of these organisms can be found in th introductory remarks. A brief glossary, a two-page bibliography, and a combined subjec common-name, and scientific-name index complete the volume. This source represents goo general information regarding these vertebrates found in New Zealand; the colored phot graphs are excellent.

NORTH AMERICA

940a. **Catalogue of American Amphibians and Reptiles**, No. 1- . [s.1.]: American Soc ety of Ichthyologists and Herpetologists, (1963-1970); Society for the Study of Amphibiar and Reptiles (1971-), 1963- . Published irregularly. Price varies.

Published as individual monographic reports, these reports provide an adequate amou of information on individual species of amphibians and reptiles found throughout the Nort American continent. Each report is authored by a different person, is published in loosele format, and represents a specific species of amphibian or reptile. For each species repo published, the scientific name, person who first named it, date of naming, common name, an alternative scientific names are provided. In addition alternative scientific names, persor who named them, and date of naming are also included. Information found in the descriptiv account of the species considers subspecies, size, coloration, basic and specific extern anatomical characteristics, a listing of publications where color illustrations can be foun geographical distribution, fossil record, a bibliography, and in some cases a key to the specie The amount of information varies from one monographic report to another. Each of th monographs of the species average from two to three pages in length. Geographical rang maps are usually found in each of the monographs. Indexes are provided for a cumulation c these monographic reports. For example, there is a scientific-name index for the first 40 reports that deal with the amphibians and reptiles.

941. Leviton, Alan E. **Reptiles and Amphibians of North America**. New York: Doubleda 1971. 250p. LC 70-147351.

Abounding with a large number of black-and-white and colored photographs, this sourc considers the various families of amphibians and reptiles found throughout North Americ The information presented in the descriptions of the families varies as does its lengt Descriptions run from one paragraph to several pages. Such parameters as habitat, breedin patterns, geographical range, behavior, coloration, size, vocal anatomy, egg-laying, and th like are presented in many of the accounts. In some cases, species are mentioned with th family account in order to illustrate a particular point. A number of selected references an a separate but combined common/scientific-name index for amphibians and reptiles conclud the volume. The work is written in an easy-going style. It presents a number of interestin facts about these vertebrates.

OUTH AMERICA

42. Duellman, William E., ed. **The South American Herpetofauna: Its Origin, Evolu- on, and Dispersal**. Lawrence, Kans.: Museum of Natural History, University of Kansas, 979. 485p. (Monograph of the Museum of Natural History, The University of Kansas, no.). $10.00pa. ISBN 0-89338-008-3.

Divided by regions of South America, this source provides the reader with information n the fossil record, geography of the region, number of genera and species, geographical nge maps, and a list of genera found throughout the particular region. In some cases, eographical distributions of the genera are also presented. A brief bibliography completes ach of the regional descriptions. Some examples of the regions contained herein are lowland opical forests, Guianan region, dry lowland regions, and so on. A few of the chapters escribing these regions are written in Spanish. In addition, several introductory chapters escribe South American herpetofauna in general as well as its relationship with the reptiles nd amphibians of Africa and Australia. Line-drawings and a few black-and-white photo- raphs depicting the geography provide the illustrative material. A separate subject and axonomic index concludes the source. It should be noted that this source does not provide n account of individual species; rather, it discusses the reptiles and amphibians of a region South America in terms of their evolution and geographical range.

INITED STATES

43. Ashton, Ray E., comp. **Endangered and Threatened Amphibians and Reptiles in e United States**. [Milwaukee, Wis.]: Society for the Study of Amphibians and Reptiles, 976. 65p. (Herpetological Circular, no. 5). LC 78-109412.

Species are arranged under their respective states of the union. For each species corded, the scientific name, common name, status (endangered, threatened, rare, periph- al), the reason for the endangerment, current protection practice, and recommendations are iven. In addition, the tables found near the end of the work provide a list of nationally reatened amphibians and reptiles. A species status sheet closes the volume. There is no dex. This work represents a joint project of the Society for the Study of Amphibians and eptiles Liaison Committee and numerous regional herpetological societies.

44. Conant, Roger. **Reptiles and Amphibians of the Northeastern States: A Non- echnical Resume of the Snakes, Lizards, Turtles, Frogs, Toads, and Salamanders of e Area**. 3d ed. Philadelphia: Zoological Society of Philadelphia, 1957. 40p.

This source considers those amphibians and reptiles found in Maine, New Hampshire, ermont, Massachusetts, Connecticut, Rhode Island, New York, New Jersey, Delaware, and e eastern portions of Pennsylvania and Maryland including Washington, D.C. The species e arranged under their common group names, such as venomous snakes, turtles, salaman- ers, and frogs. For each of the species listed, the common name, scientific name, basic entifying anatomy, coloration, and size are given. In some cases, other facts regarding the rganism are presented, such as vocal sac, behavior, and so on. A black-and-white photograph f the animal accompanies the description. Descriptions are usually several paragraphs long. addition, some general information regarding each group is presented. Other sections clude some brief material on first aid for snake bites, and caring for captive specimens. A st of selected references and a combined subject/common-name index conclude the volume.

45. DeGraaf, Richard M., and Deborah D. Rudis. **Amphibians and Reptiles of New ngland: Habitats and Natural History**. Amherst, Mass.: University of Massachusetts ress, 1983. 85p. $17.50; $9.95pa. LC 83-5125. ISBN 0-87023-399-8; 0-87023-400-5pa.

Consisting of 56 species, this source contains descriptive information on those amphibi- ns and reptiles found throughout the New England area. The states considered are Maine, ew Hampshire, Vermont, Massachusetts, Rhode Island, and Connecticut. Arranged under common group name such as salamanders, snakes, turtles, and so on, one page of descriptive matter is devoted to each organism. For each species included, the common name, scientifi name, order, and family are listed. In addition, geographical range, relative abundanc habitat, age and size at sexual maturity, breeding period and egg-laying, number of eggs lai incubation period, larval period, movement, food and feeding, and a few selected reference are presented. A geographical range map and a pen-and-ink drawing of the animal accompan the description. An eight-page bibliography, a brief glossary, and a combined common scientific-name index conclude the volume. Preceding the descriptive information can b found a checklist of New England amphibians and reptiles.

946. Martof, Bernard S., and others. **Amphibians and Reptiles of the Carolinas an Virginia**. Chapel Hill, N.C.: University of North Carolina Press, 1980. 264p. $14.95pa. L 79-11790. ISBN 0-8078-4252-4.

Each page represents a descriptive account of a species. The species are arranged unde their respective orders. Approximately 190 species are considered. For each of the organism listed, the common name, scientific name, size, coloration, basic external anatomy chara teristics, habitat, geographical distribution, egg-laying, size of egg, and diet are given. colored photograph and a geographical range map accompany the descriptive material. history and systematic list of the amphibians of North Carolina, South Carolina, and Virgini comprise the introductory information. An 11-page glossary, a 2-page list of selected refe ences, and a combined common/scientific-name index conclude the source.

947. Nussbaum, Ronald A., Edmund D. Brodie, and Robert M. Storm. **Amphibians an Reptiles of the Pacific Northwest**. Moscow, Idaho: University Press of Idaho, 1983. 332 $24.95. LC 82-60202. ISBN 0-8901-086-3.

Representing approximately 62 species, this work considers those amphibians an reptiles found throughout the states of Idaho, Oregon, and Washington. Each species arranged under its respective genus, which in turn is entered under family. Common an scientific names of the families and genera along with general information for each group ar present. For each of the organisms listed, the common name, scientific name, person who fir identified it, identifying anatomical characteristics, coloration, size, habitat, geographic distribution, variations of the species (subspecies), and life history are given. The life histor consists of breeding, courtship, egg-laying, size of eggs, clutch size, and larvae. Eac descriptive account runs several pages long. A geographical range map and, in most cases, black-and-white photograph of the organism accompany the descriptive account. Biologic keys to the different types of amphibians and reptiles of the northwest (down to species) ar also present. A series of eight colored plates depict a number of the organisms. A 2-pag glossary, an 18-page bibliography, and a combined common/scientific-name index conclud the volume.

948. Pickwell, Gayle. **Amphibians and Reptiles of the Pacific States**. New York: Dove 1972. 235p. LC 72-76868. ISBN 0-486-21686-1.

Offering less species information than *Amphibians and Reptiles of the Pacific Northwe* (see entry 947), this source considers the amphibians and reptiles found within the states o Washington, Oregon, and California. The source is divided into several sections. The basi descriptive accounts can be found in the sections of amphibians of the Pacific states an reptiles of the Pacific states. The species included in these two sections are arranged unde their respective genera, which in turn are listed under families, suborders, and orders. Som general anatomical information is given for each family. The descriptive information on th species consists of scientific name, common name, coloration, geographical range, and, i some cases, egg-laying. The average length of the descriptive account is usually no more tha one paragraph. Subspecies are also included as entries. Two other sections dealing with lif habits of amphibians and life habits of reptiles in these states considers several species, an a discussion of their life histories ensues. A small chapter on the collecting, handling, an care of amphibians and reptiles is present. A series of 64 plates depicting these vertebrates i black-and-white photos, a set of biological keys (down to species) of the amphibians an reptiles, a brief glossary, a seven-page bibliography, and a combined common/scientific-name index complete the volume.

WEST INDIES

949. Schwartz, Albert, and Robert W. Henderson. **Amphibians and Reptiles of the West Indies: Descriptions, Distributions, and Natural History**. Gainesville, Fla.: University of Florida Press, 1991. 720p. $75.00. LC 90-48025. ISBN 0-8130-1049-7.

The 585 species, 157 of which are amphibians and 428 of which are reptiles, are arranged under their respective orders. It is important to note that the number of species of herpetofauna of the West Indies far exceeds the number of the mammals (60 species) and the birds (277 species). For each of the organisms listed, the scientific name, person who first named it, date of naming, holotype, size, identifying external anatomical features, coloration, references to publications that provide illustrations of the animals, geographical distribution, subspecies (if any), and the natural history (breeding, diet, vocalization, egg production, etc.) are given. The references to the publications providing illustrations allow access to either line-drawings, black-and-white photographs, or colored photographs. Unfortunately, the references do not indicate which type of illustration is present in each publication. A range map of the appropriate island(s) accompanies each of the descriptive accounts. The accounts average one-half of a page in length; a few cover several pages. There are no illustrations of the species contained within the volume. A systematic index and a general list of references can be found in the introductory section. A 41-page bibliography consisting of those references cited in the text, and a scientific-name index complete the reference work. Although this is a comprehensive source, the absence of illustrations reduces the utility of the volume.

Amphibia (Geographic Section)

AUSTRALIA

950. van Kampen, Pieter Nicolaas. **The Amphibia of the Indo-Australian Archipelago**. Leiden, Netherlands: E. J. Brill, 1923. 304p. LC agr24-71.

Geographically, this source includes those amphibians (254 species) found throughout the Dutch Archipelago as well as Borneo, Timor, New Guinea, and Pelawan. Species are arranged under their respective genera, which in turn are entered under families and orders. A brief description of the order and family is presented. For each of the species listed, the scientific name, person who first named it, citations to the literature, identifying external anatomy, size, coloration, habitat, and geographical distribution are given. Tadpole information is also present for a number of the species. This information consists of external anatomy, size, coloration, and habitat. Descriptions run from several paragraphs to several pages long. Biological keys to families, subfamilies, genera, and species can be found interspersed among the descriptive information. Black-and-white drawings of a few of the species are presented, 29 in all. A listing dealing with the distribution of species on the islands of the Indo-Australian Archipelago, and a scientific-name index conclude the work.

Although this work is old and devoid of extensive illustrative matter, it does provide very useful anatomical descriptions of these organisms.

CANADA

951. Logier, E. B. S. **The Frogs, Toads and Salamanders of Eastern Canada**. [n.p.]: Clarke, Irwin, 1952. 127p. LC 54-1238.

The species are arranged under their respective families. For each amphibian listed, the common name, scientific name, person who first named it, geographical range, geographical distribution in eastern Canada, size, basic external anatomy, coloration, habits, and habitat are given. Descriptive information is usually two pages in length. A small black-and-white drawing of the organism can be found accompanying most descriptions. Picture keys (down to species) can be found throughout the text. Keys indicate by common name the page on which the description can be found. General biological information concerning amphibians can be found in the first six sections of the source. Such areas as anatomy, physiology, senses, instinct, intelligence, life histories, and the like are covered. An eight-page glossary, a three-page bibliography, and a combined subject, common-name, and scientific-name index complete the volume.

EUROPE

952. Boulenger, George Albert. **The Tailless Batrachians of Europe**. London: Printed for the Ray Society, 1897-1898. Reprint, New York: Arno Press, 1978. 2 pts. in 1. (Biologists & Their World Series). $38.50. LC 77-81096. ISBN 0-405-10679-3.

The species are arranged under their respective genera, which in turn are entered under families, suborders, and orders. General anatomical information is provided for the family, suborder, and order. In addition, keys to the genera are also present. For each of the species listed, the scientific name, citations to the literature, alternative scientific names, detailed external anatomical characteristics, coloration, male and female differences, geographical variations within the species, detailed size measurements, geographical distribution, skeletal anatomy, behavior, egg information, tadpole considerations, and habitat are given. Descriptive accounts can run from 10 to 15 or more pages. Line-drawings of the specimens in whole or in part, as well as colored plates of the species, and geographical range maps provide the illustrative matter. Introductory information consists of the biology of amphibians. A list of specimens preserved in the British Museum, a 13-page bibliography, and a scientific-name index conclude this two-part set. Although this work is outdated, it does, however, provide a good deal of information on the species listed therein.

953. Steward, J. W. **The Tailed Amphibians of Europe**. New York: Taplinger, 1970. 180p. LC 79-97191. ISBN 0-8008-7540-0.

Each of the chapters represent a family. The species/subspecies are listed under their respective genera, which in turn are entered under their families. Preceding the species account is some general information on the family and genus as a whole. For each of the amphibians listed, the scientific name, person who first identified it, alternative scientific names with dates of naming and people who first named them, common name, geographical distribution, habitat, coloration, general external anatomy, size, reproduction (mating, gestation period, information on the young, metamorphosis, etc.), diet, and enemies are given. Black-and-white photographs and line-drawings provide the illustrations. A series of 16 geographical range maps; common names of these species in English, French, German, Dutch, Italian, Polish, Czech, and Russian; a four-page bibliography; and a separate common/scientific-name index conclude the volume. A total of 59 species and subspecies are contained within the book.

JAPAN

954. Okada, Yaichiro. **Anura (Amphibia)**. Tokyo: Biogeographical Society of Japan, 1966. 234p. (Fauna Japonica).

The species and subspecies are arranged under their respective genera, which in turn are entered under their families. Preceding the descriptive account, general anatomical information on the family and genus, bibliographic citations to genus literature, geographical distribution, and taxonomic keys to the species are presented. For each of the species/subspecies listed, the scientific name, person who first named it, and alternative scientific name with taxonomic references to the literature are given. In addition, comprehensive external anatomy, coloration, skeletal information, sexual differences, tadpole information including measurements, variation within species, detailed geographical distribution, habitat, egg and egg-mass, and a variety of detailed tables dealing with anatomical measurements and egg-laying are presented. The length of the accounts varies from several to 15 or more pages. Colored and black-and-white photographs of the amphibians as well as line-drawings depicting external and internal (skeletons) anatomical parts provide the illustrations. A 10-page bibliography and a scientific-name index conclude the volume. Much of the material contained within this work can be found in *The Tailless Batrachians of the Japanese Empire* (see entry 955).

955. Okada, Yaichiro. **The Tailless Batrachians of the Japanese Empire**. Nishigahara, Tokyo: Imperial Agricultural Experiment Station, 1931. 215p. LC agr33-397.

Containing 58 pages of plates not including the 215 pages set forth in the citation, this source considers those frogs and toads endemic to Japan and adjacent islands. A total of 59 species and subspecies are included. The species are arranged under their respective genera, which in turn are entered under families. In addition to the listing of the species, a list of citations to the literature, brief anatomical characteristics, and a key to the species can be found under the genus name. For each species/subspecies considered, the scientific name, person who first identified it, common name in Japanese, a list of citations, external anatomical characteristics along with measurements, coloration, skeletal information, differences between the sexes, tadpole information and measurements, geographical distribution, habitat, egg information, detailed size measurements, and diet are given. Descriptive accounts usually run from three to five pages in length. A series of colored and black-and-white plates depict the organisms in the adult and tadpole stages. In addition, line-drawings depicting anatomical parts of these amphibians can be found interspersed among the text. The introductory section considers the various geographical locations making up the Japanese Empire; geographical maps accompany the description. A two-page bibliography and a combined common-name (Japanese), and scientific-name index can also be found near the end of the volume. Although dated, a wealth of information is presented on the species included.

MALAWI

956. Stewart, Margaret M. **Amphibians of Malawi**. Albany, N.Y.: State University of New York Press, 1967. 163p. LC 67-63247.

The species are arranged under their respective genera. General information is given for each of the genera. For each of the species listed, the geographical distribution, size in males, females, and juveniles, basic external anatomical characteristics, coloration, similar type species, sexual dimorphism, variations in adult and juvenile coloration, voice patterns, breeding time, habits, and habitats are given. Line-drawings and colored photographs of the species provide the illustrative matter. A key for the identification of the species precedes the descriptive accounts. The accounts run two to three pages in length. A two-page bibliography, a brief glossary, and a combined common/scientific-name index conclude the volume.

NEW GUINEA

957. Menzies, J. I. **Handbook of Common New Guinea Frogs**. Wau, Papua New Guinea: Wau Ecology Institute, 1976. 75p. (Wau Ecology Institute Handbook No. 1).

Representing the families Bufonidae, Leptodactylidae, Ranidae, Hylidae, and Microhylidae, this source provides general information on the common species that inhabit New Guinea. The frogs are arranged under their respective families. Preceding the species account some general information on the family is given. For each of the species presented, the scientific name, common name, habitat, geographical distribution, size, basic external anatomy, and breeding information are given. The descriptive accounts range from one-half to one full page in length. Approximately 50 species are contained within the volume. Colored plates provide the illustrations. Introductory information consists of general frog biology, coloration, the composition and origin of the frog fauna of New Guinea, bibliographic references representing the families, a guide to the identification of frogs occurring on Ne Guinea, and a systematic account of the species. The appendix consists of a listing of fro species under their family names. Such facts as the person who named it and date of namin are given. A combined subject and scientific-name index concludes the work.

NORTH AMERICA

958. Bishop, Sherman C. **Handbook of Salamanders: The Salamanders of the Unite States, of Canada, and of Lower California**. 3d printing. Ithaca, N.Y.: Comstock; div. c Cornell University Press, 1974. 555p. (Handbooks of American Natural History). ISBN 0-8014-0038-4.

The species of salamanders are arranged under their respective genera, which in tur are entered under their families. A brief introduction to the family and a key to the species o the genera are considered before the account of the species. For each salamander entered, th common name, scientific name, person who first named it, geographical distribution, habita size, general external anatomy, coloration, breeding, egg information (number, size), incuba tion period, and external anatomy of the larvae are presented. Descriptions are usually severa pages long. In addition, a geographical range map and a black-and-white photograph of th animal accompany each entry. Introductory information consists of relationships to othe groups, general habits, courtship patterns, fertilization and care of the eggs, larvae descrip tions, how to find and preserve salamanders, and equipment needed for collecting thes creatures. A 66-page bibliography, geographically arranged, as well as a common/scientific name index conclude the volume.

959. Cope, E. D. **The Batrachia of North America**. Ashton, Md.: Eric Lundberg, 1963 525p.

Originally published as U.S. National Museum Bulletin, no. 34, this source consider the frogs and toads of North America. For each of the species presented, the scientific name person who first named it, references to the literature, extremely detailed external anatom and size measurements, coloration, diet, habitat, and geographical distribution are given. Th length of the accounts run from 1 page to over 10 pages. Discussions of the families an genera are also present. In addition, biological keys to families, genera, and species can b found throughout the text as well as line-drawings of the heads and mouth. Eighty-six plate consisting of line-drawings (many of them labeled) of the internal systems of frogs and toad can be found near the end of the work. The introductory pages considers the general anatom of the frogs and toads. A combined subject and scientific-name index concludes the volume This is a detailed work and it is not written for the novice.

960. Dickerson, Mary C. **The Frog Book: North American Toads and Frogs, with a Study of the Habits and Life Histories of Those of the Northeastern States**. New York Dover, 1969. 253p. $9.95pa. LC 69-15901. ISBN 0-486-21973-9.

The species are arranged under their respective families. For each of the frogs and toad listed, the common name, scientific name, person who first named it, detailed coloration various measurements, external anatomy, geographical distribution, vocal anatomy, behavior habitat, and other interesting facts are given. A large number of black-and-white and colore plates of these organisms are present. Many of them are life-size. Introductory informatior considers development, metamorphosis, phylogeny, temperature and hibernation, voice regeneration, poisonous characteristics, coloration, ethology, and geographical distribution A nine-page bibliography and a combined subject, common-name, and scientific-name index conclude the source. The volume presents many interesting facts about these species in a writing style easier to read than that of *The Batrachia of North America* (see entry 959).

Reptilia (Geographic Section)

AFRICA

961. Fitzsimons, Vivian F. M. **The Lizards of South Africa**. Pretoria, South Africa: Published by Order of the Trustees of the Transvaal Museum, Pretoria, Union of South Africa, 1943. Reprint, Amsterdam: Swets & Zeitlinger, 1970. 528p. (Transvaal Museum Memoir, no. 1).

Consisting of approximately 276 species of lizards, this source provides information regarding these reptiles. The species are arranged under their respective genera, which in turn are entered under their families. Anatomical information, geographical distribution, habitat, diet, and so on, as well as references to the literature and biological keys are compiled for the family and genus. The biological keys are to the genera, species, and subspecies. For each of the species listed, the scientific name, person who first named it, common names, and references to the literature are given. In addition, external anatomy, coloration, size, behavior, and geographical distribution are given. References to the plate numbers are also listed. The descriptive accounts are usually no more than two pages in length. Line-drawings of various anatomical parts of certain species of lizards are provided. A series of black-and-white plates depicting the entire species, as well as the heads of many of the lizards and the pads of their appendages can be found at the very end of the work. A list of South African place names and their locations, an 11-page bibliography, and a scientific-name index proceed the black-and-white plates.

962. Isemonger, R. M. **Snakes of Africa: Southern, Central, and East**. Johannesburg: Thomas Nelson, 1962. Reprint, Cape Town: Books of Africa, 1983. 284p. ISBN 0-09-499562-1.

The species are listed under broad common-name groups such as blind burrowing snakes, front-fanged snakes, and so on. For each of the species entered, the common name, scientific name, coloration, size, geographical distribution, diet, habitat, and other interesting facts are given. The descriptive accounts usually vary from one-half to one full page in length. A series of line-drawings and black-and-white and colored photographs provide the illustrations. The introductory material consists of myths and superstitions, observations, snake-catching, venom, snake characteristics, and general information on the groups of snakes present within the source. A combined subject, common-name, and scientific-name index completes the volume.

ASIA

963. Lim, Francis Leong Keng. **Fascinating Snakes of Southeast Asia: An Introduction**. Kuala Lumpur, Malaysia: Tropical Press, 1989. 124p. ISBN 967-73-0045-8.

Abounding with a large number of magnificent photographs of these creatures, this source provides some general information on the snakes inhabiting the Malaysian region, including Sabah and Sarawak. The species are listed under their respective genera, which in turn are entered under subfamilies and families. A few facts regarding the families, subfamilies, and genera are given. For each species listed, the common name, scientific name, size, habitat, behavior, egg-laying, diet, and whether the snake is venomous or nonvenomous are considered. Descriptive accounts are usually no more than two paragraphs in length. Colored photographs of the snakes accompany the descriptions. Information on snakes and snakebites is presented in the introductory section. A list of snakes found in peninsular Malaysia (scientific name, common name, size), a distribution chart and list of common names of snakes of southeast Asia, a two-page bibliography, and a combined subject, common-name, and scientific-name index conclude the volume.

AUSTRALIA

964. Bustard, Robert. **Australian Lizards**. Sydney: Collins, 1970. 162p. LC 78-880537 ISBN 0-00-211420-8.

Divided into three parts, this source considers the biology and species information o the many lizards inhabiting the Australian continent. Part 1 discusses the Australian environ ment and the biology of lizards. Such areas as vision, hearing, smelling, temperature regula tion, defense behavior, reproduction, autotomy, and coloration changes are briefl considered. Part 2 represents an account of the various species of lizards. Each species i arranged under the common name of the family; scientific name of the family is also provided Examples of these common-name groups are the monitor lizards, the geckos, the drago lizards, and so on. General information is given for the species such as coloration, egg-laying diet, size, geographical distribution, and other facts. The species are discussed within th context of the family chapter; more information is given on some species than on others. Par 3 discusses conservation, research, and lizards in captivity. A large number of colored an black-and-white photographs, depicting the species, provide much of the illustrative materia A four-page bibliography and a combined subject, common-name, and scientific-name inde conclude the volume.

965. Cogger, Harold. **Australian Reptiles in Colour**. Sydney: A. H. & A. W. Reed, 1967 112p. LC 67-66291.

Containing a series of colored photographs of representative reptiles found in Australia this source provides information on the many families contained within the Reptilia. For eac of the families listed, the common name, scientific name, behavioral patterns, coloration habitat, geographical distribution, representative species, diet, and other interesting facts ar given. The descriptive accounts of the families are usually two to seven or more pages i length. A common-name index concludes the volume.

966. De Rooij, Nelly. **The Reptiles of the Indo-Australian Archipelago**. Leiden: E. J Brill, 1915. Reprint, Vaals: A. Asher, 1970. 2 vols. in 1. LC 71-595749.

Descriptive accounts and taxonomic keys are given for those reptiles found in the are known as the Dutch East-Indies. This region contains those islands between Sumatra west ward, and New Guinea eastward, and includes Borneo, Timor, and New Guinea. Some of th islands east of New Guinea are also considered. The work consists of 306 species of lizards 35 turtles/tortoises, and four crocodiles. The species are arranged under their respectiv genera, which in turn are entered under families. Some general anatomical information i given for the families and genera; keys to the species are entered under the genera. For eac of the species presented, the scientific name, person who first named it, citations to th literature, basic identifying external anatomy, size, coloration, and geographical distributio are considered. Black-and-white drawings of many of the species listed can be foun accompanying the descriptive account. A list of the species showing their distribution amon the Indo-Australian Archipelago, and a scientific-name index conclude each volume. Ir addition, a systematic index can be found at the beginning of each volume.

Volume I. *Lacertilia, Chelonia, Emydosauria.*

Volume II. *Ophidia.*

967. Frauca, Harry, and Claudy Frauca. **Harry Frauca's Book of Reptiles**. Brisbane Australia: Jacaranda Press, 1966. 100p.

Containing many large, black-and-white and several colored photographs depicting specific species, this source provides information on the reptiles inhabiting the continent o Australia. Chapters represent groups of reptiles such as dragons, skins and pygopodids geckos, monitor lizards, and so on. Information in the chapters discusses the groups as whole and considers species within the context of the discussions. Such parameters as behavior brief identifying anatomical characteristics, diet, habitat, geographical distribution, colora- tion, and other interesting facts can be found throughout the chapters. A list of selected references concludes the volume. There is no index. This source provides general information on these vertebrates; no detailed information is provided.

968. Goode, John. **Freshwater Tortoises of Australia and New Guinea (In the Family Chelidae)**. Melbourne: Landsdowne, 1967. 154p. LC 68-100786.

Consisting of 15 species, this source provides some basic information on each of the tortoises found throughout Australia and New Guinea. Species are arranged under their respective genera. Some general anatomical information is given for each genus. For each of the tortoises listed, the common name, scientific name, person who first identified it, geographical distribution, basic external anatomy, coloration, reproductive information, size, habitat, and taxonomic relationships are given. Descriptive accounts are usually two pages in length. In addition to the species accounts, such information as breeding, physiology, behavior, structure, geographical distribution, captivity considerations, and fossils are presented. A large number of colored as well as black-and-white photographs of the species provide much of the illustrative matter. Line-drawings (some labeled) of the carapace and skulls are also present. A checklist (taxonomic references included) of tortoises of Australia and New Guinea, an eight-page bibliography, a brief glossary, and a combined subject, scientific-name, and common-name index complete the volume.

969. Gow, Graeme F. **Snakes of Australia**, rev. ed. London: Angus & Robertson, 1983. 166p. LC gb85-16449. ISBN 0-207-14437-0.

Providing more information than Kinghorn's *The Snakes of Australia* (see entry 970), this source represents a compilation of information on the snakes found throughout Australia. Species are arranged under their respective genera, which in turn are entered under families. For each of the snakes listed, the common name, scientific name, person who first named it, date of naming, geographical distribution, scalar measurements, size, coloration, egg-laying, size of young, diet, and behavior are given. In addition, cross-references to the colored plates found at the end of the volume are contained with the descriptive account. A total of 48 colored plates are present. There is also a separate section that provides descriptive information on the dangerous species. Introductory information consists of general habits, symptoms and treatments of snakebites, and the care of snakes in captivity. A two-page bibliography and a combined common and scientific-name index can be found preceding the colored plates.

970. Kinghorn, J. R. **The Snakes of Australia**, rev. ed. Sydney: Angus and Robertson, 1964. 197p.

This revision in collaboration with H. Cogger considers general information regarding the snakes that inhabit the Australian continent. Species are arranged under their respective families. After some brief information on the family, the descriptive accounts of the species ensue. For each of the snakes listed, the common name, scientific name, person who first named it, coloration, scalar measurements, size, and geographical distribution are given. In addition, a colored picture of the anterior portion of the snake accompanies the descriptive information. Each descriptive account, with the picture, is one page long. The introductory material consists of the behavior of snakes, snakebites and treatment, biological keys to the snakes (down to species), and the origin of Australian reptiles. A one-page bibliography and a combined common/scientific-name index conclude the handbook.

971. Worrell, Eric. **Reptiles of Australia: Crocodiles, Turtles, Tortoises, Lizards, Snakes: Describing Their Appearance, Their Haunts, Their Habits**. 2d ed. Sydney: Angus and Robertson, 1970. 169p. ISBN 0-207-94741-4.

Containing over 330 illustrations, a few in full color, this source provides species information on the various reptiles inhabiting Australia. The species are arranged under their respective genera, which in turn are entered under families. General anatomical information is presented for the family and genus. For each of the species presented, the common name, scientific name, person who first named it, geographical distribution, anatomical description with measurements, size, coloration, and behavioral patterns are given. It should be noted that the amount of information for each species varies. In some cases, only geographica distribution and basic external anatomy are considered. Taxonomic keys to the species ca be found interspersed throughout the text. The descriptive accounts of the species provid cross-references to the plates. Accounts generally run one-half of a page or less. A brie glossary, and a combined common/scientific-name index conclude the volume. This source unlike *Harry Frauca's Book of Reptiles* (see entry 967), contains a large number of species and an adequate amount of information on each one.

CENTRAL AMERICA

972. Campbell, Jonathan A., and William W. Lamar. **The Venomous Reptiles of Lati America**. Ithaca, N.Y.: Comstock; division of Cornell University Press, 1989. 425p. $59.5C LC 88-47934. ISBN 0-8014-2059-8.

Comprising those reptiles of Mexico, Central America, Caribbean Islands, and Sout America, this source presents 145 species that are considered venomous. For each reptil included, the common name, scientific name, name of person who first named it, and date o naming are listed. In addition, the family, geographical distribution, length, basic externa anatomy, teeth information, scales, coloration, behavior, and, in many cases, symptoms o bite attacks are given. Other interesting facts about these creatures are also present. Geo graphical range maps, 568 colored plates of the species (more than one picture for som species), a series of colored plates representing topographic and vegetation maps of Lati America, and a number of line-drawings provide the illustrative material. Taxonomic key are provided throughout the work. Introductory information consists of snakebite treatmen and first aid, an annotated bibliography of snakebite in Latin America, and a list of producer of antivenoms for venomous snakes of Latin America. An adequate glossary, a 24-pag bibliography, and a combined common/scientific-name index complete the source. Th colored plates are excellent. This is a first-rate publication.

CEYLON (See Sri Lanka)

CHINA

973. Pope, Clifford H. **The Reptiles of China: Turtles, Crocodilians, Snakes, Lizards** New York: American Museum of Natural History, 1935. 604p. (Natural History of Centra Asia, vol. X). LC 35-8277.

After some general information discussing the collecting of amphibians and reptiles i China, the accounts of the species ensue. Species (22 turtles, 1 crocodile, 130 snakes, 6(lizards) are arranged under their respective genera, which in turn are entered under families Keys for the identification of families, subfamilies, genera, and species can be foun throughout the various sections. For each of the species listed, the scientific name, perso who first identified it, citations to the literature, detailed external anatomical measurements coloration, geographical distribution, habits, habitats, and other pieces of information ar given. The length of the accounts of turtles and snakes run from one page to several pages The accounts of the lizard species can be found in the section "Annotated Check List o Chinese Lizards." The information given for each lizard species consists of scientific name person who first named it, citations to the literature, geographical distribution, and numbe of specimens examined. Taxonomic keys are also present within this section. Other areas considered in the source are habitat preferences of snakes in China, number of maxillary teet found in species of Chinese snakes, a list of species and subspecies by provinces, a map o China and list of localities, a 28-page bibliography, and a series of black-and-white plates depicting many of the species. A combined subject and scientific-name index completes the volume.

EUROPE

974. Boulenger, G. A. **The Snakes of Europe**. London: Methuen, [1913]. 269p. LC a15-2150.

The species are arranged under their respective genera, which in turn are entered under the families. A brief anatomical discussion is presented for the family and genus. For each of the species listed, the scientific name, person who first named it, common name, size, anatomical identifying characteristics, coloration, geographical distribution, behavior, diet, and reproductive information are given. Line-drawings representing the anterior portion of the snakes provide the illustrative material. In addition to the descriptive accounts, other sections within the book contain material on the external anatomy, coloration, skeleton, dentition, poison apparatus, nervous system, internal organs, behavior, parasites, and geographical distribution of snakes. A combined subject, common-name, and scientific-name index concludes the volume. This work, as one might suspect because of its imprint, is not nearly as comprehensive as Steward's *The Snakes of Europe* (see entry 975).

975. Steward, J. W. **The Snakes of Europe**. Newton Abbot, England: David and Charles, 1971. 238p. LC 76-581513. ISBN 0-7153-5199-0.

Considering the Typhlopidae, Boidae, Colubridae, Viperidae, and Crotalidae, this source provides general information on those snakes found throughout the European continent. The snake species/subspecies are arranged under their respective genera, which in turn are entered under families; each family represents a chapter. Some general information on the families and genera precedes the account of the species. For each of the organisms listed, the scientific name, person who first named it, common name, alternative scientific names (dates and person who named it), geographical distribution, size, basic external anatomy, diet, and, in some cases, litter size are given. Descriptive accounts usually run from one-half of a page to several pages in length. Black-and-white photographs depicting the species, as well as some labeled line-drawings, provide the illustrative material. Also present within the source is a key to the identification of the European snakes, diagrams depicting head scalation patterns, 36 geographical distribution maps, and a list of common names of these snakes, used in European countries. A five-page bibliography and a separate common/scientific-name index round off the volume.

976. Street, Donald. **The Reptiles of Northern and Central Europe**. London: B. T. Batsford, 1979. 268p. ISBN 0-7134-1374-3.

Consisting of 25 species, this work provides a great deal of information on the lizards, snakes, and tortoises contained therein. For each species included, the common name, scientific name, person who first named it, date of naming, and size are considered. In addition, detailed scalation information, coloration including differences between male and female as well as juvenile coloration, coloration found in different varieties, habitat, behavior, diet, enemies, reproduction (mating, birth of young, growth and maturity, oviparity), hibernation, ecdysis, behavior in captivity, geographical distribution, and names of the species in a number of different European languages are presented. References to the literature can be found throughout these accounts. The lengths of these accounts run from 3 to over 10 pages. A large number of colored and black-and-white photographs provide the illustrative material. General information on the reptiles as a whole can be found in the introductory section. An 11-page bibliography consisting of the references cited within the text, and a combined common/scientific-name index complete the volume.

HONDURAS

977. Wilson, Larry David, and John R. Meyer. **The Snakes of Honduras**. 2d ed. Milwaukee, Wis.: Milwaukee Public Museum, 1985. 149p. $29.95pa. LC 85-28353. ISBN 0-89326-115-7.

Consisting of approximately 70 species, this work provides information on snakes foun throughout the country of Honduras. For each of the species listed, the scientific name, perso or persons who first named it, alternative scientific names with accompanying citations geographical distribution, basic external anatomy, size, coloration, taxonomic informatior ecological distribution, and locality records are given. A large number of geographical rang maps accompany the descriptive accounts. Biological keys to families, genera, and specie can be found throughout the text. Colored and black-and-white photographs depicting som of the snake species as well as the geographical regions provide the illustrations. Geographica information on Honduras is provided in the introductory material. A number of table indicating geographical distribution patterns of the snakes, such as their relation to elevation vegetational formations, and so on can be found at the end of the volume. An eight-pag bibliography consisting of those references that were cited within the text closes the source There is no index.

INDIA

978. Deoras, P. J. **Snakes of India**. 2d ed., rev. New Delhi: National Book Trust, 1970 148p.

The species are arranged under their respective families. For each species listed, th scientific name, common name, Hindi name, Gujarati name, Marathi name, geographica distribution, size, coloration, anatomical identifying characteristics, behavior, and, in mos cases, egg-laying information are given. The descriptive accounts average one page in length A biological key down to family and species as well as the distribution of different specie of snakes recorded in India precede the descriptive accounts. General information on snakes including venoms, is discussed in the introductory section. Black-and-white and colore photographs of the snakes provide the illustrations. A two-page bibliography, and a separat scientific index and a combined subject and common-name index can be found at the clos of the volume.

979. Gharpurey, K. G. **The Snakes of India & Pakistan**. 5th ed. Bombay, India: Popula Prakeshan, 1962. 156p. LC sa63-4052.

Previously published under the title *The Snakes of India*, this source provides genera information on the variety of snake species found throughout India and Pakistan, along wit a few species found in Britain, South Africa, and Australia. The species are arranged unde their common-name groups, such as the kraits, the cobras, and so on. Preceding the descriptiv account of the species is some general information regarding the group as a whole. For eac of the species included, the common name, scientific name, geographical distribution coloration, size, behavior, egg-laying, size of eggs, scalar measurements, common names i different parts of India, venom information, and a number of other interesting facts ar presented. The amount of descriptive material varies depending on the snake considered Descriptive accounts run from one page to several pages in length. A series of black-and-whit photographs provide the illustrations. Introductory information considers teeth and poison glands, snake poison, snake scales, classification of snakes, and poisonous snakes. A com bined subject, common-name, and scientific-name index can be found at the end of the source

LATIN AMERICA (see Central America)

MALAWI

980. Sweeney, R. C. H. **Snakes of Nyasaland**. Zomba, Nyasaland: The Nyasaland Society and The Nyasaland Government, 1961. Reprint, **Snakes of Nyasaland: With New Adde Corrigenda and Addenda**. Amsterdam-Vaals: Asher, 1971. 203p. $37.50. LC 78-585333 ISBN 90-6123-242-2.

The species are arranged under their respective families. Preceding the descriptive account is general information pertaining to the family as a whole. For each of the snake listed, the scientific name, person who first named it, common name, African names, size, basic external anatomy, coloration, markings, scalar measurements, geographical distribution, habitat, behavioral patterns, diet, and longevity are given. The accounts vary in length from one page to several pages. Also included within the source are biological keys to the families and species, and a table giving scale counts. Introductory information consists of the classification, identification, and other general facts regarding snakes. A few black-and-white photographs depicting certain snake species, and a number of line-drawings showing the anterior portion of the snakes provide the illustrations. A separate English-name, African-name, and scientific-name index can be found near the end of the work.

MIDDLE EAST

981. Joger, Ulrich. **The Venomous Snakes of the Near and Middle East**. Wiesbaden, Germany: L. Reichert, 1984. 115p. (Beihefte zum Thubinger Atlas des Vorderen Orients. Reihe A, Naturwissenschaften, no. 12). LC 85-146136. ISBN 3-88226-199-4pa.

Considering the region from Egypt and Turkey to Pakistan and to the Transcaspian Soviet Republics, this source provides general information on those venomous snake species found therein. The species/subspecies (approximately 31 of them) are arranged under their respective families. For each species entered, the scientific name, person who first named it, common name, other scientific names, persons who named them and the dates of naming, taxonomic information such as the number of scale rows and ventrals, size, coloration, habitat, and geographical distribution are given. The length of a descriptive account is usually one page long. A checklist and biological keys, down to species, of the venomous snakes of the Near and Middle East are also provided. A 12-page bibliography and an appendix consisting of the maximum snake sizes, diets, and references can be found near the end of the volume. A series of line-drawings depicting scalar formation on the heads of the snakes, and 20 geographical range maps conclude the work.

NICARAGUA

982. Villa, Jaime. **The Venomous Snakes of Nicaragua: A Synopsis**. Milwaukee, Wis.: Milwaukee Public Museum, 1984. 41p. (Contributions in Biology & Geology Series, no. 59). $6.95. ISBN 0-89326-107-6.

The species and subspecies are arranged under their respective families. For each of the species listed, the scientific name, person who first named it, citations to the literature, etymology of the name, local names, basic external anatomy, geographical distribution, habitat, habits, coloration, karyotype, and a number of other pieces of factual information are given. Geographical range maps provide the illustrative material. Other sections include biological keys down to species of the venomous snakes in Nicaragua (in English and Spanish), and a five-page bibliography. There is no index.

NORTH AMERICA

983. Carr, Archie. **Handbook of Turtles: The Turtles of the United States, Canada, and Baja California**. Ithaca, N.Y.: Comstock; div. of Cornell University Press, 1952. 542p. (Handbooks of American Natural History). $55.00. LC 52-9126.

Consisting of 79 species and subspecies of turtles that can be found throughout the United States, Canada, and Baja, this work provides basic information on the turtles listed. Species are arranged under their respective genera, which in turn are entered under families and suborders. A brief anatomical description of the suborder, family, and genus, along with keys to the families, genera, and species are provided. For each of the species considered, the common name, scientific name, person who first identified it, geographical distribution, anatomical identifying characteristics, size (variety of anatomical measurements), coloration, habitat, and behavior are given. In many cases, information on breeding and diet is also presented. Geographical range maps as well as black-and-white photographs depicting the species provide the illustrative material. A 68-page bibliography (arranged by author) and a combined subject, common-name, and scientific-name index can be found at the end of the source. In addition, the bibliography is indexed using a geographical approach (states and provinces).

984. Ernst, Carl H. **Venomous Reptiles of North America**. Washington, D.C.: Smithsonian Institution Press, 1992. 236p. $35.00. LC 91-3535. ISBN 1-56096-114-8.

Save for the Gila monster, this source considers those poisonous snakes found in the families Elapidae and Viperidae. A total of 20 species of snakes and one species of lizard comprise the volume. Each species comprises a chapter. For each of the species listed, the scientific name, person who first identified it, date of identification, and common name are given. In addition, the descriptive account considers size, coloration, anatomical measurements, differences between male and female, karyotype, fossil record, geographical distribution, geographical variation, similar species, habitat, behavior, reproduction, growth and longevity, food and feeding, venom and bites, predators and defense, populations, conservation status, and other interesting facts. A geographical range map accompanies each of the descriptions. These accounts are detailed but very readable. The lengths of these accounts usually run anywhere from 4 to 10 or more pages. A general introduction is also presented for each of the families. The introductory section consists of some general information on the external anatomy of snakes as well as general information concerning venom. Medical treatment for venomous bites is also considered. Colored as well as black-and-white photographs of these animals provide the illustrative material. The photographs are excellent. A 51-page bibliography, a glossary of scientific terms, and a combined common/scientific-name index conclude the work.

985. Ernst, Carl H., and Roger W. Barbour. **Snakes of Eastern North America**. Fairfax, Va.: George Mason University Press, 1989. 282p. $75.75. LC 88-34205. ISBN 0-913969-24-9.

Consisting of 58 species of snakes, this source considers all snake species found east of the Mississippi River to its juncture with the St. Croix River, and northwest to the western border of Ontario. Preceding the species accounts some general information on the family is given. Common and scientific family names are considered. For each snake presented, the scientific name, person who first named it, date of naming, and common name are included. In addition, a three- to five-page descriptive account follows. It consists of size, coloration, external anatomy, karyotype, fossil record, geographical distribution, geographical variation, confusing species, habitat, behavior, reproduction, growth, food and feeding, predators and defense, populations, and other interesting facts. References to the literature are cited throughout this textural account. Geographical range maps and black-and-white photographs of the snakes accompany the descriptive material. In addition, a number of colored plates of these creatures can be found in the center of the book. The captions under the plates consider the scientific and common names. A taxonomic key, down to the species level, can be found in the introductory material. A brief glossary containing pronunciation and definitions, a 60-page bibliography, and a combined common/scientific-name index can be found at the end of the volume. This work provides a good deal of information on each of the 58 species of snakes included.

986. Pope, Clifford H. **Turtles of the United States & Canada**. New York: Knopf, 1971. 343p.

Originally published in 1939, this source is in its sixth printing. The book is divided into two basic sections. Section 1 considers the biology of turtles while section 2 represents a descriptive account of the 62 species/subspecies inhabiting the United States and Canada. The biological information considers structure, relationships, size, sexes, reproduction, egg, young, longevity and growth, hibernation, sunning, food and feeding, enemies and defenses intelligence, and the care of baby turtles. The species in section 2 are arranged under their respective genera, which in turn are entered under families. Each of the family chapters are introduced by their common-name group, such as the musk and mud turtles, the map turtles, the snapping turtles, and so on. Preceding the descriptive accounts of the species, some general information on the family as a whole is considered. For each of the turtles listed, the common name, scientific name, person who first named it, brief anatomical description, a variety of anatomical measurements, information on the young, the sexes, longevity, geographical distribution, habitat, behavior, reproduction, mating, nesting, nests, enemies, and information on how to deal with the species in captivity are given. The descriptive accounts run from 2 to 10 or more pages. A large number of black-and-white photographs depicting the species provide the illustrative material. A 12-page bibliography, a checklist of the turtles of the United States and Canada, and a combined subject, common-name, and scientific-name index complete the volume.

987. Smith, Hobart M. **Handbook of Lizards: Lizards of the United States and Canada**. Ithaca, N.Y.: Comstock, 1946. 3d printing, 1967. 557p. (Handbooks of American Natural History, vol. 6).

The lizards included within this source are arranged under their respective genera, which in turn are entered under the families. A description of the families and genera as well as keys to the genera and species precede the descriptive accounts. For each lizard included, the common name, scientific name, person who first named it, geographical range, size, coloration, scalation, habitat, behavior, egg information, diet, and a number of references to the literature are presented. The length of the account is usually several pages. The illustrations consist of black-and-white photographs of the vertebrates, geographical range maps, and labeled line-drawings depicting certain areas of external anatomy. Introductory information consists of characteristics, distribution, major groups, fossil history, anatomical structures, habitats, life history, behavior patterns, folklore, economic importance, and methodologies related to the collecting and preserving of these animals. A number of geographical range maps, a listing of species and accompanying references arranged by specific states of the United States, an additional six-page bibliography, and a combined common/scientific-name index conclude the source.

988. Wright, Albert Hazen, and Anna Allen Wright. **Handbook of Snakes of the United States and Canada**. Ithaca, N.Y.: Comstock; division of Cornell University Press, 1957. 2 vols. (Handbooks of American Natural History). $79.50 set. LC 57-1635. ISBN 0-8014-0463-0.

The snakes are listed under their respective genera, which in turn are listed under families. A description of the families and genera as well as a key to the genera can be found preceding the descriptive accounts of the various species. For each snake included, the common name, scientific name, person who first named it, references to the taxonomic literature, other common names, geographical range, coloration in great detail, habitat, period of activity, breeding, eggs (number, size, color), type of nest, information on the young, food ingested, and field notes are given. Descriptions are usually several pages long. Natural history quotes from naturalists are scattered throughout the descriptive material. Black-and-white photographs of the snakes, detailed line-drawings of the heads, and geographical range maps provide the illustrative material. The introductory matter consists of snake naming, geographical range, size, longevity, distinctive characteristics, color, habitat, period of activity, breeding considerations, ecdysis, food, venom and bite, and enemies. A brief glossary, a five-page bibliography, and a combined person, common-name, and scientific-name index can be found at the end of volume 2.

NYASALAND (see Malawi)

SOUTH AMERICA

989. Peters, James A., and Roberto Donosco-Barros. **Catalogue of the Neotropical Squamata**. Washington, D.C.: Smithsonian Institution Press, 1986. 2 pts. in 1. 347p., pt. 1; 293p., pt. 2. $32.50. LC 86-600220. ISBN 0-87474-757-0.

Representing a reprint of United States National Museum Bulletin, no. 297, published in 1970, this source provides taxonomic information on snakes and lizards, and amphisbaenians found throughout Central and South America. For each of the species contained within, a specific genus taxonomic key, citations to the literature, and geographical distributions are provided. The genera, in turn, can be found under their respective families. Except for some line-drawings depicting body forms of the amphisbaenia, there are virtually no illustrations. In essence, this work represents very detailed anatomical information in order to determine one species from another. A scientific-name index can be found at the end of each part. In addition, an addenda and corrigenda to parts 1 (snakes) and 2 (lizards and amphisbaenians) can be found preceding the taxonomic keys.

SRI LANKA

990. Deraniyagala, P. E. P. **The Tetrapod Reptiles of Ceylon**. London: Dulau, 1939. 412p. (Colombo Museum Natural History Series). LC 74-235968.

This work, which represents volume 1, considers the turtles and crocodiles of Sri Lanka. The species are arranged under their respective genera, which in turn are entered under families. An anatomical description, list of citations, and taxonomic keys to genera can be found under the families listed. Material under the genera consists of a series of references, and anatomical descriptions. For each of the species listed, the scientific name, person who first named it, list of citations to the literature, local common names, external anatomy, coloration, size, skeleton, food, reproduction, embryology, behavior of young, scute changes, and geographical distribution are given. These are very detailed accounts and can run 20 to 30 or more pages in length. Black-and-white drawings as well as black-and-white photographs provide the illustrative material. Introductory information consists of zoogeography, economics, collecting, classification, and the use of keys. A nine-page bibliography, and an index of common English names, an index of common Sinhalese names, a scientific-name index, and an author index conclude the volume. Although somewhat outdated, this source provides a great deal of information on the species included.

UNITED STATES

991. Babcock, Harold Lester. **The Turtles of New England**. Boston: Printed for the Society with aid from the Gurdon Saltonstall Fund, 1919. 106p. (Memoirs of the Boston Society of Natural History, vol. 8, no. 3); (Monographs of the Natural History of New England).

Abounding with 16 large plates (mostly colored) depicting the species, this source provides information on the 17 species listed. The descriptive accounts of the species vary from several pages to 17 pages in length. The species are arranged under their respective families. For each of the turtles listed, the scientific name, person who first named it, common names, references to the literature, size, coloration, anatomical information, egg-laying, size measurements of different anatomical parts in males and females over a period of years, geographical distribution, habits, habitats, diet, enemies, and economic importance are given. It should be noted that not all of the species listed contain all of the informational parameters listed above. A five-page bibliography is also present. The plates appear at the end of the volume. There is no index; the table of contents can suffice as an access to the species contained within the source.

992. Ernst, Carl H., and Roger W. Barbour. **Turtles of the United States**. Lexington, Ky.: University Press of Kentucky, 1972. Reprint, Ann Arbor, Mich.: Books on Demand, UMI. 347p. $99.60. LC 72-81315. ISBN 0-8131-1272-9; 0-8357-8593-9.

The majority of the chapters each represent a particular family of turtles or tortoises. Preceding the accounts of the species a brief discussion of the family is presented along with the scientific and common names of the group. For each species considered, the scientific name, person who named it, common name, size, external anatomy, coloration, weight, geographical variation and distribution, habitat, behavioral patterns, reproductive information including material on nests and eggs, growth patterns and aging, food and feeding, predators, movement behavior, populations, and other incidental facts are presented. The lengths of these descriptions vary from several pages to 10 or more pages. References to the literature can be found throughout the descriptive material. Geographical range maps, black-and-white and colored photographs, and a number of tables provide the illustrations. A key to the turtles of the United States can be found preceding the accounts of the species. Other sections include the origin and evolution of North American turtles, basic descriptions on how to care for turtles in captivity, and a list of parasites, commensals, and symbionts found in certain turtles. A glossary of scientific names (pronunciations included), a 47-page bibliography consisting of those references found throughout the text, and a combined subject, common-name, and scientific-name index conclude the work. It should be mentioned that although species contained within this work are also present in *Turtles of the World* (see entry 1000), the accounts here tend to be longer and more informative.

993. Holbrook, John Edwards. **North American Herpetology; or, A Description of the Reptiles Inhabiting the United States**. Philadelphia: J. Dobson, 1842. Reprint, Lawrence, Kans.: Society for the Study of Amphibians and Reptiles, 1976. 5 vols. in 1. LC 76-6229 (regular); 76-6766 (patrons). ISBN 0-916984-02-8 (regular); 0-916984-03-6 (patrons).

This work represents the first attempt to treat comprehensively all the amphibians and reptiles that were known to be present in the United States at the time. The work began in 1836; 25 states belonged to the Union. The species are arranged under their respective genera. Preceding the species account a brief description of the genera is given. For each vertebrate listed, the external anatomy in some detail, coloration, size, habits, geographical distribution, and other interesting facts are considered. In addition, a number of references to the taxonomic literature accompany the descriptive account. These accounts are usually four to five pages in length. The majority of the plates are black-and-white; a few colored plates can be found throughout these volumes. A scientific index listing the current name, first-edition name, and second-edition name can be found in the beginning of the work. This index leads to the volume and page number for that particular species. Although somewhat outdated, this represents an historical work.

994. Shaw, Charles E., and Sheldon Campbell. **Snakes of the American West**. New York: Knopf, 1974. 330p. LC 73-7304. ISBN 0-394-48882-2.

Considering the states of Washington, Oregon, California, Nevada, Idaho, Utah, Arizona, Montana, Wyoming, Colorado, and New Mexico, this source provides information on approximately 70 species of snakes inhabiting this above geographical range. Preceding the descriptive accounts of the species, general characteristics of these reptiles and information on keeping them as pets are presented. The species are arranged under broad common-name groups such as farm snakes, rear-fanged snakes, kingsnakes, and so on. After a brief introduction to the group as a whole, the snakes are presented. For each of the species listed, the common name, scientific name, geographical range and range map, size, coloration, egg-laying, hatching, habitat, defensive mechanisms, behavior, and a variety of other interesting facts are given. Colored plates depicting each of the 70 species can be found at the end of the source. A table considering the life-span of the species, and a checklist of the snakes of the American West can be found in the appendices. A 10-page bibliography, and a combined subject, common-name, and scientific-name index proceed the appendices.

VENEZUELA

995. Pritchard, Peter C. H., and Pedro Trebbau. **The Turtles of Venezuela**. Athens, Ohi Society for the Study of Amphibians and Reptiles; Oxford, Ohio: Purchased from Publication Secretary, Dept. of Zoology, Miami University, 1984. 403p. (Contributions to Herpetolog no. 2). LC 83-51450. ISBN 0-916984-11-7 regular edition; 0-916984-12-5 patron's editio

Consisting of 23 species/subspecies, this source provides in-depth information on th turtles found in Venezuela. The species are arranged under their respective families. Gener anatomical information on the family and the genera is present under the family entry. F each of the turtles included, the scientific name, person who first named it, date of namin common name, and list of references to the literature are given. In addition, detaile information on size, external anatomy (shell, head, extremities), growth parameters (size an weight), geographical distribution, geographical variations, habitat, feeding, reproductio economic importance, and vernacular names are discussed. Many of the descriptive accoun run from 10 to 15 pages in length. Introductory information includes material on th distribution and zoogeography of South American turtles, shell nomenclature, and biologic keys to the turtles of Venezuela (down to species). The biological keys are written in Englis and Spanish. A 23-page bibliography (arranged by author), a list of locality records, and discussion on turtle conservation regulations in Venezuela can be found near the end of th source. Colored plates depicting the species and a series of geographical range maps complet the volume. There is no formal index in the book. The table of contents found at the beginnin of the work along with the biological keys can serve as the index.

VIETNAM

996. Campden-Main, Simon M. **A Field Guide to the Snakes of South Vietnam**. Wash ington, D.C.: Smithsonian Institution, United States National Museum, Division of Reptile and Amphibians, 1970. Reprint, Lindenhurst, N.Y.: Herpetological Search Service & Ex change, 1984. 114p. $9.95pa. ISBN 0-9612494-0-4.

Although billed as a field guide, this work is not published or presented as a guid Rather, it represents an account of the species of snakes found throughout South Vietnan For each of the species presented, basic external anatomy, coloration, habit, habitat, geo graphical range, and venomous status are given. A geographical range map and a black-and white line-drawing of the head of the snake accompany the description. The drawing, rang map, and descriptive account is usually one page in length. There are no colored photograph and the arrangement of the book would make it difficult to use as a traditional field guide. biological key identifying down to species, a two-page bibliography, and a scientific-nam index conclude the source.

Reptilia (Systematic Section)

TESTUDINES AND RHYNCHOCEPHALIA (Turtles, Tortoises, and Tuataras)

997. Alderton, David. **Turtles & Tortoises of the World**. New York: Facts on File, 198 191p. $24.95. LC 88-16240. ISBN 0-8160-1733-6.

Basically divided into four sections, this work provides the reader with basic informatio regarding the biology of major turtles and tortoises inhabiting the planet. The section (chapters) deal with form and function, reproduction, evolution and distribution, and a account of the families. For each of the families listed, the geographical range, species withi the family (common and scientific names given), behavior, basic external anatomy, size weight, geographical range maps, and a number of other interesting facts are given. A larg number of colored and a few black-and-white photographs depicting the species provide th illustrative material. A list of turtle species (common and scientific names) arranged systematically, a brief glossary and bibliography, and a combined common/scientific-name index conclude the volume. This work represents a good starting place for acquainting oneself with the biology of these fascinating creatures.

998. Boulenger, George Albert. **Catalogue of the Chelonians, Rhynchocephalians, and Crocodiles in the British Museum (Natural History)**. London: Printed by Order of the Trustees, 1889. Reprint, Codicote, Herts.; Lehre, Germany: Wheldon & Wesley; Verlag J. Cramer, 1966. 311p. $45.00. ISBN 3-7682-0443-X.

The species are arranged under their respective genera, which in turn are entered under families, superfamilies, suborders, and orders. Anatomical identifying characteristics and citations to the literature are presented for the order, suborder, superfamily, family, and genus. Biological keys to the families, genera, and species are also listed. For each of the species mentioned, the scientific name, citations to the literature, anatomical identifying characteristics, size, and geographical distribution are given. Line-drawings of skulls and carapaces provide most of the illustrations. A scientific-name index can be found near the end of the work. A series of six black-and-white plates depicting several turtle species completes the volume. This work is not for the novice. It represents a systematic account of those anatomical features used to distinguish one species from another.

999. Cobb, Jo. **A Complete Introduction to Turtles and Terrapins**. Neptune City, N.J.: T. F. H. Publications, 1987. 125p. $9.95; $5.95pa. ISBN 0-86622-275-8; 0-86622-280-4pa.

Based on the theme of caring for these creatures, this work considers a few dozen species of turtles and terrapins. The species are arranged under broad common-name groups, such as hardy tortoises, emydid turtles, and tropical tortoises. For each of the species listed, the common name, scientific name, geographical range, size, coloration, diet, and other facts are given. Lengths of the descriptions range from one-third of a column to two columns. A large number of colored photographs depicting these animals can be found throughout the work. Brief discussions on evolution, classification, anatomy, housing, health and hygiene, and reproduction in captivity are also present. There is no index. It should be noted that the information given in this source is very general and very basic.

1000. Ernst, Carl H. **Turtles of the World**. Washington, D.C.: Smithsonian Institution Press, 1989. 313p. $45.00. LC 88-29727. ISBN 0-87474-414-8.

Encompassing the variety of turtles and tortoises endemic to the planet, this work represents an up-to-date account of the species as well as the taxonomic changes that have occurred over the years. Preceding the accounts of species (257 of them) a discussion of the order, family, and genus is presented. For each turtle and tortoise entered, the scientific name, person who first named it, date of naming, and common name are given. In addition, coloration, external anatomy, size, geographical distribution and variation, habitat, behavioral patterns including courtship and mating, egg and nest information, food and feeding, and other interesting facts are considered. References to the literature can be found throughout the descriptive material. Black-and-white as well as colored photographs provide most of the illustrative material. An occasional labeled line-drawing representing shell anatomy is also present. A 19-page bibliography representing those references found throughout the text, a glossary of scientific names, and a combined common/scientific-name index conclude the work. It should be noted that this book was written as a direct result of the author's initial publication entitled *Turtles of the United States* (see entry 992).

1001. Marquez M., Rene. **Sea Turtles of the World: An Annotated and Illustrated Catalogue of Sea Turtle Species Known to Date**. Rome: Food and Agriculture Organization of the United Nations, 1990. 81p. (FAO Species Catalogue, vol. 11; FAO Fisheries Synopsis, no. 125, vol. 11). ISBN 92-5-102891-5.

Consisting of lengthy descriptive accounts, this source provides information on the sea turtles known throughout the world. The majority of the accounts are preceded by a one-page

description of the Cheloniidae. Such topics as basic external anatomy, geographical distrib
tion, habitat, reproductive material, predation, interest to fisheries, and the number of gene
and species within the family are considered. A second family, Dermochelyidae, is als
present without an account. A total of seven species are included under Cheloniidae, while
is included in Dermochelyidae.

Each species is arranged under its respective genus, which in turn is entered und
family. The scientific name, person who named it, date of naming, citations to the literatur
and synonyms are presented for each of the genera. For each of the species listed, the scientifi
name, person who named it, date of naming, citations to the literature, alternative scientifi
names, and subspecies are given. In addition, basic external anatomy, geographical distrib
tion, habitat, biology, reproductive material (eggs, nesting, incubation, etc.), feeding beha
ior, size, local species names, and bibliographic references are given. Black-and-whi
drawings depicting the turtle, diagrams showing general external anatomy, and a geographic
range map accompany the descriptions. The length of the descriptions vary from four to eig
pages. An illustrated glossary of technical terms and measurements, and an illustrated key
the families and genera can be found in the introductory pages. A 13-page bibliography, an
a combined index of common and scientific names conclude the handbook.

1002. Obst, Fritz Jurgen. **Turtles, Tortoises and Terrapins**. New York: St. Martin's Pres
1986. 231p. $19.95. LC 85-61662. ISBN 0-312-82362-2.

Abounding with a large number of life-sized colored and black-and-white photograp
of these reptiles, this work considers a variety of biological aspects pertaining to the turtle
tortoises, and terrapins of the world. Encyclopedic information including geographic
distribution, habitats, taxonomy, anatomy, physiology, sexual behavior, life-spans, intell
gence, diseases, parasites, diet, and much more is presented. In addition to the photograph
line-drawings and geographical range maps add to the illustrative material. A list of enda
gered species of turtles, a checklist of turtles with appropriate bibliographic references, an
a separate common/scientific-name index round off the work. This volume provides a goo
introduction to these fascinating reptiles.

1003. Pritchard, Peter C. H. **Encyclopedia of Turtles**. Neptune City, N.J.: T. F. H. Publ
cations, 1979. 895p. LC 81-168290. ISBN 0-87666-918-6.

Abounding with large numbers of black-and-white and colored photographs, this wo
provides information to all turtle species of the world. The various species are discussed und
their genus headings. Material discussed includes number of species within a genera, commo
and scientific names of the species, geographical range, coloration, size, diet, nesting info
mation, and habitats. These descriptions within each genera vary from a few pages to 10
more pages. The illustrations accompany the descriptive matter. Introductory informatio
consists of keys to families and genera, structure and function of turtles, and turtle evolutio
and fossil history. The genera of these vertebrates can be found under chapter headings suc
as emydid turtles, land tortoises, mud, musk, and snapping turtles, soft-shelled turtles, se
turtles, and so on. Appendix 1 consists of a checklist (scientific names arranged under famil
and name of person who first named it as well as date of naming), while appendix 2 lis
names of both living and extinct genera. A brief glossary and a combined common/scientifi
name index conclude the work. This is a good starting place for general information regardin
these creatures.

SQUAMATA; SAURIA (Lizards)

1004. Boulenger, George Albert. **Catalogue of the Lizards in the British Museu (Natural History)**. 2d ed. London: Printed by the Order of the Trustees, 1885-1887. Reprin
1965. 3 vols. in 2. $280.00. ISBN 3-7682-0239-9.

With each entry arranged under its respective genus, which in turn is entered under i
family, this work gives a detailed anatomical account of each species considered. Coloratio
and an accompanying table detailing size are also present. Sizes of total length, head, widt

of head, body, fore limb, hind limb, and tail are included. In addition, keys to the genera (down to species), and a number of references to the taxonomic literature can be found accompanying each description. Illustrations preceding the accounts are absent. A systematic index and a scientific-name index preceding the accounts are present for each volume. These three volumes represent an extremely detailed, sophisticated, and scholarly account of these vertebrates. Those wishing for lighter-type information would do well to avoid this set.

Volume 1. *Geckonidae, Eublepharidae, Uroplatidae, Pygopodidae, Agamidae.*

Volume 2. *Iguanidae, Xenosauridae, Zonuridae, Anguidae, Anniellidae, Helodermatidae, Varanidae, Xantusiidae, Teiidae, Amphisbaenidae.*

Volume 3. *Lacertidae, Gerrhosauridae, Scincidae, Anelytropidae, Dibamidae, Chamaeleontidae.*

1005. Mattison, Chris. **Lizards of the World**. New York: Facts On File, 1989. 192p. $24.95. LC 89-1237. ISBN 0-8160-1900-2.

The descriptive accounts of the species is preceded by a series of chapters involved with lizard biology. These sections consider form and function, ecology, diet, mechanisms of feeding, defense mechanisms, reproduction, geographical distribution, habitat, and some pointers on how to raise lizards in captivity. Information on the species is found in chapter 9, "The Classification of Lizards." Adequate descriptions of families are considered. Within this descriptive material, the species are presented. For each of the families represented, the scientific and common names, number of species, coloration, geographical distribution, general external anatomy, size, behavioral patterns, egg and nest information, and other interesting facts are considered. Black-and-white as well as a large number of colored photographs provide most of the illustrative matter. A two-page bibliography and a combined common/scientific-name index conclude the work. This volume represents an informative and readable treatise on these vertebrates.

1006. Sprackland, Robert George. **Giant Lizards**. Neptune, N.J.: T. F. H. Publications, 1992. 288p. $79.95. ISBN 9086622-634-6.

Arranged under their respective families, this source consists of all lizards of the world that are considered large; that is, approaching one meter or more in length and having considerable bulk. No legless species are considered based on these parameters. For each of the lizards present, the common name, scientific name, person who named it, date of naming, etymology of the name, additional common names, geographical distribution, habitat, size (in cm and in.), and a variety of other pieces of information are given. The lengths of the descriptions and information contained therein vary from one-half of a page to several pages. Beautiful colored photographs of these beasts accompany the descriptive information. A small section on the medical care of these animals is present near the end of the work. The pages of the book are glossy, which enhances its value and adds additional interest to the subject matter. A brief glossary, a four-page bibliography, and a combined subject, common-name, and scientific-name index conclude the source.

SQUAMATA; SERPENTES (Snakes)

1007. Boulenger, G. A. **Catalogue of Snakes in the British Museum (Natural History)**. London: Printed by Order of the Trustees, 1893. Reprint, Weinheim, Germany: J. Cramer; New York: Hafner, 1961. 3 vols. in 2.

Follows the same format and relates the same type of information as was found in *Catalogue of the Lizards in the British Museum (Natural History)* (see entry 1004).

Volume 1. *Typhlopidae, Glauconiidae, Boidae, Ilysiidae, Uropeltidae, Xenopeltidae, Columbridae, Aglyphae.*

Volume 2. *Colubridae, Aglyphae.*

Volume 3. *Colubridae (Opisthoglyphae and Proteroglyphae), Amblycephalidae, and Viperidae.*

1008. Coborn, John. **The Atlas of Snakes of the World**. Neptune City, N.J.: T. F. H Publications, 1991. 591p. $100.00. LC 91-223356. ISBN 0-86622-749-0.

Abounding with large numbers of colored photographs depicting these creatures, thi source attempts to provide information on all snakes inhabiting the planet. The familie Typhlopidae, Anomalepidae, Leptotyphlopidae, Aniliidae, Acrochordidae, Boidae, Uropelti dae, Xenopeltidae, Colubridae, Elapidae, Hydrophiidae, Viperidae, and Crotalidae are cov ered. Each species is arranged under its respective genus, which in turn is entered unde subfamily, family, and infraorder. General biological information such as size, habita number of groups within a group, very basic external anatomy, and, in some cases, geographi cal distribution, reproduction, and behavior are given for infraorder, family, subfamily, an genus. In addition, the person who named it and date of naming are given for the genera listec For each of the species mentioned, the scientific name, person who first named it, date o naming, geographical range, size, and, in some cases, basic external anatomy, coloratior behavior, diet, and terrarium needs are considered. The large colored photographs of th snakes can be found interspersed among the textural accounts. The descriptive accounts rang in length from one sentence to several paragraphs.

In addition, such topics as the evolution, classification, and general biology of snakes housing and care of snakes in captivity, general husbandry and care, diseases and treatmen and reproduction and propagation can be found in the first 100 pages of the work. A smal section on snake venoms, antivenins, and treatment of snake bites can be found near the en of the volume. A three-page bibliography, a brief glossary, and a combined common/scientific name index conclude the reference source.

1009. Mehrtens, John M. **Living Snakes of the World in Color**. New York: Sterling, 1987 480p. $55.00. LC 87-9932. ISBN 0-8069-6461-8.

Arranged under broad snake categories, this work provides an excellent opportunity t discover the beautiful variety of colors of these types of reptiles. The species are entered unde the categories of primitive snakes, typical snakes, and venomous snakes. For each specie included, the common name, scientific name, habitat, geographical range, behavior, size food, gestation period, and care are considered. In addition, general information concernin a snake group is presented and precedes the species account. A brief glossary and a separat common/scientific-name index conclude the book. It should be noted that the informatio presented is not detailed; however, the work provides a good starting point for those wantin general knowledge of these vertebrates.

1010. Phelps, Tony. **Poisonous Snakes**. rev. ed. London: Blandford, 1989. 237p. $19.95pa LC 89-189375. ISBN 0-7137-2114-6; 0-7137-0877-8pa.

The species are discussed under their respective genera, which in turn are entered unde broad common names. Examples are the vipers, elapids, and so on. Such facts as size scientific name of species, geographical range, habitat, coloration, and the degree to whic the snake is dangerous are presented for the species contained within each of the gener sections. Other sections of the volume include classification and distribution (list of scientifi names of snakes and their geographical distributions), habits, behavior, venom, snakebites and information on how to care for poisonous snakes that are in captivity. Colored an black-and-white photographs of the snakes as well as charts and graphs provide the illustrativ material. Appendix 1 lists principal antivenin sources, while appendix 2 considers emergenc procedures. A brief glossary, a two-page bibliography, and a combined subject, common name, and scientific-name index conclude the work.

1011. Underwood, Garth. **A Contribution to the Classification of Snakes**. London: Britis Museum (Natural History), 1967. 179p. (Publication British Museum, Natural History). LC 67-91980.

By basically focusing on families and subfamilies, this source provides detailed ana tomical information that distinguishes one group from another. Species are considered withi these discussions. Introductory information considers those anatomical structures useful i the classification of snakes. A few labeled line-drawings depicting skeletons and musculature provide the illustrations. A summary classification of families and subfamilies, a seven-page bibliography, and a combined subject, common-name, and scientific-name index complete the volume.

CROCODYLIA (Alligators and Crocodiles)

1012. Alderton, David. **Crocodiles & Alligators of the World**. London: Blandford, 1991. 190p. $32.87. LC gb91-289. ISBN 0-7137-2145-6.

In addition to a description of the species of crocodiles and alligators inhabiting the planet, this work provides solid information on the biology of these creatures. Such topics as form and function, reproduction, and evolution and distribution of crocodilians are found in as many chapters. The form and function section considers anatomical features, feeding, diet and digestion, hunting, mobility, breathing, thermo-regulation, social interactions, communication, predation, mortality, and other assorted facts. The species are arranged under broad common names, such as alligators and crocodiles. For each of the organisms presented, the common name, scientific name, geographical distribution, geographical range maps, size, population, habitat, mating, nesting, hatching, coloration, and information related to their endangered status are given. A large number of colored as well as a few black-and-white photographs provide the illustrations. A three-page bibliography and a combined subject, common-name, and scientific-name index conclude the volume. This is a well-balanced work and represents a good starting point for information on these interesting creatures.

1013. Grenard, Steve. **Handbook of Alligators and Crocodiles**. Malabar, Fla.: Krieger, 1991. 210p. $49.50. LC 89-71337. ISBN 0-89464-435-1.

Basically divided into four parts, this work considers the biology of alligators and crocodiles as well as providing descriptive accounts of the species found throughout the world. Part 1 discusses generally the anatomy and physiology of the organ systems found in these creatures. Part 2 (Alligatorinae), part 3 (Crocodylinae), and part 4 (Gavialinae and Tomistominae) provide the descriptive accounts. For each of the species listed, the common name, scientific name, person who first named it, date of naming, status, geographical distribution, habitat, diet, nesting and reproductive information, conservation measures, and a number of other interesting facts are presented. A geographical range map accompanies each of the descriptions. The amount of information varies from species to species as does the length of the description (1 to 10 or more pages). Black-and-white and colored photographs depicting the animals provide the majority of illustrative matter. A 26-page bibliography appears near the end of the work. An illustrated identification key to the crocodilia, a measurement conversion table, and a combined subject, geographic, common-name, and scientific-name index conclude the volume. This publication represents a well-balanced approach and a good starting point in order to gain an understanding of these reptiles.

1014. Penny, Malcolm. **Alligators & Crocodiles**. London: Boxtree, 1991. 128p. LC gb91-36762. ISBN 1-85283-132-4.

Heavy on photographs and light on text, this source considers some general information on many of the species of crocodiles and alligators inhabiting the planet. The species are arranged under their subfamilies. For each species listed, the common name, scientific name, geographical distribution, habitat, appearance and size, reproduction, and diet are given. Other information provided within other sections of the work consists of reproduction, birth and young, metabolism and movement, hunting and prey, and location and habitat. It should be noted that the above parameters are presented mainly in magnificently colored photographs of these creatures with appropriate captions. This source does not come close to the amount of information presented in *Crocodiles and Alligators* (see entry 1015). In any event, the photographs are fantastic and one can gain some general knowledge about these creatures. A combined subject and common-name index concludes the work.

1015. Ross, Charles A., ed. **Crocodiles and Alligators**. New York: Facts on File, 1989. 240p. $35.00. LC 89-30416. ISBN 0-8160-2174-0.

This source provides a myriad of biological facts dealing with these incredible reptiles. Such parameters as evolution, anatomy, physiology, behavior, diet, reproduction, habitats, economic importance, and conservation and management are presented in an interesting and appealing manner. Worldwide coverage is given. In addition, a section on living crocodilians is present. This section provides a brief descriptive account of the many crocodiles present throughout the world. The species are presented under their subfamilies. For each of the crocodiles listed, the common name, scientific name, coloration, size, habitat, geographical distribution, reproduction, and diet are given. A small geographical range map as well as a large colored painting of the species accompany the page-length description. A very large number of colored photographs can be found interspersed throughout the textural material. The photographs are worth the price of the book. This is an excellent work and provides a great introduction to these reptiles.

1016. Steel, Rodney. **Crocodiles**. London: Christopher Helm, 1989. 198p. $34.95. LC 90-179707. ISBN 0-7470-3007-3.

Consisting of worldwide coverage, this work provides a wealth of information on alligators and crocodiles. These reptiles are arranged under broad common-name groups. Such chapters as man-eaters, rarities and dwarfs, alligators and caimans, and gavials and fals gavials are included. For each of these sections, general information on the group as a whole, along with the names of species contained therein are presented. Such biological parameters as external anatomy and measurements, size, predatory behavior, diet, behavior in general, habitat, nesting information, egg size, hatchlings, coloration, fossil information, geographical distribution, and other interesting facts are presented. In addition to the above, other sections include information on fossils and conservation measures. Line-drawings and colored photographs depicting these creatures provide the illustrative material. A checklist of principal crocodilian genera (names of genera, family, suborder, and order), a checklist of living crocodilian species and their common names, a one-page bibliography, and a combined subject, common-name, and scientific-name index conclude the volume. This work contains a good deal of information on these beasts. It is a good place to start for information on these interesting reptiles.

FIELD GUIDES

Amphibia and Reptile (General)

1017. Capula, Massimo. **Simon & Schuster's Guide to Reptiles and Amphibians of the World**. New York: Simon & Schuster, 1989. 256p. $10.95pa. LC 89-21671. ISBN 0-671-69136-8; 0-671-69098-1pa.

A translation of *Tutto Anfibi e Rettili*, this work serves as a field guide to the many amphibians and reptiles found throughout the planet. A total of 80 amphibians and 121 reptiles are included. For each species mentioned, the scientific name, common name, classification (order, family), geographical distribution, brief external anatomy, coloration, size, habitat and breeding information are given. A colored photograph of the vertebrate can be found on the opposing page of the descriptive matter. Two species are contained on each page. In addition to general information on each group, a taxonomic table (down to family) is presented as well. A brief glossary, a two-page bibliography, and a combined common/scientific-nam index conclude the guide.

Amphibia and Reptilia (Geographic Section)

NORTH AMERICA

1018. Behler, John L. **The Audubon Society Field Guide to North American Reptiles and Amphibians**. New York: Knopf, 1979. 743p. $15.95. LC 79-2217. ISBN 0-394-50824-6.

Published with a flexible cover for easy use, this guide presents information necessary to help identify the many species of reptiles and amphibians found throughout the North American continent. After some brief introductory information on each of the families, the accounts of the species ensue. For each vertebrate listed, the common name, scientific name, size (in. and cm), coloration, brief external anatomy, voice (if appropriate), breeding information (seasons, eggs, etc.), habitat, geographical range, and other interesting facts are included. Subspecies are listed when appropriate. A small geographical range map accompanies each description. The colored pages of plates can be found at the beginning of the guide. The amphibians and reptiles portrayed on these plates are arranged under their common-name groups. The common name and size of the animal as well as cross-references to the text appear under each picture. In addition, a thumb pad found on the left-hand side of the page shows a small general outline of the body form of the animal, thereby making identification less painful. A guide to these thumbnail body forms can be found preceding the colored plates. A brief glossary and a combined common/scientific-name index conclude the field guide.

1019. Brown, Vinson. **Reptiles & Amphibians of the West**. Healdsburg, Calif.: Naturegraph, 1974. 79p. LC 74-3204. ISBN 0-87961-029-8; 0-87961-028-Xpa.

Species are listed under the common names of the families, such as iguanid lizards, common frogs and tailed frogs, and so on. The common and scientific names of the family are presented. For each of the species mentioned, the common name, scientific name, size, geographical range, habitat, coloration, and, in some cases, vocal patterns are given. A list of colored plates and black-and-white drawings provide help in the identification process. Procedures on catching and taking care of these animals, a picture key to the basic amphibian and reptile groups, and a checklist of the species can be found in the introductory information. A one-page list of selected readings and a separate common/scientific-name index conclude the field guide.

1020. Cochran, Doris M., and Coleman J. Goin. **The New Field Book of Reptiles and Amphibians: More than 200 Photographs and Diagrams**. New York: G. P. Putnam, 1970. 359p. LC 69-18168.

This field guide considers those reptiles and amphibians that occur in all 50 of the United States. Species are arranged under their respective genera, which in turn are entered under families. A brief description of the family and genus is presented. For each of the species listed, the common name, scientific name, external anatomical features, size, breeding information, habitat, and geographical range are given. Black-and-white photographs of the organisms can be found accompanying the descriptions. In addition, 16 pages of colored plates can be found in the center of the guide. The main sections of the source consist of sirens and salamanders, frogs and toads, crocodiles and alligators, turtles, lizards, and snakes. A list of selected references, and a combined common/scientific-name index complete the volume. The information given for many of the species is substantial for a field guide. This work could substitute for a type of miniencyclopedia.

1021. Conant, Roger, and Joseph T. Collins. **A Field Guide to Reptiles and Amphibians: Eastern and Central North America**. 3d ed. Boston: Houghton Mifflin, 1991. 450p. (The Peterson Field Guide Series; 12). $24.45; $15.95pa. LC 90-21053. ISBN 0-395-37022-1; 0-395-58389-6pa.

Species are arranged under their respective families, which in turn are entered und orders. Basic common-name groups appear at the top of the pages. In addition to the speci account, brief information is given on the order and family. For each species entered, t common name, scientific name, size (in. and cm), weight (lbs. and kgs), brief extern anatomical characteristics, habitat, diet, similar species, geographical range, and subspeci are given. The colored plates of these vertebrates are positioned in the center of the gui with cross-references from plates to text and vice versa. Each plate is a two-page spread wi common and scientific names and brief anatomical descriptions on one page and color pictures of species on the other. Line-drawings provide other illustrative matter. A sma section on how to identify tadpoles can be found near the rear of the work. A brief glossar list of references, geographical range maps, and a combined common/scientific-name ind conclude the field guide. Introductory sections consider how to catch and transport the animals, how to care for them in captivity, and what to do in case of a snakebite.

1022. Conant, Roger, Robert C. Stebbins, and Joseph T. Collins. **Peterson First Guide Reptiles and Amphibians**. Boston: Houghton Mifflin, 1992. 128p. $4.95pa. LC 91-3301 ISBN 0-395-62232-8.

Based upon Conant and Collins' *A Field Guide to Reptiles and Amphibians: Easte and Central North America*, 3d ed. (see entry 1021), and Stebbins' *A Field Guide to Weste Reptiles and Amphibians* (see entry 1025), this work provides for the novice, gener information necessary to identify the more common species of amphibians and reptiles fou throughout the North American continent. The species included are arranged under bro common-name groups, such as giant salamanders, mole salamanders, true frogs, chorus frog blind snakes, water snakes, and so on. For each of the species listed, the common name, siz coloration, habitat, and brief behavioral characteristics are given. Colored pictures of the creatures can be found on the opposing page of text. Beginning with the salamanders a ending with the snakes, over 350 species are considered. A common-name index complet the field guide. This is a very useful identifying tool for the beginner.

1023. Schmidt, Karl Patterson, and D. Dwight Davis. **Field Book of Snakes of the Unit States and Canada**. New York: G. P. Putnam, 1941. 365p. LC 41-25160.

The snakes are listed under their common-name groups, such as bull snakes, blac headed snakes, black swamp snakes, scarlet snakes, and so on. For each of the speci included, the common name, scientific name, coloration, brief external anatomy, geographic range, habitat, size, diet, breeding habits, and references to the literature are given. In additio keys to the species, line-drawings of the heads of many snakes, and geographical range ma accompany the descriptions. Thirty-four pages of plates (mostly black-and-white phot graphs) of snakes can be found near the end of the source. The introductory material extensive and covers 62 pages. It deals with folklore of snakes, definition and classificatio external characteristics, coloration, poison apparatus, treatment of snakebites, habitats, behavior patterns, and the collection, preservation, and study of snakes. A combined subject, commo name, and scientific-name index concludes the volume.

1024. Smith, Hobart M., and Edmund D. Brodie. **Reptiles of North America: A Guide Field Identification**. New York: Golden Press, 1982. 240p. (Golden Field Guide Series $10.50pa. LC 81-83000. ISBN 0-307-13666-3pa.

Abounding with a large number of colored drawings, this work provides information order to identify turtles, lizards, snakes, amphisbaenids, and crocodiles of North Americ Two hundred seventy-eight species and over 500 subspecies are contained within this guid The species are listed under their respective genera, which in turn are entered under subfam lies and families. General information is given for each of the families and subfamilies. F each species presented, the common name, scientific name, and coloration are given. geographical range map along with a colored picture of each species accompany the descri tive account. The colored pictures can be found on the opposing pages of the description The descriptions average one to two paragraphs in length. Biological keys down to speci n be found throughout the text. Other sections briefly describe anatomy, physiology, production, enemies and defensive mechanisms, and the care of reptiles. A two-page bliography and a combined subject, common-name, and scientific-name index conclude the ld guide.

25. Stebbins, Robert C. **A Field Guide to Western Reptiles and Amphibians: Field arks of All Species in Western North America, Including Baja California**. 2d ed., rev. oston: Houghton Mifflin, 1985. 336p. (The Peterson Field Guide Series; vol. 16). $17.95; 2.95pa. LC 84-25125. ISBN 0-395-38254-8; 0-395-38253-Xpa.

Each species is arranged under its respective family/ subfamily. The scientific and mmon names of the family and subfamily are presented along with a brief description of ese groups. For each animal entered, the common name, scientific name, size, general ternal anatomy, habitat, behavior, egg information, diet, geographical range, and subspecies considered. The illustrative material consists of black-and-white as well as colored plates. ese plates are represented by a two-page spread. The common name, scientific name, and loration are present on one page, while pictures of the representative animals are found on other. Introductory information consists of how to capture amphibians and reptiles, caring r them, methodology and procedures carried out in the field, and a list of identification keys. brief glossary, a set of geographical range maps that correspond to the text, and a combined mmon/scientific-name index conclude the guide.

26. Tyning, Thomas F. **A Guide to Amphibians and Reptiles**. Boston: Little, Brown, 90. 400p. (Stokes Nature Guides). $19.95; $11.95. LC 89-28444. ISBN 0-316-81719-8; 316-81713-9pa.

Not arranged in quite the same manner as most of the field guides, this volume presents ecies found throughout North America. Amphibians are listed first, followed by the reptiles. addition to providing the common and scientific names of each organism, each of the ecies consists of sections on how to recognize the animal, distinguishing the sexes, tadpole ormation (when appropriate), egg information, how to find the animal, and descriptive ormation on what to expect in observations, such as territorial behavior, vocalization, ting, eggs and egg-laying, overwintering, and so on. A quick-reference chart proceeding ch description considers length of breeding season, breeding habitat, number of eggs posited, time necessary for hatching, life-span of adults, and other facts depending on the ecific organism. A geographical range map and a line-drawing of the species accompany descriptive material. The descriptive account can be approximately 5 to 10 pages long. ere are no colored or black-and-white plates of the amphibians and reptiles. A selected list references can be found closing this guide. These references are arranged under common-me groups as well as individual species mentioned in the text. There is no index.

This guide can serve as a miniencyclopedia due to the large amount of information esented for each of the amphibians and reptiles covered. This work is more useful for oking for particular reptiles and amphibians. The descriptive information can verify correct entifications.

27. Zim, Herbert S., and Hobart M. Smith. **Reptiles and Amphibians**. Racine, Wis.: olden Press/Western, 1987. 160p. (Golden Guides). $3.95. LC 88-134520. ISBN 0-307-057-6pa.

Consisting of 212 species, this easy-to-use guide focuses mainly on those species found oughout North America. The species are arranged under their broad common-name groups. e groups are turtles, lizards, snakes, alligators and crocodiles, frogs and toads, and amanders. For each of the species listed, the size, habitat, geographical distribution, diet, d, in some cases, egg information are given. A geographical range map accompanies many the descriptive accounts. In addition, a large colored picture of the animal also accompanies account. Each page is devoted to a species. In most cases, the colored picture encompasses proximately two-thirds of a page while the account makes up the other one-third. In dition, some general information on the collecting and captivity of amphibians and reptiles is present. A brief list of selected references, a list of scientific names arranged chronolog cally by page number, and a common-name index conclude the guide.

This guide includes the more familiar types of amphibians and reptiles. It is not comprehensive guide; rather, it represents a good first start in the identification of the vertebrates.

TEXTBOOKS

1028. Duellman, William E. **Biology of Amphibians**. New York: McGraw-Hill, 198 670p. $59.95. LC 85-14916. ISBN 0-07-017977-8.

Divided into four parts, this text represents a well-balanced look at all facets amphibian biology. The four parts are life history, ecology, morphology, and evolution. Pa 1 (life history) considers such areas as reproductive strategies, courtship and mating, vocal zation, eggs and development, larvae, and metamorphosis, while part 2 (ecology) dwells relationships with the environment, food and feeding, enemies and defense, populations, a community ecology and species diversity. Part 3 (morphology) presents the various amphi ian systems in terms of their external and internal anatomy, and part 4 (evolution) discuss origins of amphibians; the cytogenetic, molecular, and genomic facets of evolution; phyl geny; and biogeography. Electron micrographs, labeled line-drawings, graphs, charts, table photographs, and geographical range maps provide the illustrative material. A 53-pa bibliography consisting of those references cited within the text, and a combined subje common-name, and scientific-name index conclude the text.

1029. Porter, Kenneth R. **Herpetology**. Philadelphia: W. B. Saunders, 1972. 524p. L 75-188390. ISBN 0-7216-7295-7.

This work represents a basic text in the study of amphibians and reptiles. Approximate 40% of the volume considers the anatomy and physiology of these creatures. The major org systems of each group are covered. In addition, both amphibians and reptiles are consider in relation to origins and phylogeny, geographical distribution patterns, coloration, di reproductive adaptations, and population dynamics, as well as moisture and temperatu regulations. The illustrations consist of labeled line-drawings of internal features, lin drawings of specimens, geographical range maps, and an occasional black-and-white phot graph. A list of references appear at the end of each chapter. A separate scientific-name a subject index concludes the textbook. This is a well-balanced text as it covers all of the maj biological disciplines associated with amphibians and reptiles.

1030. Seigel, Richard A., Joseph T. Collins, and Susan S. Novak, eds. **Snakes: Ecolo and Evolutionary Biology**. New York: Macmillan, 1987. 529p. LC 86-31212. ISBN 0-0 947830-8.

Divided into three sections, this text focuses on many aspects of snake biology. Sectio 1 considers the taxonomy and fossil history of snakes, as well as geographical distributio problems in phylogeny and zoogeography, and functional morphology. Section 2 conce trates its efforts on the collecting and life histories of these reptiles, as well as how to be maintain them in captivity. Section 3, which encompasses the majority of the work, is devot to the life history and ecology of snakes. Such areas as behavior, reproduction, populati studies, foraging, communities, spatial patterns and movements, activity patterns, physiolog status, and conservation and management are considered. A series of graphs, charts, a labeled line-drawings provide the illustrative material. At the end of each chapter is bibliography of those references cited within the chapter. A scientific-name index conclud the work. This volume provides a good amount of information on the biology of snakes.

031. Zug, George R. **Herpetology: An Introductory Biology of Amphibians and Rep-les**. San Diego: Academic Press, 1993. $50.00. LC 92-30758. ISBN 0-12-782620-3.

Divided into six parts, this work provides a diversified amount of information on the nphibians and reptiles. It is geared to upper-level undergraduates. The first five parts of the xtbook consider the biological principles governing reptiles and amphibians. Areas included e general anatomy, diet and feeding, defense and escape, reproduction and development, oming and migration, homeostasis, and population studies. The sixth part is devoted to the assification and systematics of the two groups. Black-and-white photographs, geographical nge maps, line-drawings, and graphs and charts provide the illustrative material. ibliographic references for each of the 18 chapters can be found near the end of the volume. he citations contained within the bibliography represent papers that were used by the author writing the chapters as well as review articles on the topic. A separate subject and ientific-name index concludes the work. Common names are provided in the subject portion f the index.

JOURNALS

)32. **The British Herpetological Society. Bulletin**. Vols. 1- , nos. 1- . London: British erpetological Society, 1980- . Quarterly. $40.00/yr. (includes **The Herpetological Jour-al**). ISSN 0260-5805.

Serving in some respects as a newsletter of the British Herpetological Society as well a supplement to *The Herpetological Journal* (see entry 1034), this serial considers the ociety's meetings, reports on symposia, and a number of informative articles dealing with e biology of amphibians and reptiles. The number of articles varies with each issue. Book views are present in many issues. Black-and-white and colored photographs depicting erpetofauna species make up the majority of the illustrations.

)33. **Herpetologica**. Vols. 1- , nos. 1- . Austin, Tex.: The Herpetologists' League, 1936- . uarterly. $70.00/yr. ISSN 0018-0831.

From the location of ribosomal DNA in reptiles, to the effect of predation on sympatric dpole species, this journal presents research articles dealing with all aspects of amphibian d reptile biology. Approximately 10 to 15 articles ranging in length from 5 to 20 or more ges each can be found in each issue. Other features of the journal consist of book reviews, formation on the Herpetologists' League itself, and various announcements. Graphs, charts, ght microscopy plates, line-drawings, and photographs provide the illustrative material.

)34. **The Herpetological Journal**. Vols. 1- , nos. 1- . London: British Herpetological ociety, 1985- . Semiannually. $40.00/yr. (includes a subscription to **The British Herpeto-gical Society. Bulletin**). ISSN 0268-0130.

Consisting of review articles, full-length research papers, and short notes, this journal devoted to all aspects of amphibian and reptilian biology. Such areas as ecology, behavior, stematics, physiology, and so on provide examples. The research papers vary in length from proximately 4 to 10 or more pages. The papers appearing in the short-notes section are tween two and three pages, while the length of the reviews varies considerably. A section book reviews and an occasional obituary can also be found at the end of the issue. Graphs, arts, line-drawings, and photographs provide the illustrative material.

)35. **Herpetological Review**. Vols. 1- , nos. 1- . [Athens, Ohio]: Society for the Study of mphibians and Reptiles, 1967- . Quarterly. $12.00/yr. ISSN 0018-0831.

Serving, to some degree, as a supplement to the *Journal of Herpetology* (see entry 1036), is serial presents a variety of information pertinent to those interested in the field; in essence, serves as a newsletter. Such items as information regarding the annual meeting of the ociety for the Study of Amphibians and Reptiles (SSAR), the Society's business report, herpetological meeting announcements, informational notes on specific species/subspecie techniques used in research, herpetological husbandry, population studies, geographic distribution on certain species, and book reviews are presented. Photographs, charts, graph and line-drawings provide the illustrations. Each issue is approximately 30 pages in length

1036. **Journal of Herpetology**. Vols. 1- , nos. 1- . [Athens, Ohio]: Society for the Stud of Amphibians and Reptiles, 1968- . Quarterly. $60.00/yr. (institutions). ISSN 0022-1511.

This journal consists of research papers dealing with all aspects of amphibian an reptilian biology. Examples are ecology, behavior, systematics, population studies, and s on. Approximately 25 to 30 articles are contained within each issue. The length of the pape vary from 2 to 10 pages. It should be noted that the shorter papers are placed under a sectio of notes. Graphs, charts, tables, line-drawings, and photomicrographs depicting light as we as electron microscopy provide the illustrations.

TAXONOMIC KEYS

Amphibia and Reptilia (General)

1037. Ballinger, Royce E., and John D. Lynch. **How to Know the Amphibians an Reptiles**. Dubuque, Iowa: Wm. C. Brown, 1983. 229p. (The Pictured Key Nature Series $15.00. LC 82-83544. ISBN 0-697-04786-5.

As the series title suggests, this work includes biological keys (down to species) as we as drawings of the amphibians and reptiles in order to readily identify them. A pictured ke to larval amphibians has also been inserted. Species endemic to Canada and the United State have been included. In addition to the keys, descriptions regarding orders, families, an species are presented. These descriptions include size, coloration, and geographical distrib tion maps. A classification outline as well as a combined common/scientific-name inde conclude the guide.

1038. Savage, Jay, and Jaime Villa R. **Introduction to the Herpetofauna of Costa Ric Introduccion a la Herpetofauna de Costa Rica**. [Athens, Ohio]: Society for the Study c Amphibians and Reptiles, 1986. 207p. (Contributions to Herpetology, no. 3). LC 86-6114 ISBN 0-916984-16-8.

This source presented in English and Spanish considers the taxonomic keys to the grou of amphibians and reptiles found in Costa Rica. Genera and species keys are represented. Th work includes keys to the caecilians, salamanders, frogs and toads, turtles, lizards, snake and crocodiles. A very basic line-drawing representing each group of animals can be foun preceding the keys to the group. Also present is a bibliographic index, which lists maj references for each family and genus; an annotated bibliography, which consists of th references cited; and a scientific-name index.

1039. Schwartz, Albert, and Robert W. Henderson. **A Guide to the Identification of th Amphibians and Reptiles of the West Indies Exclusive of Hispaniola**. Milwaukee, Wis Milwaukee Public Museum, 1985. 165p. $29.95. LC 85-4902. ISBN 0-89326-112-2.

Taxonomic keys down to the species level are presented within this volume. These key are arranged under genera for a variety of geographical locales. Therefore, a key to the Cuba Anolis, a key to the Jamaican and Caymanian Anolis, a key to the greater Puerto Rican Anoli and so on are listed. A bibliographic citation, person who first named the species, an geographical distribution are considered under each of the species. In addition, family name are listed above the genera. Line-drawings as well as black-and-white and colored plates the animals provide the illustrations. The appendix consists of a list of the scientific name of the species under their geographical locales. A brief glossary, a 42-page bibliograph nsisting of citations found within the keys, and a scientific-name index to the illustrations in be found closing the volume. This is a very detailed work.

)40. Taylor, Edward Harrison. **The Caecilians of the World: A Taxonomic Review**. awrence, Kans.: University of Kansas Press, 1968. 848p. LC 67-14429.

The species are arranged under their respective genera, which in turn are entered under milies and orders. Preceding the taxonomic accounts of the organisms, brief descriptions the orders; taxonomic keys to the orders, families, genera, and species; and taxonomic formation concerning the genera are presented. Bibliographic references to the taxonomic erature of the genera are also listed. For each of the species mentioned, the scientific name, rson who first named it, taxonomic reference(s), detailed anatomical measurements, den-ion numbers, coloration, and, in many cases, geographical distribution are given. Black-id-white photographs and line-drawings of the caecilians, as well as tables representing tailed anatomical measurements, accompany the descriptions. Introductory information scusses the anatomy of the animals and how these anatomical structures are related to their xonomies. There are 166 living species and two fossil species included within the work. A 7-page bibliography and a combined subject and scientific-name index conclude the volume. his is a detailed taxonomic work.

)41. Terentev, P. V., and S. A. Chernov. **Key to Amphibians and Reptiles. Opredelitel' resmykayuschchishsya i Zemnovodnykh**. 3d ed., enl. Jerusalem: Israel Program for ientific Translations, 1965. 315p. LC he66-30.

Translated from the Russian by L. Kochva, this source represents taxonomic keys to the ecies as well as species accounts of those organisms found in the USSR. Species are ranged under their respective genera, which in turn are entered under families, orders, and bclasses. General anatomical information is presented for the genus, family, order, and bclass. For each animal listed, the scientific name, person who first identified it, date of entification, size, detailed external anatomical information, coloration, geographical distri-tion, habitat, and breeding information are presented. Some descriptive accounts contain ore information than others. Black-and-white drawings depicting the animal as a whole as ell as drawings depicting its parts provide an aid in determining one species from another. 10-page bibliography consisting of literature on the fauna and taxonomy of amphibians id reptiles of the USSR, a systematic index, a series of geographical range maps, and a ientific-name index complete the volume. Although this source has been placed under xonomic keys, it could just as easily be considered a handbook.

CHECKLISTS AND CLASSIFICATION SCHEMES

Amphibia and Reptilia (General)

42. Cope, Edward D. **Check-List of North American Batrachia and Reptilia: With a stematic List of the Higher Groups, and an Essay on Geographical Distribution**. ashington, D.C.: U.S. Government Printing Office, 1875. 104p. (Bulletin of the United ates National Museum, no. 1). LC s13-114.

Based on the specimens contained in the U.S. National Museum, this work provides a t of amphibians and reptiles endemic to the North American continent. Each species is ranged under its respective genus, which in turn is entered under family, order, and class. e person responsible for naming the families are present. For each of the species listed, the ientific name, citation to the taxonomic literature, and geographical distribution are given. total of 358 species are considered. An essay on the geographical distribution of the amphibians and reptiles of North America, a four-page bibliography, and a famil genera-name index complete the checklist.

1043. Henderson, Robert W. **A Checklist and Key to the Amphibians and Reptiles Belize, Central America**. Milwaukee, Wis.: Milwaukee Public Museum Press, 1975. 63 (Contributions in Biology and Geology, no. 5). LC 77-376963.

Comprising 134 species, this checklist provides a listing of the amphibians and reptil inhabiting Belize. Each species is arranged under its respective genus, which in turn is enter under family, order, and class. For each of the species entered, the scientific name, pers who first named it, references to the literature, synonyms, subspecies, habitat, geographic distribution, and vernacular names are given. In addition, biological keys to the famili genera, and species can be found throughout the text. A seven-page bibliography consisti of those references cited within the checklist can be found at the end of the work. Access the names of the classes, orders, families, and genera of these amphibians and reptiles in t checklist can be located in the table of contents.

1044. Logier, E. B. S., and G. C. Toner. **Check List of the Amphibians and Reptiles Canada and Alaska**. Toronto: The Royal Ontario Museum, 1961. 92p. (Toronto: Roy Ontario Museum, Life Sciences Division. Contribution No. 53).

The species/subspecies (105 of them) are arranged under their respective families, whi in turn are entered under orders and classes. For each of the amphibians and reptiles liste the scientific name, person who first named it, common name, geographical range, Canadi locality records, and bibliographic citations to the literature are presented. Geographical ran maps (77 of them) accompany many of the species listings. A table of contents, which ac as a systematic index, can be found on the opening pages. This table presents the scienti and common names of the species under their families with appropriate page numbers to t text. A nine-page bibliography concludes the checklist. There is no scientific- or commo name index. It should be noted that this work represents a revision of Contribution No. 53

1045. Loveridge, Arthur. **Check List of the Reptiles and Amphibians of East Afri (Uganda; Kenya; Tanganyika; Zanzibar)**. Cambridge: Museum of Comparative Zoolog 1957. [151]-362p. (Bulletin of the Museum of Comparative Zoology, vol. 117, no. 2). L a57-7603.

Each species (527 of them) is arranged under its respective genus, which in turn entered under family, suborder, order, subclass, and class. For each of the genera listed, t person who first named it, alternative generic names, as well as citations to the literature a considered. For each of the species listed, the scientific name, person who named it, comm name, alternative scientific names, citations to the taxonomic literature, and geographic distribution are given. A three-page annotated bibliography and a scientific-name ind complete the checklist.

1046. Marx, Hymen. **Checklist of the Reptiles and Amphibians of Egypt**. Cairo: Unit States Naval Medical Research Unit Number Three, 1968. 91p. (Special Publication [Unit States. Naval Medical Research Unit, no. 3]).

Based on the collections made by the United States Naval Medical Research Unit N 3, this work provides information on the reptiles and amphibians inhabiting Egypt. T species/subspecies (6 amphibians and 87 reptiles) are arranged under their respective famili Keys to the species within the family precede the list of species. For each of the organis considered, the scientific name, person who first named it, common name, geographic distribution, and citations to the taxonomic literature are given. A number of black-and-wh photographs depicting these animals can be found throughout the pages. A three-pa bibliography and a series of 37 maps depicting Egypt and its environs conclude the checkl Each of the maps shows geographic distribution of specific species.

)47. Mertens, Robert, and Heinz Wermuth. **Die Amphibien und Reptilien Europas**. rankfurt Am Main, Germany: Verlag Waldemar Kramer, 1960. 264p.

Species and subspecies are arranged alphabetically under their respective genera. For ıch species/subspecies entered, a list of alternative scientific names is presented. For each f these, the person who named it, and citations to the literature are given. A scientific-name ıdex concludes the checklist. A few black-and-white line-drawings of external anatomical ırts provide the illustrations. This source provides the researcher with the current scientific ıme, as well as those names, based on the taxonomies, considered to be valid in the past.

)48. Miyata, Kenneth. **A Check List of the Amphibians and Reptiles of Ecuador, With Bibliography of Ecuadorian Herpetology**. Washington, D.C.: Division of Reptiles and mphibians, National Museum of Natural History, Smithsonian Institution, 1982. 70p. Smithsonian Herpetological Information Service, no. 54).

The species (682 of them) are arranged under their respective families, which in turn e entered under orders and classes. For each of the species listed, the scientific name, person ho first named it, and date of naming are given. The bibliography, which consists of 50 ıges, lists bibliographic citations to the taxonomic literature. These citations are arranged phabetically under the authors. There is no index.

)49. Schmidt, Karl P. **A Check List of North American Amphibians and Reptiles**. 6th l. Chicago: American Society of Ichthyologists and Herpetologists, 1953. 280p. LC a54- 504.

Species and races are entered under their respective families, which in turn can be found ıder orders, and classes. For each species/race listed, the scientific name, person who named , references to the taxonomic literature, geographic range, and common name are given. A vo-page list of forms introduced into North America and a combined common/scientific- ıme index conclude the checklist. This work is based in part on Cope's *Check-List of North merican Batrachia and Reptilia*... (see entry 1042).

)50. Schwartz, Albert, and Robert W. Henderson. **West Indian Amphibians and Rep- les: A Check-List**. 2d ed. Milwaukee, Wis.: Milwaukee Public Museum, 1988. 264p. Contributions in Biology and Geology Series, no. 74). $14.95pa. ISBN 0-89326-156-4.

The species are listed under their respective orders. For each of the species considered, e scientific name, person who first named it, citations to the literature, and geographical stribution are given. A series of geographical range maps and a scientific-name index onclude the checklist.

)51. Society for the Study of Amphibians and Reptiles. Committee on Common and cientific Names. **Standard Common and Current Scientific Names for North American mphibians and Reptiles**. Edited by Joseph T. Collins. 3d ed. Athens, Ohio: The Society, 990. 41p. $5.00.

Species and subspecies are listed under their respective genera. For each amphibian and ptile listed, the scientific name, person who first named it, date of naming, and common ıme are given. Appendix 1 consists of a list of Hawaiian amphibians and reptiles, while ppendix 2 considers species that are now known to be present and breeding within the ontinental United States. A one-page bibliography closes the circular.

)52. Stuart, L. C. **A Checklist of the Herpetofauna of Guatemala**. Ann Arbor, Mich.: niversity of Michigan, Museum of Zoology, 1963. 150p. (Miscellaneous Publications, Museum of Zoology, University of Michigan, no. 122). LC 63-63501.

Consisting of 88 species of amphibians and 231 species of reptiles, this source provides list of these vertebrates found throughout Guatemala. The species/subspecies are arranged ıder their respective genera, which in turn are entered under families, orders, and classes. or each of the genera listed, the person who first named it and a reference to the taxonomic terature are given. For each of the species presented, the person who first named it, references to the taxonomic literature, alternative scientific names, type locality, and geo graphical range are considered. In addition, biological keys to the families, genera, and specie can be found throughout the text. A scientific-name index completes the checklist, along wit a map of Guatemala.

1053. Villa, Jaime, Larry David Wilson, and Jerry D. Johnson. **Middle American Herpe tology: A Bibliographic Checklist**. Columbia, Mo.: University of Missouri Press, 198 131p. $35.00. LC 87-19115. ISBN 0-8262-0665-4.

Consisting of 782 species of amphibians and reptiles found in Mexico and Centra America, this work provides a checklist for these organisms. The geographical area consist of southeast Mexico, Yucatan, Guatemala, Belize, El Salvador, Honduras, Nicaragua, Cost Rica, and Panama. The scientific names of the species are arranged alphabetically under the respective families. In addition to the scientific name, the person who first named it, an citations to the literature are provided. These citations represent references to the majo papers, as well as those papers providing illustrations and information on geographica distribution. A few black-and-white photographs are interspersed within the list. A series o 91 colored plates of species can be found preceding the checklist. A 22-page bibliograph consisting of cited references within the text, and a scientific-name index conclude th volume.

1054. Welch, Kenneth R. G. **Herpetology of Africa: A Checklist and Bibliography of th Orders Amphisbaenia, Sauria and Serpentes**. Malabar, Fla.: Robert E. Krieger, 1982. 293p $26.50. LC 81-17233. ISBN 0-89874-428-8.

Each chapter represents a family. The species are entered under their respective gener within the family. For each species listed, the scientific name, person who first named it, dat of naming, geographical distribution, and a reference(s) to the literature are given. Appendi 1 represents a one-page list of references to the families, appendix 2 considers regiona references, while appendix 3 is a list of crocodilian and chelonian species arranged unde genera. A 61-page bibliography, an index to the genera, and a scientific-name index to th species and subspecies conclude the checklist.

1055. Welch, Kenneth R. G. **Herpetology of Europe and Southwest Asia: A Checklis and Bibliography of the Orders Amphisbaenia, Sauria and Serpentes**. Malabar, Fla Robert E. Krieger, 1983. 135p. $17.50. LC 82-12645. ISBN 0-89874-533-0.

The species are listed under their respective genera, which in turn are entered unde family; each family represents a chapter. For each of the species listed, the scientific name person who first named it, date of naming, and geographical distribution are given along wit citations to the literature. A 35-page bibliography containing those references cited in th text, an index to the genera, and an index to the scientific names of the species and subspecie complete the checklist.

Amphibia (General)

1056. Frost, Darrel R., ed. **Amphibian Species of the World: A Taxonomic and Geo graphical Reference**. Lawrence, Kans.: Allen Press and The Association of Systematic Collections, 1985. 732p. LC 85-9220. ISBN 0-942924-11-8.

This work presents the first attempt to compile an annotated checklist of world amphibi ans in this century: 4,014 species have been included. Each species is arranged under its genus which in turn is entered under its subfamily, family, and order. For each species included, th scientific name, person who first identified it, date of identification, taxonomic reference type species, geographical distribution, and a comment (taxonomic reference) are presented A list of journal abbreviations, museum abbreviations, and a scientific-name index conclud the checklist. This work is considered to be a summary of the state of the literature o amphibian taxonomy, the definitive work up to this time.

1057. Gorham, Stanley W. **Checklist of World Amphibians Up to January 1, 1970. Liste des Amphibiens du Monde D'Apres L'etat du 1er Janvier 1970**. St. John's, Canada: New Brunswick Museum, 1974. 172p. LC 78-371191.

Each species is arranged under its respective genus, which in turn is entered under family, order, and class. For each of the genera included, the scientific name, person who first named it, date of naming, and synonyms are listed. For each of the species considered, the scientific name, person who first named it, date of naming, and synonyms are given. In many cases, the geographical range by continent is included. A selected list of taxonomic references can be found in the introductory pages. An index to the families concludes the checklist. There is no scientific-name index. It should be noted that this list contains the names of recent amphibian species up to the year 1970.

1058. Harding, Keith A. **Catalogue of New World Amphibians**. Oxford, New York: Pergamon Press, 1983. 406p. $135.00. LC 83-12131. ISBN 0-08-028899-5.

Species are listed alphabetically under their respective families. Families are arranged phylogenetically. For each of the species presented, the scientific name, person who first named it, date of naming, references to the literature, and geographical distribution are given. Amphibians found throughout North and South America are covered along with the Hawaiian Islands and the West Indies. Section 2 lists the scientific name of all species under their order. The orders and the species are found under the geographical headings North America, Mexico and Central America, South America, and the West Indies. These broad geographical headings are further subdivided by name of country. Therefore, if one wished to know the name of amphibian species belonging to the order Caudata in El Salvador, one would go to the Mexico and Central America section, look up "El Salvador," and then look for the order under El Salvador. An alphabetical list of synonyms, a 56-page bibliography of references cited within the text, an author index, and a scientific-name index conclude the checklist. This list is considered to be the first checklist of amphibian species/subspecies found in the Americas.

1059. Smith, Hobart M., and Edward H. Taylor. **An Annotated Checklist and Key to the Amphibia of Mexico**. Washington, D.C.: U. S. Government Printing Office, 1948. 118p. (United States National Museum, Bulletin 194). LC 48-46438.

Each species (161 of them) is arranged under its respective genus, which in turn is entered under family, order, and class. For each of these above groups (class, order, family), the scientific name, the person who first named it, and a citation to the taxonomic literature are given. For each genus listed, the scientific name, person who first named it, a citation to the literature, genotype, and geographical range are considered. Keys to the orders, suborders, genera, and species can be found interspersed among the checklist. A table of contents lists class, orders, suborders, families, and genera as well as the biological keys. A geographical index, listing the names of the species under the states in Mexico, and a scientific-name index complete the volume.

Reptilia (General)

1060. Harding, Keith A., and Kenneth R. G. Welch. **Venomous Snakes of the World: A Checklist**. Oxford, New York: Pergamon Press, 1980. 188p. LC 80-40162. ISBN 0-08-025495-0.

Originally published as supplement number 1 to the journal *Toxicon* (1980), this source lists those species of snakes found throughout the world that are considered to be venomous. The scientific names of the species/subspecies are entered under their respective genera, which in turn are entered under subfamilies and families. The genera and species are listed in alphabetical order. In addition to the scientific name, the person who named it, alternative scientific names, person who named alternative name, date of naming, and geographical distribution are given. In addition to the above arrangement, a geographical arrangement is also present. This arrangement provides access by name of country; scientific names of snakes are listed under their respective countries. A nine-page taxonomic bibliography precedes th geographical arrangement. An author index and a combined subject, common-name, and scientific name index conclude the checklist. There are no illustrations.

1061. Iverson, John. **Checklist of the Turtles of the World with English Common Name** Oxford, Ohio: Society for the Study of Amphibians and Reptiles, 1985. 14p. $3.00. (Herpe tological Circulars, Society for the Study of Amphibians and Reptiles; no. 14).

Species and races of turtles and tortoises are listed under their genera, which in turn ar entered under the families. For each species presented, the scientific name and common nam are given.

1062. Iverson, John B. **A Checklist With Distribution Maps of the Turtles of the Worl** Richmond, Ind.: Paust Printing, 1986. 282p. ISBN 0-9617431-0-7pa. LC 87-108814.

The species/subspecies (246 species, 49 species with subspecies, 164 subspecies) ar arranged under their respective families. For each species listed, the scientific name, perso who first named it, date of naming, common name, holotype, type locality (location wher specimen was collected), geographical distribution, number of subspecies (if any), an comments on taxonomic status are presented. The above information is not given for ever species listed. In some cases, the scientific name, person who named it, date of naming, an common name is all that is given. In addition, a large geographical distribution map accom panies each species listed. Each encompasses approximately two-thirds of a page. A 20-pag bibliography, and a scientific-name index complete the checklist.

1063. King, F. Wayne, and Russell L. Burke, eds. **Crocodilian, Tuatara, and Turtl Species of the World: A Taxonomic and Geographic Reference**. Washington, D.C Association of Systematics Collections, 1989. 216p. $29.00pa. LC 89-18155. ISBN (942924-15-0.

This work represents a checklist of those 271 species of crocodiles, tuataras, and turtle found throughout the world. Each species is listed under its respective genus, which in tur is entered under subfamily, family, order, and class. Geographical distribution and taxonomi comments with citations to the literature are presented for the class, order, family, subfamily and genus. For each of the species listed, the scientific name, person who first named it, da of naming, taxonomic citation to the literature, original scientific name, type species, typ specimen, type locality, geographical distribution, taxonomic comments with citations to th literature, contributors and reviewers, status with citations to the literature, geographic code (countries, states, etc. where species occur), and common names, if any, are given. Black and-white sketches of a tuatara and many of the turtle species provide the illustrative materia A 53-page bibliography consisting of references to the literature cited within the checklis and a combined common/scientific-name index conclude the volume.

1064. Smith, Hobart M., and Edward H. Taylor. **An Annotated Checklist and Key to th Reptiles of Mexico Exclusive of the Snakes**. Washington, D.C.: United States Governmer Printing Office, 1950. 253p. (United States National Museum, Bulletin 199). LC 50-61446

Each species (approximately 679 of them) is arranged under its respective genus, whic in turn is entered under family, suborder, order, subclass, and class. For each subclass, orde and suborder, person who named it, citations to the taxonomic literature, and number c subgroups under each of the groups are given. Family information consists of person wh named it, citations to the literature, number of genera, and geographical range. For each c the genera listed, the person who named it, citations to the literature, genotype, geographic range, and number of species/subspecies are considered. For each of the species listed, th scientific name, person who named it, citations to the literature, alternative scientific name type locality, and geographical distribution are given. In addition, keys to the subclasse orders, suborders, families, genera, and species can be found scattered throughout th checklist. A listing of the scientific names of the reptiles under the states of Mexico, and scientific-name index conclude the volume.

1065. Smith, Hobart M., and Edward H. Taylor. **An Annotated Checklist and Key to the Snakes of Mexico**. Washington, D.C.: United States Government Printing Office, 1945. 239p. (Smithsonian Institution, United States National Museum, Bulletin 187). LC 45-37567.

The names of the species are arranged under their respective genera, which in turn are entered under families, suborders, orders, and classes. Information on the genera consists of person responsible for naming the group, bibliographic citation, genotype, geographical range, and number of species. For each of the species listed, the person who named it, alternative scientific names, citations to the literature, type locality, and geographical range are given. Biological keys to the genera and species can be found throughout the book in the genera and species entries. A geographic index of species, and a scientific-name index conclude the checklist.

1066. Welch, Kenneth R. G. **Lizards of the Orient: A Checklist**. Malabar, Fla.: Robert E. Krieger, 1990. 162p. $21.50. LC 89-34220. ISBN 0-89464-327-4.

Each species/subspecies is arranged under its respective genus, which in turn is entered under family. Each family constitutes a chapter. A total of 11 families are considered. For each of the genera listed, the scientific name, person who named it, date of naming, and citations to the literature are considered. For each of the species presented, the scientific name, person who named it, date of naming, type locality, geographical distribution, and a citation to the literature are given. A classification scheme of the lizards, down to family, can be found in the introductory section. A 22-page bibliography, which consists of the references cited within the text, and a scientific-name index conclude the checklist.

1067. Welch, K. R. G. **Snakes of the Orient: A Checklist**. Malabar, Fla.: Robert E. Krieger, 1988. 183p. $28.50. LC 86-27298. ISBN 0-89464-203-0.

Each species and subspecies is arranged under its respective genus, which in turn is entered under its family. Each family represents a chapter. Genus information consists of person who named it, date of naming, and citations to the literature. For each organism included, the scientific name, person who named it, citations to the taxonomic literature, type locality, and geographical distribution are given. A 22-page bibliography, an index to the genera, and an index to the species and subspecies conclude the checklist.

1068. Williams, Kenneth L., and V. Wallach. **Snakes of the World**. Malabar, Fla.: Robert E. Krieger, 1989- . 2 vols. to date. $31.50 vol. 1; $33.50 vol. 2. LC 88-1. ISBN 0-89464-215-4 vol. 1; 0-89464-216-2 vol. 2.

These two volumes represent a list of genera as well as living and extinct species of snakes found throughout the world. The names of the genera and species are associated with references to the taxonomic literature. A comprehensive bibliography can be found at the end of each volume. According to the authors, other volumes in this set will include illustrated identification keys to the snakes of the world, a summary of the literature on snake biology, and a revised classification scheme.

Volume 1. *Synopsis of Snake Generic Names.*
Volume 2. *Synopsis of Living and Extinct Species.*

ASSOCIATIONS

1069. **American Society of Ichthyologists and Herpetologists**. Dept. of Zoology, Southern Illinois Univ., Carbondale, IL 62901-6501. (618) 453-4113.

1070. **Herpetologists' League**. Texas Natural Heritage Program, Texas Park and Wildlife Dept., 4200 Smith School Rd., Austin, TX 78744. (512) 448-4311. FAX (512) 389-4394.

1071. **Scandinavian Herpetological Society (Nordisk Herpetologisk Forening)**. Esthersvej 7, DK-4600 Koge, Denmark. 53663023.

1072. **Society for the Study of Amphibians and Reptiles**. Miami Univ., Dept. of Zoology, Oxford, OH 45056. (513) 529-4901.

CHAPTER 7

THE BIRDS

Consisting of over 8,600 extant species, the birds are believed to have evolved from reptilian stock. Anatomical features include modified scales known as feathers, wings, hollow bones, and air sacs that enable them to exhibit flight. As in the case of the mammals, they are warm-blooded and therefore able to regulate their temperature internally. They also possess the inherent ability to build nests in which their eggs are laid. In addition, save for a few of the more complex reptilian species, the birds and mammals are the only vertebrates exhibiting a four-chambered heart. This type of pump is more efficient than the wo-chambered heart found in fish or the three-chambered heart found in amphibians and most reptiles. The incredible variety in coloration schemes exhibited by his magnificent group makes them a joy to study and observe.

A well-laid scheme doth that small head contain,
At which thou work'st, brave bird,
With might and main ...
In truth, I rather take it thou hast got
By instinct wise much sense about thy lot ...

—Jane Welsh Carlyle (1801-1866)
To a Swallow Building Under Our Eaves

DICTIONARIES

1073. Campbell, Bruce. **The Dictionary of Birds in Color**. New York: Viking Press, 1974
352p. $22.50. LC 73-17954. ISBN 0-670-27225-6.

Over half of this work is devoted to colored photographs of birds, and the remainde considers a descriptive account of 1,200 species. Contained within each description ar common name, habitat, coloration, morphological characteristics, behavior, diet, and breed ing patterns. The species are arranged alphabetically by their genera. Each description range in length from one-quarter to one-half of a column. The number, in bold print, at the end o the discussion leads to the colored photograph of the organism. In addition, there is a introductory chapter that describes faunal regions, origins and species, anatomy, and classi fication, as well as a description on bird families. The textural material on families can b found under their respective orders.

This is not what one would consider to be a typical dictionary. It could easily be argue that this volume belongs under handbooks, or encyclopedias. It was placed here as an arbitrar decision.

1074. Choate, Ernest A. **The Dictionary of American Bird Names**. rev. ed. Boston Harvard Common Press, 1985. 226p. $9.95. LC 84-28975. ISBN 0-87645-121-0.

Arranged by sections; then alphabetically, within each section, are listed common name and scientific names of birds. The lengths of the definitions vary depending upon the term For each genera mentioned within the scientific name section, the name of the person wh first identified the genus, the Greek or Latin derivation of the term, and other taxonomic piece of information are given. Definitions of the birds compiled within the common names sectio expound upon the derivations of the terms. The lengths of the explanations vary in this sectio as well, from two pages for the tern to two lines for the teeter bird. The terminology containe therein is based upon the sixth edition of the American Ornithologists' Union's *Check-Lis of North American Birds* (see entry 1302).

In addition to the main portion of the work, there is a biographical appendix that list Ornithologists and those associated with Ornithology. For each person listed, the birth an death dates are considered along with a thumbnail biographic sketch. The length of the sketc is usually no more than one paragraph. A five-page bibliography and an English/Lati glossary complete the work.

1075. Gotch, A. F. **Birds—Their Latin Names Explained**. New York: Blanford Press 1981. 348p. $22.50. LC 81-670192. ISBN 0-7137-1175-2.

Arranged by order and then family, this volume explains the Latin meaning of the order family, and species of 1,850 of the 8,600 bird species known. Each chapter represents a order. Inclusion was based upon popularity of the particular bird. In addition, the number o species found in each family are listed, along with habitat information on the organisms Popular names are listed first followed by their scientific names. Although the work is entitle *Latin Names Explained*, in many instances the scientific names of these creatures wer obtained from Greek roots.

At the end of the work can be found a general index that includes technical terms, name of zoologists, and animals other than birds; an index of English names (English names of al birds listed in the book); and an index of Latin names (orders, families, subfamilies, an scientific names).

1076. Jobling, James A. **A Dictionary of Scientific Bird Names**. Oxford, New York Oxford University Press, 1991. 272p. $36.11. LC 91-7675. ISBN 0-19-854634-3.

This work, arranged in alphabetical order, attempts to present the derivation and efinition of all genera names as well as the specific names of birds. In other words, the cientific name of the bird has been broken up into its two parts (genus/species) and located eparately within its alphabetical listing. No common names are listed. The genera terminology ppears with its first letter capitalized, while there is no capitalization of the specific species ame. Entries for each term consist of its derivation, usually Latin or Greek, and its meaning. n 11-page bibliography and an appendix listing genera and species names not included in e main portion conclude the dictionary. It should be noted that this is not a typical dictionary; is only useful for genera and species names of birds.

)77. Leahy, Christopher. **The Birdwatcher's Companion: An Encyclopedic Handbook f North American Birdlife**. New York: Hill and Wang, 1982. 917p. LC 82-11867. ISBN -8090-3036-5.

Although the terms *handbook* and *encyclopedia* are found in the title, this work is more f a dictionary in character and format. Terminology listed deals with the art of birding as ell as birds. Such areas as bird families, common names of birds, the people for whom they ere named, ecological principles, geographical locations, and other miscellaneous terms are cluded. Definitions can be as short as one sentence or as long as several pages. A few ack-and-white drawings as well as a number of colored plates provide the illustrations. olored plates depict certain birds, while the black-and-white drawings, in addition to presenting certain birds, also represents a number of geographical range maps. Appendix consists of all species of birds known to occur in North America, arranged phylogenetically, hile appendix 2 is a list of 140 species that are not normally found on the North American ontinent. Appendix 3 provides migratory information on a number of bird groups. A 94-page bliography completes the work.

)78. Lockwood, W. B. **The Oxford Book of British Bird Names**. Oxford, New York: xford University Press, 1984. 174p. $18.95. LC 84-4362. ISBN 0-19-214155-4.

Arranged in dictionary format, this volume contains approximately 1,500 words that late to 257 species of birds found in Britain. The listing is comprised of common and ternative names of these birds originating in Britain. The definitions, which in most cases onsider a history of the word, vary in length from one or two sentences to one column. The troductory section considers a brief history of the English language (including dialect and andard language), linguistic evolution, Germanic consonant changes, motivation and anges in bird names, the age of some bird names, standardization of these names, and lemics. A three-page bibliography follows the introduction.

Examples of some of these terms are Geir, Marrot, Hoopoe, Gowk, Brook Ouzel, onechat, etc.

)79. Weaver, Peter. **The Birdwatcher's Dictionary**. Calton, Waterhouses, Staffordshire, ngland: T & A D Poyser, 1981. 155p. $16.00. LC 78-4628. ISBN 0-85661-028-3.

This work represents a list of terms and phrases that birdwatchers in Britain are likely hear when practicing the art of birding. Slang, scientific terms, bird names, and a number f associations are included. The definitions for the most part are brief, running from one to everal sentences in length; some definitions are one paragraph. Line-drawings of a number f birds as well as some geographical range maps are interspersed throughout the pages. In ddition to the dictionary itself, there are a number of appendices. The appendices consist of n acronym dictionary, a list of bird species found in Europe and North America (arranged y North American name, accompanied by the British and scientific name), the birdwatcher's ode of conduct, and a list of species, arranged phylogenetically, that are on the British and ish lists.

ENCYCLOPEDIAS

1080. Austin, Oliver Luther. **Birds of the World: A Survey of the Twenty-Seven Orde and One Hundred and Fifty-Five Families**. New York: Golden Press, 1961. Reprint, [s.l Spring Books, 1988. LC 61-13290; gb89-18731. ISBN 0-600-55727-8.

Abounding with large colored paintings, this work represents the more common a familiar species of birds inhabiting the earth. The birds are arranged under their commo name groups, below which are stated the orders and families. A description of each bird grou follows. Size, geographical range, coloration, egg-laying, nesting attributes, incubati period, and other interesting facts are presented, in most cases, within the descriptive materi The colored paintings and geographical range maps accompany the description. The amou of information considered for each bird group varies from several paragraphs to several page A two-page bibliography and a combined common/scientific-name index conclude this wor

1081. Brooke, Michael, and Tim Birkhead, eds. **The Cambridge Encyclopedia of Orn thology**. New York: Cambridge University Press, 1991. 362p. $49.50. LC 91-214229. ISB 0-521-36205-9.

Arranged by broad categories dealing with the form, function, and behavior of bird this source represents a first-rate publication. It is not typical of other bird encyclopedias th for the most part list and describe particular species of birds. Rather, its focus is on t biological principles endemic to the bird group. It covers sections on migration, flight, fo and feeding, geographical distribution, nests, eggs, courtship displays, anatomy and physic ogy of systems, and much more. Bird groups and species are introduced in these sections illustrate biological principles. In essence, this work could easily serve as a supplement to text in ornithology. In addition, a section on modern birds lists and describes families a presents the number of species present within each family. Line-drawings, geographical ran maps, graphs, and colored photographs provide the illustrative matter throughout the informative sections. A list of ornithological organizations, a brief glossary, a list of refe ences for each of the 11 chapters, and a separate subject index, common-name index, a scientific-name index complete the source.

1082. Campbell, Bruce, and Elizabeth Lack, eds. **A Dictionary of Birds**. Calton, Englan T & A D Poyser, 1985. 670p. $75.00. LC 84-72101. ISBN 0-931130-12-3.

Although titled as a dictionary, this work takes on more of the characteristics of encyclopedia and, as a result, was placed in this category. Basically, the entries consist general subjects, common names of birds, and bird families. It is similar in format to *A Ne Dictionary of Birds* (see entry 1093). The lengths of the descriptions vary from one senten to several pages. Well-known bird species and important subjects relating to birds are give more space. For example, the section on penguins is five pages and considers, characteristic distribution, movements, food, behavior, breeding, and genera. The use of photograph drawings, and, in some cases, diagrams enhances the text. Brief bibliographies can be fou at the end of major entries. A table of bird classification, down to family, can be found on t opening pages. This classification acts as an index and leads the reader to articles concerni specific types of birds. There is no index to the volume.

1083. Gilliard, E. Thomas. **Living Birds of the World**. New York: Doubleday, 1958. 400 LC 58-10729.

Although somewhat outdated, this work nevertheless presents informative accounts bird groups. A large number of colored and black-and-white photographs grace the page The bird types are arranged under their orders. For each species mentioned, scientific na of the family, geographic range, size, breeding and nesting habits, and basic external anatom are given. A number of species are described as representative birds of a group. A bri bibliography and a combined scientific/common-name index close the work.

1084. Gooders, John. **The Great Book of Birds**. New York: Dial Press, 1975. 352p. LC 74-23534. ISBN 0-8037-3110-8.

Another of the many books containing excellent photographs of birds; many of these photographs take up an entire page. Descriptions, which consider major attributes of the family, are found under the family names of the birds. For each family mentioned, the scientific and common names of the family, geographical habitat and range, behavioral habits, nesting and breeding, and certain species of the bird group are included. Each of the family groups are represented by large, colored photographs that depict representative species. A brief bibliography and a combined common/scientific-name index are present.

1085. Grzimek, Bernhard, ed. **Grzimek's Animal Life Encyclopedia: Birds, I, II, III, Volume 7, 8, 9, 1974**. New York: Van Nostrand, 1974. 579p., 620p., 648p. LC 79-183178.

Abounding with a large number of colored plates, many of them devoted to a particular bird, this classic work considers all orders of birds found throughout the world. For each of the orders, a myriad of facts about the order, family, and specific species of birds are covered. Types of information include behavior, coloration, courtship display, nesting, egg-laying, size, geographical distribution, food, and other interesting facts. Geographical range maps and line-drawings accompany the text. A systematic classification of Aves, and an English, German, French, and Russian polyglot incorporate the species contained within each of the volumes of birds. A brief bibliography and a combined common/scientific-name index complete the work. The indexes are not cumulative; each index represents its own specific volume. This three-volume set of birds is a great place to start in gathering information about the types of birds endemic to the planet.

1086. Hanzak, J. **The Pictorial Encyclopedia of Birds**. New York: Crown, 1967. 582p. LC 68-11449.

Abounding with black-and-white photographs (some colored), this volume includes all the major bird types found throughout the world. The birds are arranged under their respective orders. For each species included, the family name, scientific name, basic external anatomy, geographical distribution, size, weight, and, in some cases, nesting and breeding information are given. A combined common/scientific-name index can be found at the end of the book. This volume is short on text and long on photographs. Some general information on each species is given as listed above. However, the bulk of this work is taken up by the photographs.

1087. Martin, Brian P. **World Birds**. Enfield, Middlesex, England: Guinness, 1987. 192p. LC 88-158045. ISBN 0-85112-891-2.

Composed of seven basic chapters, this work follows the same idea as the Guinness book of records except in this case it's for the birds (no pun intended). For example, it answers questions, with descriptive information, such as what is the rarest species of all ivory-billed woodpeckers, what birds are rarest in Britain, what is the smallest bird in the world, what is the largest tree nest in the world, and so on. Black-and-white, as well as colored photographs enhance these interesting pieces of information. Although this would not be considered your typical type of encyclopedia, it does offer some interesting facts about birds. In addition, it will answer certain questions that would be extremely difficult, if not impossible to answer through traditional means. A common-name index completes the work.

1088. Perrins, Christopher M., ed. **The Illustrated Encyclopaedia of Birds: The Definitive Guide to Birds of the World**. London: Headline, 1990. 420p. $50.00. LC gb90-17490. ISBN 0-7472-0277-X.

Following an extremely well-written and illustrated introduction, this work, by means of colored plates and brief textural material, attempts to consider all major birds of the world.

The introduction consists of topics such as anatomy, evolution, geography, ecology, migration, endangered species, and conservation. The birds, themselves, are arranged by their respective orders, followed by families and species (common-name entries). The orders, families, and species are not listed alphabetically; the index will need to be consulted in most cases. For each family of birds, there are two pages devoted to colored plates of the organisms, followed by brief descriptions of the species. The description includes common name, scientific name, range, habitat, size, and some interesting facts, such as breeding and egg-laying. Descriptions are usually no more than two paragraphs in length. Besides a world checklist of species, there is a scientific-name index and a common-name index.

This is an excellent work; the colored plates alone are worth the price.

1089. Perrins, Christopher, ed. **The Illustrated Encyclopedia of Birds: The Definitive Reference to Birds of the World**. New York: Prentice Hall Editions, 1990. 420p. $50.00. LC 90-34400. ISBN 0-13-083635-4.

Abounding with many large, colored photographs and consisting of 1,200 bird species, this volume is divided phylogenetically into three major sections. Ostriches to button quails comprise the first, plovers to woodpeckers, the second, and passerines, the third. Although interesting to browse through, there is no alphabetical arrangement, and in order to locate a particular bird or bird type, one must use the index. For each group of Aves considered, the order, family, geographical range maps, size, plumage color, voice, eggs, diet, and species (scientific and common names) are included within a rectangular box (fact box) that appears before the text of the entry.

The textural material, lengthy in many cases, consists of a myriad of facts about the group including but not limited to habitat, behavior, diet, breeding patterns, and body size. Colored photographs, showing the individual members of the bird group in a variety of poses, are interspersed within the text. Colored and black-and-white drawings can also be found scattered throughout the volume.

There is an introductory chapter on what is a bird that discusses many of the groups' characteristics, including external and internal anatomy. A two-page bibliography, a brief glossary, and a subject index complete the work. The index will indicate major sections of the text as well as the fact boxes.

1090. Robiller, Franz. **Birds Throughout the World**. Old Working, England: Gresham Books, 1979. 218p. $60.00. LC 80-481542. ISBN 0-905418-39-5.

A translation of *Vhogel in Aller Welt*, this book consists of some of the more common types of birds. They are arranged under broad geographical areas of the world. The regions are the Holaractic, Ethiopian, Oriental, Australian, Neotropic, and the barren region of Antarctica. Each section (region) discusses a number of bird species in a story-like manner. Throughout the text, geography, past history of the region, ecology, and the like are mentioned. A variety of facts are presented for the bird groups as well as particular species that have been included. Colored photographs provide the illustrative material. A four-page bibliography concludes the volume. Although this makes for interesting reading, it is not set up as a reference work. The only access given for a particular type of bird is through the table of contents. This is not the type of book for quickly finding a particular fact about a bird species.

1091. Scott, Peter, ed. **The World Atlas of Birds**. London: Mitchell Beazley, 1974. Reprint, Canada: B. Mitchell; Distributed, New York: Crown, 1989. 272p. LC 74-8575. ISBN 0-517-32159-9.

Arranged by continents, this volume includes a number of common species found within the land mass. Each of the birds included is presented under its respective geographical habitat. Therefore, areas such as tropical rain forests and pampas and llanos can be found as subheadings under South America. After a brief description of the habitat, a number of representative birds are listed. For each species included, the common name, scientific name, geographical range, nesting habits, and incubation period are considered. In addition, a photograph of the ecological niche, a colored painting of the bird, and a map of the continent showing the representative ecological niches are given. A classification scheme, a one-column glossary, and a combined common/scientific-name index complete the volume. The classification scheme is fairly comprehensive and considers orders, families, number of species and characteristics within the family, behavior, nesting and breeding habits, habitat, feeding, and food source of the species contained within the work. A pen-and-ink drawing of each bird is also included.

1092. Terres, John K. **The Audubon Society Encyclopedia of North American Birds**. New York: Knopf, 1980. 1109p. $75.00. LC 80-7616. ISBN 0-394-46651-9. Reprint, New York: Wings Books/Outlet Book, 1991. 1109p. $39.99. LC 91-21877. ISBN 0-517-03288-0.

This work considers information on all birds that either nest or have been sighted in the lower 48 states, Alaska, Canada, Greenland, Bermuda, and Baja California. In addition to entries describing particular bird species (847 birds) and families, as well as brief biographical sketches of 126 naturalists, this voluminous source considers many of the physiological, behavioral, and anatomical concepts of the group. Articles, alphabetically arranged, vary in length from one sentence to several pages. Such information as common name, scientific name, feeding habits, incubation period of eggs, coloration of the birds and their eggs, age, flight speed, hybrids, and geographical range are given for many of the species entries. Many fine line-drawings and labeled diagrams, a multitude of excellent, colored photographs (875 of them), including full-paged ones, and a number of geographical range maps enhance the discussions of the entries. A selected reading list, and a 47-page bibliography arranged by authors complete the work. This bibliography represents over 4,000 entries that were used in researching the encyclopedia. Although there is no index, the large number of cross-references within the work will prove to be useful in accessing the information. This work represents a significant contribution to the ornithological world.

1093. Thomson, Sir A. Landsborough. **A New Dictionary of Birds**. New York: McGraw-Hill, 1964. 928p. LC 64-18267.

Many of the entries contain within this source are encyclopedic in nature and therefore it has been placed in this category. Entries include subjects, common names, and scientific names of birds. Scientific entries can be found down to the family level. Most of the birds are described in at least one column; many are given several pages. General characteristics, behavior, reproduction, social organization, distribution, taxonomy, and types of species are considered for the better-known birds. Major articles on a number of general subjects can also be found. Pen-and-ink drawings, reproductions of colored paintings, and black-and-white photographs provide the illustrations.

A list of major articles on general subjects as well as on groups of birds can be found within the opening pages. An index of generic names can be found rounding off the volume. This work, unfortunately, is out of print. For a need for a source such as this, *A Dictionary of Birds* (see entry 1082), would fit the bill.

1094. Whitfield, Philip, ed. **Longman World Guide to Birds**. London: Longman, 1986. 219p. LC 87-106752. ISBN 0-582-89354-2.

Arranged by common names of the bird groups, this beautifully illustrated work encompasses many of the major birds found throughout the world. Under each bird group, the order, family, and species are given. For each species included, the common name, scientific name, geographical range, habitat, size in centimeters and inches, and a brief description dealing, in most cases, with its coloration, courtship behavior, number of eggs in a clutch, and incubation periods are considered. Colored plates of the birds described can be found on each opposing page of the text. A classification scheme of Aves, down to family, and a combined scientific/common-name index close the work.

BIBLIOGRAPHIES

1095. Coues, Elliott. **American Ornithological Bibliography**. New York: Arno Press, 1974. 1066p. (Natural Sciences in America Series). $43.00. LC 73-17794. ISBN 0-405-05704-0.

This work represents the second and third installments of *American Ornithological Bibliography*, which was reprinted from the 1870-1880 edition issued by the U.S. Government Printing Office. It was originally published in the U.S. *Geological and Geographical Survey of the Territories. Bulletin* (vol. 5, no. 2, pp. 239-330, and no. 4, pp. 521-1066). The second installment contains faunal works and papers that deal with Latin America and the West Indies, while the third installment contains citations that cover avian taxonomy, emphasizing species, genera, and families.

The citations in the third installment are arranged under the bird families and are listed by dates, oldest to most recent, while the references found in the second installment are listed by dates only. The references contained within these two installments are dated from the seventeenth, eighteenth, and nineteenth centuries.

An author and geographic index can be found at the second installment while the third installment has no index.

1096. Zimmer, John Todd. **Catalogue of the Edward E. Ayer Ornithological Library**. Chicago: Field Museum of Natural History, 1926. Reprint, New York: Arno Press, 1974. 2 vols. in 1. (Natural Sciences in America). $55.00. LC 73-17850. ISBN 0-405-05773-3.

This catalog represents a listing of those ornithological books found in the Edward E. Ayer Ornithological Library. The majority of the works (books and periodicals) listed were published in the nineteenth century; some eighteenth century and early twentieth century volumes are included. Citations to the works are arranged alphabetically by author. All necessary bibliographic information is presented for each entry. A list of periodicals included in the list can be found at the end of the bibliography. Unfortunately, there is no subject index.

INDEXES AND ABSTRACTS

1097. **Recent Ornithological Literature**, 1986- . Washington, D.C., American Ornithologists' Union, the British Ornithologists' Union, and the Royal Australasian Ornithologists Union, 1986- . Quarterly. $8.75/issue (members), $10.00/issue (nonmembers), $15.00/issue (institutional subscribers). sn 87-28298.

Although published as an individual serial, this work represents supplements to *The Emu, The Auk*, and *Ibis*. The only exception was in 1986, when it reflected supplements to *The Auk*, and *Ibis*. This work indexes approximately 900 national and international journals and provides access to those articles contained within that would interest ornithological enthusiasts in the wild-bird literature. A list of the scanned journals appears annually in the fourth supplement.

The citations are arranged alphabetically by author under broad subject headings such as behavior and vocalizations, diseases and parasites, evolution and genetics, and the like. The general biology and the taxonomy, systematics, and faunistics headings are broken down into a geographical arrangement. A few citations provide a sentence or two describing the contents of the article. New journal, revised journal, and renamed journal titles can also be found in the issues. No author or subject index can be found save for the broad subject-heading approach.

1097a. **Zoological Record**. Vols. 1- , nos. 1- . Philadelphia: BIOSIS. Zoological Society of London, 1865- . Annual. $2,200.00/yr. ISSN 0084-5604.

For a description of this index, see entry 55. The section relating to this chapter is section 18: Aves.

DIRECTORIES

1098. Ornithological Societies of North America. **The Flock: Membership Directory**. Lawrence, Kans.: Ornithological Societies of North America, Allen Press, 1991. 127p. $30.00.

This listing represents the members of The American Ornithologists' Union, The Association of Field Ornithologists, The Cooper Ornithological Society, and The Wilson Ornithological Society. For each member entered, the address, phone number, and fax number are listed (not all members will have an entry for phone number or fax number). In addition, there are codes that let the reader know what organization the member belongs to as well as the status of the member within the organization. Examples of status are benefactor, life member, patron, fellow, student, and so on. Besides the main alphabetical listing of names, there is a geographic listing found at the end of the directory. Geographic listing is by state followed by the member's zip code.

HANDBOOKS

Aves (General)

1099. Dorst, Jean. **The Life of Birds**. New York: Columbia University Press, 1974. 2 vols. $135.00. LC 74-8212. ISBN 0-231-03909-3.

This two-volume set considers in detail the characteristics of birds as a group. It discusses such parameters as anatomy and physiology, behavior, feeding habits, geographical ranges, ecology, and the like. Each chapter considers a different parameter. Among some of the chapters are colors of birds, food and feeding habits, vocalizations, courtship and sexual displays, nest building, eggs and young, deserts, tropical savannas, polar environment, island avifaunas, and so on. Volume 1 concentrates its chapters on anatomy, physiology and behavior, while volume 2 depicts the various ecological niches that birds inhabit. Specific genera and species are used to illustrate the salient points. Black-and-white photographs, line-drawings, graphs, tables, and a number of geographical range maps provide the illustrative material. Each volume lists bibliographic citations near the rear of the work. These references were cited within the text. Volume 2 contains a combined subject and common-name index to both volumes. Scientific names can be found next to the common names within the index.

1100. Dunning, John B., ed. **CRC Handbook of Avian Body Masses**. Boca Raton, Fla.: CRC Press, 1993. 371p. $65.00. LC 92-20884. ISBN 0-8493-4258-9.

The species, which are presented by their scientific names, are entered under their respective families, which in turn are entered under their orders. The book is arranged phylogenetically beginning with the order Tinamiformes and ending with the Passeriformes. Approximately 75% of the birds found throughout the world (6,283 species) are compiled within this reference source. For each of the species listed, the sex of the samples, the number of bird species within the sample, the mean of the body mass, the standard deviation, the range of body mass, the collecting season (breeding, postbreeding, spring migration, etc.), and the geographic location are given. The above information is given in a tabular format. In addition to the body masses of the 6,283 species birds found in part I of the work, the body masses and composition of migrant birds in the eastern United States can be found in the second portion of the work.

A bibliography consisting of 703 entries, and a scientific-name index conclude the handbook.

1101. Fuller, Errol. **Extinct Birds**. New York: Facts on File, 1988. 256p. $40.00. LC 87-9073. ISBN 0-8160-1833-2.

Abounding with excellent colored plates of the birds, this work considers all species that have become extinct since 1600. For each animal entered, the scientific name, common name, person who first named it, date of naming, the length, wing measurements of male and female, and an external anatomical description, when known, are given. In addition, general information regarding habitat, suggested reasons for extinction, and so on are provided. A seven-page bibliography, and a combined subject, common-name, and scientific-name index conclude this beautiful work. A good deal of information is considered within these pages, and although it presents itself as a coffee-table book, it is not. It is a scholarly work.

1102. **Handbook of the Birds of the World**. Cambridge, England; Washington, D.C.: International Council for Bird Preservation, 1992- . 10 vols. $165.00 vol. 1. ISBN 84-87334-10-5 vol. 1.

This projected 10-volume work attempts to provide information and illustrations on every bird species that exists throughout the world. Volume 1 of this set was published in 1992 and covers ostrich to ducks.

In addition to bird descriptions, volume 1 considers such avian topics as evolution, anatomy and physiology, migration, systematics, and the like. Each species is arranged under its respective genus, which in turn is entered under family. Each of the families constitute a chapter within the source. For each of the families listed, the scientific name, common name, generalized information, including taxonomic considerations, are presented. The descriptive account of each species consists of common name; scientific name; common names in French, German, and Spanish; additional common names; person who first named it; date of naming; size; coloration; habitat; food and feeding; breeding; behavior; status and conservation; geographical distribution; and bibliographic references. A geographical range map accompanies each of the descriptive accounts. Approximately three descriptive accounts can be found on each page. Colored plates and photographs depicting the birds, and geographical range maps provide the majority of the illustrations.

Volume 1 presents 640 pages, 50 colored plates of birds, 17 colored plates of anatomy, over 350 colored photographs, 550 distribution maps, and more than 5,000 bibliographical references. This is sure to be one of the major classics in avian publication history. Listed below are the first volume and the volumes that are to be published.

Volume 1. *Ostriches to Ducks.*
Volume 2. *Raptors to Bustards.*
Volume 3. *Jacanas to Parrots.*
Volume 4. *Turacos to Hummingbirds.*
Volume 5. *Mousebirds to Woodpeckers.*
Volume 6. *Broadbills to Tyrant Flycatchers.*
Volume 7. *Sharpbills to Thrushes.*
Volume 8. *Warblers.*
Volume 9. *Tits to Tanagers.*
Volume 10. *New World Warblers to Crows.*

As of this time, no dates have been established as to when each of the 10 volumes will be published.

1103. **Handbuch der Zoologie: Eine Naturgeschichte der Stämme des Tierreiches**. Siebenter Band, Zweite Hälfte: **Sauropsida: Aves**. Gegrundet von Willy Kükenthal. Hrsg. von J. G. Helmcke and others. Berlin: W. de Gruyter, 1927-1934.

This work belongs to a classic German set described in entry 68 in chapter 1. The second half of the seventh volume is devoted to the birds. The subject material concentrates mainly on the morphology, physiology, and taxonomy of Aves. Line-drawings (many of them abeled) and black-and-white photographs provide the illustrative material. A 28-page bibliography and a combined subject and scientific-name index conclude the work.

104. King, A. S., and J. McLelland, eds. **Form and Function in Birds**. London, New York: Academic Press, 1980-1989. 4 vols. $153.00 vol. 1; $195.00 vol. 3; $195.00 vol. 4. LC 79-5023. ISBN 0-12-407501-0 vol. 1; 0-12-407502-9 vol. 2; 0-12-407503-7 vol. 3; 0-12-407504-5 vol. 4.

This four-volume set, in essence, deals exclusively with the anatomy and physiology of the birds. Its depth of coverage is far greater than Dorst's *The Life of Birds* (see entry 1099). With the use of labeled diagrams, transmission electron microscopy photographs, graphs, and tables, this work discusses the various systems that are found in Aves. A lengthy bibliography can be found at the end of each chapter, representing those references cited in the text. Bird species and genera are employed to illustrate anatomical and physiological points. A subject index appears at the end of each chapter.

Volume 1 is concerned with avian morphology, coelomic cavities, digestive system, urinary organs, female genital organs, the blood cells, and the autonomic nervous system, while volume 2 concentrates on male genital organs, the cloaca, phallus, endocrine glands, cardiovascular system, lymphatic system, the cranial nerves, and the functional anatomy of the avian jaw apparatus. Volume 3 deals with the integument, locomotor system, somatic peripheral nerves, nasal cavity and olfactory system, external and middle ears, inner ear, eye, structure and function of avian somatosensory receptors, and the structure and function of avian taste receptors. Volume 4 considers the mainstreams in the evolution of vertebrate respiratory structures, the larynx and trachea, functional anatomy of the syrinx, functions of the syrinx and the control of sound production, anatomy of the lungs and air sacs, the blood supply to the lung, the morphometry of the avian lung, respiratory mechanics and air flow, physiology of gas exchange, control of breathing, and the central nervous system. Although they read like texts, they are much too comprehensive to be used in a one-semester course. They are extremely useful as supplementary material, and for those zoologists engaged in anatomical and physiological bird research.

105. King, Warren B., comp. **Endangered Birds of the World: The ICBP Bird Red Data Book**. Washington, D.C.: Smithsonian Institution Press, 1981. 1 vol. (unpaged). (Red Data Book, Vol. 2: Aves). $22.50; $12.50pa. LC 81-607796. ISBN 0-87474-584-5; 0-87474-583-7pa.

Originally published in loose-leaf format, this work lists all species of birds that are considered endangered, rare, vulnerable, or indeterminate (not sure of status at the time). Birds that are now considered out of danger are also presented. For each species listed, the common and scientific names, the order, family, status, distribution, population, habitat, conservation measures taken, conservation measures proposed, and general remarks are given. In addition, the person who first named the species is listed next to the species name. A brief bibliography accompanies each listing.

Arrangement is by families; scientific names are then alphabetically arranged within each family. Although the work does not have an index per se, there are ample lists, found in the introductory chapters, to allow one to find a specific organism. Within these introductory chapters, known as preambles, there are other lists. The arrangement of species on a zoogeographical/geopolitical basis, and a list of birds known or thought to have become extinct since 1600 are examples.

106. Lever, Christopher. **Naturalized Birds of the World**. London: Longman Scientific & Technical; New York: John Wiley, 1987. 615p. $230.00. LC 86-28727. ISBN 0-470-20789-2 (USA only).

Arranged by common name, this work attempts to list all birds that were naturalized in other parts of the world. For each avian species considered, the scientific name, a map and geographic range of its natural distribution, a map and geographic range of its naturalized distribution, and the story behind the introduction of the bird to its alien environs are included.

A pen-and-ink drawing of each species is incorporated within the above description. A sho
list of references appears at the end of each tale.

A large number of selected references dealing with naturalization, a geographical inde
and an index of vertebrate species can be found completing the work. This volume makes fo
some interesting reading.

1107. Sibley, Charles G., and Jon E. Ahlquist. **Phylogeny and Classification of Birds: Study in Molecular Evolution**. New Haven, Conn.: Yale University Press, 1990. 976p. $100.00. LC 90-35938. ISBN 0-300-04085-7.

This work compliments Sibley's *Distribution and Taxonomy of Birds of the World* (se entry 1281), in that it, among other things, discusses the various procedures and principle dealing with DNA hybridization. Essentially, the work is divided into two parts. Part considers such matters as structure and properties of DNA, gene structure and functio genetic regulation, DNA reassociation and thermal stability, sequence organization of th genome, comparative evolution, and the like. Part 2 presents the evidence taken from th DNA studies to explain the relationships between certain birds within a taxonomic grou *Distribution and Taxonomy of Birds of the World* lists all bird species in a classificatio scheme based upon these studies, while this work provides the understanding as to how th classification scheme was pieced together. In addition, graphical representations are prese in order to show the DNA hybridization relationships. A 96-page bibliography and a com bined subject, common-name, and scientific-name index conclude the work.

If interested in the classification scheme itself, then *Distribution and Taxonomy of Bir of the World* would be the choice. If interested in the methodology and DNA hybridizatio evidence that shows the relationships between the birds, then this volume is needed.

1108. **Traité de Zoologie: Anatomie, Systematique, Biologie**. Tome XV: **Oiseaux**. E by Pierre-Paul Grassé. Paris: Masson, 1950. (various fascicules).

A general description of this work appears in chapter 1 (see entry 69). Chapter 15 devoted to Aves. Such broad areas as the anatomy, biology, and taxonomy of birds a discussed. Photomicrographs, line-drawings (many of them labeled), and black-and-whi photographs provide the illustrations. References to the literature appear at the end of eac section. A combined subject and scientific-name index concludes the volume. This work written in French.

Aves (Geographic Section)

AFRICA

1109. Brown, Leslie H., Emil K. Urban, and Kenneth B. Newman, eds. **The Birds of Afric** London, New York: Academic Press, 1982- . 4 vols. to date $160.00/vol. LC 81-69594.

Representing all the birds of Africa, this projected seven-volume set outdoes its pred cessor, *African Handbook of Birds* (see entry 1111), in presentation, layouts, and illustrativ material. In addition, it contains more recent information regarding these species. Eac section is introduced by a brief description of the order, family, subfamily, and genus to whic the bird belongs. The scientific names and common names are given for the family an subfamily. For each of the species included under its respective genus, the scientific nam person who first named it, common names, bibliographic citation to the taxonomic literatur geographical location, and a geographical range map are considered. In addition, detaile descriptions of both male and female (coloration, size, weight, etc.), voice patterns, behavio food, breeding habits (eggs, laying dates, laying behavior, development and care of young and a list of references are given. Each species description is usually two or more pages lon Colored plates representing all species, as well as line-drawings provide the illustrativ material. A lengthy bibliography arranged by family, and a separate common/scientific-nam index conclude each of the volumes. It should be noted that this is a first-rate publicatio filled with facts and illustrations presented in an interesting and informative manner. It is sure to be a classic. At this time, four volumes of the set have been published. They are listed as follows:

Volume 1. *Ostriches to Birds of Prey*. ISBN 0-12-137301-0.
Volume 2. *Gamebirds to Pigeons*. ISBN 0-12-137302-9.
Volume 3. *Parrots to Woodpeckers*. ISBN 0-12-137303-7.
Volume 4. *Broadbills to Chats*. ISBN 0-12-137304-5.

1110. Collar, N. J., and S. N. Stuart. **Threatened Birds of Africa and Related Islands: The ICBP/IUCN Red Data Book, Part 1**. 3d ed. Gland, Switzerland: ICBP; Cambridge, U.K.: IUCN, 1985. 761p. $75.00. LC 85-214791. ISBN 2-88032-604-4.

For each threatened species entered, the common name, scientific name, person who first identified it, date of identification, order, family, and status are given. In addition, the descriptive account of each bird consists of distribution, population, ecology, threats to the species, conservation measures taken, and conservation measures proposed. References can be found at the end of the description.

Appendix A consists of a list of threatened bird species in Africa and related islands arranged by country. Appendix B presents a list of threatened bird species in Africa and related islands arranged by category of threat, while appendix C gives a list of near-threatened bird species in Africa and related islands. Appendix D lists a number of species that will be treated in the proposed Europe and Asia volume of the ICBP/IUCN Red Data Book; appendix E considers those species at risk or requiring monitoring in Africa; appendix F contains some notes on subspecies; and appendix G presents a list of candidate bird species for treatment as threatened. A common/scientific-name index that considers species found in the descriptions as well as appendices C, D, and E concludes the volume.

1111. Mackworth-Praed, C. W., and C. H. B. Grant. **African Handbook of Birds**. London, New York: Longmans, Green, 1952-1973. 2 vols. in 6.

This multivolume work presents the many and varied species of birds found throughout the African continent. Bird species are arranged under their respective families. The scientific name, common name, names of genera, and a brief description of the family are presented. In addition, keys to the family are included. For each species recorded, the common name, scientific name, person who first named it, and a bibliographic citation to the taxonomic literature are considered. Also, coloration, size, geographical distribution, habits, nest and egg information, breeding times, food, voice pattern, and distribution of other races of the species are given. A geographical range map accompanies each of the descriptions. Line-drawings and colored plates provide the illustrative material. A separate common/scientific-name index completes each volume. The titles of the volumes are as follows:

Volume 1, series I. *Birds of Eastern and North Eastern Africa.*
Volume 1, series II. *Birds of the Southern Third of Africa.*
Volume 1, series III. *Birds of West Central and Western Africa.*
Volume 2, series I. *Birds of Eastern and North Eastern Africa.*
Volume 2, series II. *Birds of the Southern Third of Africa.*
Volume 2, series III. *Birds of West Central and Western Africa.*

Volume 2 represents additions to the species and to the descriptive information, and is considered a second edition. Although the published form of this work is confusing, it does represent a major work in bird literature.

AMERICAS

1111a. Collar, N. J., and others. **Threatened Birds of the Americas: The ICBP/IUCN Re Data Book, Part 2**. 3d ed. Washington, D.C.: Smithsonian Institution Press in cooperatio with International Council for Bird Preservation, 1992. 1150p. $75.00. LC gb93-4257. ISBI 1-56098-267-5.

This work represents a continuation of the ICBP/IUCN Red Data Books, part 1 bein *Threatened Birds of Africa and Related Islands* (see entry 1110). Taking into account Nortl Central, and South America as well as adjacent islands, this work provides informatio regarding species of birds that are threatened within the above geographical range. For eac bird listed, the common and scientific name are given. In addition, the descriptive accou considers geographical distribution, population, ecology, threats, measures taken and prc posed to combat the threat, and general remarks. These descriptions are usually from one t three pages in length. Numerous references are cited within the account. These references ca be found in the bibliography section near the end of the volume. A black-and-white drawin of some of the bird species can be found within the description. Appendices found near th end of the volume enhance the utility of the text. Such arrangements found in these appendice consist of the list of birds arranged by category of threat and by country. Near-threatene birds of the Americas are also considered. An 80-page bibliography and a combine common/scientific-name index conclude the work.

ANTARCTICA

1112. Parmelee, David Freeland. **Antarctic Birds: Ecological and Behavioral Approaches Exploration of Palmer Archipelago**. Minneapolis, Minn.: University of Minnesota Pres 1992. 203p. $38.95. LC 91-12378. ISBN 0-8166-2000-8.

Consisting of approximately 30 species of birds from 11 families, this source concer trates on those creatures found throughout Antarctica. Species listed are entered under thei respective families. Preceding the descriptive accounts, general ecological and behavior facts on the family including geographical distributions are given. For each of the birds liste the common name, scientific name, overall breeding distribution, and current status in th Palmer Archipelago are given. In addition, a number of species accounts also conside information regarding the colonies, number of birds in winter, molting, predation, breedin biology (nest sites, nests, density, copulation, egg-laying, clutch size, hatching, guard an creche periods, synopsis of the annual cycle), feeding behavior, colorations, predator behavior, and mortality. The lengths of the descriptive accounts vary from less than 1 pag to 14 or more pages. Black-and-white as well as a number of magnificent colored photograph depicting the birds provide most of the illustrations. A number of geographical range map are also present. The introductory material consists of the geography of the continent and checklist of the Palmer Archipelago birds.

An appendix listing the coastal observations beyond the Palmer study area, a seven-pag bibliography, and a separate geographic index as well as a combined common/scientific-nam index conclude the volume.

1113. Watson, George E. **Birds of the Antarctic and Sub-Antarctic**. Washington, D.C. American Geophysical Union, 1975. 350p. $18.00. LC 75-34547. ISBN 0-87590-124-7.

This work includes those species that are found in and around Antarctica, the fifth larges of the continents. After some general information concerning the avifauna, land and se environment, and climate of Antarctica, the list of bird species begins. Species are arrange under their respective families. For each bird included, the common name, scientific name whether it is a resident or a vagrant, size (in cm and in.), coloration in adult and juveniles behavioral patterns, vocalization patterns, food, eggs (size, color, weight), nesting, incuba tion, hatching (time of year, incubation and brooding period), arrival month, departure molting, predation and mortality, ectoparasites, and geographical distribution are given. I ddition, a description of the family is also considered. Line-drawings of many species, olored plates of birds, and geographical range maps provide the illustrative matter.

In addition to the species accounts listed above, there is a section on geographical ccounts. These accounts describe the various land areas and islands making up the continent. 22-page bibliography, a list of variant names of the species, and a separate common/scientific-ame index complete this work. This source represents a major endeavor in bird literature.

ASIA

114. Gould, John, and A. Rutgers. **Birds of Asia**. New York: Taplinger, 1968. 321p. LC 7-79857.

Follows the same format as the *Birds of Europe* by John Gould (see entry 1126).

AUSTRALIA

115. Blakers, M., S. J. J. F. Davies, and P. N. Reilly. **The Atlas of Australian Birds**. arlton, Victoria, Australia: Melbourne University Press, 1984. 738p. $34.95. LC 84-229290. SBN 0-522-84285-2.

Using range maps for each species entered, this work attempts to encompass all birds ound in Australia. In addition to the maps that depict sites of observation and breeding, the xtural material considers common and scientific names, small drawings of the birds, opulation ranges, and migratory and breeding patterns. Observations that record colony size re included. There is one page devoted to each species mentioned. The organisms are not rranged alphabetically by their common or scientific names; it is important that one uses the dex in order to gain access to the species.

The introductory chapter discusses how the book was compiled. The closing sections clude a list of uncommon and vagrant species (scientific and common names are given as ell as geographic range); a series of historical maps showing the number of species recorded roughout the century; a nine-page bibliography for the historical-map section; a procedure r the calculation of reporting rates for field atlas maps; acknowledgements, contributors, nd artists; a bibliography containing 1,929 entries; and an index combining both scientific nd common names.

116. Marchant, S., and P. J. Higgins, eds. **Handbook of Australian, New Zealand, & ntarctic Birds**. Melbourne, New York: Oxford University Press, 1990- . 1 vol. in 2 1400p.). $295.00. LC 91-209529. ISBN 0-19-553244-9.

This work represents volume 1 of a multivolume set yet to be published. According to e preface, volume 2 (*Birds of Prey to Terns*) is being written at this time.

The species are arranged under their respective families. Included is a good description f each family, and a few bibliographic references are given before the species themselves. or each of the birds considered, the scientific name, common name, taxonomic citation to e literature, other common names, derivations of the scientific and common names, and a ynopsis of the subspecies are given. In addition, size, coloration, similar species, habitat, istribution, population, movements, food, social organization, breeding dispersion, social ehavior, voice patterns in the adult and young, nesting information such as nest materials, ggs, clutch size, laying and incubation, plumages in the adult and young, bare parts, molts, easurements, weights, structure, sexing, aging, geographical variations, and an adequate list f references are presented. This descriptive information varies in length from a few to seven r more pages. It is detailed but highly readable. An elliptical geographical range map ccompanies each of the descriptions. Colored plates representing each of the various bird roups abound throughout the text.

The appendices consist of ectoparasites of Australian, New Zealand and Antarctic birds, mily affiliations of genera mentioned in the sections on food (food eaten by the birds is iscussed), aboriginal names of the species, maori names, and foreign names. A separate scientific/common-name index concludes volume 1. It should be pointed out that this is a first-class publication. Listed below are the parts that compose volume 1.

Volume 1, part A. *Ratites to Petrels.*
Volume 1, part B. *Pelicans to Ducks.*

Look forward to other volumes in this set. It is sure to be a classic.

1117. Simpson, Ken, and Nicolas Day. **The Birds of Australia: A Book of Identification 760 Birds in Colour**. Wolfeboro, N.H.: Tanager Books, 1986. 352p. $45.00. LC 84-8812 ISBN 0-88072-059-X.

Unlike *The Atlas of Australian Birds* (see entry 1115), this source abounds with colored paintings of species. The birds are arranged by families. For each animal listed, the common name, scientific name, range map, and a brief description of its coloration, external characteristics, races (when applicable), and habitat are given. The colored paintings complement the text. In addition to the main portion of the source, there is a section entitled "The Handbook." This part of the work considers the life cycle of birds, avifaunal regions, prehistoric birds, vegetation and landform habitats of Australia, hints for bird watchers, and further explanations of the families (food intake, breeding, etc.). This handbook section complements family descriptions found in the main portion. A glossary and a separate common/scientific-name index can be found concluding the book. The paintings, themselves, are worth a thousand words.

BAHRAIN

1117a. Nightingale, Tom, and Mike Hill. **Birds of Bahrain**. London: Immel, 1993. 283p. $76.50. ISBN 0-907151-79-5.

General information is provided for the birds of Bahrain in three major sections of the work. The "Birds Throughout the Year" section is arranged by month of the year followed by the types of birds (common names given) encountered during each of the months. The "Breeding Birds of Bahrain" section considers approximately 27 species that breed within Bahrain. For each species listed, the common name, scientific name, population numbers, nesting sites, breeding season, clutch size, incubation period, feeding habits of the young, fledging period, age of first breeding, and other general information is considered. Amount of information varies from one to two columns in length. The "Systematic List of the Birds of Bahrain" section includes approximately 290 species. The species are listed under their respective family. For each species listed, the common name, scientific name, breeder and migrant information, site encounters, and geographical distribution is given. In many cases coloration is also considered. A wealth of colored plates depicting the bird species can be found throughout the entire work; especially within the three sections listed above. Introductory chapters consist of the physical geography of Bahrain, migration of the birds, and where to look for birds throughout Bahrain. Appendices consist of species recorded as having probably occurred in Bahrain, rejected species (those not occurring in the Systematic List) and a list of species in the collection of Al-Areen Wildlife Park at the beginning of 1990. A brief bibliography, a common-name, and a scientific-name index conclude the source.

BRAZIL

1117b. Sick, Helmut. **Birds in Brazil: A Natural History**. Princeton, N.J.: Princeton University Press, 1993. 703p. $95.00. LC 92-19971. ISBN 0-691-08569-2.

Originally published in 1984 as *Ornitologia Brasileira—Uma Introducao*, this work represents approximately 16% of all bird species found throughout the world and about 55% of all bird species found within South America. Approximately 1,635 species, representing 91 families and 23 orders found within Brazil, its territorial waters and Atlantic islands have been included.

Each species is arranged under its respective families, which in turn is entered under its order. The scientific and common names are given for the families along with general information describing the family as a whole. Such family information as morphology, special adaptations, behavioral patterns, breeding, migration, mortality and the like are considered in most cases. In some instances, information on the order, suborder, and superfamily is also presented. The amount of information and the length of the description varies greatly from one species to another. The length of these descriptions range from several sentences to a page. For each species listed, the common name, scientific name, and person who first named it are given. Information such as size, coloration, weight, taxonomic considerations, habitat, behavioral characteristics, and geographical distribution is also considered in many cases. Black-and-white drawings are depicted for a number of the species. Geographical range maps and illustrative diagrams are interspersed within the species account section. Bibliographic lists pertaining to families are also present within this section. Colored plates of many of the birds can be found following the index.

Introductory information consists of the climate and terrain of Brazil, habitats of Brazilian birds, a short history of Ornithology in Brazil, conservation in Brazil, biogeography and speciation of the birds, and general information relating to birds of the neotropics, fossil birds, categories of Brazilian birds, population analyses and biodiversity. An illustrated guide to the orders and families of Brazilian birds separates the introductory section from the species accounts.

A seven-page bibliography, and a separate scientific/common name-index can be found near the end of the volume.

CANADA

1118. Godfrey, W. Earl. **The Birds of Canada**. rev. ed. Ottawa: National Museum of Natural Sciences, National Museums of Canada, 1986. 595p. $39.95. LC C85-097101-2. ISBN 0-660-10758-9.

Species listed herein (578 of them) can be found under their respective orders, followed by families. A brief description of the order and family are given before the species are discussed. For each species entered, the common name, French name, scientific name, person who first named it, an external anatomical description, measurements of adult male and female, distinguishing marks, habitat, nesting information, geographical range, a list of subspecies, behavioral patterns, and a geographical range map are given. Color plates of these birds abound throughout the book. A glossary, a list of selected references arranged under provinces, and a combined common/scientific-name index can be found closing the volume. This is a first-class work.

CENTRAL AMERICA

119. Skutch, Alexander F. **Birds of Tropical America**. Austin, Tex.: University of Texas Press, 1983. 305p. (The Corrie Herring Hooks Series; 5). $29.95. LC 82-8597. ISBN -292-74634-2.

Arranged by the common names of the birds, this source attempts to include all those irds found in the tropics of Central and Latin America. For each species included, a lengthy escription of the bird follows (anywhere from 2 to 20 or more pages). Descriptive informa-on contains size, coloration, nesting and feeding habits, behavioral patterns, incubation, and ne like. The descriptive material does not contain many subheadings. As a result, one may ave to read through a portion of it to find the information desired. A four-page bibliography nd a combined common/scientific-name index complete the book.

EYLON (see SRI LANKA)

CHINA

1120. De Schauensee, Rodolphe Meyer. **The Birds of China**. Washington, D.C.: Smit sonian Institution Press, 1984. 602p. $37.00pa. LC 83-10314. ISBN 0-87474-363-X.

Arranged under their respective families, this work considers all the birds endemic China. For each species included, the common name, scientific name, coloration, geograph cal distribution, and references to the colored plates of birds are given; plates are located front of the work. In addition, a brief description of each family is presented. Common an scientific names of the families are provided. 1,195 species of birds, belonging to 88 familie are presented.

The colored plates are arranged by common-name group and provide references to th descriptive information. In addition, a number of black-and-white drawings of many bir can be found interspersed throughout the text. A five-page bibliography, a list of varia names, a checklist of the birds of China, and a separate common/scientific-name index ca be found ending this comprehensive guide.

1121. Tso-Hsin, Cheng. **A Synopsis of the Avifauna of China**. Beijing, China: Scien Press, 1987. 1222p. $163.00. LC 88-194315. ISBN 0-685-17938-9.

The species are arranged under their respective genera, which in turn are entered und families and orders. In addition to English names, the Chinese names are given for the orde family, genus, and species. For each species entered, the scientific name, person who name it, common name (in English and Russian), citation to the literature, breeding habita geographical range, and status are given. Geographical range maps accompany the speci account. A total of 1,186 species and 953 subspecies are listed. The geographical range ma provide the only illustrative matter. Near the end of the source can be found a distributi table, a gazetteer (English and Chinese phonetic alphabets, and Chinese characters), a listin of geographical place names including mountains and rivers that were cited in the text, seven-page bibliography listing citations in Chinese, a 30-page bibliography listing citatio in other languages, and separate indexes to the birds written using Chinese characters, Englis names, and scientific names.

COLUMBIA

1122. Hilty, Steven L., and William L. Brown. **A Guide to the Birds of Colombi** Princeton, N.J.: Princeton University Press, 1986. 836p. $99.50; $45.00pa. LC 84-1821 ISBN 0-691-08371-1; 0-691-08372-Xpa.

Arranged by family, this work provides a compendium of bird species found in Colon bia. After a brief description of the family in general, the species are listed. For each bi included, the common name, scientific name, identifying characteristics, similar specie voice, behavior, breeding, status and habitat, and range are listed. This information is bri and usually takes up no more than one-third to one-half of a column. The middle section the book contains mainly colored plates of the birds mentioned within the text. The introdu tory chapter considers the topography, climate, vegetation, habitat descriptions, migrant conservation, and a list of national parks of Colombia. The ending of the book contains series of appendices, among which are information on finding birds in Colombia, a list birds of Isla San Andres and Isla Providencia, a listing of the species illustrated on the plate a somewhat lengthy bibliography, a compilation of small range maps (20 per page) associate with the common names of the birds, and a separate index to common names and scientifi names.

Although twice as large, this source follows the same basic format as *Birds of Ne Guinea* (see entry 1136).

:OSTA RICA

123. Stiles, F. Gary, and Alexander F. Skutch. **A Guide to the Birds of Costa Rica**. Ithaca, [.Y.: Comstock; division of Cornell University Press, 1989. 511p. $65.00; $35.00pa. LC 8-43444. ISBN 0-8014-2287-6; 0-8014-9600-4pa.

Arranged by orders followed by families, this work considers over 830 species of birds ›und in the country of Costa Rica. The birds, under their respective families, are entered by ıeir common names. For each species listed, the common name, scientific name, description ;ize and color), habits, voice, type of nest and number of eggs, status, geographical range, nd plate number are given. In addition, a description of the family itself is given. An lustrated glossary of anatomical terms used in the descriptions can be found in front of the ird listings. The colored plates can be found in the center of the book, consisting of common ames and a brief description of the species on one side of the page and the colored picture f the birds on the opposing page. Besides the bird listings, other pieces of information :garding Costa Rica include geography and climate, avian habitats, avifauna, conservation, nd birding and birding localities. A three-page bibliography, and a combined common/ :ientific-name index conclude the guide. The plate numbers are included in the index.

:GYPT

124. Goodman, Steven M., and others. **The Birds of Egypt**. Oxford, New York: Oxford niversity Press, 1989. 551p. $125.00. LC gb88-49585; 88-22546. ISBN 0-19-857644-7.

Arranged in two parts, this work considers those bird species found throughout Egypt. art 1 considers such areas as geography, conservation, bird hunting, and environmental hanges as they pertain to Egypt. Part 2 is a listing of the species. The birds are listed under ıeir respective families, which in turn are listed under the orders. For each animal listed, the ommon name, scientific name, local name (in Arabic), subspecies, habitat, and geographical ınge are given. Colored plates of the birds as well as geographical range maps provide the lustrations. It should be mentioned that this work emphasizes the geographical range of each pecies. A 26-page bibliography, and separate indexes of scientific names, common names n English), official Egyptian Arabic names (transliterated), and official Egyptian Arabic ames (in Arabic script) conclude this work. It should be noted there are only seven pages of olored plates; that is the entire set of illustrations, save for the range maps.

:UROPE

125. Avon, Dennis, and Tony Tilford. **Birds of Britain and Europe in Colour**. London: lanford Press; New York: Sterling, 1989. 176p. $12.95pa. LC 90-159894. ISBN 0-7137-067-0.

Abounding with magnificently colored photographs, this work considers 54 species of irds found throughout Europe, England, and Ireland. For each species included, the common ame, scientific name, family name, size, coloration, basic external anatomy, voice patterns, eographical distribution, habitat, nesting, and egg information are given. In addition, a olored photograph of the bird and a geographic range map are present. A combined ommon/scientific-name index concludes the work. For a more comprehensive treatment, one ıay wish to consider *The Breeding Birds of Europe* (see entry 1127).

126. Gould, John, and A. Rutgers. **Birds of Europe**. London: Methuen, 1966. 321p. LC 6-8383.

A two-page spread is devoted to each of the birds listed in this work. One page presents escriptive information while the other contains a colored painting of the species. The escription information consists of common name, scientific name, coloration, size, geo- raphical range, nesting habits, egg-laying, color of eggs, and voice. Not all these parameters vill be given for each species contained within this volume. This is an easy read and only erves as an introduction to a number of the birds found in Europe. There is no index and the table of contents must be used to access these species. The birds are listed under the commo name of their families within the table.

1127. Pforr, Manfred, and Alfred Limbrunner. **The Breeding Birds of Europe**. Londo Croom Helm, 1983. 2 vols. $48.00 set. LC 81-208928. ISBN 0-88072-024-7 set.

Translated by Richard Stoneman and edited by Iain Robertson, this source presen breeding information on birds endemic to the European continent. The birds are entered unde their common names, each being represented by a two-page spread. For each species liste the common name, scientific name, geographical distribution, breeding information, an other facts such as length, wing length, weight, voice, breeding period, size of clutch, colc of eggs, size of eggs, incubation, and fledgling period are given. In addition, a full-pag colored photograph of each bird found in its natural habitat can be found on the opposin page of the descriptive material. Smaller photographs of birds and their eggs are present o the page containing the description. A separate common/scientific-name index conclude each volume. The volumes are as follows:

Volume 1. *Divers to Auks.*
Volume 2. *Sandgrouse to Crows.*

This work represents a goodly amount of information regarding breeding behavior. Th colored photographs are excellent. Colored geographical range maps accompany each specie discussed.

GREAT BRITAIN

1128. Hollom, P. A. D. **The Popular Handbook of British Birds**. 5th ed., rev. Londo H. F. & G. Witherby, 1988. 486p. $34.95. LC 89-164612. ISBN 0-85493-169-4.

Based on the five-volume set *The Handbook of British Birds* (see entry 1129), this wor considers the various types of birds found in England, Scotland, Wales, and Ireland. It is shortened version of the above set. Species are arranged under their common names. For eac bird included, the scientific name, coloration, size plumage differences, habitat, behavio food, nesting information (color and number of eggs, incubation period), status, and distr bution are given. References are provided to guide the reader to the colored plate of the bir and the colored plate of the eggs. Approximately two pages are devoted to each bir Black-and-white drawings of many of the species, as well as a series of colored plates of th birds and a series of colored plates of the eggs provide the illustrations. A systematic list c species, allowing access by common name, can be found on the opening pages. A combine scientific/common-name index completes the work.

1129. Witherby, H. F., ed. **The Handbook of British Birds**. London: H. F. & G. Witherby 1938-1941; Reprint, 1965. 5 vols.

See *The Popular Handbook of British Birds*, 5th ed. (see entry 1128), for basi information contained within this set.

Volume I. *Crows to Firecast.*
Volume II. *Warblers to Owls.*
Volume III. *Hawks to Ducks.*
Volume IV. *Cormorants to Crane.*
Volume V. *Terns to Game-Birds: Systematic List and Indices.*

HAITI

1130. Wetmore, Alexander, and Bradshaw H. Swales. **The Birds of Haiti and the Domini can Republic**. Washington, D.C.: U.S. Government Printing Office, 1931. 483p. (Smith sonian Institution, United States National Museum, Bulletin 155). LC 31-26685.

The species are arranged under their respective families, which in turn are entered under orders. For each of the species listed, the scientific name, person who first named it, common names, additional scientific names, citations to the taxonomic literature, habitat, behavioral patterns, size, diet, coloration, egg information, and geographical distribution are given, in most accounts. The amount of information present in each description varies; some of the above information is absent. The accounts are usually one page in length. Black-and-white plates depicting a number of the birds provide the illustrative material. The introductory chapters consist of the geography of Haiti, ornithological investigations, and a general discussion of the birds found throughout the island chain. An 11-page bibliography, and a combined common/scientific-name index conclude the volume.

INDIA

1131. Ali, Salim, and S. Dillon Ripley. **Compact Handbook of the Birds of India and Pakistan: Together with Those of Bangladesh, Nepal, Bhutan and Sri Lanka**. 2d ed. Oxford, New York: Oxford University Press, 1987. 737p. $98.00. LC 89-101301. ISBN 0-19-562063-1.

Contains the text of the second editions of volumes 1-3 and the first editions of volumes 4-10 of *The Handbook of the Birds of India and Pakistan: Compact Edition*.

Species are arranged under their respective families. For each species covered, the common name, scientific name, person who first identified it, citation to the literature, and additional common names are considered. In addition, the description of the bird consists of size, coloration, status, distribution, habitat, behavioral patterns, food, voice, color of bare parts, and breeding. Pictures of the heads of birds, geographical range maps, and 104 plates of birds (most are colored) provide the illustrative matter. A bibliography, common-name index, and scientific-name index complete the work. Because this is the compact edition, four pages can be found on each page within this book.

JAPAN

1132. Brazil, Mark A. **The Birds of Japan**. London: Christopher Helm, 1991. 466p. $49.95. LC 90-62321. ISBN 1-56098-030-3.

Interspersed within a series of line-drawings, this work considers those birds found in Japan. A total of 583 species are considered. Species are arranged under their respective families. For each bird mentioned, the common name, scientific name, and Japanese name are given. A brief description considering habitat, geographic range, and breeding characteristics follows. The lengths of the accounts differ from one-quarter to one full page. In addition, a series of colored plates of many of the birds can be found throughout the source. Range maps for each species mentioned are contained in the appendix. A bibliography, consisting of 1,286 entries, a Japanese-name index, and a combined common/scientific-name index conclude the work. Except for the drawings and plates, no identifying anatomical characteristics are considered in the text. Introductory chapters dealing with climate, ocean currents, avifauna, distribution, migration, and bird watching in Japan are present.

KENYA

1133. Lewis, Adrian, and Derek Pomeroy. **A Bird Atlas of Kenya**. Rotterdam, Brookfield, Vt.: A. A. Balkema, 1989. 620p. ISBN 90-6191-716-6.

Arranged under family names, this source includes species found in Kenya. For each bird listed, the common name, scientific name, and a brief description of its habitat and geographic location are considered. Breeding is mentioned in some species. A series of range maps and an occasional black-and-white photograph can be found interspersed within the text. A useful bibliography, and a separate common/scientific-name index conclude the book. Although some works are titled with the word *atlas*, they are in the encyclopedic category.

This source is indeed an atlas as it shows the distribution of Kenya's birds in fine detai Identifying characteristics, such as color, external anatomy, and so on are not presented.

MADAGASCAR

1134. Langrand, Olivier. **Guide to the Birds of Madagascar. Guide des Oiseaux d Madagascar**. New Haven, Conn.: Yale University Press, 1990. 364p. $50.00. LC 90-3066 ISBN 0-300-04310-4.

After a lengthy introduction (70 pages), of which can be found sections on the natur habitat, avifauna, bird community analysis, protected areas, and potential species in th vicinity of the coast of Madagascar, and so on, the listings of the birds commence. The specie are arranged under their respective orders, followed by classes. For each organism considere the common name, scientific name, French name, name of person who first identified it, da of identification, identifying characteristics, habitat, behavior, diet, nesting, distribution, an status are given. This descriptive material usually takes up one column on a page. Colore plates can be found in the center of the book. A list of distribution maps, an index of Malagas names, and a separate index of common and scientific names can be found concluding th volume.

NEPAL

1135. Inskipp, Carol, and Tim Inskipp. **A Guide to the Birds of Nepal**. 2d ed. Washingto D.C.: Smithsonian Institution Press, 1991. $55.00. 400p. LC 91-60470. ISBN 1-56098-097-

The 836 species contained herein are arranged by common group names, such as grebe game birds, ducks, kingfishers, parrots, and so on. For each species listed, the common nam scientific name, name of person who first recorded it, other common names, subspecie geographical range, and habitat are given. Each description is usually no more than one-qua ter of a column. Geographical range maps of Nepal and line-drawings of the specie accompany the descriptive information. The range maps contain symbols that are used t represent where the specimen has been recorded, as well as where it was recorded in th breeding season. In addition, possible breeding and proved breeding locales are considere Eight pages of colored plates can be found in the middle of the source. Each page of plate contains a number of different species.

The opening section of this source, which encompasses about 80 pages, conside topography, climate, vegetation, bird distribution and conservation, protected areas, migra tion, ornithological history, bird-watching areas, identification section, and a key to the map The identification section covers some of the more difficult groups of species and is illustrate with line-drawings. Species are grouped under common-name groups. Facts given in th section are common and scientific names, length, coloration, and, in many cases, wing lengt and span.

A bibliography containing 819 entries, and a separate common/scientific-name inde conclude the book. Material cited in the text can be found in the bibliography.

NEW GUINEA

1136. Beehler, Bruce M., Thane K. Pratt, and Dale A. Zimmerman. **Birds of New Guine** Princeton, N.J.: Princeton University Press, 1986. 293p. (Wau Ecology Institute, Handboo No. 9). $80.00; $39.50pa. LC 85-42673. ISBN 0-691-08385-1; 0-691-02394-8pa.

Beginning with a section on Papuan natural history, this source contains the various bir species of New Guinea. The birds are arranged by families under their common names. Afte a brief description of the family, the organisms are listed. For each organism included, th common name, scientific name, description, similar species, habits, voice, and range a listed. This descriptive material is brief and usually takes up no more than one-third of column. Line-drawings enhance the description. Most of the plates, found in the center of th lume, are colored and depict the species mentioned in the text. A gazetteer of New Guinea, brief bibliography, and a combined common/scientific-name index conclude the volume.

EW ZEALAND

37. Harvey, Bruce. **A Portfolio of New Zealand Birds**. Wellington, New Zealand: A. H. A. W. Reed, 1970. 61p. LC 71-138066. ISBN 0-80480-666-7.

Not nearly as comprehensive in the number of species contained or the length of the scriptive information as *Buller's Birds of New Zealand* (see entry 1138), this volume nsiders a representative number of bird species found throughout New Zealand, a total of . Birds are arranged by their common names. For each of the species included, the scientific me, geographical range, food, breeding and nesting habits, voice pattern, and coloration e supplied. A magnificent painting of the bird can be found on the page opposing the scriptive material. The birds contained within are the northern blue penguin, grey-faced trel, pycroft's petrel, flesh-footed shearwater, pied shag, blue heron, australiasian harrier, uthern black-backed gull, red-billed gull, white-fronted tern, new zealand pigeon, kaka, llow-crowned parakeet, long-tailed cuckoo, morepor, kingfisher, rifleman, pipit, grey arbler, north island fantail, silvereye, pied tits, bellbird, tui, and the north island saddleback. e table of contents serves as the index.

38. Turbott, E. G., ed. **Buller's Birds of New Zealand**. Christchurch, New Zealand: hitcombe & Tombs, 1967. 261p. LC 67-20253.

Reproducing the 48 stone-plate lithographs from the second edition (1988), this work s been edited and brought up to date by E. G. Turbott. It was originally published as *A story of the Birds of New Zealand* and written by Sir Walter Lawry Buller. The birds are ted under their common names. The scientific names of the species and subspecies, rson(s) who first identified them, dates of identification, and geographical distributions are ated. In this edition, the descriptive material is separated into two sections, that of the editor this work and that of Buller's original text. General taxonomic relationships, geographical nges, status, colorations, voice patterns, courtship displays, diets, and nesting behaviors can discerned from the two sections of text. The colored plates representing the species are cellent. In addition, there is a table of contents listing the common and scientific names of e birds, as well as a list of the plates. A short bibliography and a combined common/ ientific-name index close this folio-sized volume.

ORTH AMERICA

39. American Ornithologists' Union and Academy of Natural Sciences of Philadelphia. **rds of North America: Life Histories for the 21st Century**. Washington, D.C.: American nithologists' Union, 1992- . Bimonthly. $175.00/vol. $1,875.00 (complete set, 18 vols.). C 93-641491. ISSN 1061-5466.

Beginning in 1992 and finishing in 2003, this monumental monographic series will nsider over 700 species of birds that breed in the United States (including Alaska and awaii) and Canada. A total of 14,000 pages is expected when the work is completed.

Each monograph within the series represents a detailed scientific account of a species bird. In essence, each monograph represents a detailed reference source for that particular ecies. For each species of bird considered, the scientific name, French name, Spanish name, mmon name, order, and family are given. A colored photograph of the bird and a colored ographical distribution map (breeding and wintering ranges throughout the hemisphere) pear on the first page of the monograph.

The species account contained within each monograph includes a general introduction, loration, basic identifying external anatomy, size of males and females, geographical stribution, historical changes in distribution, geographic variation, subspecies, migrational tterns and behavior, habitat, feeding behavior, diet, nutrition and energetics, metabolism d temperature regulation, drinking, pellet-casting, defecation, food selection and storage, vocalization patterns, various forms of behavior (agonistic, sexual, locomotion, predatior etc.), breeding (physiology, phenology, nest site, nest construction, size of eggs, egg-laying incubation, hatching), growth and development of the young, parental care, demography an populations, genetic structure, population status and regulation, conservation and management, molts and plumages, and bare parts. A number of line-drawings provide the illustration within each of the monographs. For example, when discussing vocalization patterns, sonogram of the voice accompanies the description. Bibliographic references to the literatur as well as several appendices complete the monograph. The appendices consist of table providing information on mass measurements and linear measurements of the species.

Although it is impossible to relate the size of each monograph, the monograph on th European starling consists of 24 pages, of which 3 pages are devoted to the bibliography Each monograph is being written by a leading North American avian biologist.

Not since A. C. Bent's monumental work *Life Histories of North American Birds* (se entry 1144), begun in 1912, has anything of this magnitude been attempted. This work, whic contains the latest information on each North American species, is sure to become the classi work on the birds of North America.

1140. Audubon, John James. **The Birds of America**. New York: Macmillan, 1937. Reprin New York: Macmillan, 1985. 1 vol. (unpaged). LC 85-3030.

Short on description, this classic work is known for the fantastic colored paintings c the birds by John James Audubon. The species, which are endemic to the North America continent, are arranged under their common names. For each bird included, a full-size paintin and a short description are given. Each of these entries occupies one full page. In addition t the scientific name, the description consists of range, habitat, size, coloration, food, and voic Not all these parameters are given for each bird. The descriptions are no more than sever sentences in length and appear at the bottom of the page. There are a total of 435 plate several species appear on many of the plates. A common-name index concludes this famou collection.

1141. Baird, Spencer F. **The Birds of North America: The Descriptions of Species Base Chiefly on the Collections in the Museum of the Smithsonian Institution**. Philadelphi Lippincott, 1860. Reprint, New York: Arno Press, 1974. 2 vols. in 1. (Natural Sciences i America Series). $73.00. LC 73-17799. ISBN 0-405-05715-6.

Species are entered under their respective genera, which in turn can be found und families and orders. For each bird entered, the common name, person who first named it, da of naming, and citations to the literature are presented. In addition, the descriptive accoun discuss coloration, plumage, size of nostrils, length of the wing, length of the tail and th length of the bird itself, as well as geographical range. As in the case of *Key to North America Birds* (see entry 1270), the descriptions are very detailed, although *Key to North America Birds* provides more specific details than this source. A comprehensive bibliography and separate common/scientific-name index complete the volume. The bibliography contain those citations appearing throughout the text. The citations refer to the original taxonomy an identification of each species. Unfortunately, there are no illustrations.

It should be noted that this volume only includes those species that were part of th collection of the Smithsonian Institution, while *Key to North American Birds* attempts include all species occurring in North America.

1142. Baird, Spencer F., T. M. Brewer, and R. Ridgway. **The Water Birds of Nort America**. Boston: Little, Brown, 1884. Reprint, New York: Arno Press, 1974. 2 vols. in (Natural Sciences in America Series). $74.00. LC 73-17800. ISBN 0-405-05716-4.

Originally issued as volumes 12-13 of *Memoirs of the Museum of Comparative Zoolog Harvard College*, this comprehensive and authoritative work considers all those birds en demic to North America that are found in a specific ecological niche (i.e., water). Detaile taxonomic descriptions can be found for the families and species of these birds. Species ar arranged under their respective families. Detailed information on size and coloration considered in order to identify one species from another. Examples are size of wing, depth of bill, size of tarsus, and size of middle toe. Geographical ranges and habitats are presented as well. Citations to the taxonomic literature are given for every species mentioned within the text. The illustrations consist mainly of black-and-white drawings depicting the heads of these creatures; some drawings of the entire bird are included. A separate common/scientific-name index can be found at the end of the work. This is a very comprehensive, detailed work.

1143. Bendire, Charles Emil. **Life Histories of North American Birds: With Special Reference to Their Breeding Habits and Eggs, with Twelve Lithographic Plates**. Washington, D.C.: Government Printing Office, 1892. (Smithsonian Institution. United States National Museum. Special Bulletin No. 1). Reprint, New York: Arno Press, 1974. 2 vols. in 1. (Natural Sciences in America). $68.50. LC 73-17802. ISBN 0-405-05720-2.

Based largely on the species found in the U.S. National Museum, this work considers a variety of facts dealing with these birds. The species are entered under their respective families, which in turn are entered under orders. The common and scientific names of the family can also be found. For each bird included, the scientific name, common name, person who first named it, and citations to the taxonomic literature are presented. In addition, the descriptive account consists of geographical range, nesting and egg information, migration patterns, breeding ranges, and diet. Quotes from the diaries of naturalists are included within the descriptions. Each volume concludes with a number of black-and-white photographs of the eggs of the birds along with a combined common/scientific-name index. There are no photographs or drawings of the birds themselves.

1144. Bent, Arthur Cleveland. **Life Histories of North American Birds**. Washington, D.C.: G. P. O., 1919-1950. (Smithsonian Institution United States National Museum Bulletins). Reprint, New York: Dover.

This series of books is considered to be one of the most comprehensive works dealing with the behavior of bird species found throughout North America. Each book within this series considers a particular group or grouping of birds. The species are entered under their respective families, which in turn are entered under orders. The scientific and common names of the families are given. For each bird included, habits, and, in some cases, geographical range are considered. The habits (behavior) of the bird make up the greater part of each description; 20 or more pages is not unusual. The author, in his descriptive material, cites and includes the behavioral descriptions penned by a number of people who have observed and written about the birds' behavioral patterns. Such areas as egg information, plumage, food, voice patterns, diseases, courtship display, winter vs. summer habitats, and so on are included to some degree or another in many of the entries. This type of information varies from entry to entry depending upon what was known at the time. For the most part, the entries read like natural histories. A very large number of black-and-white photographic plates depicting the birds and their nests can be found at the end of each book. A lengthy bibliography as well as a combined author, subject, common-name, and scientific-name index are also present. Following is a list of the books within this series:

Life Histories of North American Birds of Prey. 1958. 2 vols. $10.95pa. vol. 1; $10.95pa. vol. 2. ISBN 0-486-20931-8 vol. 1; 0-486-20932-6 vol. 2.

Life Histories of North American Blackbirds, Orioles, Tanagers & Their Allies. 1958. $11.95pa. ISBN 0-486-21093-6.

Life Histories of North American Cuckoos, Goatsuckers, Hummingbirds and Their Allies. $12.95pa. ISBN 0-486-26029-1.

Life Histories of North American Diving Birds. 1986. $6.95pa. ISBN 0-486-25095-4.

Life Histories of North American Flycatchers, Larks, Swallows and Their Allies. $12.95pa. ISBN 0-486-25831-9.

Life Histories of North American Gallinaceous Birds. 1932. $10.95pa. ISBN 0-486-21028-6.

Life Histories of North American Gulls and Terns. $8.95. ISBN 0-486-25262-0.

Life Histories of North American Jays, Crows & Titmice. 1988. $12.95pa. ISB 0-486-25723-1.

Life Histories of North American Marsh Birds. 1927. $9.95pa. ISBN 0-486-21082-0

Life Histories of North American Nuthatches, Wrens, Thrashers & Their Allies. 194 $10.95pa. ISBN 0-486-21088-X.

Life Histories of North American Petrels and Pelicans and Their Allies. $11.95pa. ISB 0-486-25525-5.

Life Histories of North American Shore Birds. 2 vols. 1927-1929. $9.95pa. vol. $9.95pa. vol. 2. ISBN 0-486-20933-4 vol. 1; 0-486-20934-2 vol. 2.

Life Histories of North American Thrushes, Kinglets & Their Allies. 1949. $9.95p ISBN 0-486-21086-3.

Life Histories of North American Wagtails, Shrikes, Vireos & Their Allies. 195 $9.95pa. ISBN 0-486-21085-5.

Life Histories of North American Wild Fowl. $15.95. ISBN 0-486-25422-4.

Life Histories of North American Wood Warblers. 2 vols. $8.95pa. vol. 1; $8.95pa. vo 2. ISBN 0-486-21153-3 vol. 1; 0-486-21154-1 vol. 2.

Life Histories of North American Woodpeckers. 1939. $9.95. ISBN 0-486-21083-9.

1145. Ehrlich, Paul R., David S. Dobkin, and Darryl Wheye. **Birds in Jeopardy: Th Imperiled and Extinct Birds of the United States and Canada, Including Hawaii an Puerto Rico**. Stanford, Calif.: Stanford University Press, 1992. 261p. $45.00; $17.95pa. L 91-29555. ISBN 0-8047-1967-5; 0-8047-1981-0pa.

Employing the use of the USFWS (U.S. Fish and Wildlife Service) and the Nation Audubon Society lists, as well as other information, this source provides information on tho birds throughout most of North America that are endangered, threatened, and the like. F each of the birds listed, the common name, scientific name, nesting information, die geographical range, conservation status, and recovery plans are given. A full page is devote to each of these accounts. A colored drawing of the head of each bird accompanies th description.

The birds are arranged under continental species and subspecies, Hawaiian species an subspecies, and Puerto Rican species and subspecies that are threatened or endangered, an bird species that are candidates for the list of those that are threatened or endangered a given.

In addition, a list, along with descriptions of birds not on the official lists that exhib vulnerability as well as birds that have become extinct since 1776 is also included. Som general information on the conservation of birds along with environment threats can be foun near the end of the volume. A seven-page bibliography and a combined common/scientifi name index conclude the work.

1146. Jones, John Oliver. **Where the Birds Are: A Guide to All 50 States and Canad** New York: William Morrow, 1990. 400p. $24.95; $15.95pa. LC 89-49399. ISBN 0-68 09609-3; 0-688-05178-2pa.

This unique and interesting volume indicates the best places to go within a state province where it is most likely to encounter these fine, feathered friends. The work arranged by states and by provinces. For virtually every state and province listed, birdin hotlines, local birding groups, a description of geographical locales, maps of the state, an maps of the locales are presented. Descriptive matter includes name, address, and telephor number of locale, general directions, habitat, and housing information.

The second part of the book (bird charts) consists of a list of birds arranged under th common-name groups, geographical locations of where the birds can be found, frequency birds within the locales, and what seasons are the best for viewing the species. All birds a entered under their common names. In addition, there is a two-page supplement that lists ra and limited species.

1147. Mayr, Ernst, and Lester L. Short. **Species Taxa of North American Birds: A Contribution to Comparative Systematics**. Cambridge, Mass.: Nuttall Ornithological Club, 1970. 127p. (Publications of the Nuttal Ornithological Club, no. 9). $7.00. LC 78-24266. ISBN 1-877973-19-X.

Basically, this work is divided into three parts: an analytical table of species-level taxa, a section concerned with taxonomic comments, and a discussion section considering taxonomic problems of the species. The analytical table consists of a listing of scientific names of birds under their families. Taxonomic species data such as monotypic, uncomplicated polytypic, strongly differentiated polytypic, member of superspecies, and member of species group is considered. The taxonomic comment portion lists genera and species and briefly discusses taxonomic relationships and problems within these groupings. Citations to the taxonomic literature are referenced throughout this section. The taxonomic section considers such areas as problems of species delimitation, hybridization, evolutionary problems, and so on. A 19-page bibliography concludes the work.

1148. Palmer, Ralph S., ed. **Handbook of North American Birds**. New Haven, Conn.: Yale University Press, 1962- . 5 vols. LC 62-8259.

Arranged by common name, this voluminous work describes each species in considerable detail. For example, 14 pages are devoted to the description of the short-tailed hawk. Information includes scientific name, coloration, color phases, geographical range, measurements and weights of males and females, hybrids, geographical variations, subspecies, anatomical descriptions needed for field identification, voice characteristics for both male and female, habitats, distribution records, migration patterns, banding status, reproduction (includes displays, copulation, nest building, location of nests, dates when eggs are laid, clutch size, incubation and incubation period, development of young, age of bird when first able to fly, breeding success), survival information, habits (includes daily routine, hunting techniques, prey capture), and food getting. Geographical range maps, pen-and-ink drawings, and colored plates provide the illustrative material. A bibliography listing the cited literature found in the text, and a combined common/scientific-name index conclude each volume. The volumes published up to this point are the following:

Volume 1. *Loons through Flamingos.*

Volume 2. *Waterfowl (first part); Whistling Ducks, Swans, Geese, Sheld-Ducks, Dabbling Ducks*. $120.00. ISBN 0-300-02078-3.

Volume 3. *Waterfowl (concluded); Eiders, Wood Ducks, Diving Ducks, Mergansers, Stifftails*. $120.00. ISBN 0-300-02078-3.

Volume 4. *Family Cathartidae (New World Condors and Vultures). Family Accipitridae (first part); Osprey, Kites, Bald Eagle and Allies, Accipiters, Harrier, Buteo and Allies)*. $45.00. ISBN 0-300-04059-8.

Volume 5. *Family Accipitridae (concluded); Buteos, Golden Eagle. Family Falconidae; Crested Caracara, Falcons*. $45.00. ISBN 0-300-04060-1.

1149. Reilly, Edgar M., and Olin Sewall Pettingill, eds. **The Audubon Illustrated Handbook of American Birds**. New York: McGraw-Hill, 1968. 524p. LC 68-22765.

This volume represents another of the many types of bird books that illustrate certain species. The birds (875 species) found throughout the United States (all 50) and Canada are presented. The birds are arranged under their respective families (common and scientific names are given). After a brief introduction to the family, one or more species are listed. The descriptive matter relating to the bird consists of scientific name, common name, appearance, coloration, voice, range and status, habitat, seasonal movements, and nesting and breeding. In addition, a bibliographic citation is also present. The majority of the descriptions are usually a half page long. Black-and-white and colored photographs as well as pen-and-ink drawings provide the illustrative material. The photographs depict the bird in its natural setting; many show birds within their nests and feeding their young. An index to the illustrators and a combined common/scientific-name index can be found at the end of the book.

1150. Ridgway, Robert. **The Birds of North and Middle America: A Descriptive Catalogue of the Higher Groups, Genera, Species, and Subspecies of Birds Known to Occur in North America, from the Arctic Lands to the Isthmus of Panama, the West Indies and Other Islands of the Caribbean Sea, and the Galapagos Archipelago**. Washington, D.C.: U.S. Government Printing Office, 1911-50. 11 vols. (Smithsonian Institution. United States National Museum, Bulletins).

The species are arranged under their respective genera, which in turn are entered under families. Taxonomic considerations, bibliographic references to the taxonomic literature, and keys to the genera can be found under the families, while taxonomic references, general anatomical descriptions, coloration, geographical ranges, and keys to the species can be found under the genera headings. For each of the species listed, the scientific name, person who first named it, detailed external anatomical measurements in the adult male and female as well as the immature male and female, coloration in adult and immature males and females, geographical range, and bibliographic citations to the taxonomic literature are given. Descriptive accounts are usually no more than one to two pages in length. Black-and-white drawings depicting external anatomical sections of the birds provide the illustrative material. A combined scientific/common-name index can be found at the end of each volume.

1151. Terres, John K. **The Audubon Society Encyclopedia of North American Birds**. New York: Knopf, 1980. 1109p. $75.00. LC 80-7616. ISBN 0-394-46651-9. Reprint, New York: Wings Books/Outlet Book, 1991. 1109p. $39.99. LC 91-21877. ISBN 0-517-03288-0.

Because this work has numerous characteristics of a handbook as well as an encyclopedia it has been listed in the handbook section as well as encyclopedia section. For an annotation of this work, see entry 1092.

OMAN

1152. Gallagher, Michael, and Martin W. Woodcock. **The Birds of Oman**. London, New York: Quartet, 1980. 310p. $75.00. LC 81-17173. ISBN 0-685-01041-4.

Arranged under their respective families, this work encompasses those species found in Oman. For each bird entered, the common name, scientific name, geographical range, seasonal population consideration, size, coloration, and habitat are given. In addition, a description of the family is considered and includes number of species in the world, number of species in Oman, external anatomy, coloration, and identification parameters. Excellent colored plates of these birds can be found on the opposing page of the descriptive information. Such facts as climate, rainfall, physiographic regions, desert environment, breeding, migration, and the observation of birds in Oman are considered in the introductory chapters. Appendices consist of a checklist of the birds of Oman, additional species not found in the main portion of text, species requiring confirmation as to their presence in Oman, a list of birds reported to be escaped captives, and ringed birds recovered in Oman. A three-page bibliography and a combined common/scientific-name index conclude the volume.

PAKISTAN

1153. Roberts, T. J. **The Birds of Pakistan**. Karachi, New York: Oxford University Press 1991. 2 vols. $85.00 vol. 1; $85.00 vol. 2. LC 91-930173. ISBN 0-19-577404-3 vol. 1 0-19-577405-1 vol. 2.

Species are arranged under their respective families; the orders are also recorded. For each species included, the common name, scientific name, person who first identified it coloration, size (body length, wingspan, wing length, bill length), habitat, distribution, status habits, breeding biology, and vocalization are given for the greatest majority of the species entered. A geographical range map accompanies the description. Each species description is pproximately two pages in length. Black-and-white drawings and plates as well as colored ates provide the illustrations. The appendices consist of a glossary, a lengthy bibliography citations that are cited within the text, and a gazetteer of locations in Pakistan. A combined mmon/scientific-name index concludes each volume.

ALEARCTIC REGION (Comprises all of Europe, Africa north of the Sahara, nd Asia north of the Himalayas)

154. Cramp, Stanley, ed. **Handbook of the Birds of Europe, the Middle East and North frica: The Birds of the Western Palearctic**. Oxford, New York: Oxford University Press, 977-1988. 5 vols. LC 79-42914.

Arranged by orders, followed by families, this work describes in great detail all bird ecies found in Europe, the Middle East, and North Africa. For each bird listed, the scientific me, common name, foreign-language names (Dutch, French, German, Russian, Spanish, c.), name and date of person who first named it, geographical range, external anatomical scription, habitat, distribution, population, movements, food, social patterns and behavior, ice, breeding, coloration, molting information, measurements, weights, and geographical riation are given. Citations to the literature can be found throughout the descriptive formation. In addition, brief descriptions of the order and family are considered. A large mber of geographical range maps, colored plates, and drawings can be found throughout e text providing excellent illustrations. A very comprehensive bibliography, a series of lored and black-and-white plates of eggs, and separate indexes for scientific, English, ench, and German bird names conclude this impressive set. Titles of volumes are as follows:

Volume 1. *Ostriches to Ducks*. $135.00. ISBN 0-19-857358-8.
Volume 2. *Hawks to Bustards*. $145.00. ISBN 0-19-857505-X.
Volume 3. *Waders to Gulls*. $150.00. ISBN 0-19-857506-8.
Volume 4. *Terns to Woodpeckers*. $165.00. ISBN 0-19-857507-6.
Volume 5. *Tyrant Flycatchers to Thrushes*. $175.00. ISBN 0-19-857508-4.

Along with volume 5, volumes 6 and 7 will include all passerine birds.

55. Harrison, Colin. **An Atlas of the Birds of Western Palaearctic**. Princeton, N.J.: inceton University Press, 1982. 322p. $49.50. LC 82-800069. ISBN 0-691-08307-X.

Species are arranged under the common names of the families. A brief description of e group, along with the scientific name of the family are considered before the species are ted. For each bird mentioned, the common name, scientific name, breeding and seasonal bitats, and geographical range are included. Geographical range maps in color depict the ographical distribution of these animals. Descriptions are short and usually take up e-third of a column of text. Small drawings of the adult bird can be found at the bottom of ch page. There is a one-page bibliography and a separate common/scientific-name index. e introductory chapters consist of interpreting the maps found within, climatic zones, getation, and recent climatic changes.

ANAMA

56. Ridgely, Robert S., and John A. Gwynne. **A Guide to the Birds of Panama with osta Rica, Nicaragua, and Honduras**. 2d ed. Princeton, N.J.: Princeton University Press, 89. 534p. $60.00; $29.95pa. LC 88-29309. ISBN 0-691-08529-3; 0-691-02512-6pa.

Arranged under families with their common-name equivalents, this guide considers the rious bird species in many parts of central America. For each species listed, the common me, scientific name, brief anatomical description (size, coloration, etc.), similar species, atus and distribution, habits, and geographical range are given. A brief description of the family is also included. Numbered plates are present that lead the reader to the center portion of the work containing the colored plates of the birds identified within the text. Information on climate, migration and local movements, recent developments in Panama ornithology and conservation, and a checklist of the birds of Southern Middle America can be found in the introductory chapters. The guide closes with a section on how to find the birds in Panama, a brief bibliography, and separate indexes for scientific names and common names.

1157. Wetmore, Alexander. **The Birds of the Republic of Panama**. Washington, D.C.: Smithsonian Institution, 1965-1984. 4 vols. LC 66-61061. ISBN 0-87474-122-X (vol. 3).

After the name of the order, the family is listed using both the scientific and common names. A brief description of the family ensues, followed by a taxonomic key to the species within the family group, and, finally, the listing of the species. For each bird included, the scientific name, common name (in English and Spanish), person who named it, and bibliographic citations are given. In addition, the descriptive account contains size, coloration (adults, chicks, juveniles), habitat and geographical range, clutch size, nesting habits, and other interesting facts. Subspecies are included as well. Black-and-white drawings of a number of the birds mentioned can be found scattered throughout the volumes. A combined common/scientific-name index is present at the end of each book. Common names are in English and Spanish. The titles of the volumes are as follows:

Volume 1. *Tinamidae (Tinamous) to Rynchopidae (Skimmers).*

Volume 2. *Columbidae (Pigeons) to Picidae (Woodpeckers).*

Volume 3. *Passeriformes: Dendrocolaptidae (Woodcreepers) to Oxyruncidae (Sharpbills).*

Volume 4. *Passeriformes: Hirundinidae (Swallows) to Fringillidae (Finches).*

PHILIPPINES

1158. DuPont, John Eleuthere. **Philippine Birds**. Greenville, Del.: Delaware Museum of Natural History, 1971. 480p. (Monograph Series No. 2). $45.00. LC 70-169119. ISBN 0-913176-03-6.

Species are arranged under their respective families. Scientific names and common names of the families are given. For each species listed, the common name, scientific name, person who first named it, date of naming, bibliographic citation to its taxonomy (description), coloration, measurements (wing, bill, tarsus), and geographical range are presented. The written descriptions are short and to the point. The colored plates are scattered throughout the work and are in close proximity to the descriptions of the birds. A 15-page bibliography and a combined common/scientific-name index conclude this volume.

SAUDI ARABIA

1159. Bundy, G., R. J. Connor, and C. J. O. Harrison. **Birds of the Eastern Province of Saudi Arabia**. London: H. F. & G. Witherby, 1989. 224p. $75.00. LC gb89-32926. ISBN 0-85493-180-5.

After a 40-page introduction to the landscape, climate, vegetation, and origins of the birds endemic to this area, the list of species begins. The species are arranged under their respective families. For each bird included, the common name, scientific name, and descriptions are given. The descriptive matter varies according to the species. Geographical range, nesting habits, number of eggs laid, incubation periods, and populations are the main points considered. Excellent colored photographs are scattered throughout the text. The appendices consist of a systematic list of species and their status, passage migrants in the eastern province of Saudi Arabia, geographic coordinates, and bar charts depicting average monthly relative humidity and temperature. A five-page bibliography and a separate common/scientific-name index complete the work.

SOUTH AMERICA

1160. Blake, Emmet R. **Manual of Neotropical Birds. Volume 1: Spheniscidae (Penguins) to Laridae (Gulls and Allies)**. Chicago: University of Chicago Press, 1977- . 674p. $100.00. LC 75-43229. ISBN 0-226-05641-4.

Geographically speaking, this work is concerned with all bird species found on the mainland of Central and South America. Mexico, West Indies, Galapagos, and Falkland Islands are excluded. Species are listed under their respective genera, which in turn are listed under families and orders. Each of the families comprises a chapter and includes a synopsis of the genera, a key to each of the genera, and a listing of the species contained within the genera. The description for each species includes scientific name, person who first named it, date of naming, measurements (wings and tails), coloration, distinguishing features (between adult, immature, and juvenile), geographical distribution, and a citation to the literature. A series of black-and-white and colored plates, drawings, and geographical range maps can be found throughout the book. A combined common/scientific-name index is present. A brief synopsis of each of the families included can also be found in the front portion of the volume. As of this writing, only volume 1 has been published.

1161. De Schauensee, Rodolphe Meyer. **A Guide to the Birds of South America**. Wynnewood, Pa.: Published for the Academy of Natural Sciences of Philadelphia by Livingston Publishing, 1970. Reprint, with additions, Wynnewood, Pa.: Pan American Section, International Council for Bird Preservation, 1982. 470p. $25.00pa. LC 84-158599. ISBN 0-317-04625-X.

Arranged under orders, followed by families, this work gives an account of the many species of birds found throughout the South American continent. According to the author, this continent supports the largest number of species of any continent, no less than 2,926. For each species listed, the common name, scientific name, size, coloration, and geographical range are given. A brief description of the family is also incorporated at the beginning of each section. Pen-and-ink drawings as well as black-and-white and colored plates constitute the illustrations. The plates can be found in the center portion of the book. The appendix consists of a list of species that are considered casual, accidental, or doubtful. A combined common/scientific-name index completes the guide.

1162. Ridgely, Robert S., and Guy Tudor. **The Birds of South America**. Austin, Tex.: University of Texas Press, 1989. 516p. $65.00. LC 88-20899. ISBN 0-292-70756-8.

The work represents the first volume of a projected four-volume set. Based on De Schauensee's *A Guide to the Birds of South America* (see entry 1161), this treatise simplifies and adds to that work in the ecological and behavioral areas. The birds are arranged under their respective genera, which in turn are entered under families. In addition, a brief description of the family and genus are considered before the species are listed. For each bird included, the common name, scientific name, other common names, size (in cm and in.), coloration, habitat, behavior, voice patterns, and geographical range are given. A geographical range map accompanies each description. The colored plates, which are referenced in the descriptive matter, can be found at the beginning of the book. Bibliographic citations are considered throughout the work. The introductory sections include chapters on habitat, biogeography, migration, and conservation. The appendix consists of casual migrants from North America. A seven-page bibliography, and a separate common/scientific-name index conclude this volume.

Volume 1. *The Oscine Passerines.*

SOUTH PACIFIC

1163. DuPont, John Eleuthere. **South Pacific Birds**. Greenville, Del.: Delaware Museum of Natural History, 1976. 218p. (Monograph Series -Delaware Museum of Natural History, No. 3). $34.95. LC 75-23917. ISBN 0-913176-04-4.

General information is provided on those bird species inhabiting Fiji, Tongan, Samoan, Cook, Society, Tuamotu, Marquesas, Austral, Pitcairn, and Henderson groups of islands. The birds are listed under their respective families. The scientific and common names are given for each of the families listed. For each of the species considered, the common name, person who first named it, date of naming, citation to the taxonomic literature, coloration (male and female), soft anatomical parts of the bird, size (male and female) of wings, tails, bills, and tarsi, and geographical range are given. The accounts are usually one half page in length. Colored plates (31 plates) of these birds can be found throughout the text. A variety of species are contained on a plate. References to the plates are in the descriptive accounts. A 10-page bibliography, a list of island names, and a combined common/scientific-name index conclude the source.

SRI LANKA

1164. Henry, G. M. **A Guide to the Birds of Ceylon: With 30 Half-Tone Plates of Which 27 Are Coloured and 136 Black-and-White Drawings**. 2d ed. New York: Oxford University Press, 1971. 457p. LC 76-24812. ISBN 0-19-217629-3.

A seven-page glossary, a checklist of the birds of Sri Lanka, and a list of the plates begin this comprehensive treatment. The birds are arranged under their respective families. Families are identified by their common and scientific names; a brief description of each family is included. In addition, the families are arranged under their orders. A brief description of each order is also considered. For each species mentioned, the common name, scientific name, person who first named it, coloration, geographical range, behavior, breeding, eggs (size and color), and other interesting tidbits are offered. Black-and-white drawings and colored plates provide the illustrations, which can be found throughout the text matter.

TRINIDAD

1165. French, Richard. **A Guide to the Birds of Trinidad and Tobago**. 2d ed. Ithaca, N.Y.: Cornell University Press, 1991. 426p. $72.50; $34.50pa. LC 91-6396. ISBN 0-8014-2567-0; 0-8014-9792-2pa.

The accounts of the species are arranged under their respective families, which in turn are arranged under orders. A brief description of each family is considered. For each species included, the common name, scientific name, habitat, status, range, subspecies, size, coloration, measurements (wing size and weight), voice patterns, food, nesting, eggs, incubation, and behavior are given. Black-and-white drawings of a number of these birds are found throughout the text. The colored plates can be found in the center of the work. Cross-references from text to plates are provided. Appendix I consists of a list of species recorded on Tobago, while appendix II offers a list of Tobago species not recorded on Trinidad. Introductory information consists of the environment of Trinidad and Tobago, as well as the ecology and distribution, breeding, migration, and conservation of the species. A 13-page bibliography and a separate common/scientific-name index can be found closing out the book.

TOBAGO (see TRINIDAD)

FORMER SOVIET UNION (USSR)

1166. Flint, V. E., and others. **A Field Guide to Birds of the USSR: Including Eastern Europe and Central Asia**. Princeton, N.J.: Princeton University Press, 1984. 353p. $85.00; $29.95pa. LC 83-42558. ISBN 0-691-08244-8; 0-691-02430-8pa.

Because this work follows the same basic format as many of the handbooks listed in this section, it has been included within this grouping. After a very brief discussion of the order, the bird species are listed under their respective genera. For each animal considered, the common name, scientific name, size, coloration, habitat, egg-laying and nesting behavior, geographical range, and the names of similar species are given. Geographical range maps and black-and-white drawings of many of the birds can be found throughout the descriptive material. The center of the book consists of 48 colored plates of bird species contain within the text. Cross-references within the description identify the plate number as well as the geographical map number. A separate common/scientific-name index can be found at the end of the book. There are two common-name indexes (English and Russian). In addition, there is a list of scientific names, in the same order as they appear in the descriptive account. The names are arranged under the families and the common names; plate and range map numbers are given for each listing. This scientific-name listing is the cross-reference list and can also be found near the completion of the work.

1166a. Knystautas, Algirdas. **Birds of Russia**. London: HarperCollins, 1993. 256p. $36.00. LC gb93-58178. ISBN 0-00-219913-0.

Consisting of over 800 species of birds found throughout former republics of the USSR, including Russia, this work provides some general information on these animals. The species are arranged under their respective orders. Common and scientific name is given for each order listed. For each of the species considered, the common and scientific name is given. The amount of descriptive information for each of the species listed varies. In most cases, the geographical distribution, habitat, diet, and breeding information (incubation time, number of eggs in a clutch, etc.) is presented. The length of these accounts vary from a few sentences to a several paragraphs. Colored photographs of many of the birds described within the text can be found in the middle of the accounts. In addition, a number of black-and-white photographs depicting bird species as well as the geography of Russia can also be found scattered throughout the text. Information dealing with bird migrations and bird conservation in Russia is presented in the introductory chapters. A checklist of the birds of Russia, providing common and scientific names and breeding information can be found near the end of the work. A combined common/scientific-name index completes the book.

VANUATA

1167. Bregulla, Heinrich L. **Birds of Vanuata**. Oswestry, Shropshire, England: Anthony Nelson, 1992. 294p. $44.00. ISBN 0-904614-34-4.

Consisting of 121 species, this source considers those birds endemic to a group of volcanic islands in the southwest Pacific known as Vanuata. These islands were the former French and British condominium of the New Hebrides. The birds are arranged under their respective families, which in turn are entered under orders. The common and scientific names of the family, along with a brief description of the family and order are presented. For each of the species listed, the common name, scientific name, other common names, and the French and Bislama names are given. In addition, the size, coloration, distribution, status, conservation, habitat, behavior, voice patterns, food, nest, and breeding are also present within the descriptive account. The accounts vary in length from two to three pages. A series of 24 colored plates as well as a number of black-and-white photographs depicting the birds and the islands provide the illustrations.

Other sections of the book contain maps of the southwest Pacific and Vanuata, a list of the islands, the avifauna of Vanuata along with conservation considerations, a checklist of the birds, morphology of the birds and shapes of the eggs, and a series of hints for birdwatchers in the islands. A brief glossary, an eight-page bibliography, and a combined common/scientific/geographic-name index conclude the volume. This is a well-presented source.

VENEZUELA

1168. De Schauensee, Rodolphe Meyer, and William H. Phelps, Jr. **A Guide to the Birds of Venezuela**. Princeton, N.J.: Princeton University Press, 1978. 424p. $90.00; $35.00pa. LC 76-45903. ISBN 0-691-08188-3; 0-08205-7pa.

Arranged under families, this work considers the common and scientific names, size, coloration, nesting habits, and status of these creatures. The description for each species is about one-third of a column. In addition, there is a brief identifying description of the family. Line-drawings are interspersed throughout the text. The center of the book finds a series of black-and-white and colored plates of the species contained within the text. A species index to the plates, a brief bibliography, and a separate common/scientific-name index complete the volume.

Aves (Ecological Section)

1169. Cross, Theodore. **Birds of the Sea, Shore and Tundra**. New York: Weidenfeld & Nicolson, 1989. 1 vol. (unpaged). $65.00. LC 88-092284. ISBN 1-55584-385-9.

This work represents a compilation of magnificent photographs of birds, many of which take up one or two pages. The text is no more than a few sentences for each of the birds included. The brief information varies from one bird to another. Some of the birds included are egrets, herons, boobys, puffins, ospreys, geese, owls, hawks, gulls, and loons. There is no index and no table of contents. This work should only be purchased for the magnificent colored plates.

1170. Haley, Delphine, ed. **Seabirds of Eastern North Pacific and Arctic Waters**. Seattle, Wash.: Pacific Search Press, 1984. 214p. $22.95. LC 83-19411. ISBN 0-914718-86-X.

Teeming with a large number of colored photographs, this work considers the Procellariiformes, Pelecaniformes, and Charadriiformes of the region. Chapters are comprised of specific family groups within these orders. For each family, the common name, scientific name, behavioral patterns, geographical ranges, breeding, egg-laying, nesting, incubation information, bird rearing, food, and feeding are considered. In addition, a list of the pertinent species within the family are listed along with a description. The species descriptions consist of size, coloration, diet, and geographical range. Geographical range maps are included throughout the various chapters. The work concludes with a classification of eastern North Pacific and Arctic seabirds (down to species), a nine-page bibliography, and a combined subject, common-name, and scientific-name index. This is an excellent publication.

1171. Hancock, James. **The Birds of the Wetlands**. London: Croom Helm, 1984. 152p. £13.95. LC gb84-16423. ISBN 0-7099-1287-0.

Covering North America, South America, Africa, India, China, Japan, Indonesia, Australia, and Europe, this book considers all those bird species endemic to wetland ecosystems. The ecosystems included are the Florida Everglades, Northern Argentina, Tana River, Kenya, Bharatpur, Zhalong Reserve, Shinhama Reserve, Chiba, Pulau Dua, Darwin and the South

Alligator River, and Coto Donana. The descriptive information is arranged under respective continents. Various bird species are considered within this descriptive information, along with colored photographs of same. There are no subheadings within each section that allows the reader to easily find an entry; the index is necessary. A short bibliographic list arranged under the wetland ecosystems, and a combined common/scientific-name index complete the work. Page numbers in bold type refer to the illustrations.

1172. Hosking, Eric, and Ronald M. Lockley. **Seabirds of the World**. New York: Facts on File, 1984. 159p. $27.95. LC 83-1751. ISBN 0-87196-249-7.

Abounding with a large number of colored-action photographs, this work includes penguins, petrels, cormorants, gannets, boobies, skuas, gulls, terns, and auks. Each of the chapters depicts and discusses a major bird group. In addition to the colored photographs, each chapter discuss the behavior, habitat, geographical range, home life, distribution, longevity, and many other interesting facts about the group. Certain species are mentioned in order to illustrate some of the salient points. A section on the seabird as an individual can be found in the first chapter. A combined common/scientific-name index concludes the work. Most of the chapters are 20 to 30 pages in length.

1173. Line, Les, Kimball L. Garrett, and Kenn Kaufman. **The Audubon Society Book of Water Birds**. New York: Harry N. Abrams, 1987. 256p. LC 87-1434. ISBN 0-8109-1863-3.

Abounding with magnificent colored photographs of these birds, this work introduces the reader to the many and varied types of species inhabiting this ecological niche. The photographs show the birds in their natural habitats and comprise, in many cases, one- or two-page spreads. The birds, themselves, are arranged by common-name groups. For example, "Dancing Cranes and Other Waders," "Of Rockhoppers and Kings," and "Fishing with Pouches, Spears, and Hooks" are representative of the types of chapter headings. The descriptions, which are written to appeal to the layperson, consider behavioral patterns, food getting, coloration, geographical distribution, nesting practices, and so on. Scientific names as well as common names are present for those species used within the chapter. Some general notes on the photographers, and a combined common/scientific-name index conclude this volume.

1174. Pledger, Maurice, and Charles Coles. **Game Birds**. New York: Dodd, Mead, 1983. 112p. LC 83-1821. ISBN 0-396-08171-1.

Although this folio-sized volume considers only 24 of the wildfowl species, it does it with the ardent bird-lover in mind. Descriptive information can be found on one page and a full-page colored painting of the bird on the opposing page. The description consists of a history of the bird, number of species within its family, habitat, feeding behavior, breeding, egg-laying, number of eggs within a clutch, color of eggs, incubation time, and voice characteristics. The scientific name is also included within the text. Types of birds considered are pheasant, quail, partridge, grouse, hazel hen, ptarmigan, grouse, prairie chicken, wild turkey, bustard, mallard, teal, wigeon, geese, woodcock, and snipe. A brief bibliography of wildfowl books can be found at the end of the volume.

1175. Soothill, Eric, and Richard Soothill. **Wading Birds of the World**. London: Blandford, 1989. 334p. LC 90-186429. ISBN 0-7137-2130-8.

According to Soothill, waders are those birds that wade in shallow or deep water some time during the year in search of food. The birds are arranged under their respective families. The scientific name, common name, and number of species recorded are given for each of the families. For each species listed, the scientific name, common name, length, coloration, characteristics, behavior, habitat, food, voice patterns, courtship display, breeding, and geographical distribution are given. In addition, a geographical range map and a colored picture of the bird accompany the textural material. The amount of descriptive information varies depending on the species. In some cases, one or more pages are devoted to a particular wader. A one-page bibliography and a separate common/scientific-name index can be found at the end of the work.

Aves (Systematic Section)

SPHENISCIFORMES (Penguins)

1176. Peterson, Roger Tory. **Penguins**. Boston: Houghton Mifflin, 1979. 238p. $40.00. LC 79-10101. ISBN 0-395-27092-8.

Consisting of the 17 penguin types found throughout the world (mostly Antarctica), this work provides interesting information on the species and their behavioral and social patterns. After a brief introduction to the group as a whole, as well as their history, an account of the variety of penguins ensues. For each of the species, the common name, scientific name, geographical range, and habitat are given, and, in some cases, population and egg information are considered. A line-drawing of the head of the penguin accompanies the description. The rest of the work considers such biological parameters as their environment, their behavior and socialization patterns, and their relations with other animals and man in both a positive and a negative manner. In an additional section, their northern look-alikes such as the puffins are portrayed. A large number of colored photographs, many of them taking up one or two pages, provide much of the illustrative matter. A two-page bibliography and a combined subject, common-name, and scientific-name index conclude the source. This is a delightful book to read. It provides the reader with a good start in the understanding of these interesting creatures.

PROCELLARIIFORMES (Albatrosses, Shearwaters, Petrels, and Fulmars)

1177. Warham, John. **The Petrels: Their Ecology and Breeding Systems**. London; San Diego, Calif.: Academic Press, 1990. 440p. $59.95. LC gb90-33834. ISBN 0-12-735420-4.

The petrels include those birds known as the shearwaters and albatrosses and the storm, diving and gadfly petrels. With that in mind, this work considers all aspects of these birds. The introductory chapter considers various biological parameters of the order Procellariiformes, to which the petrels belong. Such information as external morphology, anatomy, body size, allometry, general and breeding ecology, global distribution, populations, and systematics and nomenclature are included. The book goes on to discuss in depth the families and genera contained within the order. Discussions of these families and genera consist of morphology, plumage, moulth, sexual dimorphism, hybridization, genera, walking, swimming, diving, flight, behavior, geographical distribution, foods and feeding, breeding, dispersal and migration, populations and mortality, and conservation. The remainder of the volume considers biological aspects such as breeding, pre-egg stage, the petrel egg, incubation, and chick stage. Black-and-white photographs, geographical range maps, tables, charts, and graphs provide the illustrations. A 39-page bibliography, a checklist of the Procellariiformes, and a combined common/scientific-name index, as well as a subject index complete the work.

It should be noted that this is not a typical reference book on birds. It is highly specialized and does not contain colored photographs or plates. Rather, it represents an in-depth description of these fascinating creatures.

PELECANIFORMES (Cormorants, Darters, and Pelicans)

1178. Johnsgard, Paul A. **Cormorants, Darters, and Pelicans of the World**. Washington, D.C.: Smithsonian Institution Press, 1993. 445p. $49.00. LC 92-31997. ISBN 1-560-98216-0.

Consisting of 32 species of cormorants and shags, two species of darters, and seven species of pelicans, this work provides in-depth information on these birds of the world. The species are arranged under their respective families (common and scientific names given). For each of the species listed, the common names, scientific name, person who first named it, date of naming, subspecies, coloration (in prebreeding adults, immatures, juveniles, and nestlings), detailed measurements, and weights are given. In addition, basic external anatomy, ecology, movements and migrations, competition and predation, foods and foraging behavior, social behavior, reproductive biology, population status, and evolutionary relationships are considered. Geographical range maps and simple sketches of the birds (in most cases) accompany the descriptive accounts. The lengths of the accounts vary from a few to seven or more pages. Colored plates of these magnificent birds can be found interspersed throughout the section of species account.

While section 2 considers the descriptive accounts, section 1 provides detailed information on the biology of this group of birds and composes approximately one-third of the book. Such areas as taxonomy, comparative morphology and anatomy, various types of behaviors, reproductive biology, and population dynamics and conservation biology are included.

The appendices include a biological key for species identification, a glossary of scientific and vernacular names, and head-profile-identification drawings. A 19-page bibliography, and a combined common/scientific-name index conclude the work. This reference source is considered to be the first worldwide survey of cormorants, darters, and pelicans.

CICONIIFORMES
(Herons, Bitterns, Storks, Ibises, Spoonbills, and Others)

1179. Hancock, James, and James Kushlan. **The Herons Handbook**. London: Croom Helm, 1984. 288p. $45.00. LC gb84-26798. ISBN 0-7099-3716-4.

Arranged under the common names of the herons, this work includes all herons found around the world. For each species entered, the common name, scientific name, person who named it, data of naming, and taxonomic reference are given. In addition, a two- to four-page description is given for each of the birds. Descriptive information includes identification, distribution and population, migration, habitat, behavior (feeding, breeding), and some general points on the nest, the eggs, and young. A geographical range map and a colored painting can be found for each of the birds described. Introductory material consists of classification, courtship, feeding, and the identification of herons and egrets. A six-page bibliography and a combined common/scientific-name index complete the work.

1180. Hancock, James A., James A. Kushlan, and M. Philip Kahl. **Storks, Ibises and Spoonbills of the World**. London; San Diego, Calif.: Academic Press, 1992. 328p. $139.00. LC gb92-43708. ISBN 0-12-322730-5.

Partially based on information compiled in *The Herons Handbook* (see entry 1179), this source provides detailed information on the stork, ibis, and spoonbill species found throughout the world. A total of 20 species of stork, 23 species of ibis, and six species of spoonbill are contained within. Each of the descriptive accounts of the species represent a chapter within the book. For each of the species listed, the common name, scientific name, person who named it, date of naming, citation to the literature, and additional common names are given. In addition, size, coloration, basic external anatomy, nestling information, courtship displays, voice patterns, flight information (measurements), distribution, population, ecology, breeding, taxonomy, and conservation are presented. A full-page painting of each of the birds as well as a geographical range map accompanies the descriptive account. The lengths of these accounts vary from four to six or more pages. Appendix tables for each of the species can be found near the end of the volume. Such information as size of wing, culmen, tarsus, tail, wing-span, and weight, eggs (length, width, weight), and egg-laying are presented. Introductory sections consider classification, conservation, courtship and reproduction, feeding behavior, and ecology of these birds. Colored photographs provide the illustrations for the introductory chapters. A 54-page bibliography and a combined common/scientific-name index conclude the volume. This is an excellent publication. It provides a good deal of information on these species of birds.

1181. Voisin, Claire. **The Herons of Europe**. London: T & A D Poyser, 1991. 364p. $40.73. LC gb91-99490. ISBN 0-85661-063-1.

Narrower in scope than *The Herons Handbook* (see entry 1179), this work presents a great deal of information on the nine species that are breeding in Europe at the present time. An entire chapter is devoted to each of these species. For each of the herons included, the common name, scientific name, size, and wing span are presented. In addition, plumage coloration (adult breeding, adult nonbreeding, juvenile, older young at the nest, downy young), bare parts (adult breeding, adult nonbreeding, juvenile, downy young), field characteristics (coloration, behavioral patterns, voice patterns), male and female measurements (wing, tail, bill, tarsus, toe), breeding and wintering areas, timing of breeding and movements, habitat, population size and trends, displays and calls, fear behavior, courtship displays, calls, pair formation, copulation, nest building, greeting behavior, egg-laying, incubation and hatching, development and care of the young, feeding behavior and food, and predation are included. A large number of citations to the literature are interspersed within the text. Chapters can run as long as 20 to 30 pages. Geographical range maps, graphs, line-drawings, tables, and colored plates provide the illustrative material. The species included are *Botaurus stellaris* (bittern), *Ixobrychus minutus* (little bittern), *Nycticorax nycticorax* (black crowned night heron), *Ardeola ralloides* (squacco heron), *Bubulcus ibis* (cattle egret), *Ardea cinerea* (grey heron), *Ardea Purpurea* (purple heron), *Egretta garzetta* (little egret), and *Egretta alba* (great white egret). Chapters on the classification of herons, origin of herons, general appearance and special features of herons, breeding behavior and biology, habitat, resource partitioning and species diversity, feeding behavior, food of herons, and protection of herons in Europe are also included. An 18-page bibliography reflecting those references found throughout the text, as well as a combined subject, common-name, and scientific-name index conclude the work.

PHOENICOPTERIFORMES (Flamingos)

1182. Allen, Robert Porter. **The Flamingos: Their Life History and Survival: With Special Reference to the American or West Indian Flamingo (Phoenicopterus ruber)**. New York: National Audubon Society, 1956. 285p. (Research Report of the National Audubon Society, no. 5). LC 56-58658.

This work is concerned with all the flamingos of the world; depending upon the taxonomist, there are four to six species that inhabit the planet. It discusses them within broad sections that deal with a number of biological parameters. These sections consider in great detail the distribution and migration, population numbers, food habits and ecology, breeding cycles (courtship displays, nest and eggs, incubation periods, etc.), and conservation. The various species are introduced within these sections and are used to illustrate biological principles. Black-and-white photographs and line-drawings provide most of the illustrative matter. A few color photographs are also present. A 21-page bibliography, a list of flamingo names in a variety of languages, and a combined subject, common-name, and scientific-name index complete the volume.

ANSERIFORMES (Ducks, Geese, Swans, and Screamers)

1183. Delacour, Jean. **The Waterfowl of the World**. London: Country Life, 1973. 4 vols. LC 55-600. ISBN 0-668-02970-6.

Originally published in 1954, this classic work considers all species of birds of the world that fall into the general category of waterfowl. The species are arranged under common-name groups, such as the whistling ducks, the brents, and so on. Preceding the species accounts, an adequate description of the group as a whole is presented. Such aspects as plumage, geographical distribution, behavioral patterns, and captivity are included. For each species mentioned, the common name, scientific name, references to the literature, coloration, size (wing, tail, culmen, and tarsus), juvenile features, geographical distribution, behavioral patterns, and captivity information are considered. Colored plates and geographical range maps provide the illustrations. A systematic index of the order Anseres (waterfowl) appears near the beginning of the first, second, and third volumes and reflects the species contained within those volumes. A separate common/scientific-name index appears at the end of each volume and likewise reflects same.

Volume 1.	The magpie goose, whistling ducks, swans and geese. Sheldgeese and shelducks.
Volume 2.	The dabbling ducks.
Volume 3.	Eiders, pochards, perching ducks, scoters, golden-eyes and mergansers, stiff-tailed ducks.
Volume 4.	General habits. The reproductive cycle. Ecology. Distribution and species relationships. Fowling. Conservation and management. Aviculture. Domestic waterfowl. The anatomy of waterfowl. Fossil Anseriformes. Corrections and additions.

1184. Gooders, John, and Trevor Boyer. **Ducks of North America and the Northern Hemisphere**. New York: Facts on File, 1986. 176p. $24.95. LC 86-6333. ISBN 0-8160-1422-1.

Arranged by the common name of the duck, this beautifully illustrated compendium considers the species of duck found in North America. Written in easy-to-understand language, this source includes the common name, scientific name, range maps, body and wing size, weight of males and females, egg color, egg clutch, incubation period, and fledgling period of duck species. In addition, informative descriptive information is presented, usually consisting of one to three pages. This work abounds with colored paintings of the duck species mentioned. A combined common/scientific-name index as well as a brief bibliography conclude the book.

1185. Johnsgard, Paul A. **Ducks, Geese, and Swans of the World**. Lincoln, Nebr.: University of Nebraska Press, 1978. 404p. $35.00. LC 78-8920. ISBN 0-8032-0953-3.

Arranged by tribes, followed by the common names of these avian delights, this excellent source encompasses the ducks, geese, and swans of the world. For each bird listed, the common name, scientific name, person who first identified it, date of identification, synonyms, subspecies, range, measurements, weights, external and internal features for identification, habitat, foods, reproduction, status, relationships to other groups, and suggested readings are given. A total of 148 species have been included. Pen-and-ink drawings, range maps, and a series of colored plates of some species provide the illustrations. A table of contents, list of illustrations, and an introduction to the family Anatidae (ducks, swans, and geese) can be located on the opening pages. A useful bibliography, a glossary, and a common/scientific-name index conclude this volume.

1186. Kear, Janet. **Ducks of the World**. New York: Mallard Press, 1991. 216p. $30.00. ISBN 0-7924-5636-X.

Abounding with colored photographs, this source provides general information on the various species of ducks inhabiting the planet. The species are arranged under their common-name groups, such as shelducks, steamer ducks, dabbling ducks, and so on. Some general information on the group as a whole is presented such as diet, reproduction, vocal patterns, number of species within the group, behavioral patterns, and the like. Not all the information given above applies to each group. The species follow the group accounts; one page is devoted to each of the ducks presented. Very general information is given for each of the species, such as common name, weight, size, coloration, geographical distribution, and diet. The accounts are usually no longer than one paragraph. The colored photograph of the species fills up the major portion of each page. A combined subject and common-name index concludes the volume. It should be noted that this is mainly a pictorial work with general information on the duck groups and the ducks themselves.

1187. Owen, Myrfyn. **Wild Geese of the World: Their History and Ecology**. London B T Batsford, 1980. 236p. LC 81-114096. ISBN 0-7134-1831-6.

The species are arranged under the genus *Anser* (10 species) and the genus *Branta* (species). For each bird included, the coloration, plumage, relations to other species, popula tions, geographical distribution, nesting, egg information, habitat, winter biology, an exploitation and conservation are given. The average description averages two to four page long. Geographical range maps accompany the descriptive matter. Other illustrative materia includes colored plates, line-drawings, graphs, and charts. In addition to the species accounts sections on classification, social behavior, movements and migration, summer and winte biology, population dynamics, and conservation and exploitation are present. The appendice consist of breeding data for geese, body weights of geese, and measurements of geese. A 13-page bibliography as well as a combined subject, common-name, and scientific-nam index complete the volume.

1188. Soothill, Eric, and Peter Whitehead. **Wildfowl of the World**. London: Blandfor Press, 1978; rep., 1988. 297p. £12.95pa. LC gb89-41636. ISBN 0-7137-0863-8; 0-7137 2110-3pa.

Accompanied by colored photographs of each species, this book considers the swans geese, and ducks of the world. Birds are arranged by common names, and each listing contain the scientific name, range map, description, characteristics and behavior, habitat, distribution food habits, voice, display, and breeding. In addition, a description of wetlands for wildfow is included. These are arranged by name of country. A list of wildfowl collections around th world, a short bibliography, and a separate index of common names and scientific names ca be found at the end of the source.

This work is not as comprehensive as *Ducks, Geese, and Swans of the World* (see entr 1185), as the former contains 148 species, while the later encompasses 128.

1189. Wilmore, Sylvia Bruce. **Swans of the World**. New York: Taplinger, 1974. 229p. LC 74-3669. ISBN 0-8008-7524-9.

As one would imagine, this source does not contain a large number of species. For th species included, the amount of descriptive material devoted to them is admirable; approxi mately 15 to 20 pages is given for each account. The descriptive information contains, among other parameters, the size, weight, flight, voice, behavioral patterns, coloration, nesting breeding, egg and incubation information, population studies, and much more. Black-and white photographs, line-drawings, and geographical range maps provide the illustrativ material. The species contained within this volume are the mute (*Cygnus olor*), trumpete (*Cygnus cygnus buccinator*), whooper (*Cygnus cygnus cygnus*), whistling (*Cygnu columbianus columbianus*), bewick's (*Cygnus columbianus bewickii*), black (*Cygnu atratus*), black-necked (*Cygnus melanocoryphus*), and the coscoroba (*Coscoroba coscoroba* swan. A four-page bibliography and a combined subject, common-name, and scientific-nam index conclude the work.

FALCONIFORMES
(Vultures, Hawks, Eagles, Kites, Harriers, Osprey, Falcons, and Others)

1190. Brown, Leslie. **Eagles of the World**. Cape Town, South Africa: Purnell, 1976. 224p LC 77-356676. ISBN 0-360-00318-4.

In addition to an account of the various species of eagles, this work considers othe biological parameters of these birds such as behavior, breeding, and so on. The descriptiv accounts (26 pages) of the eagles are arranged under broad common names, such as sea an fish eagles, snake eagles, and the harpy group. The account discusses the group as a whol and considers size, weight, geographical range, habitat, and egg and incubation information The various species are introduced within the account. Black-and-white photographs of th birds accompany the descriptions as well as the other sections contained in the volume. Th remainder of the source considers physical characters, senses, adaptation for predation, behavioral patterns, breeding cycles, the young eagle, and conservation of eagles. Appendix 1 presents a summary of the present knowledge of eagle species, such as general habits, detailed diurnal behavior, hunting methods, food needs and preferences, breeding biology, and survival and longevity. Appendix 2 considers nest sites, dimensions, share of sexes, and occupation time. Appendix 3 summarizes breeding data for the better-known eagles and includes clutch size, incubation period, sexes incubating, fledgling period, young per successful in the nest, and young per pair overall, while appendix 4 is a list of threatened species with main threats outlined. A six-page bibliography and a combined subject, common-name, and scientific-name index complete the volume.

1191. Cade, Tom J. **The Falcons of the World**. Ithaca, N.Y.: Cornell University Press, 1982; Reprint, 1987. 192p. $49.50. LC 81-68743. ISBN 0-8014-1454-7.

After an introductory section (55 pages) that includes anatomical attributes, classification, distribution and migration, feeding adaptations, size and flying performance, hunting success, sexual dimorphism, social behavior, and reproduction, an account of the species commences. For each falcon included, the common name(s), scientific name, and a full-page colored painting of the bird are present. In addition, a two- to four-page description of each species is included and speaks to the various subjects that were covered in the introduction. A series of maps depicting the falcons' distribution patterns, a three-page bibliography, and a combined subject, common-name, and scientific-name index conclude this magnificent work.

1192. Johnsgard, Paul A. **Hawks, Eagles, & Falcons of North America: Biology and Natural History**. Washington, D.C.: Smithsonian Institution Press, 1990. 403p. $45.00. LC 89-48558. ISBN 0-87474-682-5.

Including a large number of excellent colored plates of these beasts, this work provides the reader with an adequate amount of information on each of the species of hawks, eagles, and falcons found throughout North America. For each of the species listed, the common name, scientific name, person who first named it, date of naming, and additional common names are given. In addition, geographical distribution, North American subspecies, coloration and external anatomical characteristics in the adult male and female as well as the subadult and juvenile, detailed measurements of size and weight, habitats, ecological considerations, foods and foraging, social behavior, breeding biology, and evolutionary relationships and status are presented. A geographical range map and a line-drawing of the head of the bird accompany each of the descriptive accounts. Accounts are usually five or more pages in length. Other areas within the volume considers comparative biology, evolution, classification, zoogeography, foraging ecology and foods, comparative behavior, reproductive biology, and population biology and conservation of these birds of prey.

The appendices consist of a key to the species of North American Falconiformes, origins of vernacular and scientific names of North American Falconiformes, a glossary, and a field identification of views and anatomical drawings of these vertebrates. A 31-page bibliography and a combined common/scientific-name index conclude the reference source.

1193. Newton, Ian, ed. **Birds of Prey**. New York: Facts on File, 1990. 240p. $40.00. LC 90-33303. ISBN 0-8160-2182-1.

Arranged somewhat differently than most of the handbooks, this work provides the reader with information on the birds of prey found throughout the world as well as their biology. Basically, the work is divided into three parts. The first section considers descriptive accounts of the species, the second considers their biology, and the third discusses their relationship with man. A large number of excellent colored photographs enhance the text in all three sections.

After some general information describing raptors and their families the descriptive accounts of the species ensue. The species are listed under their families. For each of the species listed, the common name, scientific name, coloration, basic external anatomy, size, habitat, geographical distribution, reproductive information, and diet are given. A colored painting of each of the birds accompanies each account.

Part two considers the biological aspects of this group. Such major biological phenomena as habitats, populations, feeding behavior, social behavior, reproduction, mortality, and migration and movements are considered. Approximately 100 pages is devoted to the biology of these birds. Part three considers the relationships of the birds of prey with humans and discusses such areas as human impacts on the birds of prey as well as their conservation and management.

A systematic checklist of all living diurnal raptors in the order Falconiformes, a one-page bibliography, and a combined subject, common-name, and scientific-name index conclude the volume. A great deal of information, presented in an interesting manner, can be found throughout the work.

1194. Weick, Friedhelm. **Birds of Prey of the World: A Coloured Guide to Identification of All the Diurnal Species Order Falconiformes**. Hamburg, Germany: Verlag Paul Parey, 1980. 159p. $48.00. ISBN 3-490-08518-3.

Including 1,144 colored figures and 160 line-drawings, this work is an attempt to compile all the diurnal species found in the Falconiformes. Text is written in both the German and English languages. The main portion of this volume is split into three parts. There is a key for identifying these species by employing categories such as size (very large to very small), length of neck, length of legs, head and neck coverings, color of bill, shape of bill, color of plumage, color of eye, shape of crest, and so on. The scientific name of the bird lies to the left of the specific description. In this way, identification can be made by a number of characteristics. The second portion is a compendium of the order Falconiformes. Names of genera are listed, along with descriptive external anatomies, under their respective families, superfamilies, and suborders. Line-drawings of representative bird types can be found scattered throughout these two sections. The third portion of the source contains the colored plates, each of which is accompanied by a list of tables. Each table contains the scientific name, person who first identified it as such, date of naming, common name, distribution, length, wing span, tail span, weight, tarsus length, and so on. A list of scientific names, German names, and common names, as well as an annotated bibliography conclude the volume.

GALLIFORMES
(Grouse, Pheasants, Quail, Peafowl, Guinea Fowl, Turkeys, and Others)

1195. Alderton, David. **The Atlas of Quails**. Neptune, N.J.: T. F. H. Publications, 1992. 144p. $39.95. ISBN 0-86622-145-X.

This magnificent work considers all the quail species found throughout the world. A two-page spread is devoted to each of the quails mentioned; one page for the description and the opposing page for the colored paintings of the birds. For each species listed, the common name, scientific name, person who first named it, date of naming, large geographical range map, length of the bird, geographical distribution, coloration (young and adult), breeding success in captivity, nest building, number and color of eggs in a clutch, incubation period, and a number of other interesting facts are given. All of the above information is not considered for each species presented. Colored paintings of the male and female accompany the description and can be found on the opposing page of the descriptive material. The descriptions are usually no more than one to four paragraphs in length; the coverage given to each species differs considerably. It should be noted that this work is heavier on illustrations than on descriptive accounts.

In addition, other sections consider such subjects as the keeping, housing, feeding, breeding, and health care of quails. These sections are illustrated with colored photographs. In addition, a checklist of the genera and species of quail is present. Scientific and common names are present in the checklist. A combined subject, common-name, and scientific-name index concludes the volume.

1196. Bergmann, Josef. **The Peafowl of the World**. Surrey, England: Saiga, 1980. 99p. ISBN 0-904558-51-7.

The work is divided into five parts and consists of the history, general description, species, and management of peacocks as well as a section on peacocks in art. The general description information considers courtship behavior, nesting information (type of nest, number of eggs in clutch, etc.), brief anatomical description including feathers, hatchling information, and the molting process. For each of the eight species presented, the common name, scientific name, geographical distribution, and the coloration of the crown, head, neck, breast, belly, thighs, back, shoulders, wing coverts and tertiaries, secondaries, primaries, tail, bill, legs and the feet as well as the total length, length of wing, tail, culmen, tarsus, thigh, and spur are given. This information is presented separately for the male and the female. The management section deals with natural living habits, housing for peacocks, type of aviary, feeding, breeding and rearing, transport and selling, and illness and disease. The history section presents how peacocks were looked upon down through the ages, while the art section considers how peacocks were used in different art mediums. A large number of black-and-white drawings as well as full-page colored plates of the peacocks provide the illustrative material. A combined subject, common-name, and scientific-name index, and a one-page bibliography complete the work.

1197. Johnsgard, Paul A. **The Grouse of the World**. Lincoln, Nebr.: University of Nebraska Press, 1983. 413p. $42.50. LC 82-21922. ISBN 0-8032-2558-X.

Species are entered under their common names. For each bird included, the scientific name, person who first named it, date of naming, other vernacular names, geographical range, subspecies, measurements, identification parameters, field marks, age and sex criteria, distribution, habitat, population density, habitat requirements, nesting and brooding requirements, food and foraging behavior, mobility and movements, reproductive behavior, nesting behavior, and evolutionary relationships are given. Large geographical range maps accompany the descriptive information. Descriptions can run 16 or more pages. Colored and black-and-white photographs as well as black-and-white drawings provide the illustrations. Photographs show the birds in their natural habitats.

In addition to the individual accounts, there are sections on evolution and taxonomy, physical characteristics, molts and plumages, physiological traits, hybridization, reproductive biology, population ecology and dynamics, social behavior and vocalizations, aviculture and propagation, and hunting, recreation, and conservation. One hundred pages are devoted to these disciplines.

Name derivations of grouse and ptarmigans, a key to identification of grouse and ptarmigan species, and a hunter harvest and population status estimates of grouse and ptarmigans (arranged by state and country) make up the appendices. A 34-page bibliography, and a combined common/scientific-name index conclude the volume.

1198. Wayre, Philip. **A Guide to the Pheasants of the World**. London, New York: Country Life, 1969. 176p. LC 71-506394. ISBN 0-600-43350-1.

Arranged by common-name groups, this book attempts to give an up-to-date account of pheasants. For each group of birds, the genus followed by particular species of that genus are described. Genera descriptions include external anatomy, habitat, coloration, courtship display, size of eggs, and so on. For each species listed, the common name, scientific name, and distinguishing characteristics are mentioned. Colored paintings of the species provide the illustrations. A short bibliography, a checklist of pheasants of the world (arranged by genera), and a combined common/scientific-name index conclude the book. In addition, sections on housing and management of breeding birds, procuring and transport of pheasants, incubation, rearing, and diseases of pheasants can be found in the introductory chapters.

GRUIFORMES
(Cranes, Rails, Coots, Seriemas, Bustards, and others)

1199. Johnsgard, Paul A. **Cranes of the World**. Bloomington, Ind.: Indiana University Press, 1983. 258p. $40.00. LC 82-49015. ISBN 0-253-11255-9.

Basically, arranged into two sections, this volume brings to the reader all varieties of cranes that inhabit the globe. Section 1 discusses a number of aspects of crane biology, such as classification, evolution, social behavior, vocalization, ecology, population dynamics, reproduction, hybridization, endangered species, conservation, and cranes found in myth and legend. A number of line-drawings, tables, and beautiful colored photographs of cranes found in their natural habitats provide the illustrations for this section. Section 2 consists of the species arranged by common names. For each species considered, the common name, scientific name, person who named it, date of naming, other vernacular names, geographical range, a list of subspecies or semispecies, measurements, weight, and anatomical and identification descriptions are given. In addition, there is a great deal of information on each crane regarding distribution and habitats, foods and foraging behavior, migrations and movements, sociality, interspecific interactions, all aspects of breeding (including courtship), population status, and evolutionary relationships. Each species account also includes a fine pen-and-ink drawing of the bird, and a geographical range map. The volume also contains a number of black-and-white photographs of many species. Origins of scientific and vernacular names of cranes, a key to the species and subspecies of cranes, an adequate bibliography, and a combined common/scientific-name index can be found at the close of the book. The descriptions of these species are lengthy and contain a good deal of information regarding the cranes of the world.

1200. Ripley, S. Dillon. **Rails of the World: A Monograph of the Family Rallidae**. Boston: David R. Godine, 1977. 406p. $400.00. LC 75-619273. ISBN 0-87923-199-8.

Illustrated with 41 paintings by J. Fenwick Lansdowne, this folio-sized book presents the rails that inhabit the earth. The opening chapters consist of characteristics, distribution, and evolution and speciation of rails. The species, which comprise the greatest portion of the work, are arranged under their respective genera. In addition to a description of each species, a taxonomic key to each genera is given. For each bird included, the common name, scientific name, alternative names, and a bibliographic citation are given. In addition, the descriptive matter consists of size, coloration, habitat, geographical range, food, voice, and breeding information including egg size and color. Subspecies are listed after the species. A taxonomic key to the subspecies is present. In addition to the magnificent paintings, a series of geographical range maps can be found scattered throughout the text. A description of fossil species, a bibliography, and a separate common/scientific-name index complete this excellent volume. The bibliography comprises those citations found throughout the text. The citations are arranged under authors with oldest references appearing first.

CHARADRIIFORMES
(Snipes, Plovers, Sandpipers, and others)

1201. Johnsgard, Paul A. **The Plovers, Sandpipers, and Snipes of the World**. Lincoln, Nebr.: University of Nebraska Press, 1981. 493p. $45.00. LC 80-22712. ISBN 0-8032-2553-9.

As a companion volume to *Ducks, Geese, and Swans of the World* (see entry 1185), this work includes shorebirds of the world. This source is arranged by families, followed by tribes in many cases. For each bird listed, the common name, scientific name, other vernacular names, subspecies, range (map included), measurements, weight, coloration, external anatomy, habitats, social behavior, reproduction, status and relationship to other groups, and, in some cases, a bibliographic citation are given. Brief keys to genera and/or species occur in many of the descriptions. Colored and black-and-white photographs of the birds provide many of the illustrations. Line-drawings and range maps are also included. A head-profile-identification guide (line-drawings of the heads of these birds), a listing that includes derivations of generic and specific names, a bibliography of sources cited within the text, and a separate common/scientific-name index can be found ending this volume. Opening chapters include taxonomy and evolutionary relationships, reproductive biology, and keys to families, subfamilies, and tribes.

COLUMBIFORMES (Pigeons, Doves, and Sandgrouse)

1202. Goodwin, Derek. **Pigeons and Doves of the World**. 3d ed. Ithaca, N.Y.: Cornell University Press, 1983. 363p. $49.50. LC 81-70700. ISBN 0-8014-1434-2.

The species/subspecies of pigeons and doves can be found in a series of chapters that identify the groups by their common names. For example, chapters such as the typical pigeons and the pink pigeon, cuckoo doves and long-tailed pigeons, and the doves of the genus *Leptotila* are present. For each of the species present, the common name, scientific name, alternative scientific name, and a citation to the taxonomic literature are considered. In addition, basic external anatomy, coloration (males, females, and juveniles), geographical distribution, habitat, diet, behavioral patterns, nesting information, vocalization, courtship displays, alternative common names, and a brief list of references are given. A black-and-white drawing of the bird and a geographical range map accompany the descriptive accounts. The accounts range from one to four or more pages in length.

A large number of colored plates can be found interspersed among the pages of text. Keys to the plates are also present and provide the reader with an outline of the birds as well as their common and scientific names.

Introductory information consists of a wide variety of biological principles pertinent to these bird species. Such areas as display and social behavior, nesting, parental care, clutch size and egg color, plumage, evolution, and the like are considered. This introductory material consists of approximately 50 pages of the work. A separate common/scientific-name index concludes the handbook.

PSITTACIFORMES (Parrots, Lories, Cockatoos, and Macaws)

1203. Forshaw, Joseph M. **Parrots of the World**. 3d ed., rev. Willoughby [N.S.W.]: Lansdowne Editions, 1989. 672p. ISBN 0-7018-2800-5.

Illustrated by William T. Cooper, this work brings to life the myriad of parrots that can be found in all the niches of the world. The introductory section consists of fossil history, classification of parrots, physical attributes (external and internal anatomy with labeled diagrams), distribution, longevity and mortality, habitats, feeding, flight, voice, nesting habits (including egg-laying and incubation), and molting.

Each bird species is arranged under its respective genus, which in turn is entered under subfamily, family, and order. In addition, a brief description of the family/subfamily, tribe, and genus is considered. For each species mentioned, the common name, scientific name, person who first identified it, size, weight, coloration in adults and juveniles, and geographical distribution are given. Descriptive information for each subspecies consists of scientific name, person who first identified it, length of wings and tails in males and females, coloration, and geographical distribution. In addition, diet, vocal patterns, nesting information, and number and size of eggs are provided for the species as a whole. Geographical range maps accompany the descriptive matter. The descriptive accounts are usually one to two pages in length. The colored paintings of these birds are fantastic and depict the species recorded within the text. A 15-page bibliography and a separate common/scientific-name index concludes the volume. The bibliography consists of citations found throughout the book. This work represents a very comprehensive treatment of parrots found throughout the globe.

STRIGIFORMES (Owls)

1204. Burton, John A., ed. **Owls of the World: Their Evolution, Structure and Ecology** New York: E. P. Dutton, 1973. 216p. LC 73-8271. ISBN 0-525-17432-X.

Divided into three parts, this source considers the various species of owls inhabiting th planet. Part I considers owls as a symbol down through the ages, the evolution of owls, an the characteristics of owls. Part II presents the owl species. The owls can be found under thei common name groups, such as barn and bay owls, wood owls, hawk owls, screech owls, an so on. For each of these groups, the number of species within the group (family), size, externa anatomy, geographical ranges, habits, habitat, behavioral patterns, diet, and information o eggs, nests, and the young are given. In addition, the species are listed under their geographica locations. For each species listed, the scientific name, person who first named it, date o naming, and brief external anatomy and coloration are considered. A number of geographica range maps are present for each of the groups mentioned. Magnificent colored photograph and colored paintings abound throughout the text. A number of these take up a full page. Par III considers a checklist of owl species, vocalization patterns of owls, a brief glossary, one-page listing of books, and a combined subject, common-name, and scientific-name index This is a marvelous book and should be enjoyed by many.

1204a. Freethy, Ron. **Owls: A Guide for Ornithologists**. London, England: Bishopsgat Press, 1992. 134p. £12.95. LC gb94-3316. ISBN 1-85219-042-6.

Although not as encompassing as Burton's *Owls of the World* (see entry 1204), no Hume's *Owls of the World* (see entry 1205), this work represents a good introduction to th many species of owls found throughout the planet. The birds are presented under thei common name group which represents chapters. Such groups as the barn, otus, bubo, snowy fish and strix owls, among others, are represented within these pages. For many of the specie of owls depicted, the common name, scientific name, size, external anatomical characteristics coloration, reproductive information such as size of clutch, incubation time, etc., behaviora characteristics, and geographical range is given. It should be noted that the length and amoun of information given for each of the species varies widely. These descriptive accounts ca range from a few sentences to several pages depending upon the species. Colored an black-and-white illustrations depict many of the owls discussed within the text. Classificatio tables of the Strigiformes, down to subspecies, can be found in the many tables near the en of the volume. A two-page bibliography completes the work. There is no index.

1205. Hume, Rob. **Owls of the World**. Philadelphia: Running Press, 1991. 192p. $40.00 LC 91-52606. ISBN 1-56138-032-6.

Encompassing the families Tytonidae and Strigidae, this source considers the variou owl species found throughout the world. The species are presented under their respectiv families. The common and scientific names of the family are listed along with the number o species found within the family. For each owl listed, the common name, scientific name, size coloration, geographical distribution, habitat, and status are given. In addition, the textura material discusses the basic external anatomy, and considers in more detail its ecologica niche. In some species, a great deal more information is given including number of eggs in clutch, voice patterns, behavioral patterns (hunting, etc.), diet, breeding information, an taxonomic considerations. The length of a description varies from one-third of a page t several pages. A beautiful colored picture of each owl accompanies the descriptive materia A series of 151 geographical distribution maps, which are cross-referenced from the descrip tion; a table containing the name of the owl (common and scientific names), the length, an distribution; a selected list of references; and a combined common/scientific-name inde complete the volume. This is a beautifully published book which provides general informatio regarding the owl species of the world.

1206. Voous, Karel H. **Owls of the Northern Hemisphere**. Cambridge, Mass.: MIT Pres 1988. 320p. $55.00. LC 88-13367. ISBN 0-262-22035-0.

Karel Voous, a Dutch ornithologist, and Ad Cameron, an avian artist, have produced a beautifully illustrated and highly descriptive work on the 47 known owl species occurring north of the tropics. Besides numerous drawings of the species, there is a superb full-color plate for each of the owls mentioned.

For each species included, a detailed account follows. Descriptions are based on the most current data available and cover faunal type, distribution, climatic zones, habitat, geographical variation, related species, fossil species, structure, vision and hearing, behavioral characteristics, ecological hierarchy, breeding, food and feeding, population dynamics, and relationships to man.

The remainder of the volume includes a map section depicting distribution of each species, an appendix listing common and scientific names of the species, name of person who named the species and year it was named, extensive bibliography, and a subject index including general subjects, and scientific and common names. This is a rich and rewarding work on owls. A great deal of research has been incorporated into this volume as evidenced by the lengthy descriptions and extensive bibliography.

1207. Wardhaugh, A. A. **Owls of Britain and Europe**. Poole, Dorset: Blandford Press, 1983. 128p. £7.95. LC 83-108605. ISBN 0-7137-1260-0.

Not as glitzy or as large as many of the bird books listed, this work nevertheless covers the many species of owls found in Britain and Europe. Two of the six chapters consider the species and make up the greatest portion of the volume. They are entitled "Owls resident in Britain" (six species), and "European owls not resident in Britain" (seven species). The birds are arranged by their common names and entries include their scientific names and lengthy descriptions discussing color, size, behavior, geographical range, feeding, breeding, and status. There are a number of interesting color photographs as well as some black-and-white photographs illustrating the birds. The other portions of this work delve into the origin of owls and their place among the birds, their characteristics, life-style comparisons, and points on owl watching. A short bibliography and a combined subject, common-name, and scientific-name index complete the book.

PICIFORMES (Woodpeckers, Toucans, Honeyguides, Puffbirds, and Others)

1208. Short, Lester L. **Woodpeckers of the World**. Greenville, Del.: Delaware Museum of Natural history, 1982. 676p. (Monograph Series No. 4). $99.95. LC 79-53793. ISBN 0-913176-05-2.

The 198 species are arranged under their respective genera, which in turn are entered under tribes and subfamilies. A brief description of the genera is offered. For each of the species included, the common name, scientific name, geographical range, weight, size, coloration, anatomical considerations, habitat, foraging habits, voice patterns, courtship display, breeding, and taxonomy are given. References to the 101 colored plates are contained within the descriptive matter. Descriptions are usually two to five pages in length. The introductory chapters consider the family Picidae, plumage and structure, behavior, and zoogeography, evolution, and systematics of the woodpeckers. A 22-page bibliography and a combined common/scientific-name index conclude the source. This is an excellent book filled with a good deal of information on the biology of woodpecker species, presented in a semitechnical writing style.

PASSERIFORMES (Perching Birds)

209. Austin, Oliver L. **Song Birds of the World**. New York: Golden Press, 1967. 318p. .C 67-8899.

Written with the novice in mind, this book considers the many passerine birds (perching irds) that have come to be known as the songbirds. The descriptions of these birds are rranged under their respective families. The common and scientific names of the families are given. Each descriptive account considers the number of species in the group, general size and coloration, habitats, diet, clutch size, incubation time, nest building, and a number of specific species within the group. Colored paintings of the birds accompany the family descriptions. A combined common/scientific-name index concludes the volume. This work is an easy read and presents a good introduction to these types of birds.

Hirundinidae (Swallows)

1210. Turner, Angela, and Chris Rose. **Swallows & Martins: An Identification Guide and Handbook**. Boston: Houghton Mifflin, 1989. 258p. $35.00. LC 89-080262. ISBN 0-395-51174-7.

Divided into two major sections, this work provides information on the swallow and martin species of the world. The first section is similar to a field guide. The birds are arranged by their common-name groups. For each species included, the common name, scientific name, habitat, coloration, geographic range maps, and cross-references to the descriptive information found in section two are given. For each bird considered in section one, a color plate of the organism can be found on the opposing page.

Section two contains the descriptive information on these species and considers common name, scientific name, person who first identified it, taxonomic reference, field characters, habitat, distribution, status, migration, food, behavior, breeding, voice, size, coloration, and external anatomical parts. Approximately two pages are devoted to each bird. A 21-page bibliography consisting of references present throughout the text, and a combined common/scientific-name index conclude the volume.

Thraupidae (Tanagers)

1211. Isler, Morton L., and Phyllis R. Isler. **The Tanagers: Natural History, Distribution, and Identification**. Washington, D.C.: Smithsonian Institution Press, 1987. 404p. $70.00; $49.95pa. LC 85-11747. ISBN 0-87474-552-7; 0-87474-553-5pa.

Arranged under their respective genera, this work includes tangers found throughout the world. For each species recorded, the scientific name, common name, reference to plate, length (in cm and in.), and subspecies are considered. In addition, the geographical range, elevational range, habitats, behavioral patterns, vocalization patterns, breeding information, and references are included. The species descriptions (242 species listed) are usually two pages in length. A geographical range map accompanies each description. Some general information on the genus is also given. The colored plates of birds can be found in the center of the book. For each colored plate there is an opposing page that further describes the birds contained on the plate. Such parameters as common name, geographical range, and coloration are described. Introductory information consists of a glossary and a section discussing size and appearance, distribution and habitat, social behavior, feeding and food, vocalizations, and the breeding behavior of tanagers. A 17-page bibliography and a combined common/scientific-name index conclude the work.

Passeridae (Sparrows)

1212. Summers-Smith, J. Denis. **The Sparrows: A Study of the Genus Passer**. Calton, England: T & AD Poyser, 1988. 342p. $48.00. LC 88-157958. ISBN 0-85661-048-8.

The majority of the chapters contained herein are devoted to a specific type of sparrow. The information contained on each of these birds includes common name, scientific name, subspecies, synonyms, and citations to the literature. In addition, after a brief description describing its taxonomic relationship, the various parameters are presented. Among them are size, coloration, biometrics, geographical distribution, habitat, behavior, breeding biology, survival, molting process, voice patterns, and food ingested. The males and females are discussed separately when deemed necessary, such as in the case of coloration. Geographical ange maps, graphs, line-drawings, and colored plates of the birds provide the illustrations. he lengths of the chapters depicting the swallows vary; they can run from approximately 6 o 15 pages. It should be noted that this book represents very complete accounts of these pecies. Other chapters include characteristics and interrelationships, origins and evolution, nd the systematic position of the sparrows. The appendices consist of a key to the names of ne sparrows (bibliographic citations to species), and a gazeteer. A 20-page bibliography, and combined subject, common-name, and scientific-name index conclude the work.

Estrildidae (Waxbills, Grass Finches, and Mannikins)

213. Goodwin, Derek. **Estrildid Finches of the World**. Ithaca, N.Y.: Comstock; division f Cornell University Press, 1982. 328p. LC 81-70708. ISBN 0-8014-1433-4.

Arranged by common-name groups, such as the blue-bills, the seed-crackers, the winspots, and so on, this work includes the estrildid finches of the world. For each of these ommon-name groups, a description of the group and the species contained within the group re given. Each species is accessed by common name, followed by scientific name, name of erson who first named it, date of naming, bibliographic citation, a length description, field haracters, distribution, habitat, nesting information, display, and social behavior. In addition, list of references, range maps, and colored paintings of the groups are included. There are ntroductory chapters dealing with nomenclature, distribution and adaptive radiation, plum- ge and coloration, behavior and biology, and estrildids in captivity. An index of common ames and an index of scientific names can be found ending the volume.

Corvidae (Crows, Jays, and Magpies)

214. Goodwin, Derek. **Crows of the World**. 2d ed. Seattle, Wash.: University of Wash- ngton Press, 1986. 299p. $45.00. LC 87-671143. ISBN 0-565-00979-6.

The crows are arranged under their common-name groups. For each of these birds, the ommon name, scientific name, person who named it, date of naming, bibliographic citation o the taxonomic literature, coloration, size, field characters, distribution, habitat, feeding and eneral habits, voice patterns, display and social behavior, other common names, and a short st of bibliographic citations are given. A geographical range map accompanies each descrip- on. Some of the parameters listed above may not be given for each species. Black-and-white rawings and a number of colored plates provide the illustrations. Drawings and/or plates are ot provided for every species entered.

In addition to the accounts of the species, there are sections dealing with nomenclature, daptive radiation, plumage and coloration, maintenance behavior, social behavior and ourtship display, nesting and parental care, voice and vocal mimicry, and antipredator ehavior. These sections encompass 60 pages and are found in the front portion of the work. separate scientific/common-name index concludes the volume.

FIELD GUIDES

Aves (General and Ecological Section)

215. Bologna, Gianfranco. **Simon & Schuster's Guide to Birds of the World**. New York: imon & Schuster, 1981. 511p. LC 80-39507. ISBN 0-671-42234-0; 0-671-42235-9pa.

Edited by John Bull, this source provides general information in identifying a variety f bird species found throughout the world. A total of 424 species are included. The species re arranged alphabetically by scientific names. For each of the species listed, the scientific ame, common name, order, and family are given. In addition, size, weight, coloration, habitat, geographical distribution, egg-laying information, and a small black-and-whit sketch of the bird are presented. A colored picture depicting the species can be found on th opposing page of descriptive material. The accounts are usually one half page in length. Th introductory information considers such avian matters as flight, plumage, coloration, ana omy, physiology, reproduction, migration, and ecology. A brief glossary, a three-pag bibliography, and a common-name index conclude the field guide.

Because the birds are arranged alphabetically by scientific names, the use of the guid can be limited (i.e., birds within a certain group can not be found together).

1216. Cerny, Walter. **A Field Guide in Color to Birds**. London: Cathay, 1975. 343p. ISB 0-904644-10-3.

Birds are arranged under orders, followed by families. Common and scientific name are presented for these two groups. For each species mentioned, the common name, scientifi name, descriptive information necessary for identification purposes, geographical range ma and a colored picture of the bird are given. Bird pictures appear on the opposing pages of th descriptive material. Species throughout the world are included.

A series of pictures depicting birds in flight, a number of colored plates showing eg broods, a description of eggs and nests of a number of species, a list of birds that nest i marginal regions, and a one-page bibliography can be found near the end of this guide. I addition, there is a separate common/scientific-name index. The colored plates of the bird are extremely well done.

1217. Chandler, Richard J. **The Facts on File Field Guide to North Atlantic Shorebirds A Photographic Guide to the Waders of Western Europe and Eastern North Americ** New York: Facts on File, 1989. 208p. $19.95. LC 88-046179. ISBN 0-8160-2082-5.

The species are entered under their common names. For each bird included, the commo name, scientific name, size, coloration, external anatomical parameters, food, juvenile co oration, coloration differences between summer and winter when appropriate, voice patter status, habitat, distribution, racial variation, and a consideration of similar species are give A colored photograph or photographs of the birds accompany the descriptive material. series of black-and-white photographs depicting shorebirds in flight, a two-page biblio raphy, and a combined common/scientific-name index complete the guide. In additio sections on North Atlantic shorebird species, shorebird plumages and molts, and methods an equipment necessary to photograph shorebirds can be found in the introductory chapters.

1218. Harrison, Peter. **A Field Guide to Seabirds of the World**. Lexington, Mass.: Stephe Greene, 1987. 317p. LC 86-31984. ISBN 0-317-60437-6.

Designed as a companion volume to *Seabirds: An Identification Guide* (see entry 1219 this work differs in that it contains colored photographs to depict the 320 species; a total c 741 photographs are included. The work is divided into two sections. Section 1 contains th colored photographs, giving common name, scientific name, and references to section which contains the descriptive information. For each bird included, the common nam scientific name, size (length and wing; cm and in.), coloration (summer, winter, and juvenile habits, geographical distribution, and similar species are given. A geographical range ma accompanies each of the descriptions. The textural information references the reader back t the colored photographs. In addition, a set of tubenose identification keys (drawings of bird within bird groups with their wings spread out) can be found near the end of the volume. one-page bibliography and a separate common/scientific-name index conclude the guid Introductory information provides a brief description of the various orders of seabirds.

1219. Harrison, Peter. **Seabirds: An Identification Guide**. rev. ed. London: Croom Hel 1985. 448p. $35.00; $24.95pa. LC 85-17158. ISBN 0-395-33253-2; 0-395-60291-2pa.

Consisting of seabirds from around the world, this guide is useful in the identificatio of these species. Basically, the work is divided into two sections. Section 1 contains th colored plates, along with brief descriptions of the species. The plates can be found on th pposing pages of the descriptive matter. The species are arranged under their common-name roups (e.g., large and medium-sized penguins, crested penguins, etc.). For each species ntered, the common name, scientific name, size in centimeters, alternative name, brief istinguishing features between the adult, juvenile, and chick, and cross-references to the aps, as well as additional descriptive information are given. Section 2, which contains the dditional descriptive information, has the species arranged under their respective families. fter a brief discussion of the families and genera, the species are listed. For each bird entered, e common name, scientific name, length in centimeters and inches, coloration and size of e chick, juvenile, and adult, and geographical range are considered. This descriptive formation provides cross-references to the plates and the maps. In addition, there are 312 eographical range maps, which again provide cross-references to the plates and the text. A ibliography, and a separate common/scientific-name index conclude the volume.

220. Kaufman, Kenn. **A Field Guide to Advanced Birding: Birding Challenges and low to Approach Them**. Boston: Houghton Mifflin, 1990. 299p. (The Peterson Field Guide eries; 39). $22.95; 14.95pa. LC 89-71668. ISBN 0-395-53517-4; 0-395-53376-7pa.

Not formatted like the other field guides, this work is based on the premise that a general utline of the birds' features make identification easier. Line-drawings provide these general utlines, such as the heads of birds, or the bills, or very basic drawings of the birds. The ne-drawings are pen-and-ink; there are no colored photographs of the species. The birds are ranged by common group names, such as the medium-sized terns, the winter loons, the dark ises, the screech-owls, and so on. Each of these groupings contains the common and cientific names of the species entered, a general anatomical description (including age ariations), what anatomical feature to look for first, and an anatomical description of each f the birds. The line-drawings accompany these descriptions. An adequate bibliography and combined subject, scientific-name, and common-name index conclude the guide. This work presents a useful and different approach to bird identification.

221. Madge, Steve, and Hilary Burn. **Waterfowl: An Identification Guide to the Ducks, eese, and Swans of the World**. Boston: Houghton Mifflin, 1988. 298p. $35.00. LC 7-26186. ISBN 0-395-46727-6.

Basically arranged into two sections, this work provides the bird enthusiast with formation necessary to identify the various Anseriformes of the world. Section 1 consists f 47 plates and is arranged by common group names, such as the whistling ducks, the tropical rest ducks, and so on. Each plate contains, on one side of the page, a list of bird species that ve common name, scientific name, coloration, geographical range, differences between the lult male, female, and juveniles, and a geographical range map, while the opposing page nsists of colored plates of these birds. In addition, the page number to a more complete escription of the bird is given. These descriptions can be found in section 2. Arranged by mily, section 2 lists common name, scientific name, alternative names, identification formation, voice, measurements, geographical variation, habitat, behavior, distribution, pulation, and a reference. This section acts as a miniencyclopedia. A brief glossary, a ree-page bibliography, and a combined common/scientific-name index conclude the guide.

222. Marchant, John, and Tony Prater. **Shorebirds: An Identification Guide**. London: room Helm, 1988. 412p. $24.95pa. LC gb89-33848. ISBN 0-7470-1403-5.

Concentrating on plovers, sandpipers, and the like, this work considers such information scientific name, common names, description, voice, habits, movements, breeding, age, sex, ces, and measurements. The first portion of this volume is devoted to the colored plates of e species, including common and scientific names, range maps, and basic external characristics. In addition, it refers the reader to the page in the second portion of the work that cludes the various types of information as mentioned above. The introductory chapters plain how to make the best use of this field guide. An adequate bibliography, and a mbined common/scientific-name index complete the volume. The descriptive area in the cond part of the book could serve as brief encyclopedic entries.

Aves (Geographic Section)

AFRICA

1223. Hollom, P. A. D., and others. **Birds of the Middle East and North Africa: Companion Guide**. Calton, England: T & A D Poyser, 1988. 280p. $32.50. LC gb88-2031 ISBN 0-85661-047-X.

The authors intended this work to be a companion to the field guides dealing wit Europe, specifically with *Jim Flegg's Field Guide to the Birds of Britain and Europe* (se entry 1227). Countries included are Morocco, Algeria, Tunisia, Libya, Egypt, Turkey, Syri Iraq, Iran, Lebanon, Israel, Jordan, Saudi Arabia, Kuwait, Qatar, UAE, N. Yemen, S. Yeme and Oman. The birds are grouped by orders, although the orders and families are n mentioned. For each bird included, the common name, scientific name, size, coloration, voic pattern, status, geographical range, and habitat are given. Geographical range maps an black-and-white drawings accompany the descriptive text. The center of the guide contain the colored plates of the birds. Cross-references from the descriptions to the plates are liste A two-page bibliography, a checklist of all species recorded (common and scientific name given), and a separate common/scientific-name index can be found concluding the guide.

ASIA, SOUTHEAST

1224. King, Ben F., and Edward C. Dickinson. **A Field Guide to the Birds of South-Ea Asia: Covering Burma, Malaya, Thailand, Cambodia, Vietnam, Laos and Hong Kon** Boston: Houghton Mifflin, 1975. Reprint, **The Collins Field Guide to the Birds of Sout East Asia: Covering Burma, Malaya, Thailand, Cambodia, Vietnam, Laos, and Hon Kong**. Lexington, Mass.: Greene Press, 1988. 484p. $24.95pa. LC 87-14840. ISBN 0-828 0650-5.

Species are arranged under their respective families. Family names are presented in the common and scientific vernaculars. For each bird considered, the common name, scientifi name, identification characteristics, geographical range, habitat, and plate number are give Colored as well as black-and-white plates are found in front of each family listing. In additio general information concerning the family is presented, such as number of species foun throughout the world as compared to southeast Asia. Bird groups found in each family a described. A list of the scientific and common names of birds found in Taiwan that are n present in the descriptive portion of the guide, a six-page bibliography, and a combine common/scientific-name index conclude the volume. Maps of the area are located inside th front and back covers of the field guide.

AUSTRALIA

1225. Simpson, Ken, ed. **Simpson and Day Field Guide to the Birds of Australia: A Boo of Identification**. Ringwood, Victoria, Australia: Penguin Books Australia; New Yor Viking Penguin, 1989. 352p. ISBN 0-670-90072-9.

First published by Lloyd O'Neil Pty in 1984 as *The Birds of Australia*, this guid provides the reader with the type of information necessary to identify those birds inhabitin the Australian continent. The species (758 of them) are arranged phylogenetically und common-name groups such as the petrels, the albatrosses, the gulls, and so on. For eac species listed, the common name, scientific name, coloration, basic identifying extern anatomical parts, size, and habitat are given. In addition, a small geographical range map Australia accompanies each of the accounts. The descriptions are usually one-third of a pag in length. Colored plates of the birds can be found, in most cases, on the opposing pages their respective accounts. An illustrated key to the families is present in the introducto section. General information on the families of birds can be found near the end of the volum A brief glossary and a separate scientific/common-name index conclude the field guide.

UROPE (see GREAT BRITAIN, EUROPE, nd IRELAND)

ALAPAGOS ISLANDS

226. Harris, Michael Philip. **The Collins Field Guide to the Birds of Galapagos**. rev. ed. exington, Mass.: Stephen Greene Press, 1989. 160p. $19.95pa. LC 88-30788. ISBN 0-8289-726-9.

Originally published as *A Field Guide to the Birds of Galapagos*, this work represents revised edition of same. The species are arranged under their respective families. The mmon name and scientific name of each family is given. For each species recorded, the mmon name, scientific name, local name, size, coloration, flight behavior, food, voice tterns, breeding, and geographical distribution are presented. Line-drawings as well as ack-and-white and colored plates provide the illustrations. Introductory information in-udes climate, breeding seasons, ecology, migrants, and other pertinent information as it lates to the birds of the Galapagos. A combined common/scientific-name index completes e field guide.

REAT BRITAIN, EUROPE, and IRELAND

227. Flegg, Jim. **Jim Flegg's Field Guide to the Birds of Britain and Europe**. Ithaca, .Y.: Cornell University Press, 1990. 256p. $43.50; $19.95pa. LC 90-1558. ISBN 0-8014-742-6; 0-685-35106-8pa.

The birds are arranged under their common names and grouped under their orders. For ch species entered, the scientific name, size (in cm), coloration, voice patterns, habitat, ographical distribution, and status are given. In addition, a geographical range map and a heeled calendar (designating the months of the year) can be found accompanying each scription. The wheeled calendar is color-coded and designates in what months the popula-on of the bird is most abundant, moderately abundant, and least abundant. Colored photo-aphs of the species described can be located on the opposing pages of the guide. Introductory aterial considers bird biology, bird habitats, bird names and classification, and family aracteristics. A listing of bird clubs and societies, a one-page bibliography, and a separate mmon/scientific-name index conclude the field guide.

228. Gooders, John. **Field Guide to the Birds of Britain & Ireland**. London: Kingfisher ooks, 1986. 288p. LC gb85-42316. ISBN 0-86272-139-3; 0-86272-143-1pa.

Each page is devoted to a species and provides a colored picture and factual information. ecies are arranged by orders and families, although neither is mentioned. For each bird corded, colored pictures are provided as well as a tabular list of identification attributes. ich attributes as type, size, habitat, behavior, flocking, flight voice, coloration on different rts of the body, type of nest, eggs (number and color), incubation period, date of broods, od, and population are given in the table. In addition, status, similar species, and a ographical range map are provided. The pages are coded at the top with a small, colored ctangle. These colors represent groups of birds, such as owls, waders, herons, and so on. If e bird watcher has some knowledge as to the type of bird, they need only browse these lor-coded pages. A checklist of the species by common name, as well as a separate mmon/scientific-name index complete the guide.

229. Harrison, Colin. **A Field Guide to the Nests, Eggs and Nestlings of British and ıropean Birds**. New York: Quadrangle/New York Times, 1975. 432p. $29.95. LC 74-435. ISBN 0-8129-0553-9.

Provides the same basic information as *A Field Guide to the Nests, Eggs and Nestling of North American Birds* (see entry 1247).

1230. Jonsson, Lars. **Birds of Europe: With North Africa and the Middle East**. Londor Christopher Helm, 1992. 559p. £25.00. ISBN 0-7136-8096-2.

Comprising those birds that can be found throughout Europe, North Africa and th Middle East, this work provides general identifying information on same. Beginning with th red-throated diver and ending with the northern oriole, this source is arranged phylogenticall For most of the species listed, the common name, scientific name, size, coloration, basi external anatomy, voice patterns, habitats, seasonal spottings, and migration patterns ar considered. A geographical range map accompanies the descriptions. Colored pictures of th birds can be found on the opposing pages of text. The lengths of the accounts can run fro several sentences to one-half page; not all the above information is given for each specie listed. Introductory information briefly considers the biology of birds, such as plumag wings, size, and so on. A two-page bibliography, a list of ornithological journals and societie and a combined common/scientific-name index conclude the field guide.

1231. Perrins, Christopher. **New Generation Guide to the Birds of Britain and Europ** Austin, Tex.: University of Texas Press, 1987. 320p. (Corrie Herring Hooks Series; no. 8 $16.95. LC 86-51382. ISBN 0-292-75532-5.

Containing all species that breed in Europe, the birds are arranged under their respectiv families, which in turn are entered under orders. The common and scientific names of th family are given. For each species included, the common name, scientific name, size (in cm male and female weights, geographical distribution, coloration, nesting information, foo and the number, color, and size of eggs in a clutch are given. A geographical range ma accompanies each description. Colored paintings of the birds can be found on each opposin page of the descriptive matter. The introductory section comprises a thumb nail sketch of th anatomy and physiology of birds. A combined common/scientific-name index completes th guide.

1232. Peterson, Roger Tory, Guy Mountfort, and P. A. D. Hollom. **A Field Guide to th Birds of Britain and Europe**. 4th ed., rev. and enl. Boston: Houghton Mifflin, 1983. 239 LC 83-10807. ISBN 0-395-34416-6.

The birds are arranged under their respective families. The common and scientific name of the family are entered. For each bird included, the common name, scientific name, commo name in other languages (German, Dutch, French, Swiss), size, coloration, voice patter habitat, and references to the geographical range maps (at end of work) are given. Th illustrations consist of colored plates of the birds and can be found in the center of the guid The descriptive information leads the reader to the plate. Introductory information contain a section on how to identify birds, and a European checklist. A listing of accidentals (specie not normally found in the area), and a combined common/scientific-name index complete th field guide.

HAWAII

1233. Pratt, H. Douglas, Phillip L. Bruner, and Delwyn G. Berrett. **A Field Guide to th Birds of Hawaii and the Tropical Pacific**. Princeton, N.J.: Princeton University Press, 198 409p. $65.00; $24.95pa. LC 86-4993. ISBN 0-691-08402-5; 0-691-02399-9pa.

Arranged by the common group names of birds, this work provides access to thos species found in many of the islands of the tropical Pacific including Hawaii. In addition t the Hawaiian Islands, areas such as Micronesia, Polynesia, Palau, Mariana Islands, Tru Islands, Samoa, Fiji, Tonga, and the Cook, Society, and Marquesas Islands are included.

The species are listed under their orders, followed by their respective families (scientifi and common names given). For each bird included, the common name, scientific name, brie anatomical description, coloration, habits, means of identification from similar species, an ccurrence are given. In addition to many lined pen-and-ink drawings of the birds, there are 3 colored plates found at the end of the guide. Within the descriptive material a reference to ıe proper plate can be found.

The introductory material consists of pointers on how to use this guide as well as ıformation on the types of islands located in the Pacific. The appendices consist of a list of ird species that are questionable regarding their established place among these islands, as ell as a bird checklist, and a series of regional maps. The bird checklist contains a list of the ommon names of the birds arranged under island groups. The table indicates the birds found n specific islands in that group. For example, the section on Hawaiian Islands states that the rctic loon can be found on Oahu. The island groups mentioned are Hawaiian Islands, licronesia, Central Pacific Islands, Central Polynesia, Southeastern Polynesia, and Fiji. A seful bibliography and a combined common/scientific-name index conclude the guide. The st of colored plates can be found following the index.

RELAND (see GREAT BRITAIN, EUROPE, and IRELAND)

APAN

234. Wild Bird Society of Japan. **A Field Guide to the Birds of Japan**. Tokyo, New York: odansha International, 1986. 336p. $24.95. LC 85-40112. ISBN 0-87011-746-7.

Arranged by common names of bird groups, this guide contains the swimmers (loons, ucks, geese, etc.), aerialists (albatrosses, petrels pelicans, boobies, gulls, terns, etc.), large aders (herons, storks, ibises, cranes, rails, etc.), birds of prey (hawks, ospreys, eagles, kites, urriers, accipiters, falcons, owls, etc.), nonpasserine land birds (pigeons, cuckoos, kingfish-s, rollers, woodpeckers, etc.), and the passerine birds (larks, swallows, shrikes thrushes, ioles, etc.). For each species mentioned, the common name, scientific name, size (in cm), oloration in adult and immature birds, similar species, status, and geographic range maps e given. In addition, there is a brief discussion of the family to which these species belong. olored plates of the birds considered can be found on the opposing pages.

The guide closes with a listing and description of the various places in Japan for bird atching, and a combined common/scientific-name index.

AVA and BALI

235. MacKinnon, John. **Field Guide to the Birds of Java and Bali**. Yogyakarta, Indone-a: Gadjah Mada University Press, 1990. 391p. $47.50. LC 88-948479. ISBN 979-420-092-1.

Abounding with a large number of colored plates, this work provides information needed order to identify those bird species found on the islands of Java and Bali. The first part of e work presents the colored plates, while the second portion deals with the species accounts. he species are arranged under their respective families. General information on the family well as keys to the species within the family precede the account. For each of the species sted, the common name, scientific name, the Indonesian name, size, coloration, voice ttern, geographical distribution, status (very rare, rare, etc.), habitats, generalized behavior tterns, general breeding information, and race are given. The lengths of the descriptive counts vary from one page to one-third of a page. In addition, the plate number, leading the ader to the first part of the section, is present in the account. A total of 488 species are scribed.

The plates provide colored illustrations of 474 of the 488 species contained within the iide. For each plate presented, the scientific name, common name, and coloration of the rds can be found on the opposing page. Approximately 10 to 15 birds can be found on each ate. General information on bird watching as well as a guide to Javan bird families can be und in the introductory section. A two-page glossary and a three-page bibliography can be und near the rear of the guide.

The six appendices consider such data as the birds of Java's offshore islands, a list of sident birds whose status in Java and Bali is a concern, endemic birds of Java and Bali, and so on. Separate indices to common names in English, scientific names, and common names in Indonesian complete the field guide.

1235a. MacKinnon, John, and Karen Phillips. **A Field Guide to the Birds of Borneo, Sumatra, Java, and Bali: The Greater Sunda Islands**. Oxford, New York: Oxford University Press, 1993. 491p. $85.00; $39.95pa. LC 92-30340. ISBN 0-19-854035-3; 0-19-854034-5pa.

Consisting of 820 species, this field guide describes the avian fauna found throughout the Greater Sunda Islands. Each species is arranged under its respective families. General information, including the common and scientific names, is given for each of the families. For each of the species listed, the common names, scientific names, size, coloration of feathers, iris, and bill, vocal pattern, geographical range and distribution, status, and general behavioral characteristics are considered. These descriptions are usually one-third of a page in length. In addition, cross-references to the colored plate of the bird can be found in the species account. A number of species can be found on each of the 88 colored plates. The colored plates are located in the first half of the work, followed by the accounts.

Introductory information includes geography, biogeography, conservation, field techniques for bird watching, as well as general information on when and where to see birds in this area of the world.

The appendices, which can be found near the end of the volume, consists of endemic threatened and endangered species in the main reserves; endangered and threatened species by island; land birds found on offshore island groups; Bornean montane birds by mountain group; an annotated list of birds of the Malay Peninsula not described in the test; graphical representation of vocal characteristics; and regional ornithological clubs, journals, and museums. A 10-page bibliography and a combined common/scientific-name index conclude the guide.

It should be noted that this guide basically includes some of the material found in MacKinnon's *Field Guide to the Birds of Java and Bali* (see entry 1235). Because this work has been expanded geographically, the number of species is almost double that of MacKinnon's *Field Guide to the Birds of Java and Bali.*

MEXICO

1236. Edwards, Ernest P. **A Field Guide to the Birds of Mexico: Including all Birds Occurring from the Northern Border of Mexico to the Southern Border of Nicaragua**. 2d ed. Sweet Briar, Va.: Ernest P. Edwards, 1989. 118p. $23.50. LC 88-83374. ISBN 0-911882-11-1.

Species are arranged under family names (common and scientific names given). For each bird listed, the common name, scientific name, person who first named it, coloration, and habitat are considered. Coloration information is also presented in Spanish. Within the descriptive portion a plate number is provided that leads to the colored plates found at the end of the guide. A combined scientific/common-name index concludes the work. Common names are listed in English and Spanish in the index.

1237. Peterson, Roger Tory, and Edward L. Chalif. **A Field Guide to Mexican Birds: Field Marks of All Species Found in Mexico, Guatemala, Belize (British Honduras), El Salvador**. Boston: Houghton Mifflin, 1973. 298p. $21.95. LC 73-4970. ISBN 0-395-17129-6.

Species are arranged under their respective families. The families are presented with their scientific and common names. After a brief description of the family, the species are presented. For each of these birds, the common name, scientific name, coloration, size, habitat, geographical range, and voice characteristics are presented. Colored plates of many of the species occur in the center of the guide. A brief bibliography as well as a combined common/scientific-name index can be found at the close of the volume.

IIDDLE EAST (see AFRICA)

'EW ZEALAND

238. Chambers, Stuart. **Birds of New Zealand: Locality Guide**. Hamilton [New Zealand]: run Books, 1989. 511p. ISBN 0-473-00841-6.

The species are arranged under their common-name groups, such as the kiwis, alba-osses, herons, and so on. For each of the species listed, the family, scientific name, common ıme, ecological importance, breeding season, habitat, geographical range, coloration, exter-ıl anatomical identifying features, vocal pattern, and areas in New Zealand where the bird ın be found are provided. Although the basic information on the bird is adequate, the thrust the guide is to alert the reader as to where in New Zealand a particular type of bird can be ound. A colored photograph of the species and a geographical range map accompany each escriptive account. The accounts are usually three pages in length. Some general facts on ew Zealand, along with a map of the islands, can be found in the introductory section. A ries of four planned trips (excursions) to facilitate bird watching can be found near the end the volume. A two-page bibliography, and a separate index of maps, places, and common ames of the birds complete the field guide.

'ORTH AMERICA

239. Bull, John, and John Farrand, Jr. **The Audubon Society Field Guide to North merican Birds: Eastern Region**. New York: Knopf, 1977. 775p. LC 76-47926. ISBN 394-41405-5.

Employing a flexible cover, this guide is easy to use and handle. The colored plates of e birds can be found in the first half of the book. The descriptive information that follows e plates consists of the common name, scientific name, and family of each species. In ldition, the size, coloration, voice, habitat, range, and nesting information are provided. The ısic outlines of the birds can be found on the opening pages. These outlines are contained the edge of each colored plate in the guise of a thumb print, thereby making identification sier. A brief glossary, an index giving credit to the photographers, and a combined mmon/scientific-name index conclude the guide.

240. Bull, John, and Edith Bull. **Birds of North America, Western Region: A Quick entification Guide for All Bird-Watchers**. New York: Macmillan, 1989. 144p. (Macmil-n Field Guides). $10.95. LC 88-27263. ISBN 0-02-526610-1.

Three hundred forty of the most common bird species of western North America can be und in this easy-to-use field guide; there are approximately 700 species throughout this rritory. The birds are arranged by colors and/or common-name groups. For each of the birds ted, the common name, size, coloration, habitat, and range are given. Colored plates of the rds can be found on each opposing page of the descriptive material. Some general informa-n on optical equipment, a one-page glossary, and an index arranged by common name that ves scientific name equivalents complete the guide.

41. Bull, John, and others. **Birds of North America, Eastern Region**. New York: acmillan, 1985. 159p. (Quick Reference Field Guide Series). $10.95. LC 84-26347. ISBN 02-079660-9.

Follows the same format as *Birds of North America, Western Region* (see entry 1240).

42. Clark, William S. **A Field Guide to Hawks, North America**. Boston: Houghton ifflin, 1987. 198p. (The Peterson Field Guide Series; 35). $19.95; $13.95pa. LC 87-4528. BN 0-395-36001-3; 0-395-44112-9pa.

Arranged by the common names of bird groups, this work encompasses the many types hawks seen throughout the North American continent. For each species entered, the mmon name, scientific name, anatomical description of various life stages of the bird (coloration, wing size, etc.), type of flight, behavior, status and distribution, similar specie subspecies, etymology of bird name, and measurements are given. In addition, a bri description of the family to which the bird belongs is presented. Geographical range ma are interspersed throughout the text. Illustrations comprise a series of colored plates as we as a series of black-and-white photographs. References to these appear in the descripti entries. Types of birds included are vultures, ospreys, kites, harriers, accipiters, buteone eagles, and falcons. A lengthy bibliography, an index to references by species and topics, an a combined common/scientific-name index complete the field guide.

1243. Farrand, John. **The Audubon Society Master Guide to Birding**. New York: Knop 1983. 3 vols. $47.85 set. LC 83-47945. ISBN 0-394-54121-9 set.

Covering the species of birds found on the North American continent, this three-volun set represents a major work in the identification of these animals. Each of the species arranged under its respective family, which in turn is entered under order. Common ar scientific names are assigned to the family. After a brief description of the family, the speci can be found. For each of the birds entered, the common name, scientific name, behavio anatomical description (size, coloration, etc.), voice patterns, the names of species that a similar, and geographical range are considered. Geographical range maps also accompar the description. One to three colored photographs of each bird are depicted on the opposir page of the descriptive matter. When two pictures are present, they show different aspects the bird. For example, the red-throated loon photographs show breeding plumage coloratic as well as winter plumage coloration, while the white ibis photos show the adult and juveni forms. The California condor, has three photographs, which show the juvenile in flight, t adult in flight, and the adult perched. Each volume has a combined scientific/common-nan index; volume three also contains a comprehensive index to the three-volume set. A introductory section describes basic bird anatomy, and tips on how to identify and locate the creatures. These are excellent field guides for the identification of the birds of Nor America.

Volume 1. *Loons to Sandpipers.*
Volume 2. *Gulls to Dippers.*
Volume 3. *Old World Warblers to Sparrows.*

1244. Farrand, John. **Eastern Birds**. New York: McGraw-Hill, 1988. 495p. (An Audub Handbook). $13.50. LC 87-3430. ISBN 0-07-019976-0.

This field guide includes those birds that are usually found east of the Rocky Mountai (between the Atlantic coast and western Great Plains), as well as in provinces along southe Canada. Arranged by habitat and species that look similar to one another, this guide includ the common name, scientific name, colored photographs, field marks, similar species, habit and range of each bird mentioned. A combined common/scientific-name index concludes t source. This is a true field guide; that is, it can be carried easily in the field and is not diffic to use.

1245. Farrand, John. **Western Birds**. New York: McGraw-Hill, 1988. 496p. (An Audub Handbook). $13.95. LC 87-3425. ISBN 0-07-019977-9.

Arranged exactly like *Eastern Birds* (see entry 1244), it covers all those birds usual found west of the Rocky Mountains (states between the Pacific coast and the western Gre Plains). In addition, species found in arid portions of western Texas, Oklahoma panhand and the extreme western areas of Kansas, Nebraska, and the Dakotas, as well as Saskatchewa are included. The same general type of information is given for each species as in *Easte Birds*. The guide contains a combined common/scientific-name index.

1246. **Field Guide to the Birds of North America**. 2d ed. Washington, D.C.: Nation Geographic Society, 1987. 464p. $16.95pa. LC 86-33249. ISBN 0-87044-692-4pa.

Arranged under common bird groups (families), this guide considers those species approximately 800) known to breed in North America. The scientific name of the family is presented as well as the family's common name. For each species presented, the common ame, scientific name, size in inches and centimeters of males and females, coloration, general xternal anatomical features, habitat and geographical distributions are considered. In addi-ion, each description, which is usually one-quarter of a page long, is accompanied by a olor-coded geographical range map of North America. Colored plates of the birds (males nd females when appropriate) can be found on the opposing page of text. Other colored plates ound throughout the text depict the various groups of birds in flight. The colored plates are xcellent. The introduction consists of brief information on species, families, scientific ames, plumages, plumage sequence, field marks, labeled colored drawing of a typical bird, measurements, voice, behavior, abundance, habitat, range maps, and birding. A combined ommon/scientific-name index concludes the guide. This work represents a first-rate field uide.

247. Harrison, Colin. **A Field Guide to the Nests, Eggs and Nestlings of North American Birds**. Cleveland, Ohio: Collins, 1978. 416p. LC 77-19361. ISBN 0-529-05484-1.

Species are arranged under their respective families. Families are presented with their cientific and common names. For each bird included, the common name, scientific name, reeding territories, information describing the nest, eggs, incubation, nestling (coloration), nd nestling period are given. The illustrative material is mainly in the form of colored plates howing nestlings and eggs. Drawings of nests can be found throughout the species descrip-on. A separate common/scientific-name index concludes the guide.

248. Harrison, Hal H. **A Field Guide to Western Birds' Nests: Of 520 Species Found Breeding in the United States West of the Mississippi River**. Boston: Houghton Mifflin, 979. 279p. (The Peterson Field Guide Series; 25). $17.95; $12.95pa. LC 79-11330. ISBN -395-27629-2; 0-395-47863-4pa.

For each species considered, the common name, scientific name, breeding range, habitat, est and egg descriptions are given. Nesting information usually consists of location, size in nches and centimeters, while egg characteristics comprise number, size in millimeters, and ncubation period. Black-and-white photographs as well as colored plates depicting broods f eggs in the birds' nests are contained within the guide. A brief glossary and a combined ommon/scientific-name index conclude the volume.

249. Mathews, F. Schuyler. **Field Book of Wild Birds and Their Music: A Description of the Character and Music of Birds, Intended to Assist in the Identification of Species Common in the United States East of the Rocky Mountains**. New York: Putnam, 1921. eprint, New York: Dover, 1967. 325p. LC 67-17084.

With birds arranged under their respective families, this work considers the vocalization atterns of birds east of the Rockies. For each bird included, the common name, scientific ame, size, geographical range, coloration, egg information, and vocalization patterns are iven. The vocalization is not only described in words; the actual musical score is shown. Many of these brief scores have been written for piano. A large number of black-and-white hotographs provide the illustrative material. Introductory information consists of a brief escription of bird music, a quick course on musical keys for those who do not read music, e musical scales of the thrushes, and a brief glossary defining musical terms. Six maps elating to the migration of birds, as well as a combined common/scientific-name index omplete the guide. Although this is outdated, it provides for a rather unique reference due the musical scores contained therein.

250. Peterson, Roger Tory. **A Field Guide to the Birds: A Completely New Guide to All he Birds of Eastern and Central North America**. 4d ed., completely rev. and enl. Boston: oughton Mifflin, 1980. 384p. (The Peterson Field Guide Series; 1). $17.95; $13.95pa. LC 0-14304. ISBN 0-395-26621-1; ISBN 0-395-26619-X.

As the subtitle implies, this guide includes all birds of Canada and the United States tha are endemic to Eastern and Central North America, east of the 100th meridian. The specie are arranged by broad categories, such as the ducks and their relatives, the seabirds, th long-legged wading birds, birds of prey, and so on. The organisms, entered initially b common names, are arranged under their respective families within the broader categorie mentioned above.

For each species considered, the scientific name, measurement, description, voice, an range are given. Colored illustrations are shown on the proceeding page for each bir described. In addition to a brief introduction, there is a series of color-coded range maps foun at the end of the volume. A dictionary arrangement of all birds mentioned, both by commo and scientific names, is listed in the index. Some general knowledge of these organisms i useful in order to make good use of this, and other, field guides.

1251. Peterson, Roger Tory, and Virginia Marie Peterson. **A Field Guide to Western Birds A Completely New Guide to Field Marks of All Species Found in North America Wes of the 100th Meridian and North of Mexico**. 3d ed. Boston: Houghton Mifflin, 1990. 432p (The Peterson Field Guide Series, no. 2). $17.95; $12.95pa. LC 89-31517. ISBN 0-395 51749-4; 0-395-51424-Xpa.

Each species is arranged under the common group name of the family, accompanied b the family's scientific name. For each species listed, the common name, scientific name, siz of body, size of wing spread, coloration, geographical range, habitat, and, in many cases voice patterns are presented. The colored plates (165 plates in all) of the species can be foun on the opposing pages of the descriptive information. Numerous bird species appear on eac plate. A map of the area considered and some general information on how to identify bird can be found in the introductory material. A list of geographical range maps that ar cross-referenced to the species accounts and vice versa, a systematic checklist, and combined common/scientific-name index conclude the guide.

1252. Udvardy, Miklos D. F. **The Audubon Society Field Guide to North America Birds: Western Region**. New York: Knopf, 1977. 854p. LC 76-47938. ISBN 0-394-41410-1

Follows the same basic format as *The Audubon Society Field Guide to North America Birds: Eastern Region* (see entry 1239).

1253. Walton, Richard K., and Robert W. Lawson. **Birding by Ear: Western Nort America: A Guide to Bird-Song Identification**. Boston: Houghton Mifflin, 1990. 64p. an 3 audio cassettes. (Peterson Field Guide Series; 41). $35.00. LC 90-30858. ISBN 0-395 52811-9.

Covering the western portion of North America, this kit contains a 64-page booklet an three audio cassettes. The booklet lists the birds that can be heard on each of the cassettes For each bird considered, the common name, scientific name, habitat, and voice pattern ar mentioned. In addition, there is a small black-and-white drawing of the bird, as well as a pag number to *A Field Guide to Western Birds*, 1990 (see entry 1251). A one-page bibliography a common-name index, and a phonetic index (bird sounds written out) complete the booklet The cassettes, of course, contain the sounds of the birds.

1254. Whitman, Ann H., ed. **Familiar Birds of North America: Eastern Region**. Nev York: Knopf; distr., Random House, 1986. 192p. $4.95pa. LC 86-45588. ISBN 0-394-74839 5pa.

Except for the differences in the types of birds included, this work follows the sam format as *Familiar Birds of North America: Western Region* (see entry 1255). A total of 8 familiar species are contained in this guide.

1255. Whitman, Ann H., ed. **Familiar Birds of North America: Western Region**. Nev York: Knopf; distr., Random House, 1986. 192p. $6.95pa. LC 86-45589. ISBN 0-394-74842 5pa.

Produced as an 11-by-16-centimeter booklet, this work provides general information [t]hat will enable the novice bird watcher to identify the common bird species (80 of them) [fo]und throughout the western United States. The geographical range is bounded by the Pacific [O]cean on the west and the 100th meridian on the east. The birds are presented in groups, [b]eginning with the heron and ending with the common barn owl. For each bird listed, the [c]ommon name, scientific name, habitat, nesting information, size, coloration, voice patterns, [g]eographical range, and a geographical range map are given. A full-colored picture of each [o]f the species can be found on the opposing page of the descriptive account. General [in]formation on the families of birds, a one-page glossary identifying the major external parts [o]f the bird with an accompanying diagram, and a combined common/scientific-name index [c]onclude the field guide. This guide is intended for the novice.

[P]ACIFIC OCEAN (see HAWAII)

[P]UERTO RICO and the VIRGIN ISLANDS

[1]256. Raffaele, Herbert A. **A Guide to the Birds of Puerto Rico and the Virgin Islands**. [Re]v. ed. Princeton, N.J.: Princeton University Press, 1989. 254p. $45.00; 16.95pa. LC [8]9-34781. ISBN 0-691-08554-4; 0-691-02424-3pa.

Arranged under families, this work presents 284 species that occur or have occurred in [P]uerto Rico and the Virgin Islands. The common and scientific names of the family as well [as] a description of the family are given. For each species, the descriptive account consists of [co]mmon name, scientific name, size (cm and in.), coloration, habitat, local common names, [vo]ice patterns, nesting information, and geographical distribution. The illustrations consist [of] black-and-white as well as colored plates. Each plate consists of a two-page spread. One [of] the pages lists the species under the common-names with scientific-name equivalents. Size [an]d coloration is also considered. The other page contains the pictures of the birds. A locality [ch]ecklist, which contains a list of the birds found within the guide, as well as their geographi[ca]l location, can be found near the rear of the work. A combined common/scientific-name [in]dex and a map of Puerto Rico as well as the Virgin Islands complete the guide.

[S]OUTH AMERICA

[1]257. Dunning, John S. **South American Birds: A Photographic Aid to Identification**. [N]ewtown Square, Pa.: Harrowood Books, 1987. 351p. $47.50; $35.00pa. LC 88-6299. ISBN [0-]915180-25-1; 0-915180-26-Xpa.

The species of birds are arranged under their common-name groups, such as eagles and [ha]wks, cuckoos, owls, pigeons and doves, and so on. For each bird included, the common [na]me, scientific name, coloration, and a geographical range map are given. In addition, a large [nu]mber of the descriptions are accompanied by a colored photograph of the bird; 2,700 species [ar]e considered with over 1,400 color photographs. The introductory section includes several [pa]ges of black-and-white drawings in order to help the reader learn how to identify major [bi]rd groups; the common names and descriptions are given. A combined common/scientific-[na]me index can be found closing the work. This guide represents quite a feat considering the [fa]ct that it covers a great majority of the birds of South America. According to the author, [th]ere are approximately 2,950 species.

[V]IRGIN ISLANDS (see PUERTO RICO [a]nd the VIRGIN ISLANDS)

WEST INDIES

1257a. Bond, James. **A Field Guide to Birds of the West Indies**. 5th ed. Boston: Houghton Mifflin, 1993. (The Peterson Field Guide Series). $24.95; $19.45pa. LC 93-9467. ISBN 0-395-67701-7; 0-395-67669-Xpa.

Consisting of colored plates and black-and-white drawings of the species, this work represents an identification guide to those birds found in the Bahama Islands, Greater Antilles, the Cayman and Swan Islands, Old Providence and St. Andrew in the southwest Caribbean, and the Lesser Antilles south to Barbados and Grenada. The species listed are arranged under their respective families. The common and scientific name of the family is presented along with a brief description of the family as a whole. For each species listed, the common name(s), scientific name, size, coloration, basic external characteristics, vocal patterns, habitat, nidification, diet (in most cases), and geographical range is considered. The length of the account varies from several sentences to several paragraphs. A black-and-white drawing of many of the birds accompany the description. Colored plates are depicted for those species that are not accompanied by a black-and-white drawing. Cross-references within the descriptive account of the species are given in order to locate the colored plates. A list of vagrant birds and a combined common/scientific-name index conclude the volume.

TEXTBOOKS

1258. Gill, Frank B. **Ornithology**. New York: W. H. Freeman, 1990. 660p. $49.95. LC 89-16793. ISBN 0-7167-2065-5.

Illustrated with photos, graphs, tables, drawings, labeled diagrams, and maps, this text serves as an introduction to the science of ornithology. It is divided into a series of sections, each of which contain a number of chapters to support the theme. The section names are "Origins," "Form and Function," "Behavior and Communication," "Behavior and the Environment," "Reproduction and Development," and "Populations." Each chapter concludes with a brief bibliography under the heading "Further Readings."

1259. King, A. S., and J. McLelland. **Birds: Their Structure and Function**. 2d ed. London; Philadelphia: Bailliere Tindall, 1984. 334p. $30.50pa. LC gb84-16060. ISBN 0-7020-0872-9.

Containing a large number of labeled diagrams, this book is a shortened version of King's four-volume *Form and Function in Birds* (see entry 1104). It has been greatly condensed and can serve as a text in a one-semester course. Sections of the work consider external anatomy, integument, skeletomuscular system, coelomic cavities, digestive, respiratory, female and male reproductive systems, the urinary system, cloaca and vent, endocrine system, cardiovascular, lymphatic, and nervous system, as well as a chapter on special sense organs. Species and genera of birds are used to illustrate specific points made in the text. An appendix with a listing of the common and scientific names of the birds cited within the text as well as a subject index completes the work. It should be realized, however, that this volume contains only anatomy and physiology, and, therefore, would not serve well for a course dealing with all aspects of avian biology.

1260. Sturkie, P. D., ed. **Avian Physiology**. 4th ed. New York: Springer Verlag, 1986. 516p. $79.00. LC 85-26049. ISBN 0-387-96195-X.

Encompassing 199 illustrations, this text presents the various types of physiological mechanisms present in birds. Coverage consists of the nervous system, sense organs, muscles, immune system, blood, circulatory system, respiration, regulation of body temperature, energy metabolism, digestive system, carbohydrate, protein, and lipid metabolism, excretory system, and the endocrine system. A number of chapters are devoted to the endocrine including pituitary, reproductive systems in male and female, thyroids, parathyroids, adrenals, pancreas and pineal. Specific bird species are used to illustrate the various physiological phenomena. Labeled line-drawings, graphs, tables, and charts provide the illustrative matter. At the end of each chapter there is a lengthy set of references, mainly to the periodical literature. These references have been cited in the text. The work concludes with a subject index.

Two appendices can be found at the end of the work. Appendix I gives brief descriptions on all major orders of birds. Line-drawings of the species enhance the textural material. Appendix II lists alphabetically the common names of birds with their scientific-name counterparts. Following the appendices is a 58-page bibliography in which can be found those works cited within the text. A subject/common-name index completes the volume.

This text would serve well for undergraduate and graduate courses in ornithology.

1261. Welty, Joel Carl, and Luis Baptista. **The Life of Birds**. 4d ed. New York: Saunders College Publishing, 1988. 698p. (The Saunders Series in Organismic Biology). LC 87-32143. ISBN 0-03-068923-6.

In addition to the various ethological and ecological chapters, this work presents a series of sections concerned with the physiology of organ systems. For example, there are chapters on bones and muscles; brain, nerves, and sense organs; food, digestion, and feeding habits; blood, air, and heat; and so on. Suggested readings are listed at the end of each chapter. Labeled diagrams, line-drawings, photographs, graphs, charts, tables, and maps enhance the discussions.

Near the end of the volume an 81-page bibliography is present in which those citations that are cited in the text can be found. A subject/common-name/scientific-name index completes the book.

Although both this work and Gill's *Ornithology* (see entry 1258), are ornithologically informative, Gill's text seems to be a more enjoyable read.

JOURNALS

1262. **American Birds: The Magazine of Record and Discovery**. Vols. 1- , nos. 1- . New York: National Audubon Society, 1947- . 5/yr. $35.00/yr. (institutions). ISSN 0004-7686.

Written in a magazine-type style and geared mostly to the ornithological layperson, this publication attempts to provide informative articles concerning our fine, feathered friends. Examples are the social behavior of birds, birds of a geographic area, watching birds, and a focus on a specific type of bird. A large number of color and black-and-white photographs as well as tables and maps are used to enhance the articles. Advertisements can also be found throughout these periodicals.

The last issue of the year represents a compilation of bird counts throughout Canada, the United States, Mexico, Belize, Honduras, Costa Rica, Panama, Colombia, Peru, Venezuela, Brazil, Bahamas, Puerto Rico, Bermuda, and the Virgin Islands. Common names of species are mentioned along with the number of times they have been sighted.

1263. **Ardea**. Vols. 1- , nos. 1- . Amsterdam: Netherlands Ornithologists' Union, 1912- . Semiannually. $48.76/yr. ISSN 0373-2266.

Although an occasional article in physiology can be found, the majority of papers published in this journal deal with ethology and ecology. Articles range in length from 5 to 30 pages each. Tables, graphs, maps, and photographs are interspersed among the text material. The research papers are written in English with English, Dutch, and French abstracts. In some instances, an entire issue is devoted to a symposium. For example, in volume 79 (no. 2), 1991, the second International Symposium on Western Palearctic Geese was published.

1264. **The Auk: A Quarterly Journal of Ornithology**. Vols. 1- , nos. 1- . Lawrence, Kans.: American Ornithologists' Union, 1884- . Quarterly. $60.00/yr. (institutions). ISSN 0004-8038.

From articles dealing with mitochondrial DNA variation to a miniature activity recorder for plunge-diving seabirds, this serial encompasses articles in taxonomy, ethology, ecology, physiology, reproduction, biochemistry, and the like. In addition to 20 or more research articles, spanning 10 or more pages each in length, each issue contains a number of short communication papers. Graphs, tables, and charts are used to illustrate concepts found within the text.

1265. **The Condor: A Journal of Avian Biology**. Vols. 1- , nos. 1- . Los Angeles: Cooper Ornithological Society, 1899- . Quarterly. $60.00/yr. (institutions). ISSN 0010-5422.

Although the title indicates journal articles relating to condors, this publication includes many varieties of bird species as papers are made available. The majority of articles consider ethological and ecological fields. Some examples are mating, breeding, egg-laying, migration, and population studies. In addition, articles dealing with physiology and biochemistry can be found. Maps, graphs, and tables provide the illustrative materials.

Each issue contains, on an average, 25 research papers running about 10 pages each in length. There are, in addition, a series of short communication articles averaging a few pages each in length. Book reviews can also be found in certain issues. A brief section entitled "News and Notes" (research news, meeting dates, etc.) completes the issue.

1266. **Ibis: Journal of the British Ornithologists' Union**. Vols. 1- , nos. 1- . Oxford, England: Published for the British Ornithologists' Union by Blackwell Scientific, 1859- . Quarterly. $199.50/yr. ISSN 0019-1019.

Approximately 10 to 12 research articles varying in length from 8 to 12 pages each, can be found in each issue. In some selected issues, book reviews are also present. The ornithological papers presented deal mainly with ethology, ecology, breeding, physiology, and biochemistry. Graphs, maps, and tables can be found interspersed within the text.

1267. **Journal of Field Ornithology**. Vols. 1- , nos. 1- . Lawrence, Kans.: Association of Field Ornithologists, 1930- . Quarterly. $45.00/yr. (institutions). ISSN 0006-3630.

Fifteen or more research articles can be found in each issue. Articles vary in length from approximately 5 to 10 pages each and cover mainly a wide variety of ecology and ethology papers. Nesting and breeding habits, population studies, migration patterns, and so on are some examples. Graphs and tables enhance the text.

In addition, reviews of recent literature are included within each issue and cover journal articles and books. Journal articles reviewed are arranged under broad ornithological categories, such as "Banding and Longevity"; "Migration, Orientation, and Homing"; "Population Dynamics"; and "Nesting and Reproduction."

1268. **Ornis Scandinavica**. Vols. 1- , nos. 1- . Copenhagen, Denmark: Munksgaard International, 1970 - . Quarterly. $90.00/yr. ISSN 0030-5693.

Approximately 10 articles averaging 10 pages in length each appear in most issues. Occasionally, an issue will be devoted to a symposium. For example Vol 22 (no. 3), July-September 1991, contains the Proceedings of the 5th International Symposium on Grouse.

The research articles seem to deal mainly with areas in ethology, and ecology. Some examples are population studies, migration, and sexual dimorphism. An occasional article in physiology will be present. Tables, charts, graphs, and maps compliment the text. In addition a number of short communication papers are published, a few pages each in length.

1269. **The Wilson Bulletin: A Quarterly Magazine of Ornithology**. Vols. 1- , nos. 1- Richmond, Va.: Wilson Ornithological Society, 1889- . Quarterly. $40.00/yr. ISSN 0043 5643.

Research articles discussing mainly the ecology and ethology of bird species are presented in each issue. In addition to the five or more major papers, there are approximately 10 short communication articles. The major papers are usually 10 pages each in length, while the shorter ones are usually 5 pages each. Tables and graphs can be found within the major papers. Each issue also contains a series of book reviews.

The major thrusts of these ecological and ethological papers are in the areas of migration, feeding, nesting, populations, and the like.

TAXONOMIC KEYS

1270. Coues, Elliott. **Key to North American Birds: Containing a Concise Account of Every Species of Living and Fossil Bird at Present Known from the Continent North of the Mexican and United States Boundary, Inclusive of Greenland and Lower California**. 5th ed. Boston: Page, 1872. Reprint, New York: Arno Press, 1974. 2 vols. (Natural Science in America Series). $77.00 set. LC 73-17816. ISBN 0-405-05732-6.

This work presents very detailed information regarding the order, suborder, family, subfamily, genus, and the species. Keys to the genera and subfamilies are included. The types of details that are included under each of the species entries are length of wing, tail, bill, middle toe and claw, as well as the bird itself; coloration to the finest detail; distinguishing features between male and female, geographical distribution, nesting information, number of eggs in a clutch, and size and color of eggs. In addition, the species are entered by their scientific names along with their common names. A large number of black-and-white drawings provide the illustrations. Citations to the literature appear throughout the text.

Volume 1 consists of three parts. Part I gives detailed information for collecting, preparing, and preserving birds. Part II provides a very detailed anatomical discussion in order to understand the structure and classification of birds, and Part III begins the systematic synopsis of the North American birds (listing of species). Volume 2 continues and completes the systematic synopsis. A systematic synopsis of fossil birds and a combined subject and scientific-name index conclude the source.

CHECKLISTS AND CLASSIFICATION SCHEMES

Aves (General)

1271. Clements, James. **Birds of the World: A Checklist**. 3d ed. New York: Facts on File, 1981. 562p. LC 80-26997. ISBN 0-87196-556-9.

Species are arranged under family names. The order of each bird is also given as well as the common name of the family. For each species entered, the scientific name, common name, and geographical location are presented. Subspecies and races are not included. A hierarchical coding system has been devised and a code has been assigned to each of the birds listed. The numbers in the code relate to order, family, and species. A brief bibliography and a separate scientific/common-name index conclude the checklist.

1272. Collar, N. J., and P. Andrew. **Birds to Watch: The ICBP World Checklist of Threatened Birds**. Washington, D.C.: Smithsonian Institution Press, 1988. 303p. (ICBP Technical Publication, no. 8). $19.95pa. LC 88-42876. ISBN 0-87474-301-X.

Threatened species are arranged under their respective families. The name of the order can be found above the family name, and the common group name of the bird is given with the family name. For each species entered, the common name, scientific name, geographical habitat, and reasons why the bird is threatened are included; the major reason given is destruction of the habitat. In addition, there is a 24-page bibliography, a checklist of the birds arranged by geopolitical unit, and a separate common/scientific-name index.

1273. Edwards, Ernest Preston. **A Coded List of Birds of the World**. Sweet Briar, Va.: Ernest P. Edwards, 1974. 174p. LC 73-93701. ISBN 911882-04-9.

Species are arranged under their respective families. The common name and geographic range is presented along with the scientific name of each bird. In addition to the scientific name of the family and order, common names are also given. For example, Procellariiformes (order) are albatrosses, petrels, and so on; a family under that order, Diomedeidae, are the albatrosses. The geographical ranges are listed in codes. These codes appear on a world map found at the beginning and ending of the book.

Other listings include orders and families of birds of the world, genera of birds of the world, an index of common names of orders and families, and a brief bibliography.

1274. Edwards, Ernest P. **A Coded Workbook of Birds of the World**. 2d ed. Sweet Briar, Va.: Ernest P. Edwards, 1982. 2 vols. $30.00. LC 82-82891. ISBN 0-911882-07-3 vol. 1; 0-911882-10-3 vol. 2.

These two volumes represent a revised edition of *A Coded List of Birds of the World* (see entry 1273). Following the same basic format as *A Coded List of Birds of the World*, these two volumes incorporate more species as well as an expansion of their geographical ranges. In addition, a cross-referenced list of codes are given, taking the reader from this work back to the first edition. A separate common/scientific-name index can be found at the end of each volume.

Volume 1. *Non-Passerines.*

Volume 2. *Passerines.*

1275. Gruson, Edward S. **Checklist of the World's Birds: A Complete List of the Species, with Names, Authorities and Areas of Distribution**. New York: Quadrangle, 1976. 212p. LC 72-85239. ISBN 0-8129-0296-3.

Published in small print, this work lists the various species of birds found throughout the world. Species are arranged under their respective families; scientific and common names are listed for each of the families. For each species entered, the scientific name, common name, the sources (presented in numbered codes) where information was found, and the geographical regions (presented in letter codes) are recorded. Notes on specific species, a key to the source codes, a bibliography, and separate common/scientific-name index can be found ending the book. A world map (in front and at end of the volume) indicates the geographical codes.

1276. Howard, Richard, and Alick Moore. **A Complete Checklist of the Birds of the World**. 2d ed. San Diego, Calif.: Academic Press, 1991. 622p. $41.78. ISBN 0-12-356910-9.

Based mainly on J. L. Peters' *Check-List of Birds of the World* (see entry 1279), this work attempts to list all species of birds in all geographical areas of the world. Beginning with the order Struthioniformes (ostriches) and ending with the Passeriformes (perching birds), each species and subspecies is listed under its respective family. The use of tribes, subgenera, and superspecies were not considered in order for simplification.

Common names are listed next to the families and species. For example, the family Cathartidae are known as the new world vultures, and the species *Cathartes aura* is known as the turkey vulture. Geographical ranges are also considered for each species and subspecies mentioned. The Passeriformes, largest of the orders, represents three-fourths of the checklist.

In addition to a brief introduction, there are several pages listing the arrangement of orders and families, as well as a brief list of taxonomic references for each family. There is a separate index for scientific names and common names.

1277. Lodge, Walter. **Birds Alternative Names: A World Checklist**. London: Blandford, 1991. 208p. £10.95. LC gb91-1265. ISBN 0-7137-2267-3.

Arranged by bird families, this checklist attempts to include birds having one or more alternative names in common usage. Under each family is listed the scientific name of the organism (genus/species/subspecies) along with the common name or names. The scientific ames are not arranged alphabetically under their respective families. A common-name index nd a generic index can be found at the end of the work. The listing of orders, families, and ubfamilies can be found preceding the checklist and can also be used as an index.

277a. Monroe, Burt L., and Charles G. Sibley. **A World Checklist of Birds**. New Haven, Conn.: Yale University Press, 1993. 393p. $45.00. LC 93-60341. ISBN 0-300-05547-1.

Based on Sibley and Monroe's *Distribution and Taxonomy of Birds of the World* (see ntry 1281), this checklist encompasses 9,702 species found in 2,063 genera. The species are rranged under tribes, which in turn are entered under subfamily, family, superfamily, arvorder, suborder, and order. All of the above taxonomic groups are not always present for ach of the orders considered. The orders are arranged phylogenetically beginning with the truthioniformes and ending with the Passeriformes. For each of the species listed, the cientific name, common name, and geographical distribution is given. In addition, species ot recognized by the authors but treated as such by others, are listed in italics under the pecies name proposed by Monroe and Sibley. An index to the genera as well as a common-ame index concludes the checklist.

278. Norton, John, Simon Stuart, and Tim Johnson, comps. **World Checklist of Threat-ned Birds**. 2d ed. Peterborough: Nature Conservancy Council, 1990. 274p. £20.00. LC b91-47294. ISBN 0-86139-6014.

The species (approximately 2,200) are arranged under their respective families, which n turn are entered under orders and classes. The scientific and common names are listed for he family. For each of the species presented, the scientific name, common names, geographi-al breeding, nonbreeding, vagrant areas, status (extinct, endangered, vulnerable, rare, etc.), xploitation considerations (such as feathers, hunting, live animal trade, food, etc.), and eferences to the literature (numbers are given that refer the reader to the bibliography at the nd of the work) are considered. In addition to the bibliography, a combined common/scientific-ame index completes the checklist. The table of contents, which can be found on the opening ages of the book, lists family names under their orders.

It should be noted that this checklist includes the species of birds that were listed in the CITES (Convention on International Trade in Endangered Species of Wild Fauna and Flora) ppendices up to January 1990.

279. Peters, James Lee. **Check-List of Birds of the World**. Cambridge, Mass.: Museum f Comparative Zoology, 1931- . 16 vols.

Species and subspecies are arranged under family names. A family can be found under rder, suborder, and superfamily. For each entry, the person who first named it, citations to he taxonomic information, and geographical range are given. This is a classic and authorita-ive work. A second edition of the work began in 1979. Contents of the volumes are as ollows:

Volume 1. *Struthioniformes, Tinamiformes, Procellariiformes, Sphenisciformes, Gaviiformes, Podicipediformes, Pelecaniformes, Ciconiiformes, Phoenicopteriformes, Falconiformes, Anseriformes*. 1979. $50.00. ISBN 0-910999-01-5.

Volume 2. *Galliformes, Gruiformes, Charadriiformes*. 1934. $35.00. ISBN 0-910999-02-3.

Volume 3. No name given. Orders include Columbiformes, Psittaciformes. 1937.

Volume 4. *Cuculiformes, Strigiformes, Caprimulgiformes, Apodes*. 1940. $35.00. ISBN 0-910999-04-X.

Volume 5. *Trochili, Coliiformes, Trogoniformes, Coraciiformes*. 1945. $35.00. ISBN 0-910999-05-8.

Volume 6. *Piciformes*. 1948. $35.00. ISBN 0-910999-06-6.

Volume 7. *Eurylaimidae, Dendrocolaptidae, Furnariidae, Formicariidae, Conopophagidae, Rhinocryptidae*. 1951. $35.00. ISBN 0-910999-07-4.

Volume 8. *Tyrannidae, Pipridae, Cotingidae, Oxyruncidae, Phytotomidae, Pittidae, Philepittidae, Acanthisitiidae, Menuridae, Atrichonrnithidae.* 1979. $45.00. ISBN 0-910999-08-2.

Volume 9. *Alaudidae, Hirundinidae, Motacillidae, Campephagidae, Pycnonotidae, Irenidae, Laniidae, Vangidae, Bombycillidae, Dulidae, Cinclidae, Troglodytidae, Mimdae.* 1960. $35.00. ISBN 0-910999-09-0.

Volume 10. *Prunelidae, Turdinae, Orthonychinae, Timaliinae, Panurinae, Picathartinae, Polioptilinae.* 1964. $35.00. ISBN 0-910999-10-4.

Volume 11. *Sylviidae, Muscicapidae (sensu strictor), Maluridae, Acanthizidae, Monarchidae, Eopsaltriidae, Platysteriridae.* 1986. $75.00. ISBN 0-910999-11-2.

Volume 12. *Pachycephalinae, Aegithalidae, Remizidae, Paridae, Sittidae, Certhiidae, Rhabdornithidae, Climacteridae, Dicaeidae, Nectarimiidae, Zosteropidae, Meliphagidae.* 1967. $40.00. ISBN 0-910999-12-0.

Volume 13. *Emberizinae, Catablyrrhynchinae, Cardinalinae, Thraupinae, Tersiminae.* 1970. $40.00. ISBN 0-910999-13-9.

Volume 14. *Parulidae, Drepanididae, Vireonidae, Icteridae, Fringillinae, Carduelinae, Estrilididae, Viduinae.* 1968. $40.00. ISBN 0-910999-14-7.

Volume 15. *Bubaiornithinae, Passerinae, Ploceinae, Sturnidae, Oriolidae, Dicruridae, Callaeidae, Grallinidae, Artamidae, Cracticidae, Ptilonorhychidae, Paradisaeidae, Corvidae.* 1962. $35.00. ISBN 0-910999-15-5.

Volume 16. Comprehensive index to the volumes listed above. 1987. $75.00. ISBN 0-910999-16-3.

1280. Richmond, Charles W. **The Richmond Index to the Genera and Species of Birds.** Riverside, N.J.: G. K. Hall, 1992. 72,000 cards on 92 microfiche. $495.00. ISBN 0-8161-1795-0.

Consisting of the most up-to-date index to both the genera and species, this work considers the literature of the nineteenth century to the present. The entries are by the scientific name of the genera and species of birds. Other information included provides the researcher with person who first named it, date of naming, reference to type specimens, original bibliographic reference along with other published references, and an annotation.

This is the first time this checklist is being published. The original 72,000 cards can be found in the Smithsonian Institution, National Museum of Natural History, Division of Birds.

According to Richard C. Banks, Zoologist, U.S. Fish and Wildlife Service "The Richmond Index enables a researcher to solve a problem of the source of a 'mystery' name in a few minutes rather than spending hours in a fruitless search of references to find another mention of it. Without this unique index one might never solve the problem."

1281. Sibley, Charles G., and Burt L. Monroe, Jr. **Distribution and Taxonomy of Birds of the World**. New Haven, Conn.: Yale University Press, 1990. 1111p. $125.00. LC 90-70494. ISBN 0-300-04969-2.

Based upon DNA-DNA hybridization studies of Sibley and Ahlquist in their book entitled *Phylogeny and Classification of Birds: A Study in Molecular Evolution* (see entry 1107), this comprehensive work considers the taxonomy of bird species and arranges them based on these new phylogenetic findings. The arrangement is by subclass, followed by infraclass, parvclass, order, suborder, infraorder, family, and, in some cases, tribe. Species are listed under their respective families. Species are not listed taxonomically. For each organism listed, the scientific name, person who first named it, date of naming, common name, and geographical range are included. A number of symbols are employed within the description indicating extinction, alternate genus, alternate scientific name, and so on.

The remainder of the source includes a world numbers index (based on the American Ornithologist's Union [A.O.U.] numbering system, and includes 9,881 entries), a series of maps depicting all geographical areas of the world, a gazetteer to compliment the maps, a comprehensive bibliography arranged by author, and a combined scientific/common-name index. This work will, no doubt, become one of the classics in avian taxonomy.

It should be noted that a supplement to this work was published in 1993 by Yale University Press.108p. $25.00. ISBN 0-300-05549-8.

Aves (Geographic Section)

AFRICA

1282. Clancey, P. A., and others. **Check List of the Birds of South Africa**. Cape Town: University of Cape Town, South African Ornithological Society, List Committee, 1969. 338p. LC 78-581089.

The species (828 of them) are arranged under their respective genera, which in turn are entered under families, orders, and subclasses. The scientific and common names are given for the family. Genus information includes the person who named it and citations to the literature. For each of the species listed, the scientific name, person who first named it, common name, citations to the literature, and geographical distribution are given. An index to the orders and families, as well as a map of South Africa can be found in the opening pages. A five-page bibliography can be found at the end of the work. There are no other indices save for the index to orders and families mentioned above.

1283. Gunning, J. W. B., and Alwin Haagner. **A Check-List of the Birds of South Africa: Being a Record of All the Species Known to Occur South of the Zambezi-Cunene Line (The 16th Degree of South Latitude)**. Pretoria, South Africa: The Government Printing and Stationery Office, 1910. [75]-156p. LC 20-9002.

The species (920 of them) are arranged under their respective genera, which in turn are entered under families. Genera information consists of name, person who named it, and date of naming. For each of the species listed, the scientific name, person who named it, date of naming, common name (in English and German) and citations referring the reader to Reichenow's *Vogel Afrikas* are given. There is no index. It should be noted that this work was published as a supplement to *Annals of the Transvaal Museum*, volume II.

ANGOLA

1284. Traylor, Melvin Alvah. **Check-List of Angolan Birds**. Lisboa, Angola: 1963. 250p. (Publicacoes Culturais, No. 61).

The species (859 species and 188 subspecies) are arranged under their respective families. The common name, along with the scientific name of the family are given. For each of the species/subspecies listed, the scientific name, person who first named it, bibliographic taxonomic citation, and geographical distribution are given. A gazetteer of Angola, and a scientific-name index complete the checklist.

ANTARCTICA (see NEW ZEALAND)

AUSTRALIA

1285. Condon, H. T. **Checklist of the Birds of Australia**. Melbourne: Royal Australasian Ornithologists Union, 1975- . 311p. LC 77-353344. ISBN 0-9599832-1-0 vol. 1.

The species/subspecies (393 indigenous living species, 43 fossil species, and 10 introduced species) are arranged under their respective genera, which in turn are entered under families, suborders, and orders. The scientific and common names are given for the order suborder, and family. For each of the genera listed, bibliographic citations to the taxonomi literature are presented. For each of the species listed, the scientific name, person who firs named it, common name, geographical distribution, and bibliographic citations to the taxo nomic literature are given. A table of orders and families can be found in the opening pages A three-page list of references, arranged geographically, a gazetteer, and a scientific-nam index conclude the checklist.

This work represents volume 1 of a two-volume set. Volume 1 considers the nonpas serines. As of this writing there is no indication that volume 2 has been published.

1286. Parker, S. A., and others. **An Annotated Checklist of the Birds of South Australia** Adelaide, S. Aust.: South Australian Ornithological Association, 1979- . 2 vols. to date. LC 85-229359.

The species are arranged under their respective families. For each of the species listed the scientific name, person who first named it, date of naming, common name, geographica range, status, and habitat are given. In many cases, an extensive amount of information o geographical range is present. A gazetteer and a combined common/scientific-name inde conclude each part. At this time, part 1 and a portion of part 2 have been published.

Part 1. *Emus to Spoonbills.*

Part 2A. *Waterfowl.*

According to the introductory information found in part 2A, information on the gul and terns will be found in part 2B.

BELIZE

1287. Wood, D. Scott, Robert C. Leberman, and Dora Weyer. **Checklist of the Birds o Belize**. Pittsburgh, Pa.: Carnegie Museum of Natural History, 1986. 24p. (Special Publicatio [Carnegie Museum of Natural History], no. 12). LC 89-207290.

The species are arranged under their respective families. The common and scientifi names of the family are presented. For each of the species listed, the common name, scientifi name, temporal distribution (permanent resident, winter resident, etc.), geographical distri bution, and abundance (common, uncommon, rare, very rare) are given. There is no index.

BERMUDA

1288. Wingate, David B. **A Checklist and Guide to the Birds of Bermuda**. [s.l., s.n.] 1973, (Bermuda, Island Press). 36p. LC 77-371069.

The species (320 of them) are arranged phylogenetically by common name. For each o the species entered, the common name, scientific name, status, and seasonal distribution an abundance charts are given. The entire checklist is presented in a chart form. A species accour of a few of the birds listed (20 of them) can be found in the introductory material. Suc information as breeding and coloration are given in these accounts. There is no index.

CANADA

1289. Godfrey, W. Earl. **Checklist of Canadian Birds. Liste de Reference des Oiseau du Canada**. Toronto: Royal Ontario Museum, 1986. 61p.

The species (579 of them) are arranged under their respective families, which in tur are entered under orders. For each of the species listed, the scientific name and common name are given. There is no index.

CEYLON (see SRI LANKA)

CHILE

1290. West, Steve. **Checklist to the Birds of the Republic of Chile**. Kissimmee, Fla.: Russ Mason's Natural History Tours, 1985. 12p.

The species (over 400 of them) are listed under their respective families. Scientific and common names for the families are present. For each of the species listed, the common name, scientific name, habitat, status, and geographical range are given. There is no index and no table of contents.

CYPRUS

1291. Stewart, P. F. **A Check List of the Birds of Cyprus**. [s.l., s.n., 1971]. 92 leaves.

According to the author's note, Cyprus is geographically situated in a major migratory path. Therefore, many of the birds listed are not considered permanent residents of the island. The species/subspecies are arranged under their respective genera, which in turn are entered under families and orders. For each of the species/subspecies listed, the scientific name, person who first named it, common name, and status (migratory, permanent resident, etc.) are given. In addition, the months of the year one is most likely to encounter the species are also considered. A combined common/scientific-name index concludes the checklist.

ECUADOR

1292. Butler, Thomas Y. **The Birds of Ecuador and the Galapagos Archipelago: A Checklist of All the Birds Known in Ecuador and the Galapagos Archipelago and a Guide to Help Locate and See Them**. Portsmouth, N.H.: Ramphastos Agency, 1979. 78p. LC 78-61829.

The species are entered by common name with scientific name equivalents, and are arranged under their respective families. The common and scientific names of the families are given. By means of a checking system, the habitat, frequency of sighting, birds recorded in recent years (1972-1978), and references to illustrations of the birds that can be found in other sources are listed. A number of black-and-white and colored photographs provide illustrations for this book. Illustrations are found in the front and rear of the checklist. There is no index.

ETHIOPIA

1293. Urban, Emil K., and Leslie H. Brown. **A Checklist of the Birds of Ethiopia**. Addis Ababa, Ethiopia: Dept. of Biology, Haile Sellassie I University, 1971. 143p. LC 72-169752.

The species/subspecies (827 of them) are arranged under their respective families, which in turn are entered under orders. For each of the species listed, the scientific name, person who first named it, common name, status, geographical distribution, habitat, and breeding records are presented. Geographical information on Ethiopia can be found in the introductory material. A three-page bibliography, a gazetteer of localities, a table listing changes in classification, and a separate common/scientific-name index conclude the checklist.

GALAPAGOS (see ECUADOR)

GREAT BRITAIN

1294. British Ornithologists' Union. **Check-List of the Birds of Great Britain and Ireland**. London: British Ornithologists' Union, 1952. 106p. LC a54-1836.

The species/subspecies (426 of them) are arranged under their respective genera, which in turn are entered under families and orders. For each genera listed, the person who named it and a citation to the literature are given. For each of the species listed, the scientific name, person who named it, common name, citation to the literature, geographical distribution, and status (resident, scarce, nonbreeding visitor, etc.) are given. A list of the orders and families can be found in the table of contents. A separate genus and common-name index can be found concluding the checklist.

1295. Witherby, H. F. **A Check-List of British Birds: With a Short Account of the Status of Each: Compiled from "The Handbook of British Birds,"** by H. F. Witherby. rev. ed. London: H. F. & G. Witherby, 1941. 78p. LC 43-12546.

The species (520 of them) are arranged under their respective genera, which in turn are entered under families and orders. The person who first named the genus is given. For each of the species listed, the scientific name, common name, whether the bird is a resident, rare vagrant, vagrant, occasional winter-visitor, and so on, and geographical distribution are given.

HAWAII

1296. Bryan, E. H. **Check List and Summary of Hawaiian Birds**. Honolulu, Hawaii: Books about Hawaii, 1958. 28p.

The species/subspecies (231 of them, native and introduced) are arranged under their respective genera, which in turn are entered under families and orders. For each family considered, the scientific and common names are presented. For each genera listed, the person who named it and date of naming are considered. For each of the species listed, the scientific name, person who named it, date of naming, common names (in English and Hawaiian), and geographical distribution are presented. A common-name index completes the checklist.

IRELAND (see GREAT BRITAIN)

KENYA

1297. Backhurst, G. C. **Check-List of the Birds of Kenya**. Nairobi, Kenya: Ornithological Sub-Committee, EANHS, 1981. 40p. LC 81-980471.

Compiled from G. C. Backhurst's book *Birds of East Africa: Their Habitat, Status, and Distribution*, this source provides a list of those birds found throughout the Kenyan countryside. The species are arranged under their respective families. The scientific and common names of the family are presented. For each of the species listed, the common name and scientific name are given. A common-name index is provided at the end of the checklist.

MALAWI

1298. Benson, C. W. **A Check List of the Birds of Nyasaland: Including Date on Ecology and Breeding Seasons**. Blantyre, Malawi: Nyasaland Society, 1953. 118p. LC 54-37150.

The species (609 of them) are arranged under their respective families. For each of the species listed, the scientific name, common name, habitat, and references to the literature are given. Habitats and references are given in code. Charts for the habitat code can be found at the beginning of the book, while the reference codes appear near the end of the work. A map of Nyasaland, now known as Malawi, can be found in the opening pages. An index to generic names completes the checklist.

MALAYA

1299. Gibson-Hill, C. A. **An Annotated Checklist of the Birds of Malaya: An Annotated List of the Birds Occurring, or Known to Have Occurred, in the Territories of the Federation of Malaya and the Colony of Singapore**. Singapore: V. C. G. Gatrell, government printing, 1949. 299p. (Bulletin of the Raffles Museum, Singapore, no. 20).

The species/subspecies (575 of them) are arranged under their respective genera, which in turn are entered under families. The person who first named the genus is presented. For each of the species listed, the scientific name, person who first named it, common name, alternative scientific names, citations to the literature, geographical distribution, and status (rare vagrant, occasional vagrant, occasional visitor, resident, introduced, etc.) are given. A table of contents listing the scientific and common names of the families, and maps of the Malay Peninsula can be found in the introductory pages. A 10-page bibliography and a scientific-name index conclude the checklist.

MEXICO

1300. Birkenstein, Lillian R., and Roy E. Tomlinson, comps. **Native Names of Mexican Birds**. Washington, D.C.: United States Department of the Interior, Fish and Wildlife Service, 1981. 159p. (Resource Publication [U.S. Fish and Wildlife Service], vol. 138). LC 80-606886.

The species of birds (994 species) are listed under their respective families (89 families). For each species listed, the common name in English and the common name(s) in Spanish are given. The family name in Spanish is also provided. A separate index to the common Mexican names and the common English names concludes the list.

NEW ZEALAND

1301. Checklist Committee, Ornithological Society of New Zealand. **Checklist of the Birds of New Zealand and the Ross Dependency, Antarctica**. 3d ed. Auckland, New Zealand: Random Century, in association with the Society, 1990. 247p. ISBN 1-86941-082-3.

The species (379 of them) are arranged under their respective genera, which in turn are entered under subfamilies, families, and orders. A brief description of evolutionary considerations, along with appropriate citations to the literature, are given for the order. For each of the genera included, the person who first named it and citations to the literature are considered. For each of the species listed, the person who first named it, alternative scientific names with citations to the literature, and some general information on fossil forms along with citations to the literature are presented. Fossil information consists of geological era and geographical distribution. An index of common names, an index of scientific names, and a series of maps depicting New Zealand and the Ross Dependency complete the checklist.

NORTH AMERICA

1302. American Ornithologists' Union. **Check-List of North American Birds: The Species of Birds of North America from the Arctic through Panama, Including the West Indies and Hawaiian Islands**. 6th ed. Washington, D.C.: American Ornithologists' Union, 1983. 877p. LC 84-178617. ISBN 0-943610-32-X.

Prepared by the Committee on Classification and Nomenclature of the American Ornithologists' Union, this work provides a list of bird species located in North America as well as the West Indies and Hawaiian Islands. The species/subspecies are arranged under their respective genera, which in turn are entered under families, orders, superorders, subclasses, and classes. Scientific and common names are given for the class, subclass, superorder, order, and family. The person who first named the genus is presented. For each bird species listed, the scientific name, person who first identified it, common name, and taxonomic citation to the literature are given. In addition, geographical distribution (including breeding information, in many cases), habitat, and, in many cases, taxonomic information can also be found. A combined common, scientific-name index concludes the checklist.

NYASSALAND (see MALAWI)

PAKISTAN

1303. Khanum, Zakia, Manzoor Ahmed, and Mohammad Farooq Ahmed. **A Check List of Birds of Pakistan with Illustrated Keys to Their Identification**. Karachi: Zoological Survey Dept., 1980. 138p. (Records [Zoological Survey of Pakistan, vol. 9 (1 and 2)]).

The species/subspecies (over 1,200 of them) are arranged under their respective genera, which in turn are entered under families and orders. The species/subspecies are presented by means of biological keys. Such facts as size, coloration, and basic external anatomy are employed within the keys in order to identify families, genera, and species. A second listing, near the end of the work, presents the species/subspecies under their respective families, which in turn are entered under orders. For each of the species listed, the scientific name, common name, status, and page number indicating location within the main part of the work are given. A table of contents lists the keys to the families, genera, and species. A one-page bibliography, and a series of black-and-white drawings depicting the heads, wings, and claws of many of the birds conclude the checklist.

PERU

1304. Parker, Theodore A., Susan Allen Parker, and Manuel A. Plenge. **A Checklist of Peruvian Birds**. Tucson, Ariz.: S. Parker, 1978. 54p.

The species are arranged under their respective families. Common and scientific names of the families are given. For each of the birds listed, the common name and scientific name are included. There is no index.

PUERTO RICO

1305. Leopold, N. F. **Checklist of Birds of Puerto Rico and the Virgin Islands**. Rio Piedras, Puerto Rico: University of Puerto Rico, Agricultural Experiment Station, 1963. 119p. (Bulletin [University of Puerto Rico (Rio Piedras Campus). Agricultural Experiment Station], no. 168).

The species (209 of them) are arranged phylogenetically by their common English name. For each of the species listed (table format), the English common name and synonyms, the Spanish common name and synonyms, scientific name, and resident status are given. A two-page bibliography and a common-name index (English and Spanish interfiled) complete the checklist.

SAUDI ARABIA

1306. Jennings, Michael C. **The Birds of Saudi Arabia: A Check-List**. Cambridge, England: M. C. Jennings, 1981. 109p. ISBN 0-9507405-0-0.

This source covers some 413 species of birds found within the boundaries of Saudi Arabia. Bird species are arranged under their respective families. Scientific and common names of the family are given. For each species considered, the scientific name, common name, habitat, status, geographical range, and cross-references to the geographical range maps found near the end of the list are considered. Occasional line-drawings of a few of the birds provide the illustrations. The introductory information consists of a listing and description of the geographical areas that comprise Saudi Arabia. A six-page bibliography of works consulted, a gazetteer, the geographical range maps, and a scientific-name index conclude the work.

SOUTH AMERICA

1307. Altman, Allen, and Byron Swift. **Checklist of the Birds of South America**. 2d ed. Washington, D.C.: Printed by St. Mary's Press, 1989. 82p. $6.50pa. LC 89-90962. ISBN 0-9622559-1-2pa.

Containing one-third of the bird species found throughout the world, this source provides names of those birds inhabiting the South American continent. The species are arranged under their respective families. Common and scientific names of the families are given. For each of the species entered, the common name, scientific name, status information (resident, nonbreeding migrant, etc.), and geographical distribution are given. Material is presented in a table format. A list of birds of the South American Islands (not included in the main list), a two-page bibliography, an index to the genera, and a common-name index complete the checklist.

SRI LANKA

1308. Phillips, W. W. A. **Annotated Checklist of the Birds of Ceylon (Sri Lanka)**. rev. ed. Colombo, Sri Lanka: Wildlife and Nature Protection Society of Sri Lanka in Association with The Ceylon Bird Club, 1978. 92p.

The species/subspecies (427 of them) are arranged under their respective families, which in turn are entered under orders. The scientific and common names are given for order and family. For each of the species listed, the scientific name, person who first named it, date of naming, common names, citations to the literature, geographical distribution, and status (rare vagrant, breeding resident, etc.) are given. A systematic classification scheme, down to family, appears in the table of contents. The introductory section considers the geography of Sri Lanka, as well as bird migrations to and from Sri Lanka. There is no index.

THAILAND

1309. Deignan, Herbert G. **Checklist of the Birds of Thailand**. Washington, D.C.: Smithsonian Institution; for sale by the Superintendent of Documents, U.S. Government Printing Office, 1963. 263p. (U.S. National Museum, Bulletin No. 226). LC 64-60323.

Consisting of 1,173 species, this source provides a listing of those birds found throughout Thailand. The species are arranged under their respective genera, which in turn are entered under families and orders. For each genera listed, the person who first identified it is included. For each of the species/subspecies, the person who first named it, citation to the taxonomic literature, and geographical range are given. A map of Thailand showing the provinces can be found in the introductory material. A scientific-name index concludes the checklist.

TURKEY

1310. Vittery, A., R. F. Porter, and J. E. Squire, eds. **Check List of the Birds of Turkey**. [Sandy, Bedfordshire, England]: Records and Editorial Committee, Ornithological Society of Turkey, 1971. 34 leaves.

The species (394 of them) are listed phylogenetically by their scientific name. For each of the species listed, the scientific name, common name, status (breeding resident, breeding summer visitor, etc.), and geographical distribution are given. Geographical information on Turkey is presented in the introductory material. There is no index.

VIRGIN ISLANDS (see also PUERTO RICO)

1311. Leck, Charles F., and Robert L. Norton. **An Annotated Checklist of the Birds of the U.S. Virgin Islands**. Christiansted, St. Croix: USVI, Antilles Press, 1991. 40p. ISBN 0-916611-01-9.

The species (208 of them) are entered under their common names, phylogenetically. Fo each of the birds listed, the common name, scientific name, habitat, and, in many cases, diet frequency of occurrence, geographical distribution, and numbers (abundant, common, un common, etc.) are given. Some general information on the Virgin Islands (St. Thomas, St John, and St. Croix) is presented in the introductory information. A two-page bibliograph concludes the checklist. There is no index.

WEST INDIES

1312. Bond, James. **Check-List of Birds of the West Indies**. 4th ed. Philadelphia: Th Academy of Natural Sciences, 1956. 214p.

Including the Bahama Islands, Greater Antilles, Lesser Antilles, the Cayman and Swa Islands, and the Colombian Islands of San Andres and Providencia, this work provides a lis of bird species inhabiting the above geographical region. The species/subspecies are arrange under their respective families, which in turn are entered under orders. The person who firs named the family is considered. For each of the species/subspecies listed, the person who firs named it, bibliographic taxonomic citation, and geographical distribution are given. A scientific-name index completes the checklist.

BIOGRAPHIES

1313. Mearns, Barbara, and Richard Mearns. **Biographies for Birdwatchers: The Live of Those Commemorated in Western Palearctic Bird Names**. London; San Diego, Calif. Academic Press, 1988. 490p. (Books About Birds). $37.50. LC gb88-22412. ISBN 0-12-487422-3.

In order to be included in this monograph, one would have to have had one or mor western palearctic birds named after them as either a common name or a scientific name. Fo each inclusion (91 in all), the name of the person, date of birth and death when appropriate common and scientific names of the bird, person who first named the bird, and a bibliographic citation referring the reader to the original article where the bird was first described are given In addition, a biographical sketch of the person is presented. These sketches are approximately two to seven pages long. Needless to say, the sketch emphasizes the person's life as a bird enthusiast. A picture of the bird in question, along with a picture of the biographee are presented (in most cases). An occasional geographical range map can be found in the biographical sketches.

Appendix 1 consists of a listing of naturalists commemorated by species of uncertain status within the western palearctic, while appendix 2 consists of a selection of naturalists that were mentioned within the text. A combined common/scientific-name index, as well as a people index conclude the work.

GUIDES TO THE LITERATURE

1314. Miller, Melanie Ann. **Birds: A Guide to the Literature**. New York: Garland, 1986. 887p. (Garland Reference Library of the Humanities). $86.00. LC 85-45116. ISBN 0-8240-8710-0.

This guide represents 1,942 sources dealing with bird literature. The majority of these sources are listed under broad subject headings such as behavior, distribution and populations, evolution, ecology, migration and banding, flight, and so on. The first nine pages are devoted to general sources such as dictionaries, encyclopedias, bibliographies, and the like. A section on children's literature, biographies, and fiction can be found near the rear of the source. The discussion of each book is listed in a tabular format. This format consists of a sentence describing the book as a whole, coverage (geographical), description (types of information contained therein), illustrations, and, in some cases, additional features. The citations to the sources consist of author, title, edition, imprint, and pagination. An appendix listing a number of the major journals, and a combined author and title index conclude this reference book.

It should be noted that this guide should be used for more complete coverage of the bird literature than could possibly be found in this chapter. For example, bird sources of each of the states of the United States are included in Miller's work; this work excludes specific state material. The states of Alaska and Hawaii are exceptions. Trying to conceive of adequate broad subject headings for the large number of sources can be a chore. Miller's choice of headings are adequate. However, this work would have been more useful if a subject index had been produced.

ASSOCIATIONS

1315. **African Love Bird Society**. P.O. Box 142, San Marcos, CA 92079-0142. (619) 727-1486.

1316. **African Seabird Group**. P.O. Box 34113, Rhodes Gift 7707, South Africa. 21-6503294. FAX 21-6503726.

1317. **American Birding Association**. P.O. Box 6599, Colorado Springs, CO 80934. (719) 634-7736. (800) 634-7736. FAX (719) 471-4722.

1318. **American Border Fancy Canary Club**. 348 Atlantic Ave., East Rockaway, NY 11518. (516) 593-2841.

1319. **American Budgerigar Society**. 1704 Kangaroo, Killeen, TX 76543. (817) 699-3965.

1320. **American Cockatiel Society**. 2213 Bermuda, West Palm Beach, FL 33406. (407) 967-5439.

1321. **American Dove Association**. P.O. Box 21, Milton, KY 40045. (502) 268-3240.

1322. **American Federation of Aviculture**. 3118 W. Thomas Rd., Ste. 713, Phoenix, AZ 85017. (602) 484-0931.

1323. **American Norwich Society**. 3799 Quails Walk, Bonita Springs, FL 33923. (813) 992-6331.

1324. **American Ornithologists' Union**. Natl Museum of Natural History, Smithsonian Institution, Washington, DC 20560. (203)357-1970. Telecommunications Services (202) 357-1932.

1325. **American Singers Club**. 3564 Loon Lake Rd., Wixom, MI 48393. (313) 624-8127.

1326. **Association of Avian Veterinarians**. P.O. Box 299, East Northport, NY 11731. (516) 757-6320.

1327. **Association of Field Ornithologists**. Broadmoor Wildlife Sanctuary, Massachusetts Audubon Soc., 280 Eliot St., South Natick, MA 01760. (508) 655-2296.

1328. **Atlantic Waterfowl Council**. Dept. of Natural Resources, Game and Fish Div., 20: Butler St., Atlanta, GA 30334. (404) 656-3523.

1329. **Australasian Seabird Group**. P.O. Box 12397, Wellington North, New Zealand.

1330. **Avicultural Society of America**. 17347 Aspenglow, Yorba Linda, CA 92686. (714 996-5538.

1331. **Bird Association of California**. 679 Prospect, Pasadena, CA 91103. (818) 795-6621

1332. **Bird Strike Committee Europe**. (Bird Problems Around Airports.) Luftfartshuset Ellebjergvej 50, Postboks 744, DK-2450, Copenhagen SV, Denmark. 36444848. FA 36440303. TELEX 27096.

1333. **Birds of Prey Rehabilitation Foundation**. RR 2, Box 659, Broomfield, CA 80020 (303) 460-0674.

1334. **British Ornithologists' Union**. c/o British Museum (Natural History), Sub-dept. o Ornithology, Tring, Herts HP23 6AP. (0442) 890080.

1335. **British Trust for Ornithology**. Beech Grove, Tring, Herts, HP23 5NR. (044282 3461.

1336. **Brooks Bird Club**. 707 Warwood Ave., Wheeling, WV 26003. (304) 547-5253.

1337. **Canvasback Society**. P.O. Box 101, Gates Mills, OH 44040. (216) 443-2340.

1338. **Central States Roller Canary Breeders Association**. 305 Grosvenor Ct., Boling brook, IL 60440. (708) 985-4416.

1339. **Colonial Waterbird Society**. U.S. Fish & Wildlife Service, Patuxent Wildlife Re search Center, Laurel, MD 20708. (301) 498-0380.

1340. **COM-U.S.A.** (Cage Bird Breeding.) P.O. Box 122, Elizabeth, NJ 07207. (201 353-0669. FAX (201) 353-2065.

1341. **Cooper Ornithological Society**. 1100 Glendon Ave., Ste. 1400, Los Angeles, CA 90024. (213) 472-7868.

1342. **Cornell Laboratory of Ornithology**. 159 Sapsucker Woods Rd., Ithaca, NY 14850 (607) 254-BIRD. FAX (607) 254-2415.

1343. **Cyprus Ornithological Society**. 4 Kararis St., Strovolos 154, Cyprus. 2-420703.

1344. **Ducks Unlimited**. One Waterfowl Way, Long Grove, IL 60047. (708) 438-4300 FAX (708) 438-9236.

1345. **Eastern Bird Banding Association**. Four View Point Dr., Hopewell, NJ 08525. (609 466-1871.

1346. **Esperantist Ornithologists' Association**. (Ornitologia Rondo Esperantlingva - ORE). Trombitas utca 12, H-4031 Debrecen, Hungary.

1347. **Federation of Field Sports Associations of the EEC**. (Federation des Associations de Chasseurs de la CEE - FACE.) (Migratory Birds, among other pursuits.) 23-25, rue de la Science, boite 16, B-1040 Brussels, Belgium. 2-2304236. FAX 2-2311049. TELEX 25816 COPA B.

1348. **Florida Keys Wild Bird Rehabilitation Center**. 233 Coral Rd., Islamorada, FL 33036. (305) 852-4486.

1349. **Hawk Migration Association of North America**. Box 3482, Rivermont Sta., Lynchburg, VA 24503. (804) 847-7811.

1350. **Hawkwatch International**. 1420 Carlisle NE, No. 202, P.O. Box 35706, Albuquerque, NM 87176. (505) 255-7622. (800) 726-4295.

1351. **Inland Bird Banding Association**. Rt. 2, Box 26, Wisner, NE 68791. (402) 529-6679.

1352. **International Association for Falconry and Conservation of Birds of Prey**. (Association Internationale de la Fauconnerie et de la Conservation des Oiseaux de Proie - IAF.) Le Cochetay, B-4140 Gomze-Andoumont, Belgium. 41-687369.

1353. **International Bird Rescue Research Center**. 699 Potter St., Berkeley, CA 94710. (415) 841-9086. FAX (415) 841-9089.

1354. **International Breeding Consortium for St. Vincent Parrot**. Jersey Wildlife Preservation Trust, Les Augres Manor, Trinity, Jersey, Channel Islands JE3 5BF, England. 534-64666. FAX 534-65161.

1355. **International Border Fancy Canary Club**. 4449 Jackman Rd., No. 5, Toledo, OH 43612. (419) 476-6686.

1356. **International Council for Bird Preservation**. 32 Cambridge Rd., Girton, Cambridge CB3 OPJ, England. 223-277318. FAX 223-277200. TELEX 818794 ICBP G.

1357. **International Council for Bird Preservation, U.S. Section**. World Wildlife Fund, 1250 24th St. NW, Washington DC 20037. (202) 778-9563. FAX (202) 293-9211.

1358. **International Crane Foundation**. E-11376 Shady Lane Rd., Baraboo, WI 53913-9778. (608) 356-9462. FAX (608) 356-9465. TELEX 297778 ICF UR.

1359. **International Dove Society**. 2507 3rd Ave. N., Texas City, TX 77590. (409) 945-4629.

1360. **International Ornithological Congress**. (Congressus Internationalis Ornithologicus.) Zoology Dept., Box 600, Victoria Univ. of Wellington, Wellington, New Zealand. 4-721000.

1361. **International Osprey Foundation**. P.O. Box 250, Sanibel, FL 33957. (813) 472-5218.

1362. **International Pigeon Federation**. (Federation Colombophile Internationale.) 39, rue de Livourne, B-1050 Brussels, Belgium. 2-5376211.

1363. **International Register for the White Eared Pheasant**. Jersey Wildlife Preservation Trust, Les Augres Manor, Trinity, Jersey, Channel Islands JE3 5BF, England. 534-64666. FAX 534-65161.

1364. **International Softbill Society**. 3120 Cedar Vale Rd., Nedrow, NY 13120. (315) 492-3329.

1365. **International Union for Applied Ornithology**. (Internationalen Union fur Angewandte Ornithologie.) Weiherallee 29, W-6229 Schlangenbad, Germany. 6129-8747.

1366. **International Waterfowl and Wetlands Research Bureau**. Slimbridge, Glos., GL2 7BX, England. 453-890624. FAX 453-890697. TELEX 437145 IWRB G.

1367. **International Wild Waterfowl Association**. Hidden Lake Waterfowl, 5614 River Styx Rd., Medina, OH 44256. (216) 725-8782.

1368. **Last Chance Forever.** (Birds of Prey.) 506 Ave. A, San Antonio, TX 78218. (512) 655-6049.

1369. **National Audubon Society**. 950 Third Ave., New York, NY 10022. (212) 832-3200. Audubon hot line on legislative issues (202) 547-9017. Rare Bird Alert (212) 832-6523.

1370. **National Bird-Feeding Society**. 1163 Shermer Rd., Northbrook, IL 60062. (708) 272-0135.

1371. **National Cage Bird Show**. 630 Lake Park Dr., Addison, IL 60101. (312) 543-3757.

1372. **National Color-Bred Association**. 11614 January Dr., Austin, TX 78753. (512) 836-7116.

1373. **National Finch and Softbill Society**. 125 W. Jackson St., York, PA 17403. (717) 854-2604.

1374. **National Gloster Club**. 58 Joanne Dr., Hanson, MA 02314. (617) 294-0340.

1375. **National Institute of Red Orange Canaries and All Other Cage Birds**. 331 E. York, West Chicago, IL 60185. (708) 293-5330.

1376. **Newburyport Birders' Exchange**. 8 Columbia Way, Plum Island, MA 01951. (508) 465-8696.

1377. **North American Bluebird Society**. Box 6295, Silver Spring, MD 20916-6295. (301) 384-2798.

1378. **Nuttall Ornithological Club**. Harvard Univ., Cambridge, MA 02138. (617) 495-2471.

1379. **Ornithological Society of the Middle East**. The Lodge, Sandy, Beds., SG19, 2DL, England.

1380. **Pacific Seabird Group**. Univ. of California, Dept. of Avian Sciences, Davis, CA 95616. (916) 752-1300.

1381. **Parrot Society**. 108 B Fenlake Rd., Bedford MK42, OEU, England. 234-58922.

1382. **Pelican Man's Bird Sanctuary**. P.O. Box 2648, Sarasota, FL 34230. (813) 955-2266.

1383. **Peregrine Fund**. World Center for Birds of Prey, 5666 W. Flying Hawk Ln., Boise, ID 83709. (208) 362-3716. FAX (208) 362-2376.

1384. **Rare Center for Tropical Bird Conservation**. 1529 Walnut St., Philadelphia, PA 19102. (215) 568-0420.

1385. **Royal Society for the Protection of Birds**. The Lodge, Sandy, Beds, SG19 2DL, England. 767-80551. FAX 767-292365. TELEX 82469 RSPB.

1386. **Scandinavian Ornithological Union**. Swedish Ornithological Soc., Postfack 14219, S-104 40 Stockholm, Sweden. 8-6626434.

1387. **Seabird Group**. Royal Soc. for the Protection of Birds, The Lodge, Sandy, Beds., SG19 2DL, England. 767-80551.

1388. **Society for the Preservation of Birds of Prey**. P.O. Box 66070, Los Angeles, CA 90066. (213) 397-8216.

1389. **Society of Parrot Breeders and Exhibitors**. P.O. Box 369, Groton, MA 01450. (603) 878-4391.

1390. **South African Bird Ringing Unit**. (Bird Migration.) Univ. of Cape Town, Rondebosch 7700, South Africa. 21-6502421.

1391. **Toledo Bird Association, Zebra Finch Club of America**. 3320 Bentley Blvd., Toledo, OH 43606. (419) 476-6685.

1392. **U.S. Association of Roller Canary Culturists**. 533 Beach Ave., Bronx, NY 10473. (212) 328-9343.

1393. **Western Bird Banding Association**. California State University, Dept. of Biology, Bakersfield, CA 93311-1099. (805) 664-3179.

1394. **Wild Bird Feeding Institute**. 1163 Shermer Rd., Northbrook, IL 60062-4538. (708) 272-0135.

1395. **Wilson Ornithological Society**. Univ. of Michigan, Museum of Zoology, Ann Arbor, MI 48109-1079. (313) 764-0457.

1396. **World Pheasant Association of the U.S.A.** 2412 Arrowmill St., Los Angeles, CA 90023. (213) 262-5143.

1397. **Yorkshire Canary Club of America**. 619 Glynita Circle, Reisterstown, MD 21136. (301) 833-5981.

1398. **Zambian Ornithological Society**. P.O. Box 33944, Lusaka 10101, Zambia.

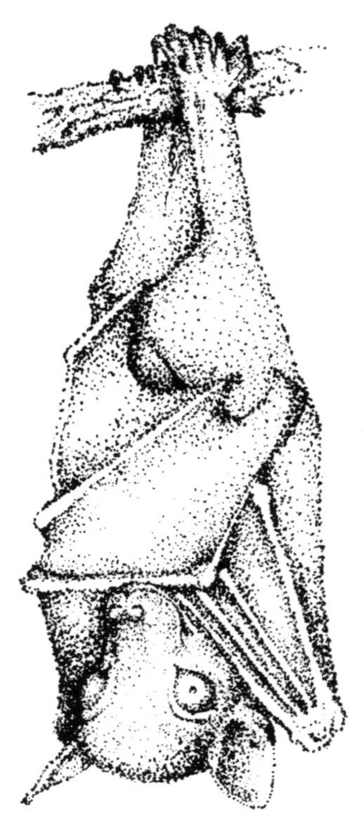

CHAPTER 8

THE MAMMALS

The mammals are a remarkable group. Although they share certain characteristics that define the group, such as a body covered with hair, specialized glands that provide nourishment for the young, a four-chambered heart, and a single bone in the lower jaw, they display an amazing diversity in form, behavior, and distribution.

In form they range from a minute bat that weighs less than an ounce, to the giant whales that weigh tons. Geographically, they can be found on every major land mass, as well as in the sea and in the air.

In the 1993 edition of *Mammal Species of the World* (see entry 1618), the number of species covered was 4,629. They range from the Monotremata, an order comprising the duck-billed platypus and the echnidnas, or spiny anteaters, which hatch their young from eggs, to the marsupials, which raise their young in pouches, to the Chiroptera (bats), which can fly, to the Mysticeta (whales), which live in the sea, to the primates, which include man's closest relatives.

This chapter covers reference works for the major groups of wild mammals. For the most part, literature sources for domesticated mammals such as pets and farm animals are not included here, although domesticated mammals are often found in feral, or wild, populations.

And when the moon gets up and night comes, he is the Cat that walks by himself, and all places are alike to him. Then he goes out to the Wet Wild Woods or up the Wet Wild Trees or on the Wet Wild Roofs, waving his wild tail and walking by his wild lone.
—Rudyard Kipling
"The Cat That Walked by Himself," Just So Stories, *1902*

DICTIONARIES

1399. Gotch, A. F. **Mammals: Their Latin Names Explained: A Guide to Animal Classification**. Poole, England: Blandford Press, 1979. 271p. LC 79-312310. ISBN 0-7137-0939-1.

Although this is not a dictionary in the strictest form, since the terms defined are not arranged alphabetically, it still serves as a source for definitions of the many taxonomic names of mammals. Each chapter is devoted to a particular order with definitions of each common-name group down to species. The derivation of the scientific name from the Latin or Greek is provided; an indication of the geographic range may also be given; and other information about the name may be included. For example, the pygmy hedgehog tenrec (*Echinops telfairi*) derives its name from the Greek *ekhinos*, "a hedgehog," "ops," from the Greek *opsis*, "aspect, appearance"; and is named after C. Telfair (1777-1833), a zoologist who founded the Botanical Gardens in Mauritius. Black-and-white illustrations of the more common species are found throughout the text as well as a few diagrams illustrating the phylogeny of an order. Three indexes provide excellent access to information in this work. The general index includes technical terms, names of zoologists, and names of animals mentioned in the text that are not mammals. The other indexes are to scientific and common names. A very engaging and useful work for all interest levels.

1400. Sokolov, Vladimir E. **A Dictionary of Animal Names in Five Languages: Mammals**. Moscow: Russky Yazyk, 1988. 349p. LC 83-183901. ISBN 5-200-00239-7.

This is volume one of a (so far) three-volume Latin, Russian, English, German, and French dictionary of animal names, this being the one for mammals. Volume two is for amphibians and reptiles and volume three for fish. The first section is a taxonomic arrangement under the Russian name with the Latin, Cyrillic, English, German and French names following. Following this are separate sections for alphabetical arrangements of Latin, Cyrillic, English, German, and French mammal names with reference numbers next to the names that refer to the description in the taxonomic (first) section. The final section is an alphabetical list of species names included in the dictionary. In all, 5,391 mammal names are included in this useful work.

1401. Stangl, Frederick B., Jr., P. G. Christiansen, and Elsa J. Galbraith. **Abbreviated Guide to Pronunciation and Etymology of Scientific Names for North American Land Mammals North of Mexico**. Lubbock, Tex.: Texas Tech University Press, 1993. 28p. (Occasional Papers, The Museum, Texas Tech University; no. 154). $3.00.

This guide offers pronunciations of those perplexing Latin- or Greek-based scientific names that have puzzled students for years. Arranged in checklist form, each scientific name is followed by a suggested pronunciation as well as its etymology (in brackets) and a literal translation. For example, *Bassariscus astutus* — b_ sär is' koos ä stoo' toos [=(Gr.) bassara (fox)] [=(L.) astutus (cunning)]. There is no index to scientific or common name, so the reader will need to know to what order and family a particular species belongs in order to find it in the checklist. Although opinions may differ on the pronunciations offered here, overall it should prove very useful to students and the professional mammalogist.

ENCYCLOPEDIAS

1402. Gould, Edwin, and George McKay, consultant eds. **Encyclopedia of Animals: Mammals**. New York: Gallery Books, 1990. 240p. ISBN 0-8317-2788-8.

This is a popular work on mammals that does not try to be truly encyclopedic but covers each major group with full-color photos and drawings and brief descriptions of the more common species within each major group. Written by 17 contributing scholars, it offers reliable and readable information in a colorful format of photos and drawings. It is arranged by order from marsupials to elephant shrews. Information on each order is written by a different author and features 2 to 10 pages of text on the characteristics of the group, a range map, a summary of the group as to number of families, genera, and species as well as size and an indication of its conservation status. Throughout are excellent photos and drawings of selected species and their habitats. Comparable to Macdonald's encyclopedia (see entry 1403), but less comprehensive.

1403. Macdonald, David, ed. **The Encyclopedia of Mammals**. New York: Facts on File, 1984. 895p. $65.00. LC 84-1631. ISBN 0-87196-971-1.

Packed with more illustrations in one volume than any other mammal encyclopedia, this work also provides an authoritative yet easily readable text. More than 170 authors from universities around the world have contributed to this text that attempts to describe all known members of the class Mammalia. It is divided into six main sections: the carnivores, sea mammals, primates, large herbivores, small herbivores, and the insect-eaters, which includes marsupials and bats. For example, the primates section first overviews the entire group, then breaks down the information into five sections: prosimians, monkeys, apes, tree shrews, and flying lemurs. Each of these individual sections begins with an outline of species and basic information, including distribution maps and drawings of skulls, all contained in a yellow header spanning the top of the first two pages. After two or more pages of general information on the group, brief species descriptions are given under the common name. Throughout the text, outstanding black-and-white and color drawings and photos decorate most pages. The appendix contains a lengthy checklist with distribution and common names provided; a bibliography arranged by group (carnivores, primates, etc.); a glossary of terms; and an index to scientific and common names. Altogether an excellent text for all libraries, and at a manageable cost for most budgets.

1404. Parker, Sybil P., ed. **Grzimek's Encyclopedia of Mammals**. English-language ed. New York: McGraw-Hill, 1990. 5 vols. $500.00 set. LC 89-12542. ISBN 0-07-909508-9 set.

This is a new edition of the mammal section (vols. 10-13) of *Grzimek's Animal Life Encyclopedia* (see entry 18). It has expanded from four to five volumes and continues the tradition of interesting, easy-to-read yet authoritative text with beautiful color illustrations and photos. It was originally published in German under the title *Grzimek's Enzyklopadie Saugetiere* in 1988. It is also available in CD-ROM form (see entry 1407).

It is arranged taxonomically, with each volume treating specific orders. Volume 1 provides an introduction and covers the Monotremata, Marsupialia, Insectivora, Macroscelidea, Chiroptera, and Dermoptera; volume 2, Scandentia, Primates, Xenarthra, and Pholidota; volume 3, Rodentia and Carnivora; volume 4, Carnivora, Lagomorpha, Cetacea, Tubulidentata, Proboscidea, Sirenia, Hydracoidea, and Perissodactyla; and volume 5, Artiodactyla. Each order is given an overview, then families are described with complete species accounts for the more common or interesting species. The text is lavishly illustrated with color drawings and photos (some that cover one or two entire pages), maps, tables and graphs. Each volume ends with a list of references, a list of the authors for that volume plus a brief biographical sketch, a list of illustration credits, and an index to scientific and common names. There is no comprehensive index for the set, however. Although the price is high, this is an excellent set for all libraries.

405. Rue, Leonard Lee, III. **Pictorial Guide to the Mammals of North America**. New ork: Thomas Y. Crowell, 1967. 299p. LC 67-12408.

This easy-to-read introduction to mammals provides information on the most common ecies of the major mammalian orders. Each chapter is devoted to an individual species. The iapter begins with the common as well as scientific names, a distribution map, black-and-hite illustrations of the tracks made by the animal, and a thumbnail sketch of the animal ize, weight, habits, habitat, food, breeding, enemies, and life-span). A two- to four-page xtual description with black-and-white photos follows. Three appendices are here: where see mammals, arranged by state; where to find more information; and a checklist of North merican mammals. Easy to use; for junior high school through college.

406. Whitfield, Philip, consultant ed. **Longman World Guide to Mammals**. Essex, ngland: Longman Group, 1985. 198p. ISBN 0-582-89211-2.

This reference work is a catalogue of the more common mammals of the world. It is ranged by family with the common names of the families covered on each page in bold tters at the top. Facing each page of description of families and representative species is a olor plate with the described species drawn in striking, colorful detail. This makes it easy to ad the brief descriptions and see an illustration at the same time. Each species account lists e common name, the scientific name, the range, habitat, size, and a paragraph or two further scribing the species. A checklist of orders and families of mammals is found at the end of e text along with an index to scientific and common names. Good source for color drawings id brief descriptions.

ELECTRONIC ENCYCLOPEDIAS

407. **The Multimedia Encyclopedia of Mammalian Biology**. New York: McGraw-Hill, 992. CD-ROM. $995.00 (single-user version); $1,250.00 (network version, up to 16 users). BN 0-07-707701-6.

This is a multimedia form of the five-volume *Grzimek's Encyclopedia of Mammals* (see try 1404), supplemented with additional articles not previously published. The database is oduced on a CD-ROM that allows for motion video sequences, animal sounds, and more an 4,000 photos and maps in addition to the text. The format is on hypertext, so that other formation may be accessed from the screen being viewed. For example, a screen on bat ight shows text on the left, and an image of a bat flying on the right, with a window for a axonomic Browser" that can access information from McGraw-Hill's *Synopsis and Clas-fication of Living Organisms* (see entry 77) and the *Dictionary of Scientific and Technical erms* (McGraw-Hill, 1989). Also, from this screen the reader may press a key for sound, a bliography, plot a graph of data collected, enlarge the image, see a movie on the topic, ask r further information such as food, habits, enemies, life cycle, or move on to other screens ing the search, find, or browse keys.

This medium allows for faster, more flexible, and more fun access to information in rzimek's five-volume work, but at more than twice the cost of the print set, which does not clude the computer equipment needed to run it. It may be the coming media however that n speedily access information not available through the printed index and entertain as well. imarily for school and public libraries.

ENCYCLOPEDIAS (Systematic Section)

Carnivora (Dogs)

1408. De Prisco, Andrew, and James B. Johnson. **Canine Lexicon**. Neptune City, N.J.: T. F. H. Publications, 1993. 896p. $79.95. LC 93-134006. ISBN 0-86622-198-0.

Although this work is titled a lexicon, it is included here for its encyclopedic treatment of the terminology and study of all aspects of dogs, both wild and domestic. It includes more than 3,500 entries arranged alphabetically from A.A.D. (Advanced Agility Dog) to Zygomatic Arch. It pictures and briefly describes more than 500 dog breeds from around the world as well as the wild species such as the coyote, wolf, and fox. Also included are articles on dog care such as dental care and grooming, as well as some medical veterinary articles on diseases such as leptospirosis and canine adenovirus. Almost any term or subject that has to do with dogs is included here. More than 1,300 striking, full-color photos illustrate this useful work. An annotated bibliography leads to further reading.

Perissodactyla (Horses)

1409. Ensminger, M. E. **The Complete Encyclopedia of Horses**. New York: Barnes, 1977. 487p. LC 74-9282. ISBN 0-498-01508-4.

This encyclopedia has a similar alphabetic arrangement like *The Encyclopedia of the Horse* (see entry 1410), but it concentrates more on information for the horse business such as medicine, breeding, showing, and racing. Articles are short, or broken into small segments. Black-and-white photos, drawings, and other illustrations as well as some color plates inserted into the front enliven the text. It also includes appendices for the Breed Registry Association, breed magazines, horse books, and colleges of agriculture in the United States and Canada. Filled with information and especially useful to the horse owner and breeder.

1410. Hope, C. E. G., and G. N. Jackson, eds. **The Encyclopedia of the Horse**. Advisory editor, William Steinkraus. New York: Viking Press, 1973. 336p. LC 72-90351. ISBN 670-29402.

Arranged in alphabetical order from Aachen (where Germany's official international horse show is held) to Zhmudky (Lithuanian heavy horse), this encyclopedia covers many interesting facts on all aspects of horse studies. The articles are one sentence to several pages, but usually less than one column in length. Many have references to further reading on the subject. Each article is signed by the author, one of more than 100 experts who contributed to this work. Color and black-and-white photos, drawings, graphs, maps, or figures are found on almost every page. Contains more historical information than *The Complete Encyclopedia of Horses* (see entry 1409). Interesting reading, packed with information for junior high school to college students or anyone with an interest in horses.

1411. Yenne, W. J. **The Pictorial Encyclopedia of Horses**. New York: Gallery Books, 1989. 191p. ISBN 0-8317-6910-6.

This tall (37 cm) book is filled with beautiful full-color photos of horse breeds from Akhal-Teke to Zemaituka. Brief, paragraph-long descriptions of the various breeds are provided, with color photos illustrating many of the breeds. The photos often cover an entire page, and many cover two. Introductory text on the history and general information on the horse, and appendices on horse racing, lists of winners of the Epsom Downs and Kentucky Derbies, steeplechase racing, and more add to the information on the individual breeds. A book for browsing and admiring more than for textual information.

BIBLIOGRAPHIES

412. Fairley, J. S. **Irish Wild Mammals: A Guide to the Literature**. 2d ed., rev. and enl. ialway, Ireland: The Author, 1992. 166p. gb92-41744.

This is a selective bibliography to books, journal articles, reports, and pamphlets ublished in the literature between 1800 and 1991 on the subject of wild mammals of Ireland. Iost references to domestic or fossil mammals are not included. This revised second edition icludes 411 more references than the first edition. The bibliography is arranged in alphabeti- al order by author, each entry being numbered (1,374 entries total). This second edition has n expanded index and is arranged by common name of the mammal.

413. Jones, Gwilym S., and Diana B. Jones. **Bibliography of the Land Mammals of outheast Asia, 1699-1969**. Honolulu: Dept. of Entomology, Bernice P. Bishop Museum; aipei, Taiwan: U.S. Naval Medical Research Unit No. 2, 1976. 238p.

This bibliography was originally compiled to aid in the investigation of medically nportant arthropods found in land mammal populations of this area. The countries covered re Burma, Thailand, Laos, Cambodia, North and South Vietnam, the Republic of the hilippines, Indonesia, Malaysia, and Hainan Island. Included in the bibliography are cita- ons to major monographic works and journal articles. The citations are arranged in alpha- etical order by author and each is assigned an entry number (5,213 total). Separate indexes subject, geographical area, and family refer to these entry numbers.

414. Magnolia, L. R., comp. **Whales, Whaling, and Whale Research: A Selected Bibli- graphy**. Long Island, N.Y.: The Whaling Museum, Cold Spring Harbor, 1977. 91p. LC 7-151069.

This bibliography on many aspects of whales, whaling, and research covers primarily e years 1946 to 1976, but does include some earlier references, one to 1820. Exactly 1,000 umbered, English-language references are arranged here in alphabetical order by author. eriodical articles, books, and other bibliographies are included. No index.

415. Ronald, K., and others. **An Annotated Bibliography on the Pinnipedia**. Charlotten- nd, Denmark: International Council for the Exploration of the Sea, 1976. 785p. LC 5-380520.

This is a bibliography of approximately 9,500 references on sea lions, seals, and alruses from the earliest time to 1975. The title suggests the bibliography is annotated, but nly the citations are given (author's name, title, date when published, and the journal or book tle). The bibliography is divided into two parts: the author section, where citations are listed alphabetical order by author; and the subject index section, where key words are listed in phabetical order to access the citations in the author section. Such indexing makes finding citation on most any subject in the bibliography an easy task.

416. Ruch, Theodore C. **Bibliographia Primatologica: A Classified Bibliography of rimates Other Than Man.** Pt. 1. **Anatomy, Embryology & Quantitative Morphology; hysiology, Pharmacology & Psychobiology; Primate Phylogeny and Miscellanea**. pringfield, Ill.: Charles C. Thomas, 1941. 241p. LC 42-2967.

This is the first volume of a projected work covering all fields of primate studies. owever, this is the only volume published. It covers the literature from the earliest times to 939 and so serves as a valuable source of retrospective literature. It includes not only ferences to journal articles but books and theses as well. It begins with lists of references the literature of the ancient world and the Middle Ages, as well as the sixteenth, seventeenth, d eighteenth centuries. After that, the bibliography is arranged under broad subject headings ich as anatomy, then under more specific subjects, such as embryology, and then to a more ecific topic such as early embryonic stages. An index to author names completes the work.

1417. Shump, Ann U., and others. **A Bibliography of Mustelids**. East Lansing, Mich.: Michigan State University, 1975- .

The purpose of this ongoing series is to compile references from journals and reports on the mustelids (skunks, badgers, martins, mink, otters, ferrets, etc.) from 1900 to the date of publication. Each issue is devoted to a particular animal in this family (i.e., part 1, ferrets and polecats; part 2, mink; part 3, skunks; part 4, weasels; part 5, otters; part 6, wolverines; part 7, badgers; part 8, martens and fishers; part 9, european mink). Each issue is individually authored by different authors, the most recent being published in 1991. Each bibliography is arranged under several main topics such as anatomy, reproduction, physiology, disease, distribution, behavior, paleontology, and evolution. An author index is included for each issue.

1418. Truitt, Deborah, comp. **Dolphins and Porpoises: A Comprehensive, Annotated Bibliography of the Smaller Cetacea**. Detroit, Mich.: Gale Research, 1974. 582p. LC 73-19803. ISBN 0-8103-0966-1.

Containing 3,549 citations, this work covers all aspects of the literature on dolphins and porpoises, in all languages, from earliest times to 1972. It is arranged within categories such as anatomy, behavior, reproduction, thermoregulation, identification keys, and museum collections. Within each category, the citations are arranged chronologically from the earliest to the most recently published. Many of the citations have very helpful descriptive annotations that tell the contents of the work and its emphasis. Three indexes (author, subject, and taxonomic name) give good access to the information contained in the bibliography.

1419. Walker, Ernest P. **Mammals of the World**. Vol. III, **A Classified Bibliography**. Baltimore, Md.: Johns Hopkins Press, 1964. 769p. LC 64-23218.

This third volume of the first and second editions of *Mammals of the World* was designed to supplement the first two volumes, by serving as a classified bibliography. Succeeding editions included a more select bibliography at the end of the second volume. The bibliography includes references to journal articles, books, and reports in all languages. It is arranged into four main groups (i.e., orders, geographical, general, and periodicals). Under each group, the references are listed alphabetically by title, with foreign language titles translated into English. This is the most comprehensive bibliography on the whole class Mammalia, especially for the older literature. Updates can be found in succeeding editions of *Mammals of the World* (see entry 1436).

1420. Wolfe, Linda D. **Field Primatology: A Guide to Research.** New York: Garland, 1987. 288p. $51.00. LC 87-23811. ISBN 0-8240-8552-3.

This is an annotated bibliography of 1,072 books and articles on field primatology. This covers primarily the recent literature, with the oldest citations from the 1960s. Each annotation is two to six sentences in length and includes information on the research site, the year(s) of study, and the major findings of the author(s). The bibliography is arranged into five chapters covering general studies, prosimians, New World monkeys, Old World monkeys, and apes. A subject index (that also contains common names), and a primate genus index are provided. An appendix includes a checklist of primates covered in the book, and is based on that provided by Jaclyn Wolfheim in *Primates of the World* (see entry 1525). A very good guide to the field primatology literature.

INDEXES AND ABSTRACTS

1421. **Current Primate References**. Vols. 1- , nos. 1- . Seattle, Wash.: Primate Information Center, Regional Primate Research Center, University of Washington, 1964- . Monthly $45.00/yr. to individuals; $50.00/yr. to institutions. ISSN 0590-4102.

This is an index to journal articles, reports, books, book chapters, and dissertations on nonhuman primates. It arranges citations under approximately 40 broad subject headings such as behavior, biological clocks, diseases, genetics, learning, neoplasia, reproductive system, toxicology, and virology. The citation is listed only once under one of these headings, but references to the numbered citation is provided under other appropriate headings. There is also a section for citations to book reviews of recent primatology books. Two indexes to each issue are provided (i.e., an index to primates by family with genus and species listed below the family, and an index to author). Yearly cumulative author index only.

1422. **Primates - Database to the Primatological Literature**. For MS-Dos machines. Lease costs are $995.00 plus shipping. A yearly subscription for 12 monthly update disks is $495.00. Available from the Primate Information Center, University of Washington, Seattle, Washington.

This database contains approximately 50,000 records on all aspects of nonhuman primate research. It begins in 1985 and can be searched by subject, species, author, title, keyword, controlled vocabulary terms, or a combination of these. The database comes with a search program, a search guide, and a complete thesaurus of controlled vocabulary terms. Updates can be received monthly with the subscription service. There are approximately 450 to 650 new citations per month.

1423. **Wildlife Review**. Vols. 1- . Fort Collins, Colo.: U.S. Fish and Wildlife Service, 1935- . 6/yr. Free to some government agencies and libraries. ISSN 0043-5511.

This index selectively indexes worldwide wildlife and natural resource literature from over 1,300 journals as well as more than 500 books and symposia proceedings. See entry 54 in chapter 1 for a full description. It includes an index for mammals as well as for birds and reptiles/amphibians.

1424. **Zoological Record**. Vols. 1- , nos. 1- . Philadelphia: Published by BIOSIS and the Zoological Society of London, 1864- . Annually. $2,200.00/yr. ISSN 0144-3607.

Zoological Record is published annually in 27 separate sections (see entry 55 for full review). One of the sections is "*Mammalia*," Section Number 19. This section indexes the literature, primarily systematic literature, on mammals. It indexes journal articles, books, some theses, and reports.

DIRECTORIES

1425. American Society of Mammalogists. **Directory of Members**. Provo, Utah: Brigham Young Univ., Dept. of Zoology, 1978- .

Issued as a supplement to the *Journal of Mammalogy* since 1978, this is an alphabetical ist of the officers and regular members of the American Society of Mammalogists. The ddress of the member follows the name, and the entry ends with the date when the member oined the Society.

426. **Directory of Marine Mammal Specialists**. Rome: Food and Agriculture Organiza-ion of the United Nations, United Nations Environment Programme, 1988. 139p. Free of harge to anyone actively involved in marine mammal conservation. LC 92-981273.

This directory of approximately 100 marine mammal specialists from around the world s a joint effort by the United Nations Environment Programme and the International Union or Conservation of Nature and Natural Resources (IUCN). It is arranged first in alphabetical rder by the last name of the specialist. The address is listed under the name and the field of pecialty listed next to the name. There follows a list of the specialists by country. Next, they re arranged under specialty by marine mammal type and by specific interest (ecology, onservation, behavior, etc.).

1427. **International Directory of Primatology**. Madison, Wis.: Wisconsin Regiona Primate Research Center, University of Wisconsin, 1992. 1 vol. (unpaged). $10.00. LC 92-644037.

This directory brings together, under one cover, material on a wide variety of primat information sources and programs that would otherwise take many hours of searching to find It is intended as a means to aid primate research and communication between the variou researchers and organizations around the world. The directory is organized into five section and four indexes. The five sections cover (1) Organizations Arranged Geographically, (2 Field Studies (locations), (3) Population Management Groups, (4) Primate Societies, an (5) Information Resources. To help find information in these sections, four indexes ar provided to organization, species, programs, and personal names. Beside each entry in th index is a section and page number where that information can easily be found in the text.

In this handy work one can find the address, mission, key personnel (with phon numbers), and species supported on 183 primate research centers such as the Red Howle Project in Guárico, Venezuela; find the address, mission, clientele, services and collection of information agencies such as the British Library of Wildlife Sounds; find out who has database on the *Lemur catta* (the Indianapolis Zoo does); and find a bibliography on th behavior of captive gorillas. For all this information and more one need pay only $10.0 (including postage and handling). A bargain price for a work that would benefit all academi and large public libraries, and especially all primate research organizations.

1428. Yates, Terry L., William R. Barber, and David M. Armstrong. **Survey of Nort American Collections of Recent Mammals**. Baltimore, Md.: American Society of Mam malogists, 1987. 76p. (Journal of Mammalogy; vol. 68, no. 2, suppl.).

Approximately every 10 years the American Society of Mammalogists conducts survey of the recent mammal collections in North America. This is the result of its fifth survey The collections are arranged alphabetically by country, state, or province, then by institutio name and collection name. For each collection listed, the following are given: address wher collection housed, total specimens, specimen types, photos or other illustrations in collectior and a contact person and phone number. Following this is a table listing mammal collection of 10,000 or more specimens, beginning with the largest collection (United States Nationa Museum, Fish and Wildlife Service). Other tables provide information on the historica development of the collections (growth from 1922) and collections with added amenities suc as a central library. A sample of the survey form used to gather this information, and bibliography of catalogs of type specimens found in North American collections complet this work.

HANDBOOKS

Mammalia (General)

1429. Anderson, Sydney, and J. Knox Jones, Jr. **Orders and Families of Recent Mammal of the World**. New York: John Wiley, 1984. 686p. LC 83-21806. ISBN 0-471-08493-X.

With descriptions of 21 orders and 131 families of living or recently extinct mammals this work provides a quick reference to information on these groups. It is a revision of *Recen Mammals of the World: A Synopsis of Families* (Ronald Press, 1967), also edited by Anderso and Knox. Twenty-one contributors wrote the various chapters on the major groups o mammals. Each chapter covers a particular order except the order Rodentia, which took fou chapters. Diagnosis, distribution, geologic range, list of families, and further remarks ar provided for each order, followed by descriptions of each family. Each family has a diagnosis description of general characters, habits, habitat, recent distribution, and a list of recer genera, their geologic range, and the major fossil groups. Black-and-white distribution map are provided as well as drawings of skulls and skeletons showing characteristic features.

68-page bibliography of references, and an index to genera and the above (no species) as well as common names of major groups complete this work. No identification keys are included. For the student and specialist seeking a more complete reference to orders and families of mammals worldwide.

1430. Broad, Steven, Richard Luxmoore, and Martin Jenkins. **Significant Trade in Wildlife: A Review of Selected Species in CITES Appendix II**. Volume 1. **Mammals**. Gland, Switzerland, IUCN; Lausanne, Switzerland: Secretariat of the Convention on International Trade in Endangered Species of Wild Fauna and Flora, 1988. 183p. ISBN 2-88032-953-1.

This work provides a summary of information for 22 mammals listed in CITES (Convention on International Trade in Endangered Species of Wild Fauna and Flora) in Appendix II, which lists species that are not currently threatened with extinction, but may be if their trade is not regulated. Each summary includes a 2- to 10-page account on the animal's conservation status, distribution, population, habitat and ecology, threats to survival, international trade, conservation measures, captive breeding programs, and a list of references for further reading. There is no index. Concise summaries, useful to students and those planning for the conservation of these animals.

1431. **Handbuch der Zoologie: Eine Naturgeschichte der Stämme des Tierreiches**. Bd. 8. **Mammalia**. Gegr. von Willy Kükenthal, fortgegührt von Thilo Krumbach, herausgegeben von J. G. Helmcke, D. Starck, H. Wermuth. Berlin: de Gruyter, 1969- . Volume 8 is up to 58 parts as of this writing. Most parts are in print.

This eighth volume of the *Handbuch* (see entry 68), is a multipart, comprehensive treatment of the class Mammalia, or Säugetiere in German. The work is an ongoing project, and is up to 58 parts as of this writing. It is primarily in German, although some parts are in English. Each part covers either an individual group of mammals or covers a particular aspect, such as reproduction or classification. It is very detailed and thorough and includes many black-and-white drawings. Lengthy bibliographies accompany each volume. Each volume also has an index to subject and scientific name.

1432. Lawlor, Timothy E. **Handbook to the Orders and Families of Living Mammals**. 2d ed. Eureka, Calif.: Mad River Press, 1979. 327p. $20.00. ISBN 0-916422-16-X.

This handbook provides keys and diagnostic descriptions of mammalian orders and families worldwide. It is larger than the little, first-edition handbook, but is arranged in much the same way. It begins with descriptions of mammalian bones, especially those of the skull that are diagnostic in identification of groups. Black-and-white line-drawings of the skulls are included. The keys are also illustrated with black-and-white drawings to aid in identification. After the student has used the keys to get to an order or family, the diagnostic characters for the group are listed. For example, the following characters may be given: the size, texture of hair, number of digits, description of teeth, skull bones and their size, vertebrae characteristics, number of genera and species, and an indication of geographical range. Includes a nice bibliography of references cited. This is a compact and useful tool for identification of mammal orders and families for students and amateurs.

1433. Lever, Christopher. **Naturalized Mammals of the World**. New York: Longman, 1985. 487p. LC 83-19527. ISBN 0-582-46056-5.

This work provides species accounts of naturalized mammals, those animals that were brought to a new geographical place to live and managed to live and reproduce in the new place. This work details many interesting stories of how a species was brought to a new place and flourished, sometimes for good, sometimes for bad. The European hedgehog, the raccoon, the domestic cat, the moose, grey squirrel, and the edible dormouse are all examples of naturalized mammals. Each species account provides a black-and-white drawing of the animal, a range map of natural and naturalized distribution, and a written account of how, when, and why the species was introduced into a new land. A lengthy bibliography of sources, and tables summarizing naturalization information complete this interesting text.

1433a. **Mammalian Species**. Nos. 1- . Provo, Utah: American Society of Mammalogists, c/o Dept. of Zoology, Brigham Young University, 1969- . Published irregularly. $10.00/yr. ISSN 0076-3519.

As the title suggests, each issue of this unique work is devoted to a particular mammal species and is authored by an expert(s) on that species. Each issue follows the same format to promote uniformity of information for each species and for ease of comparisons with other species. The format begins with a list of references to where the species was first named and other references to type locality and taxonomy. The remaining headings are context and content, diagnosis, general characters, distribution, fossil record, form and function, ontogeny and reproduction, ecology, behavior, genetics, remarks and literature cited. A black-and-white photo of the species and its skull, and a range map are provided for each species. Each species account is from 3 to 20 pages in length. Cumulative indexes to author and genera, and a systematic list are provided for every 100 species accounts. As of this writing, 443 species accounts had been published. A very valuable source for complete species descriptions and, using the lengthy list of references cited, for finding further sources of information.

1434. Morris, Desmond. **The Mammals: A Guide to the Living Species**. New York: Harper & Row, 1965. 448p. LC 65-20508.

This work provides a checklist of 4,237 living mammal species and more detailed information on 300 of the more common species. Information on the 300 species is nontechnical and covers about one page with information on size, color, habitat preference, behavior and geographical range. A black-and-white photo is placed at the top of the page of each description. Separate indexes to scientific and common name make this an easy-to-use book for the general reader.

1435. Nagorsen, D. W, and R. L. Peterson. **Mammal Collectors' Manual: A Guide for Collecting, Documenting, and Preparing Mammal Specimens for Scientific Research**. Toronto, Canada: Royal Ontario Museum, 1980. 79p. (Life Sciences Miscellaneous Publications). LC 80-510574. ISBN 0-88854-255-0.

This easy-to-read and easy-to-use manual provides basic information on collecting, documenting, and preparing mammal specimens for collections. It pictures and describes the different methods for collecting (i.e., nets, traps, baits, etc.). It also details what and how data should be recorded, complete with illustrations of a catalogue sheet, and a page from a field notebook and data sheet, as well as specimen tags. It shows how to take correct measurements of large and small specimens and how to determine sex and reproductive status. A long chapter on preparing specimens is also illustrated with helpful drawings on correct technique. Finally, special techniques such as collecting blood samples and parasites, or preserving stomach contents are discussed with methods suggested for shipping specimens. A selective bibliography suggests further reading and the appendix provides a checklist of field equipment and supplies for karyotype work. Useful information in easy-to-use form.

1436. Nowak, Ronald M. **Walker's Mammals of the World**. 5th ed. Baltimore, Md.: Johns Hopkins University Press, 1991. 2 vols. $89.95 set. LC 91-27011. ISBN 0-8018-3970-X.

This new fifth edition of a classic work is revised and updated by Ronald N. Nowak, who served as senior author with John L. Paradiso on the fourth edition. This new edition has only moderate changes from the last edition, the work that included the greatest revision and updates since the first. Specifically, there are 106 new generic accounts, 300 new photos that either replace or supplement older materials, separate accounts for each extinct genus known to have lived within the last 5,000 years, and textural revisions to reflect new information from the literature.

The arrangement of this two-volume work is the same systematic arrangement by order in the first edition. Each order is introduced with general information on the group, followed by descriptions of each family and genus in the order and including information on threatened or endangered status. Black-and-white photos accompany most descriptions. Care has been taken to include photos of live animals with photos of museum specimens or drawings used only for the rare or extinct genus. Each volume has a complete index to both volumes with entries for the scientific names in bold and the common names in ordinary type. A 113-page bibliography is included at the end of volume two.

The revisions in this new edition are not critical, so those who already own the fourth edition may consider purchase of a comparable work on mammals. One work is *Grzimek's Encyclopedia of Mammals* (see entry 1404), which is almost as comprehensive as Walker's and includes more behavior, ecology, and conservation information plus color photos and illustrations, but carries a much higher price tag ($500.00). David Macdonald's one-volume *The Encyclopedia of Mammals* (see entry 1403), although not as comprehensive, is a good choice for its overall treatment of the subject with authoritative sources, interesting reading, color photos, and a bargain price ($65.00). Walker's, however, still provides the most complete list of species (4,444) with descriptions and photos. A classic work.

1437. Thornback, Jane, and Martin Jenkins, comps. **The IUCN Mammal Red Data Book**. Gland, Switzerland: IUCN Conservation Monitoring Centre, 1982. 516p. ISBN 2-88032-600-1.

This is part of the Red Data Books compiled by the IUCN (International Union for Conservation of Nature and Natural Resources). This series of books provides comprehensive and authoritative information on the conservation status of species and serves as international registers of threatened species. Part one of this work covers the "Threatened Mammalian Taxa of the Americas and the Australasian Zoogeographic Region (Excluding Cetacea)." In this volume are listed and described 155 threatened mammals from around the world. Each species is described in one to three pages with information on the distribution, population, habitat and ecology, threats to survival, conservation measures taken, conservation measures proposed, captive breeding, and further remarks, and a list of references. At the head of the page for each species, its status is listed as either extinct, endangered, vulnerable, rare, indeterminate, out of danger, or insufficiently known. This list is updated by the *1990 IUCN Red List of Threatened Animals* (see entry 79a).

1438. **Traité de Zoologie**. Tome XVI-XVII. **Mammifères**. Pierre-Paul Grassé, ed. Paris: Masson & Cie, 1955-1982. 9 vols. LC 49-2833.

Part of a multivolume classic work, entirely in French, that summarizes the biology, anatomy, and systematics of the animal world (see entry 69). This nine-volume section deals with the entire group of mammals. Tome XVI covers, in detail, the anatomy and physiology of mammals in seven fascicles. Tome XVII (actually published first in 1955) covers the classification and description of the mammal orders in two fascicles. Many black-and-white drawings and some color plates illustrate each volume. Each fascicle is 500 to 1,100 pages in length.

Tome 16, fas. 1. *Tegument, skeleton*. Fr. 1,482. ISBN 2-225-58476-1.
Tome 16, fas. 2. *Musculature*. Fr. 1,125. ISBN 2-225-59246-2.
Tome 16, fas. 3. *Musculature*. Fr. 1,210. ISBN 2-225-29689-8.
Tome 16, fas. 4. *Nervous System, Sense Organs, Circulation, Blood and Lymph*. Fr. 1,482. ISBN 2-225-30327-4.
Tome 16, fas. 5. *Splanchnology*. Fr. 908. ISBN 2-225-36839-2.
Tome 16, fas. 6. *Reproduction*. Fr. 1,482. ISBN 2-225-59950-5.
Tome 16, fas. 7. *Embryology*. Fr. 1,712. ISBN 2-62678-2.
Tome 17, fas. 1. *Orders, Anatomie, Ethology, Systematics*. Fr. 1,170. ISBN 2-225-58487-7.
Tome 17, fas. 2. *Orders, Anatomie, Ethology, Systematics*. Fr. 1,130. ISBN 2-225-58509-1.

1439. Webb, J. E, J. A. Wallwork, J. H. Elgood. **Guide to Living Mammals**. 2d ed. London: Macmillan, 1979. 240p. LC 80-456495. ISBN 0-333-27257-9.

This unique guide briefly lists the chief characteristics of each major animal group down to family. It is arranged in classified form from the marsupials to the artiodactyles. Each major group is characterized in one page, usually, with black-and-white drawings of selected species, and range maps found throughout the text. For example, the family Tayassuidae, the Peccaries, are described with three main characteristics as follows: "1. They are medium-sized mammals with relatively short legs, thick course hair and a bristly mane. There is a gland at the centre of the back. They are omnivorous and feed on vegetation, small animals and carrion. 2. The tusks point downward and have a sharp cutting edge. The dental formula is I 2/3, C 1/1, Pm 3/3, M 3/3. 3. They are found in tropical America. There is 1 genus and 2 species." A black dot is put next to the characteristics that are especially important. A 14-page glossary helps to explain the terminology used in the descriptions. This is an easy-to-use guide that quickly summarizes the basic characteristics of the major groups of mammals.

Mammalia (Geographic Section)

In addition to this list of books covering the mammals of specific regions and countries, the following journal article may be consulted for a more expansive list of references to mammalian faunas of almost 150 countries and geographical areas.

1440. Hickman, Graham C. "National Mammal Guides: A Review of References to Recent Faunas." **Mammal Review** 11, no. 2 (1981): 53-87.

AFRICA

1441. Kingdom, Jonathan. **East African Mammals: An Atlas of Evolution in Africa**. New York: Academic Press, 1971-1982. Reprint (in paperback), Chicago: Univ. of Chicago Press, 1984-1989. 3 vols. in 7. $146.00 vol. 1; $146.00 vol. 2, pt.A; Reprint, $35.00; $138.00 vol. 2, pt.B; Reprint, $32.50; $146.00 vol. 3, pt.A; Reprint, $37.50; $211.00 vol. 3, pt.B; Reprint, $37.50; $32.50 vol. 3, pt.C.; $32.50 vol. 3, pt.D. LC 73-117136. ISBN (vol. 1) 0-12-408301-3; 0-226-43718pa; (vol. 2, pt.A) 0-12-408302-1; 0-226-43719-1pa.; (vol. 2, pt.B) 0-12-408342-0; 0-226-43720-5pa.; (vol. 3, pt.A) 0-12-408303-X; 0-226-43721-3pa.; (vol. 3, pt.B) 0-12-408343-9; 0-226-43722-1pa.; (vol. 3, pt.C) 0-12-408344-7; 0-226-43724-8pa.; (vol. 3, pt.D) 0-12-408345-5; 0-226-43725-6pa.

These volumes attempt to describe and picture the mammal fauna of the three countries that make up most of East Africa (i.e., Kenya, Tanzania and Uganda). Volume 1 contains the primates; volume 2, part A, insectivores and bats; volume 2, part B, hares and rodents; volume 3, part A, carnivores; volume 3, part B, large mammals (zebras, camels, giraffes, etc.); volume 3, part C and D, bovids. Each species is described and pictured in elegant black-and-white drawings by the author, usually with a full-page drawing and some smaller sketches, as well as a drawing of the musculature and skeleton. The accompanying descriptions are lengthy; the one for Thomson's gazelle is 12 pages, and include brief information on family order and local names as well as head and body measurements and weights, and then more lengthy information on habits, habitat, biology, ecology, and so on. A black-and-white range map is also included for each animal. Each volume includes a long bibliography, a checklist of species, a gazetteer of places mentioned in the text, and systematic and subject indexes. Of interest to all groups, even those studying art (because the drawings are remarkable).

1442. Meester, J., and H. W. Setzer, eds. **The Mammals of Africa: An Identification Manual**. Washington, D.C.: Smithsonian Institution Press, 1971. Loose-leaf. $131.30. LC 70-169904. ISBN 0-685-20923-7.

Composed primarily of keys and identification characters, this loose-leaf work was placed in the section for identification keys (see entry 1611).

1443. Smithers, Reay H. N. **The Mammals of Rhodesia, Zambia and Malawi: A Handbook**. London: Collins, 1966. 159p. LC 66-77871.

This is an easy-to-use guide to the identification and description of mammal species in these three countries of southern Africa (Rhodesia is now known as Zimbabwe). It is more of a field guide to the common species, with concise descriptions, maps, and color illustrations. Each species account covers a page or two and includes a physical description, the geographical distribution, habitat, habits, diet, breeding, and other interesting notes. A range map, a color drawing, and a drawing of the track made by the animal are included for each species. Indexes to common and scientific names are found at the end. Some of the information here is updated in Smithers's 1983 work (see entry 1444).

1444. Smithers, Reay H. N. **The Mammals of the Southern African Subregion.** Pretoria: University of Pretoria, Republic of South Africa, 1983. 736p. £95.00. LC 85-217684. ISBN 0-86979-536-8 presentation ed; 0-86979-538-4 subscriber's ed.; 0-86979-540-6 ordinary ed.

This mammoth work on the mammals of the Southern African Subregion covers the area south of the Cunene/Zambezi Rivers. This includes the countries of Botswana, Zimbabwe, part of Mozambique, the Republic of South Africa, and Namibia. It updates some of the information in the author's other books for this area. It is a very detailed and complete treatise including not only precise taxonomic information, but behavior and ecology as well. It is arranged taxonomically by order, with keys to the identification of family, genus, and species. Each species account includes the scientific, common, and Afrikaans names, the author and date where first named, taxonomic notes, a physical description with measurements and mass recorded in tabular form, the geographical distribution, the habitat, habits, food, and reproduction. A black-and-white distribution map is provided for each species. Thirty color plates with full-color drawings of selected species are found throughout the text as well as many black-and-white drawings. Drawings of some skulls are also included. A lengthy bibliography of more than 1,000 references and indexes to the scientific, English, and Afrikaans names complete the text. A major work by the authority on mammals for this area.

ARABIA

1445. Harrison, David L., and Paul J. J. Bates. **The Mammals of Arabia.** 2d ed. Sevenoaks, Kent, England: Harrison Zoological Museum, 1991. 354p. £65.00. LC gb91-29800. ISBN 0-9517313-0-0.

This is the only comprehensive work covering mammals from the whole Arabian peninsula. This second edition comes 20 years after the first edition and so includes many changes and updates. It is arranged taxonomically and includes keys to species within each family. It describes 160 species. Each species account is one column to more than one page in length and includes the external, cranial and dental measurements, external characteristics, cranial characters, dentition, distribution, and other remarks. Black-and-white distribution maps accompany each species account. Black-and-white photos or drawings, especially of skulls, are often provided with each species account as well. A glossary of terms, geographical gazetteer, 14-page bibliography, and indexes to common and scientific names complete this definitive work on the mammals of Arabia.

1446. Kingdom, Jonathan. **Arabian Mammals: A Natural History**. London, New York: Academic Press, 1991. 279p. $130.00. LC gb92-42848. ISBN 0-12-408350-1.

Although not as detailed as David Harrison's *The Mammals of Arabia* (see entry 1445), this work has more general information for the nonspecialist. The work features beautiful full-color and black-and-white drawings by the author of the many interesting mammal species in Arabia. Many of the drawings cover a full page. The first half of the work is arranged by major groups, with general information on the groups and some species. Black-and-white drawings are found here. At the end of this section is a checklist of mammals found in the Arabian Peninsula, a list of national parks and nature preserves, a short bibliography, and a subject index to the first half of the work. This is followed by a section with Arabic and English descriptions of a single species, with a full-color drawing of the species on the facing page. There is no index to this section. A good selection for those wanting nontechnical information and beautiful illustrations.

AUSTRALIA

1447. Gould, John. **The Mammals of Australia**. South Melbourne: Macmillan Company of Australia, 1977. 3 vols. ISBN 0-333-13984-4 vol. 1; 0-333-13912-7 vol. 2; 0-333-21056-6 vol. 3.

First published in 1863, this new edition of a classic work is exceptional for the beautiful color plates illustrating each species. It is published in three volumes: volume 1, Australian marsupials and monotremes; volume 2, kangaroos; and volume 3, placental mammals of Australia. The original text has been augmented by more recent information, placed in the margin, by Joan M. Dixon. The plates are reduced from their original folio size, but still convey their original beauty and style. Each volume has an index for scientific and common names as well as a bibliography of references cited in the work.

1448. Ride, W. D. L. **A Guide to the Native Mammals of Australia**. New York: Oxford University Press, 1970. 249p. LC 79-17396. ISBN 0-19-5502523.

This work describes and pictures 228 species of native Australian mammals. The species are arranged first into major groups (marsupials, rodents, bats, carnivores, and monotremes), and then into 55 smaller groups (under kangaroos are great kangaroos, large wallabies, small wallabies, etc.). The group is then described, with individual species listed under the description for the group. Descriptions of each group are a page or two in length and feature a full-page black-and-white drawing of a representative species. Descriptions are easy, nontechnical, and enjoyable to read. They cover the natural history of the group as well as distribution and habitats. A good book for students and amateurs.

1449. Strahan, Ronald, ed. **Complete Book of Australian Mammals: The National Photographic Index of Australian Wildlife**. London: Angus & Robertson, 1984. 530p. ISBN 0-207-14454-0.

Featuring beautiful full-color photos from the National Photographic Index of Australian Wildlife, this work includes descriptions of every species of native mammal and introduced mammal that can be found living in the wild in Australia today. It is arranged taxonomically with individual species accounts covering one to two pages. For each species account, a column down the first page includes information on size, concise identification, synonyms, conservation status, list of subspecies, and references for further reading. Next to this column is one page or more of textual description with information on behavior and ecology, and other interesting notes. Beautiful full-color photos, often covering a full page, give a clear picture of each species. A geographic range map showing where the animal is found in Australia is found in each species account. A beautiful book and especially useful in public libraries.

1450. Walton, D. W., and B. J. Richardson, eds. **Fauna of Australia**. Vol. 1B: **Mammalia**. Canberra, Australia: Australian Government Publishing Service, 1989. 1227p. $79.95. LC 88-125506. ISBN 0-644-06056-5.

This lengthy and comprehensive work deals with the biology of the entire mammalian fauna of Australia. It is arranged by family and follows the sequence given in the *Zoological Catalogue of Australia*, volume 5, *Mammalia* (see entry 1623). Each family is thoroughly described in separate chapters that average 20 pages. This includes information on the family's discovery, morphology and physiology, natural history, biogeography, and collection and preservation. A bibliography of literature cited finishes each family chapter

Black-and-white photos and drawings illustrate the chapters. An illustrated key to Australian genera follows the family chapters. The work concludes with a detailed index to subjects, as well as scientific and common names. Thorough enough for the specialist, but written and illustrated at a level for the student or amateur.

BORNEO

1451. Medway, Lord. **Mammals of Borneo: Field Keys and an Annotated Checklist**. Kuala Lumpur, Malaya: Printed for MBRAS by Perchetakan Mas, 1977. 172p. (Monographs of the Malaysian Branch of the Royal Asiatic Society, no. 7). LC 77-941271.

This comprehensive, though dated, work on the mammals of Borneo and adjacent islands includes land and marine mammals. There are 196 species listed and described here, 40 species being endemic or unique to Borneo. Most mammals are in the orders of Chiroptera (bats), Rodentia (rodents), and Carnivora (carnivores). After introductory text on the history of the study, diversity, distribution, and future of Bornean mammals, the annotated checklist is presented. Keys to identification down to species are provided in each order. Species accounts may be very brief (reference to where first named and distribution) to three pages. Black-and-white photos of selected species are included as well as some black-and-white drawings (especially of bats). A list of references and a selective bibliography of works on Bornean mammals completes the text.

BOTSWANA

1452. Smithers, Reay H. N. **The Mammals of Botswana**. Salisbury, Rhodesia: Trustees of the National Museums of Rhodesia, 1971. 340p. (Museum Memoir, no. 4). LC 72-183752.

This survey of the mammals in Botswana was six years in the making. An interim report titled *A Check List and Atlas of the Mammals of Botswana* by Smithers was published in 1968. The present work is taxonomically arranged by orders from Insectivora to Rodentia. Each order has keys to genus and species. Each species account covers one page or more (10 pages for the Bateared Fox) and may include such information as distribution, habitat, habits, breeding, food, and taxonomic notes. The common name and, sometimes, African name is provided as well as the reference where first named in the literature. Nice range maps are inserted for each species and line-drawings or photos are provided for some species. A four-page bibliography and indexes to scientific, common, and African names are useful additions to the text. Introductory material on the area studied, especially of the Kalahari where little collecting had previously been done, is interesting reading.

CANADA

1453. Banfield, A. W. F. **The Mammals of Canada**. Toronto, Canada: Published for the National Museum of Natural Sciences, National Museums of Canada, 1974. 438p. $47.50. LC 73-92298. ISBN 0-8020-2137-9.

Although dated, this is the only comprehensive treatment of mammals in Canada and is still in print. A new set, *Handbook of Canadian Mammals*, will eventually update this work; however, it is only up to volume 2 so far and covers marsupials, insectivores, and bats (see entry 1454). Banfield's work is meant to provide popular accounts of the mammals for students, amateurs, and specialists. In this work, 196 species are listed and described. They are listed in taxonomic order from Marsupialia to Artiodactyla. Each species is listed under its common name with the French and scientific names (and author) listed beneath it. A physical description is followed by the animal's habits, habitat, reproduction, economic status, distribution, and canadian distribution. A reference follows to where first named and a full reference or references to an article(s) on the description, biology, or ecology of the animal. Black-and-white distribution maps with shading for where the animal may be found in Canada are included for each species. Black-and-white and full-color drawings of some animals are also provided. Appendices on skull bones, a table on dental formulae, number of digits and mammae, and a glossary of terms are included before the index of common and scientific names. An excellent text for most any reader, high school and above.

1454. Van Zyll de Jong, C. G. **Handbook of Canadian Mammals**. Ottawa, Canada: National Museum of Natural Sciences, National Museums of Canada, 1983- . $19.95ea. LC 83-206387. ISBN 0-660-10328-1 vol. 1; 0-660-10756-2 vol. 2.

Now up to volume 2, this handbook is planned to go into greater detail than *The Mammals of Canada* (see entry 1453). It provides information on the life histories, systematics, and distribution of the living species of mammals in Canada. Volume one includes the marsupials and insectivores, and volume two includes the bats. Each volume is in a taxonomic arrangement with sections on orders, families, genera, and species. Each species account lists the common and French names under the scientific name followed by a list of references where first named and other descriptive references. A description, the distribution, systematics, and biology are included with each account, as well as black-and-white drawings of the skull and other distinctive features, and a distribution map. Illustrated keys to family, genus, and species are also provided. Plates of full-color drawings of representative animals are inserted in front of the text. A glossary of terms and a bibliography of references cited in the text finish the work. There is no index.

CHILE

1455. Osgood, Wilfred H. **The Mammals of Chile**. Chicago: Field Museum of Natural History, 1943. 268p. (Field Museum of Natural History, publication 542; Zoological Series, vol. 30).

Information collected on two major expeditions to Chile, in 1922 and again in 1939, form the basis of the collection of Chilean mammals now housed in the Field Museum of Natural History and from which this work was derived. Background information on the principal faunal districts and on how the specimens were collected introduce the text. A list of the 134 species covered in the text and a key to orders begin the systematic accounts of the Chilean species. Each order has keys to genera with a list of species and their descriptions following the keys. Each species account lists references to where first named and other references to the literature. The geographical range and a description as well as the number and geographical area where specimens were collected comprise the rest of the species account. Black-and-white drawings of skulls are provided for identification of most species. A bibliography of sources and an index to common and scientific names complete the text.

CHINA

1456. Allen, Glover M. **The Mammals of China and Mongolia**. New York: American Museum of Natural History, 1938-1940. 2 vols. (Natural History of Central Asia, vol. 11, pts. 1-2). LC 39-21372.

This two-volume set, containing more than 1,300 pages, reviews the mammalian fauna of this area. The information is from the Asiatic Expeditions of the American Museum of Natural History that occurred from 1916 to 1930 and so does not include the most recent information, but forms a good basis for review. After introductory material on faunal areas of China and Mongolia, collectors of these mammals, and relationships of Asian and North American mammals, the work is arranged systematically by order: volume 1, with Insectivora to the beginning of Rodentia, and volume 2, with the remainder of Rodentia, and Artiodactyla. Each order is thoroughly described and includes keys to families down to species. Species accounts are detailed and cover two to four or more pages with information on type specimen and location, description, measurements, nomenclature, occurrence and habits, the number and location of specimens examined, and references where the name was first published. Distribution maps, and striking full-page black-and-white photos of the area and various species are found throughout the text. A complete bibliography of sources is found at the end of volume 1, and a complete index to volumes 1 and 2 is found at the end of volume 2. Detailed and complete for its time of publication.

GREAT BRITAIN and EUROPE

1457. Arnold, H. R. **Atlas of Mammals in Britain**. London: HMSO, Natural Environment Research Council, 1993. 144p. (Ite Research Publication, no. 6). £12.50. ISBN 0-11-701667-5.

This work lists and describes the mammals of Britain, but more importantly, it provides extensive population survey information. This information has been gathered since 1965 and is now in a database with over 115,000 records. With this information, a detailed distribution map is included for each species, showing where the animal is found in Great Britain. A valuable tool for conservationists.

1458. Bjarvall, Anders, and Staffan Ullstrom. **The Mammals of Britain and Europe**. New York: Viking, 1987. 240p. $39.95. LC 85-41976. ISBN 0-7099-3268-5.

Providing a less technical, more popular text on the indigenous and introduced mammals of Britain and Europe, this work also features beautiful color and black-and-white drawings of many species described here. The text is arranged taxonomically from marsupials to whales, and describes each species in one paragraph to a page or two. The species accounts include a physical description and some interesting notes on their behavior, food, reproduction, and habitat. Small range maps are included for each species. Primarily for the student in junior high on up, and the general reader.

1459. Corbet, G. B, and Stephen Harris, eds. **The Handbook of British Mammals**. 3d ed. Oxford; Boston, Mass.: Published for the Mammal Society by Blackwell Scientific Publications, 1991. 588p. LC 89-1822. ISBN 0-632-01691-4.

Now into its third edition, this is the primary work to consult for species accounts of mammals of Britain, Ireland, and the Channel Isles. Introductory chapters on fossil mammals of the area, the habitats (woodlands, wetlands, farmland, etc.) where mammals live, and conservation laws precede the systematic accounts. Each species is thoroughly described in several pages with information on taxonomy, distinguishing characteristics, sign (tracks, droppings, calls, or burrows), fur, teeth, chromosomes, measurements, distribution, fossil remains, habitat, behavior and social organization, feeding, population, parasites, and even references where sound recordings of the sounds the species makes may be found. Black-and-white photos and drawings of each species are provided as well as drawings of skulls and other distinguishing characteristics. Distribution maps also accompany each species account. An appendix on extinct species and introduced or escaped species are found at the end of the text. A classic work for mammals of this area, suitable for amateur and specialist alike.

1460. Corbet, G. B. **The Terrestrial Mammals of Western Europe**. Philadelphia: Dufour Editions, 1966. 264p. LC 66-23640.

This work deals with all mammals (other than whales, seals, and bats) in the geographical area of Europe west of Russia. The first part of the book discusses the study of mammals in general, followed by a systematically arranged list and description of the mammals of this area. Keys down to species are included and each species account is usually about a page in length. Information on distribution, and a general description as well as some behaviors are in each species account. Aside from eight pages of black-and-white plates, there are no illustrations of each species. A glossary, bibliography, and index are included at the end.

1461. Orr, Richard. **Mammals of Britain and Europe**. Text by Joyce Pope. London: Pelham Books, 1983. 176p. LC 84-108223. ISBN 0-7207-1426-5.

Featuring beautiful color plates of mammals drawn by Richard Orr, this work also provides easy and interesting reading for the student and general reader. The work describes the most common mammals found in this area and pictures them in page-long, full-color plates as well as in striking black-and-white drawings. A checklist of European mammals, a bibliography for further reading, and an index to scientific and common names complete the work.

HAWAII

1462. Tomich, P. Quentin. **Mammals in Hawaii: A Synopsis and Notational Bibliography**. 2d ed. Honolulu, Hawaii: Bishop Museum Press, 1986. 375p. (Bishop Museum Special Publication 76). $32.00. LC 85-73487. ISBN 0-93087-10-2.

Although we are not including handbooks for individual U.S. states in this work, we are including Hawaii, due to its broad boundary ranges that extend far into the oceans that surround it. It therefore is considered more like a regional area. Tomich's work includes a checklist of names and origins (introduced, immigrant, indigenous, endemic, or hybrid) because most mammals were introduced to the islands by explorers and early settlers. Detailed species accounts are provided that include the reference to the original published description, the type locality, native range (where originally came from), and the range in Hawaii. Further observations of the species habits are documented, and black-and-white photos are provided for most species. An extensive 162-page annotated bibliography leads to further reading. Interesting reading for the amateur and specialist.

INDOMALAYAN (ORIENTAL) REGION

1463. Corbet, G. B., and J. E. Hill. **The Mammals of the Indomalayan Region: A Systematic Review**. Oxford: Oxford University Press, 1992. 488p. $90.00. LC 92-299. ISBN 0-19-854693-9.

This monumental work provides an account for all known living and fossil mammal species for this region. This region was once referred to as the Oriental Region and includes, roughly, west of the Indus Valley and south of the Himalaya Mountains, southern China, Southeast Asia, and adjacent islands including Taiwan, the Philippines, and the Lesser Sundas. It includes about 7,000 scientific names, keys to each taxonomic level, a 3,000-item bibliography, and five appendices with a list of recent species, new names proposed in this work, gazetteer, biographical notes (people who have significantly contributed to mammal studies in this area), glossary and abbreviations, and additions to maps. Genus and some species accounts provide a reference to where first named in the literature and synonyms, type species, geographical distribution, distinguishing characteristics, variations, and other remarks. Black-and-white drawings are included for some species as well as black-and-white distribution maps. This is a welcome up-to-date synthesis of a species-rich area.

1464. Cranbrook, Earl of. **Mammals of South-East Asia**. 2d ed. Oxford, New York: Oxford University Press, 1991. 96p. (Images of Asia). $24.95. LC 92-186974. ISBN 0-19-588568-6.

This brief, easy-to-use guide to the common mammals of Southeast Asia features many black-and-white and color photos as well as color drawings of selected species. The book is arranged taxonomically from primitive mammals to anteaters, rodents, and hares. The major families are discussed and selected species are described in an informative yet nontechnical style. A checklist of 660 species found in this area follows the text, showing the amazing diversity of mammals here. It is arranged by family common name, with the scientific name listed next to it, and, its geographical range. A selective bibliography is provided, but there is no index.

IRAN

1465. Harrington, Fred A., comp. and ed. **A Guide to the Mammals of Iran**. Tehran, Iran: Dept. of the Environment, 1977. 88p.

This is the first comprehensive guide to the living mammals of Iran. It lists and describes 148 species, with 112 drawn in full color. Although the introduction lists these mammals in taxonomic order, the arrangement of the text itself is by habitat group (e.g., Caspian forest, Caspian stream, rocky foothills, central sandy desert, high mountains, etc.). Each habitat group and the animals found there are described in one page with a full-color drawing depicting the area and the mammals on the page facing the text. The text is intended for the general public and is, therefore, easy reading to inform the public of these animals and foster better conservation. Cultural, relief, and nature-reserve maps are included as well as an index to common names only.

JAPAN

1466. Sowerby, Arthur De Carle. **The Mammals of the Japanese Islands**. Chang-Hai, China: Universite L'Aurore, 1943. 66p.

This review of the mammals of the Japanese islands was written at a time when the possessions and territories of Japan included, in addition to the three main islands of Japan, the Liu Chu Islands, Formosa, Korea, the Bonin Islands, and the Marianne, Caroline, and Marshall Islands. This extensive region is unevenly covered in this review. More recent publications on mammals of this area are in Japanese. In all 280 species in 11 orders are discussed here. This work is arranged by order from Insectivora to Sirenia. Under each heading the author discusses the various mammals from this group found in this geographical area. A few black-and-white drawings of selected species are provided as well as nine pages of black-and-white photos, found at the end of the text. There is no index or a list of references.

MALAWI

1467. Ansell, W. F. H., and R. J. Dowsett. **Mammals of Malawi: An Annotated Check List and Atlas**. Cornwall, England: Trendrine Press, 1988. 170p. £15.00. LC 88-171827. ISBN 0-9512562-0-3.

After introductory information on the physical features of Malawi, zoogeography, mammal habitats, and conservation efforts, the work begins the systematic accounts of mammals that live in this area. Each species account includes extensive taxonomy, location of specimens examined, habitat, and general notes. One hundred eighty black-and-white range maps are grouped together at the end of the text. Most species accounts refer to a range map number. Historical as well as present distribution are marked on each species map. Includes a gazetteer, a long list of references, and an index to genera, species, and subspecies and English common name.

MALAYA

1468. Medway, Lord. **The Wild Mammals of Malaya and Offshore Islands Including Singapore**. London: Oxford University Press, 1969. 127p.

A second edition of this work was reprinted with corrections in 1983 under the title *The Wild Mammals of Malaya (Peninsular Malaysia) and Singapore*, but we were unable to obtain a copy for review. Since the new edition contained only minor corrections, we reviewed the first edition for this book.

This work lists and describes 199 species of land mammals from this region. It is arranged taxonomically by order from Insectivora to Artiodactyla. Species accounts are one column to one page in length and include information on the name (scientific, English, and Malayan), distribution, identification, habits, and breeding, as well as a list of subspecies, if any. Full-color drawings of individual species are found on 15 plates placed throughout the volume. Black-and-white drawings illustrating identifying characteristics of many species are also provided. A bibliography of references cited and further works on Malayan mammals is placed at the end of the text. A concise work, good for identification, and useful to many groups from students to specialist.

MALAYSIA

1469. Tweedie, M. W. F. **Mammals of Malaysia**. Singapore: Longman Malaysia, 1978. 87p. LC 79-121662. ISBN 0-582-72424-4.

This work covers the geographical range of Malaya and Borneo, and provides brief descriptions of living species found there. It is arranged from Insectivora to Cetacea. Each species is described in one paragraph to one page, giving the common, scientific, and Malaysian names at the beginning of the description. Sixteen plates with full-color drawings of selected species and some black-and-white drawings illustrate the text. A checklist of Malaysian mammals follows the text. Provides a summary of the works by Lord Medway the *The Wild Mammals of Malaya* (see entry 1468), and *Mammals of Borneo* (see entry 1451)

MOZAMBIQUE

1470. Smithers, Reay H. N., and Jose L. P. Lobao Tello. **Check List and Atlas of the Mammals of Mocambique.** Salisbury, Southern Rhodesia: Trustees of the National Museums and Monuments of Rhodesia, 1976. 184p. (Museum memoir, no. 8).

More than a checklist, this work includes distribution, habitat notes, range maps, taxonomic notes, and keys to family, genus, and species. Introductory text on the various habitats of the area (forest, thicket, savannas, grasslands, etc.) is illustrated with black-and-white photos. A gazetteer to places mentioned in the text is also provided. Each species account lists the scientific and English as well as Portuguese common name, the author of the name and the date published, taxonomic notes, distribution, and habitat. A black-and-white range map is included with almost every species account.

NEOTROPICS

1471. Eisenberg, John F., and Kent H. Redford. **Mammals of the Neotropics**. Chicago, Ill.: University of Chicago Press, 1989- . $85.00, $34.95pa. (vol. 1); $90.00, $35.00pa. (vol. 2). LC 88-2749. ISBN 0-226-18540-6, 0-226-19539-2pa. (vol. 1); 0-226-70681-8, 0-226-70682-6pa. (vol. 2).

Planned as a three-volume set, this work provides a comprehensive survey of terrestrial and marine mammals of the Neotropical Region. The first volume, by John Eisenberg, covers the northern neotropics: Panama, Colombia, Venezuela, Guyana, Suriname, and French Guiana. The second volume, by Kent Redford and John Eisenberg, covers the southern neotropics; Chile, Argentina, Uruguay, and Paraguay.

The arrangement is the same for both volumes. An introductory chapter on the biogeography of the area is followed by individual chapters on mammalian orders arranged in taxonomic order from Marsupialia to Lagomorpha. Each family within the order is described as well as each genus and species. The species description is the most complete. For each species, a complete table of measurements is given as well as a physical description, the geographical distribution, life history, and ecology. A large black-and-white map approximately 3 by 6 inches is provided next to each description with black dots and shading indicating where the species were collected or found. Drawings of skulls for some species are also included. Black-and-white drawings are found occasionally in the text and excellent color plates are inserted in the back of the text. A lengthy list of references is found at the end of each chapter. Two final chapters in each volume summarize topics supplementing the material in the main text, such as mammalian community ecology or the effects of humans on mammals for the region. Separate indexes to scientific names and common names are at the end of each volume. This set fills a gap for a current, comprehensive reference for mammals of this region and should prove to be a standard for years to come. For all academic and large public libraries.

NEW GUINEA

1472. Flannery, Timothy. **Mammals of New Guinea**. Carina, Australia: Robert Brown and Associates, 1990. 439p. £40.00. LC 91-173838. ISBN 1-85273-029-6.

This volume covers primarily the living native mammals of the island of New Guinea and features beautiful color photos of each animal described. Offshore islands are not covered and introduced animals are not covered as well, except when wild populations have been established. Introductory material covers the geological history, vegetation, zoogeography, paleontology, and introduced mammals of New Guinea. The rest of the book is arranged taxonomically by family. The species accounts are one to four pages in length and include information on size, range, synonyms, native names, and a physical description with other information such as behavior, food, and other notes. A full-color photo or drawing (often covering an entire page) illustrates each species. A black-and-white range map is also included for each species. Fifty-two plates of skulls for identification are grouped together after the list of references. A very nice text, suitable for amateur and specialist.

NEW ZEALAND

1473. King, Carolyn M., ed. **The Handbook of New Zealand Mammals**. Auckland, New Zealand: Oxford University Press in association with The Mammal Society, New Zealand Branch, 1990. 600p. $80.00. LC 91-101373. ISBN 0-19-558177-6.

This work lists and describes a total of 46 land-breeding mammals (11 native species and 35 introduced) that are or have been established in the wild in New Zealand. The introduction explains the unusual situation of having more introduced species than native species, and other general information on the mammals of this area. An illustrated key to mammal skulls that can be found on the main islands of New Zealand is provided before the species accounts. Species accounts range from 1 page to 31 pages for the bushtail possum, and include the common and Latin name, a physical description, measurements, field sign, history of colonization, distribution in New Zealand and in the world, habitat, diet and food, behavior, reproduction, population studies, predators, adaptation to New Zealand, and significance to the ecology of New Zealand. Black-and-white photos are provided for each species, and some distribution maps are included as well. A glossary and a 69-page bibliography complete the work. An index to scientific and common names is found at the end. The fly papers have color photos of various species and skins used for identification. A thorough and interesting work for mammalogists, students, and amateurs.

NIGERIA

1474. Happold, D. C. D. **The Mammals of Nigeria.** Oxford, England: Clarendon Press, 1987. 402p. $135.00. LC 85-31056. ISBN 0-19-857565-3.

This work describes the known 247 mammal species living in Nigeria today. It is a very complete work, beginning with information on the physical environment of Nigeria and a checklist of mammal species. Keys to orders, families, genera, and species are included. Species accounts include distribution, localities where found, conservation status, physical description, ecology, reproduction, and taxonomic notes. Each account is usually one column in length. A few black-and-white drawings of distinguishing characteristics are found throughout the text, and a section of black-and-white photo plates picturing selected species and their habitats is found in the middle. General information on mammal populations in the rainforest, the savanna, the Sudan and Sahel savannas, as well as a chapter on the impact of man on the mammals in Nigeria round out the wealth of information provided here. Distribution maps are grouped at the end of the text. A lengthy bibliography and index finish this interesting text for the student and specialist alike.

NORTH AMERICA

1475. Cahalane, Victor H. **Mammals of North America**. New York: Macmillan, 1947. 682p. LC 47-4195.

This work provides a very readable account of the mammals of North America and is geared more for the general reader. Each species account is written in an engaging way, drawing the reader through each description with the ease of a popular novel. The book is arranged under 94 headings for each of the main groups found in North America (e.g., the deer family, the cats, the chipmunks and squirrels, etc.). The descriptions of each group and the individual species are two to several pages in length and include a striking black-and-white drawing of a representative species. A long list of references arranged by general group, then by animal group, finishes the text.

1476. Chapman, Joseph A., and George A. Feldhamer, ed. **Wild Mammals of North America: Biology, Management, and Economics**. Baltimore, Md.: Johns Hopkins University Press, 1982. 1147p. $75.00. LC 81-8209. ISBN 0-8018-2353-6.

This text describes the most economically important mammal species in North America north of Mexico. It was created as a reference for use in wildlife management of the animal groups biologists and naturalists were working with the most. It is arranged taxonomically with long (5 to 60 pages), detailed but very readable species accounts by individual authors. Each species account includes nomenclature, distribution, detailed description including measurements and weights, physiology, genetics, reproduction, development, sex determination, ecology, food, behavior, mortality, and economic status. The species account ends with a lengthy list of literature cited. A distribution map and black-and-white photos of skulls and the species illustrate each species account. A glossary of terms and an index by subject as well as scientific and common name complete this useful work. Of interest to all groups, especially conservationists and wildlife managers.

1477. Hall, E. Raymond. **The Mammals of North America.** New York: John Wiley, 1981. 2 vols. LC 79-4109. ISBN 0-471-05443-7 vol. 1; 0-471-05444-5 vol. 2.

This classic two volume work attempts to summarize the taxonomic studies of North American mammals from 1492 to 1977. It is arranged by order, with order, family, genus, and species descriptions provided as well as keys to family, genus, and species. Black-and-white drawings of the skulls of representative species, and drawings of selected species are found throughout the text. Distribution maps, some covering an entire page, show where individual species may be found in North America. North America here includes Greenland, Panama, and the continent between the two, as well as the Greater and Lesser Antilles south to Grenada.

A total of 3,607 species are covered here. Each species is described first by listing the scientific and common names. References to the first use of the name are included after the name as well as an indication of type locality. External measurements and marginal records are also provided for each species. Separate indexes to common and scientific names are found at the end of each volume. Volume two includes information on collecting and preparing specimens, an illustrated glossary and a comprehensive bibliography of literature cited in the texts. A very thorough and reliable work.

1478. Rue, Leonard Lee, III. **Pictorial Guide to the Mammals of North America**. New York: Thomas Y. Crowell, 1967. 299p. LC 67-12408.

This easy-to-read introduction to mammals provides information on the most common species of the major mammalian orders. Each chapter is devoted to an individual species. The chapter begins with the common as well as scientific names, a distribution map, black-and-white illustrations of the tracks made by the animal, and a thumbnail sketch of the animal (size, weight, habits, habitat, food, breeding, enemies, and life-span). A two- to four-page textual description with black-and-white photos follows. Three appendices tell where to see mammals, arranged by state; where to find more information; and provide a checklist of North American mammals. Easy to use, for junior high through college.

1479. Schmidt, John L., and Douglas L. Gilbert, comp. and eds. **Big Game of North America**. Harrisburg, Pa.: Stackpole Books, 1978. 494p. $34.95. LC 78-14005. ISBN 0-8117-0244-8.

This book summarizes wildlife management for North America's largest mammals, including elk, moose, caribou, mountain lions, pronghorns, bison, musk-ox, and so on. Fourteen chapters, each covering a particular animal, provide excellent information on taxonomy, distribution, a description, life history, population dynamics, productivity, breeding, herding characteristics, food and water requirements, and so on. Conservation issues such as estimating age and population size and how to balance hunting and conservation of the species are discussed along with other conservation problems. Each chapter is around 20 pages and includes black-and-white photos, graphs, range maps, and striking drawings by Charles W. Schwartz. The remaining 11 chapters cover broader topics such as behavior, nutrition, hunting management, and future concerns. Four appendices include pathology and necropsy techniques, care and use of the harvested animal, and an alphabetical list of plants and animals listed in the book. An interesting and factual book of interest to students and wildlife conservationists.

1480. Wrigley, Robert E. **Mammals in North America: From Arctic Ocean to Tropical Rain Forest: Wildlife Adventure Stories and Technical Guide**. Winnipeg, Manitoba: Hyperion Press; dist., New York: Sterling, 1986. 360p. $19.95. LC 90-179147. ISBN 0-920534-33-3.

Written primarily for the young-adult reader, this work describes and pictures 115 of the more common mammals found in the North American continent. It is arranged by geographical area (e.g., Mammals of the Cold Oceans, Mammals of the Tundra, Mammals of the Desert, etc.). Within each geographical area, an average of 10 species that are commonly found in that area are described and 95 of them are pictured in full-color drawings. Wildlife stories accompany some species descriptions and lead into more detailed information such as color distribution and status, food, reproduction, and physical description. Black-and-white drawings as well as distribution maps are also provided. A checklist of North American mammals and a bibliography of additional readings enhances the text. For school and public libraries, primarily.

PAKISTAN

1481. Roberts, T. J. **The Mammals of Pakistan**. London: Ernest Benn, 1977. 361p. LC 77-368354. ISBN 0-510-39900-2.

This work provides keys, illustrations, textual descriptions, and distribution maps for the 158 living species of mammals found in Pakistan. After introductory information on the area, complete with maps and color photos of the ecological zones where the mammals are found, the remainder of the text is arranged taxonomically from Insectivora to Cetacea. Each species is described in a page or more with information on the physical description, distribution and status, and biology of the animal, including references where first named in the literature. Black-and-white distribution maps and drawings of several species are also provided. A lengthy bibliography, a gazetteer of locations mentioned in the text, a glossary of terms, and an index to scientific and common names complete this notable work.

PANAMA

1482. Goldman, Edward A. **Mammals of Panama**. Washington, D.C.: Smithsonian Institution, 1920. 309p. (Smithsonian Miscellaneous Collections, vol. 69, no. 5). LC 20-26440.

Contains accounts of all the known living mammals in Panama at that time. The material is from a biological survey of the Panama Canal Zone done in 1910 by the Smithsonian

Institution and, as such, includes fascinating accounts, written in diary form, by the author of the trip through this area. Extensive descriptions of the area and the mammals, birds, and plants found there introduce the work. The rest of the work concentrates on the descriptions of the mammals arranged systematically from marsupials to primates. Each species account gives scientific and common names, a reference where first named, type location, and an interesting prose description of first-hand observations by the author. A large folded map of this area is inserted after the title page. A bibliography, black-and-white plates of skulls for identification, and an index to scientific and common names as well as subjects complete the work.

REPUBLIC OF SOUTH AFRICA

1483. Smithers, Reay H. N. **South African Red Data Book: Terrestrial Mammals**. Pretoria: Foundation for Research Development, Council for Scientific and Industrial Research, 1986. 216p. (South African National Scientific Programmes Report No. 125). ISBN 0-7988-3806.

This work updates two previous *South African Red Data Book: Mammals* (Meester, 1976, and Skinner, Fairall, and Bothma, 1977). It covers the area of the Republic of South Africa and reviews the conservation status of the species in the previous publications as well some new species. Those species deemed extinct, endangered, vulnerable, rare, out of danger, or indeterminate (may be included in one of these categories) are listed and described in this work. Information on present and former geographical distribution as well as habitat, habits, breeding in the wild and in captivity, reasons for decline, numbers in captivity, protective measures in operation and those proposed, and current research are covered for most species or for the family or order where they belong taxonomically. Some geographical range maps are included. A 14-page bibliography and an index to scientific name complete the work.

SURINAME

1484. Husson, A. M. **The Mammals of Suriname**. Leiden, Netherlands: E. J. Brill, 1978. 569p. (Zoologische Monographieen van het Rijksmuseum van Natuurlijke Historie, no. 2). LC 79-308858. ISBN 90-04-05819-2.

This very thorough work on the mammals of Suriname describes all species known to live in this country and provides guides to their identification through keys to order, family, genus, and species, with illustrations. It is arranged taxonomically with each species account providing detailed information. The scientific name is followed by a reference to where the name was originally published, the type locality, synonyms, common name in English, Dutch, and Sranantongo (official Suriname language), geographical distribution, occurrence in Suriname, detailed description with skull measurements, and a paragraph of the ecology and biology of the species. One hundred fifty-one black-and-white plates at the end of the text illustrate the skulls or provide a drawing or photo of each species mentioned in the text. A 22-page bibliography finishes the text. A very detailed and complete work.

THAILAND

1485. Lekagul, Boonsong, and Jeffrey A. McNeely. **Mammals of Thailand**. 2d ed. Bangkok, Thailand: Printed under the auspices of the Association for the Conservation of Wildlife, 1988. 758p.

This taxonomically arranged text describes all known species living in Thailand at that time. Introductory material gives background information on Thailand's geographical regions and the zoogeography of Thai mammals as well as a list of Thai species on the IUCN Red List of Endangered Species and appendix 1 of CITES. Thereafter, each order is thoroughly described. Keys to families and to species are also provided. Each species account includes the synonyms for the species, the diagnosis, size, distribution, description, ecology, and behavior. Black-and-white photos of the species, several views of the skull, as well as black-and-white drawings and distribution maps and a section of color drawings, provide excellent illustrations for identification and are a strong point of this work. A glossary and a lengthy bibliography add to the usefulness of this text.

TRANSVAAL

1486. Rautenbach, I. L. **Mammals of the Transvaal**. Pretoria, Transvaal: ECOPLAN, 1982. 211p. (ECOPLAN monograph, no.1). LC 82-185493. ISBN 0-620-05738-6.

This work, arranged systematically by order from Insectivora to Artiodactyla, lists and describes the living mammals in the Transvaal. Keys to genus and species are provided. Species accounts are usually one page in length and include information on distribution, habitat, habits, food, breeding, taxonomic notes, measurements, and records of occurrence. Range maps are given for each species. The common, scientific, and Afrikaans names are also given for each species, and separate indexes for the names are provided at the end of the text. A nice bibliography and a gazetteer of locations mentioned in the text are additional highlights of this useful work.

TRINIDAD

1487. Alkins, M. E. **The Mammals of Trinidad**. St. Augustine, Trinidad: Dept. of Zoology, University of the West Indies, 1979. 75p. (Occasional papers, no. 1). LC 81-133203.

This brief work provides information on the environment, climate, and vegetation of Trinidad as well as listing and describing the mammals found there. Each species description includes measurements, a physical description, habits, and other interesting notes. A three-page bibliography and indexes to scientific and common names finish the text.

UNITED STATES

1488. Hamilton, William J., Jr., and John O. Whitaker, Jr. **Mammals of the Eastern United States**. 2d ed. Ithaca, N.Y.: Comstock, 1979. 346p. LC 79-12920. ISBN 0-8014-1254-4.

This work covers 105 species of mammals found in the 27 states east of the Mississippi River, except Mississippi. It is arranged taxonomically from Marsupialia to Artiodactyla and provides identification keys to order, genus, and species. Each species is described in two to four pages, covering the physical description, distribution, and habits. Black-and-white distribution maps are provided for each species, and black-and-white photos accompany most species descriptions. A bibliography for further reading is arranged under a general group, then under order groups such as Chiroptera (bats). This provides a synopsis of information found in various state mammalogy texts for this area and so provides the specialist as well as the amateur with concise, readable information on a large area of the United States.

1489. Ingles, Lloyd G. **Mammals of the Pacific States: California, Oregon, and Washington**. Stanford, Calif.: Stanford University Press, 1965. 506p. $15.95pa. LC 65-10380. ISBN 0-8047-1843-1pa.

This book was originally published in 1947 as *Mammals of California*, followed by a revised edition titled *Mammals of California and Its Coastal Waters*, published in 1954. The present work includes Oregon and Washington for the first time. It is intended for use by students and other beginners in the study of mammals. It is quite thorough and covers the basics such as classification, evolution, and the range and distribution of these mammals, as well as their habitats, conservation status, and habits. After the introductory material on ecology, evolution, and on the study of mammals, the book is arranged by order from Marsupialia to Artiodactyla. Each order is completely described and includes keys to identification of species, species accounts with black-and-white drawings, and range maps. Six appendices extend the information on collecting and preparing specimens, scat (dung) identification, a handy guide to the pronunciation of many genera that are frequently mispronounced, classification, a checklist of mammals in the Pacific States, and, finally, the dental formulas of all orders except Cetacea (whales, dolphins, etc.). Although covering a limited geographical range, the information for these groups is valuable for the more general study of mammals as well.

1490. Jones, J. Knox, Jr., and others. **Mammals of the Northern Great Plains**. Lincoln: University of Nebraska Press, 1983. 379p. $35.00. LC 82-2693. ISBN 0-8032-2557-1.

The Northern Great Plains is defined in this text as the states of Nebraska, North Dakota, and South Dakota. This geographical area is also covered in another book by Jones and others titled *Guide to Mammals of the Plains States* (see entry 1491), but this work briefly describes each species in one page, whereas *Mammals of the Northern Great Plains* describes each species in more length and detail. It also provides more background information on the region such as vegetation, soils, drainage patterns, climate, and zoogeography. It is arranged taxonomically from Marsupialia to Artiodactyla with keys to identification for orders, families, genera, and species. Each species account is two or more pages in length and includes information on the origin of the name, distribution, description, and natural history, finishing with selected references for further reading. A black-and-white distribution map accompanies each species account as well as a black-and-white photo or drawing of the species. Additionally, a checklist of mammals of this area, an illustrated glossary of terms, and a bibliography of literature cited in the text add to the value of this work. Easy to read for the amateur, yet detailed enough for the specialist.

1491. Jones, J. Knox, Jr., David M. Armstrong, and Jerry R. Choate. **Guide to Mammals of the Plains States.** Lincoln, Nebr: University of Nebraska Press, 1985. 371p. $31.50; $15.95pa. LC 84-21012. ISBN 0-8032-2562-8; 0-8032-7557-9pa.

Covering the area from the Canadian Prairie Provinces south to northern Mexico, and from the Rocky Mountains east to the Mississippi, this work concentrates on the mammals found in the states of North Dakota, South Dakota, Nebraska, Kansas, and Oklahoma. It is intended to provide concise accounts of all species in this area. Arranged taxonomically from the marsupials to the even-toed ungulates, each species account is limited to one page with information on distribution, a description, habits and habitat, reproduction, and a short list of suggested references. Black-and-white range maps and photos of selected species illustrate the text. Keys to the identification of orders, families, genera, and species are also included, as well as a checklist of mammals found in this area. A good, concise, and rather recent guide.

1492. Olin, George. **Mammals of the Southwest Deserts**. Rev. ed. Globe, Ariz.: Southwest Parks and Monuments Association, 1982. 99p. $5.95. LC 81-86094. ISBN 911408-60-6.

Covering the geographical area of western Texas to the coasts of California, and north from central Utah and Colorado down to northern Mexico, this short book lists and describes 42 of the most common mammals of this area. Each species is described in one page with a full-color or black-and-white drawing of the animal on the facing page. The descriptions include information on geographical range and habitat and are written in an easy, yet fact-filled style. Especially good for public and school libraries.

1493. Van Gelder, Richard G. **Mammals of the National Parks**. Baltimore, Md.: Johns Hopkins Univ. Press, 1982. 310p. $35.00; $10.95pa. LC 81-17162. ISBN 0-8018-2688-8; 0-8018-2689-6pa.

This is a handy guide to take while visiting any of the 48 national parks. It tells where in each park to find the mammals mentioned, then gives details about each animal. Divided into two sections, the first section lists and relates the mammals found in each national park. A map of the park is also included in each two- to six-page description of each mammal found in each park. The second part of the book describes and pictures the mammals listed in the first part. Each easy-to-read, yet informative species description covers two or more pages and includes a black-and-white photo of the species, and ends with a list of national parks where that particular species may be found. An interesting and fun guide.

1494. Zeveloff, Samuel I. **Mammals of the Intermountain West**. Salt Lake City, Utah: University of Utah Press, 1988. 365p. $19.95pa. LC 88-20462. ISBN 0-87480-296-2; 0-87480-327-6pa.

This is an up-to-date guide to the mammals found between the Sierra Nevada and the Rocky Mountains and includes the states of Utah and Nevada primarily, with parts of northern Arizona, western Colorado and Wyoming, southern Idaho and Oregon, and western California. It is written in a popular style for the amateur, but is current and thorough enough for students and professionals. Introductory material describes the region and its ecological niches. The remainder of the book is arranged in taxonomic order from marsupials to even-toed hoofed animals. Each chapter covers an order, with descriptions of the various species found in the region and range maps showing where they are usually found. Each chapter ends with a list of references for further reading. Black-and-white drawings and color paintings illustrate many species described in the book. A checklist of common names of mammals in this region is printed at the beginning of the text. A very readable and current work.

FORMER SOVIET UNION (USSR)

1495. Heptner, V. G., and N. P. Naumov, eds. **Mammals of the Soviet Union**. Leiden; New York: E. J. Brill, 1989- . £50.95 vol. 1; £50.95 vol. 2, pt.2. LC 88-14592. ISBN 90-04-08874-1 vol. 1; 90-04-08876-8 vol. 2, pt.2.

Now up to two volumes, this work, originally published in Russian, is projected to update and supplement S. I. Ognev's *Mammals of the USSR and Adjacent Countries* (see entry 1496). Heptner died in 1975 and did not complete the series, but the work continues and is being translated into English. Heptner began this series with a comprehensive (1147 pages) volume on the Ungulates, a group that was not included in Ognev's work. Volume 2 covers the Carnivora (pt. 1, sea cows and dogs, bears and mustelids; pt. 2, hyenas and cats; pt. 3, pinnipeds and toothed whales, in preparation). Species included are those both living and extinct. Keys to identification are provided to species in most cases. Twenty-four species are detailed in the ungulates group. Species accounts are very detailed and complete. The account for *Sus scrofa* (wild boar), is 64 pages long. This includes a list of references where first described, a physical description, taxonomy notes, geographic distribution (including past ranges to show a change in distribution over the years), description of four subspecies, biology (population, habitat, food, behavior, migrations, reproduction, enemies, mortality), and economic importance. Black-and-white photos and drawings illustrate the text. Bibliographies of works cited and indexes to scientific and common names as well as subject finish each volume.

1496. Ognev, S. I. **Mammals of Eastern Europe and Northern Asia**. Washington, D.C.: Published for the National Science Foundation and the Smithsonian Institution by the Israel Program for Scientific translations, 1962-1967. 8 vols. LC 63-60427.

This nine-volume work is a translation of Ognev's work originally published in Russian. Beginning with volume three, the title changed to *Mammals of the USSR and Adjacent Countries*. The geographical area covered is vast and includes Poland, Finland, and the Baltic states, the USSR, and parts of eastern Turkey, Mongolia, and Manchuria. The work is arranged by taxonomic groups: Volume 1, Insectivora and Chiroptera; volume 2, Carnivora (Terrestrial); volume 3, Carnivora (rest of Terrestrial and Pinnipedia); volumes 4-7, Rodents; volume 9, Cetacea. Volume nine was published five years after the death of Ognev and was written by his student, A. G. Tomilin. Volume eight was to contain the remaining groups of rodents and the ungulates, but was not completed before the death of Ognev. The work is very detailed and complete, with keys down to species, and long species accounts. Species accounts include references to literature, common names (in French, English, German, and Russian), scientific name, local names, references to principal figures (plates, diagrams, etc.), a complete diagnosis covering up to several pages, the geographical distribution in detail (which also may be several pages), habits, and general characteristics including an account of its daily life, burrows or other living quarters, reproduction, growth and development of young, hibernation (if applicable), economic importance (pest and pest control), as well as maintenance in captivity for more common species. Hundreds of black-and-white photos and drawings illustrate the text. An extension of this work titled *Mammals of the Soviet Union* (see entry 1495) is only two volumes so far but fills in gaps unfinished in Ognev's work, notably the ungulates. A monumental work.

VIETNAM, SOUTH

1497. Van Peenen, P. F. D. **Preliminary Identification Manual for Mammals of South Vietnam**. Washington, D.C.: United States National Museum, Smithsonian Institution, 1969. 310p.

Written during the Vietnam war, this work was prepared for medical personnel to identify mammals in the area. It is arranged taxonomically with a chapter for each of the 11 orders of mammals found in South Vietnam. Keys to orders, families, genera, and species are included with species accounts following the keys. Each species account gives the scientific name, the common name, the author and date where first named, a black-and-white sketch of the species as well as its skull, a short paragraph on the appearance, measurements of the species and skull, as well as number and location of species examined. A gazetteer of place names in South Vietnam, and a map follow the species descriptions. Easy to use, with concise information.

ZAMBIA

1498. Ansell, W. F. H. **Mammals of Northern Rhodesia: A Revised Check List with Keys, Notes on Distribution, Range Maps, and Summaries of Breeding and Ecological Data.** Lusaka, Northern Rhodesia: Government Printer, 1960. 155p. LC 61-1662.

This short but comprehensive work provides keys to living species of mammals in Northern Rhodesia, now Zambia, as well as a host of other information. Introductory information on the geography and ecology of the area leads to the systematic list of mammals. The list provides keys down to species with species accounts that include information on geographical distribution, habitat, behavior, diet, and reproduction. Appendices to rejected species names (had one time been listed as occurring in Northern Rhodesia, but the records were not acceptable); a list of possible additions of species; a list of species by geographical categories such as the Ethiopian Region, and the West African Subregion; a gazetteer of localities listed in the checklist; a list of references; and an index to scientific, English, and African common names for the mammals. Distribution maps are included at the end. Color plates of selected species are found throughout the text. Updated by *The Mammals of Zambia* (see entry 1499).

1499. Ansell, W. F. H. **The Mammals of Zambia**. Chilanga, Zambia: National Parks & Wildlife Service, 1978. 126p. LC 81-980224.

This work updates and consolidates information from the author's previous work, *Mammals of Northern Rhodesia* (see entry 1498). Four years after this was published, the independent republic of Zambia was established and the name Northern Rhodesia fell to history. This new work is not actually a new edition since it has a different emphasis, so both books are necessary when studying the mammals of this country. Keys for identification to species are not given in this volume, but the author has rather concentrated more on detail of the species occurring in Zambia and their distribution. Indeed, the last 111 pages include distribution maps for species discussed in the text. The text is in checklist form with detailed taxonomic descriptions for each species. No drawings or pictures of species are included here. A detailed work primarily for the specialist.

ZIMBABWE

1500. Smithers, Reay H. N., and V. J. Wilson. **Check List and Atlas of the Mammals of Zimbabwe Rhodesia**. Salisbury, Zimbabwe Rhodesia: Trustees of the National Museums and Monuments, 1979. 193p. (Museum Memoir, no. 9). LC 81-203460.

Southern Rhodesia was named Zimbabwe in 1980, just when this book was published, hence the confusion of the name on the title and throughout the text. The work is similar to the other works by Smithers for this area. Although titled a check list, it is more than just a checklist in that it provides concise information for each species on distribution, habitat, habits, food, breeding, weights and size, as well as taxonomic notes and distribution maps. Keys to identification to genus and species are also given. An index to English and scientific names completes the volume. Updated in part by Smithers's 1983 work (see entry 1444).

Mammalia (Systematic Section)

MONOTREMATA (Platypus and Echidnas)

1501. Collins, Larry R. **Monotremes and Marsupials: A Reference for Zoological Institutions**. Washington, D.C.: Smithsonian Institution Press [for sale by the Supt. of Docs., U.S. Govt. Print. Off.], 1973. 323p. LC 73-5963.

Also submitted by the author to the University of Maryland as his master's thesis in zoology, this work was intended as a reference source for zoos in the care, maintenance, and display of the species in these two groups, the monotremes and the marsupials. Each species account includes information on its distribution, history in captivity, enclosure dimensions, dietary information (some with recipes), and may include breeding and molting seasons, development of the young, and hibernation information. Black-and-white photos of most species and some photos of enclosures for that species are found throughout the text. A 35-page bibliography and index to species and subject complete this work. An interesting text, compiling hard-to-find information in a single work.

1502. Griffiths, Mervyn. **The Biology of the Monotremes**. New York: Academic Press, 1978. 367p. LC 78-4818. ISBN 0-12-303850-2.

The platypus and the echidnas (spiny anteaters) make up this order called Monotremata. This work covers the present knowledge and research on these animals in 10 chapters. The anatomy, classification, food, feeding, physiology, reproduction, and other areas of study are covered in a technical, yet very readable writing style. Black-and-white photos, graphs, and drawings are found throughout the text. A long list of references and an index to scientific name and subject complete the text.

MARSUPIALIA (Opossums, Kangaroos, Wallabies, Koalas, etc.)

1503. Archer, Michael, ed. **Carnivorous Marsupials**. Mosman, Australia: Royal Zoological Society of New South Wales, 1982. 2 vols. ISBN 0-9599951-3-7.

Containing papers presented at the Carnivorous Marsupials Symposium held in May 1980 in Australia, this two-volume work summarizes the history and present research done on this group. It includes separately authored papers on many aspects of carnivorous marsupial studies such as reproduction and life histories, ecology, physiology, behavior, paleontology, morphology, genetics, and cytology. Each paper includes black-and-white photos, drawings, and figures to illustrate the text. Lists of references are found at the end of each paper. A taxonomic-name index is found in volume two, but there is no subject index.

1504. Archer, Michael, ed. **Possums and Opossums: Studies in Evolution**. Chipping Norton, NSW: Published by Surrey Beatty & Sons in association with the Royal Zoological Society of New South Wales, 1987. 2 vols. $112.00. LC 88-140634. ISBN 0-949324-05-1 set.

Although the title suggests the content is on possums and oppossums alone, this two-volume set actually treats the entire group of marsupials. This work originated as a symposium of the Royal Zoological Society of New South Wales in Sydney. Following the symposium, papers were submitted for publication. This work complements the phylogenetic work of *Carnivorous Marsupials*, by M. Archer (see entry 1503), and *Possums and Gliders*, edited by A. P. Smith and I. D. Hume (Surrey Beaty & Sons, 1984). The first paper of volume one is by K. P. Aplin and M. Archer and titled "Recent Advances in Marsupial Systematics with a New Syncratic Classification." This classification summarizes the new scheme of marsupial systematics. It is the one used in *Mammal Species of the World* (2d ed.) (see entry 1618). The remainder of the work provides an overview of non-Australian marsupials, followed by the largest section with papers on living and extinct Australian marsupials, grouped by family. Each chapter includes extensive references. Black-and-white photos, drawings, graphs, and maps illustrate the text, as well as two full-color plates of reconstructions of two Miocene marsupial species. A thorough work, primarily for the student or specialist.

1505. Collins, Larry R. **Monotremes and Marsupials: A Reference for Zoological Institutions**. Washington, D.C.: Smithsonian Institution Press [for sale by the Supt. of Docs., U.S. Govt. Print. Off.], 1973. 323p. LC 73-5963.

Contains information for zoos on the care, maintenance, and display of the animals in these two groups, the monotremes and marsupials. (See entry 1501.)

1506. Oldfield, Thomas. **Catalogue of the Marsupialia and Monotremata in the Collection of the British Museum (Natural History).** London: Printed by Order of the Trustees, 1888. 401p. LC 06-36238.

This work lists and describes 151 species of marsupials and three species of monotremata. Keys to family, genus, and species are provided. Each species account is a very detailed physical description, but also includes information on geographical distribution, location of type specimen, and name of collector and place where specimens were collected. References for the author of the name and where published, along with references to synonyms, as well as references to other articles on the species are included at the beginning of each description. A few black-and-white drawings are found in the text. Inserted at the end of the work are 28 plates, some hand-colored, with drawings of selected species and/or skulls. Dated, but contains very detailed anatomical characteristics that may not be found in other sources.

1507. Stonehouse, Bernard, and Desmond Gilmore, eds. **The Biology of Marsupials**. Baltimore, Md.: University Park Press, 1977. 486p. LC 76-10944. ISBN 0-8391-0852-4.

This handbook on the pouched mammals includes 24 individually authored chapters on all aspects of marsupial studies. An annotated list of the living marsupials leads into chapters on the origin and evolution of marsupials, population and behavior studies, anatomy and physiology, and reproduction of marsupials. Each chapter has a lengthy list of references. Black-and-white drawings, graphs, and photos illustrate the text.

CHIROPTERA (Bats)

1508. Andersen, Knud C. **Catalogue of the Chiroptera in the Collection of the British Museum**. 2d ed. Vol. 1, **Megachiroptera**. London: Trustees of the British Museum (Natural History), 1912. Reprint, New York: Johnson Reprint, 1966. 854p. $72.00. ISBN 0-384-01395-3.

More than just a catalog, this work lists, describes, and provides keys to species for 186 species and 228 forms of the Megachiroptera, or large bats. It is a second edition to the first section of George Edward Dobson's work published in 1878 (see entry 1510). These bats are the flying foxes and fruit bats that have a wingspan of up to 1.5 meters. Introductory material includes the distinguishing characters, anatomy, and geographical distribution of this group, followed by a key to genera. Keys to species are found in the systematic section. Each species account is detailed, especially in measurements of skulls and teeth. Taxonomic information is followed by a diagnosis, information on skull and teeth, color and color phases, range, type location, and other remarks. The species accounts are one page to several pages in length. Black-and-white drawings of skulls and distinguishing characteristics are found occasionally throughout the book. An index to scientific name finishes the volume.

1509. Barbour, Roger W., and Wayne H. Davis. **Bats of America**. Lexington, Ky.: University Press of Kentucky, 1969; Reprint 1979. 312p. $35.00. LC 73-80086. ISBN 0-8131-1186-2.

This classic text, published originally in 1969, is still in print today. It is an easy-to-use work that synthesizes the information on recent (not fossil) U.S. bats, and provides keys to their identification. Following introductory information on bats and how to identify them, a pictured key to the identification of adult bat species is given. The rest of the book is arranged taxonomically by family with information on each family, then separate species accounts. Each species account is two to six pages in length and includes information on distinguishing characteristics, variation, species that may be confused with the one described, geographic range, habitat, flight characteristics, food and feeding habits, reproduction, and further remarks. A black-and-white photo of each species, photos of distinguishing anatomical details (especially of skulls), and a range map are all placed with the species account. Twenty-four colored plates, placed in the center of the text, illustrate many of the species described. A final section on the study of bats provides information on collecting, preserving, tracking and transporting, and keeping bats in captivity. A lengthy bibliography finishes the text. Fine book for the student or amateur as well a basic text for the specialist.

1510. Dobson, George Edward. **Catalogue of the Chiroptera in the Collection of the British Museum**. London: Printed by Order of the Trustees, 1878. Reprint, Forestburgh, N.Y.: Lubrecht and Cramer, 1966. 567p. $65.00. ISBN 3-7682-0300-X.

This work contains complete accounts of 400 species of bats. The accounts of the species in the Megachiroptera (large bats) group have been updated in a second edition by Knud Andersen (see entry 1508). Although the nomenclature is dated, this source does provide detailed descriptions of several hundred bats and their geographic distribution in the world. Introductory information on bat anatomy leads into the systematic index where keys to suborder, family, genus, and species begin the descriptions. Species accounts are one page in length, usually, and include references to where first named in the literature, plus other references to descriptions and biology. This information is followed by a detailed physical description, including any varieties to the species, geographical distribution in the world (and type location), and where each species was collected and by whom. Thirty plates of black-and-white drawings of selected species, skulls, and of identifying anatomical characteristics are inserted at the end of the work. A detailed taxonomic work.

1511. Hill, John E., and James D. Smith. **Bats: A Natural History**. Austin, Tex.: University of Texas Press, 1984. 243p. $35.00. LC 83-51654. ISBN 0-292-70752-5.

This work reviews the basics of bat biology for the general reader and serious student. It includes information on the evolution of bats as well as their flight, feeding and food, thermoregulation, reproduction, echolocation, population ecology, and vocalization, and a chapter on bats of the world. A bibliography (arranged according to chapter), and three indexes to the scientific and common names as well as to subject complete the text. Many black-and-white photos and drawings are found throughout the text. A good, up-to-date work on the basics of bat biology.

1512. Miller, Gerrit S., Jr. **The Families and Genera of Bats**. Washington, D.C.: Govern ment Printing Office, 1907. Reprint, Forestburgh, N.Y.: Lubrecht & Cramer, 1966. 282p $60.00. ISBN 3-7682-0534-7.

Still in print, this classic work is dated but provides a great deal of descriptive detai useful to present-day researchers. It lists and describes 36 families and subfamilies, and 17 genera of bats of the world. Introductory material on anatomy, especially detailed denta formulas and descriptions, leads into the descriptions of the families, subfamilies, and gener of bats. Keys to suborders, families, subfamilies, and genera are provided. Information at th genus level includes references to where first described and other taxonomic references, typ species information, geographic distribution, number of forms, and rather detailed descriptiv characters. Genus accounts are usually two pages in length and sometimes include black-and white line-drawings of the skull or another anatomical detail described in the account.

1513. Robertson, James. **The Complete Bat**. London: Chatto & Windus, 1990. 160p $20.00. LC 91-142133. ISBN 0-7011-3500-X.

This popular guide to bats, especially British bats, is a good introduction for young an old. It is full of black-and-white and color photos as well as entertaining drawings. Genera information on bats is provided in such chapters as "Bat Facts," "Bats in Books," and "A Ba in the Hand: Bat Care." A good beginning book for public libraries and school libraries fron junior high on up.

1514. Schober, Wilfried. **The Lives of Bats**. New York: Arco, 1984. 200p. LC 83-45641 ISBN 0-668-05993-1.

Translated from the German, lots of black-and-white and full-color photos are used t illustrate this introductory text on bats. It is more detailed than *The Complete Bat* (see entr 1513), but is still on a beginner's level of information. After general introductory material o the evolution, distribution, and history of bats, the major families of bats are described Following this are chapters on where bats live, what they eat, echolocation, breeding hibernation, orientation, and conservation of bats. Appendices provide more information o the systematics of bats, distribution, and food. An interesting, readable text, useful in publi and school libraries.

1515. Wimsatt, William A., ed. **Biology of Bats**. New York: Academic Press, 1970-197 3 vols. LC 77-117110.

Covering all the major aspects of bat biology except systematics, this ongoing work i now up to three volumes. Each chapter is authored by an expert in the field. The chapter cover bat evolution, skeletal and muscle systems, flight, migration, hibernation, thermoregu lation, urinary and central nervous systems, the integument, taste, vision, hearing, socia organization, acoustic orientation, the cardiovascular system, blood, and care of laborator bats. Each chapter has its own list of references. Black-and-white photos and drawing illustrate the text. Each volume has a subject and author index.

PRIMATES

1516. Chiarelli, A. B. **Taxonomic Atlas of Living Primates**. London, New York: Academi Press, 1972. 363p. LC 79-129785. ISBN 0-12-172550-2.

This work provides keys to species of living primates in addition to brief informatio on each species. Each description is one page long and usually includes a black-and-whit photo of the species. The brief descriptive information includes the common names (i English, German, French, Spanish, and Italian), synonyms, geographical distribution, bod length, body weight, tail length, and coat color. Some black-and-white range maps ar included. Three separate indexes to species name, common name, and synonyms are given An easy-to-use guide.

517. Hill, W. C. Osman. **Primates: Comparative Anatomy and Taxonomy**. New York: nterscience Publishers; Edinburgh: At the University Press, 1953-1974. 8 vols. LC 54-4499.

This lengthy work covers the anatomy and taxonomy of all groups of primates in detail. 'he family then genus are described, then each species. Genus and species information ıcludes a description of external characters, dentition, internal anatomy, distribution, and ıxonomy, and may also include information on habitat, ecology, reproduction, and behavior. ome keys to identification are included. The descriptions are lengthy: for the genus *Macaca*, 84 pages; the species *Macaca mulatta* (Rhesus Monkey), 49 additional pages. Each volume overs a major group: Volume 1, Strepsirhini; volume 2, Haplorhini: Tarsioidea; volume 3, ithecoidea: Platyrrhini; volume 4, Cebidae, part A; volume 5, Cebidae, part B; volume 6, 'atarrhini, Carcopithecoidea, Cercopithecinae; volume 7, Cynopithecinae: Cercocebus, Maaca, Cynopithecus; volume 8, Cynopithecinae: Papio, Mandrillus, Theropithecus. Black-and-/hite photos of some species are included, as well as black-and-white drawings of anatomical haracteristics and skulls, and distribution maps. A comprehensive, though dated work.

518. Kavanagh, Michael. **A Complete Guide to Monkeys, Apes and Other Primates**. ondon: Jonathan Cape, 1983. 224p. ISBN 0-224-02168-0.

This work concentrates on providing easily read descriptions and color photos of all enus groups living today. After information on the evolution and origin of the primates, the ook is arranged by the major taxonomic groups (i.e., the Prosimians, monkeys of the New Vorld and the Old World, and the tailless primates). This is written in nontechnical language, asily read by the interested amateur or student. The genus descriptions are one page to six ages in length and feature full-color photos of selected species, plus black-and-white range ıaps. The appendix contains a classification of the Primates, a glossary of terms, a brief ibliography, and an index to common name.

519. Lee, Phyllis C., Jane Thornback, and Elizabeth L. Bennet., comp. **Threatened 'rimates of Africa**. Gland, Switzerland: IUCN, 1988. 151p. (The IUCN Red Data Book). 24.00. ISBN 2-88032-955-6.

Part of the Red Data Book series produced by IUCN (International Union for Conserration of Nature and Natural Resources) that lists threatened species from around the world, ıis work provides a list of the threatened primates of Africa. The categories are endangered E), vulnerable (V), rare (R), indeterminate (I), insufficiently known (K), or not threatened NT). It also reviews each species on the list, providing a summary of information, its istribution, the population, habitat and ecology, the major threats to the animal, and a list of eferences for further reading. Black-and-white photos of 14 of the species are provided. In ll, 30 species representing 50% of the primates in Africa are on this threatened list, indicating ıe terrible decline in this population.

520. Napier, J. R., and P. H. Napier. **A Handbook of Living Primates: Morphology, 'cology, and Behaviour of Nonhuman Primates**. London, New York: Academic Press, 967. 416p. LC 66-30126.

This handbook is divided into three parts. The first part covers the functional morphology of primates; the second, 56 genera descriptions; and the third part includes comparative nd supplementary data. The functional morphology section includes information on the kull, growth rates, eyes, teeth, brain, limbs, and the opposable thumbs. The genera described ı the second section are arranged in alphabetical order from *Alouatta* to *Urogale*. Each genus escription lists the subgenus, and number and name of species and subspecies. The genus escription also includes the geographical range, ecology, morphology, weights and dimenions, behavior, reproduction, adaptability to captivity, and a list of abbreviated references ıat can be found in the main bibliography at the end of the book. Nice black-and-white photos f many representative species in each genera are provided. Most of the information in section ıree is in tabular form, but it also includes a systematic list of living primates and a summary f the habitats of primates. Updated by their *Natural History of the Primates* (see entry 521).

1521. Napier, J. R., and P. H. Napier. **Natural History of the Primates**. 1st MIT Press ed. Cambridge, Mass.: MIT Press, 1985. 200p. $24.95. LC 84-28879. ISBN 0-262-14039-X.

This handbook provides several chapters of information on primate origins, anatomy, and behavior before descriptions of 57 genera of living nonhuman primates. This updates information on descriptions given in their 1967 work *A Handbook of Living Primates* (see entry 1520). It is more colorful, with many color photos and drawings to illustrate selected species, but the textual descriptions are not as detailed. A final chapter on human evolution, a glossary of terms, a list for further reading, a five-page bibliography of references, and an index to scientific and common names finish the work.

1522. Napier, P. H. **Catalogue of Primates in the British Museum (Natural History)**. London: British Museum (Natural History), 1976- . $55.00 pt. 1; $55.00 pt. 2; $65.00 pt. 3; $55.00 pt. 4; $49.95 pt. 5. (Publication nos. 744, 815, etc.). ISBN 0-565-00744-0 pt. 1; 0-565-00815-3 pt. 2; 0-565-00894-3 pt. 3; 0-565-01008-5 pt. 4; 0-565-01154-5 pt. 5.

Parts 1-3 of this catalog were authored by P. H. Napier, and part 4 by P. D. Jenkins, who took over this project after Napier left the British Museum (Natural History). These are catalogs of specimens held by the museum in its various collections. Part 1 includes families Callitrichidae and Cebidae; part 2, family Cercopithecidae, subfamily Cercopithecinae; part 3, family Cercopithecidae, subfamily Colobidae; part 4, suborder Strepsirhini, including the subfossil Madagascan lemurs and family Tarsiidae; part 5, the apes, superfamily Hominoidea. Keys to family are provided and references to where to find keys to species are given. Information on each genus includes number of species, principal synonyms, taxonomy, morphology, geographical range, ecology, behavior, and reproduction. Weights and measurements of each species in a genus are provided in tabular form. Each species account includes information on type localities, taxonomy, principal synonyms, and range. Detailed information on where each species was found, collection date, sex, age, accession number and references or remarks are listed in tabular form. Good bibliographies and an index to scientific and common names complete each volume.

1523. Preston-Mafham, Rod, and Ken Preston-Mafham. **Primates of the World**. New York: Facts on File, 1992. 191p. $24.95. LC 91-34964. ISBN 0-8160-2745-5.

Following the same pattern as other books in the " ... of the World" series, this book provides a short but good introduction to the primates with excellent full-color photos that accompany the text. The introduction is followed by two chapters on the main primate groups; the Prosimians and New World monkeys and the Old World monkeys and the apes. It also provides basic chapters on social behavior, food, habits, ecology, enemies, and defense, and a final chapter on the impact of man on the primate population. The striking photos, found on almost every other page, were taken by Ken Preston-Mafham of primates in the wild. Provides a good introduction for junior high to college level.

1524. Tuttle, Russell H. **Apes of the World: Their Social Behavior, Communication, Mentality, and Ecology**. Park Ridge, N.J.: Noyes Publications, 1986. 421p. $55.00. (Noyes Series in Animal Behavior, Ecology, Conservation, and Management). LC 86-17960. ISBN 0-8155-1104-3.

The apes include the orangutans, common chimpanzees, bonobos, gorillas, siamangs, and gibbons. This work discusses the present research and findings on this group's taxonomy, distribution, behavior, feeding, communication, sociobiology, and ecology. Each chapter discusses the general topic as it applies to each of the ape groups. A few black-and-white photos, and maps illustrate the text. A tabular comparison of the apes is provided in the last chapter. A long bibliography of sources and an index to subjects, common names, and scientific names complete the text. Primarily for the primatologist, but interesting reading for the student and the nonspecialist.

525. Wolfheim, Jaclyn H. **Primates of the World: Distribution, Abundance, and Con-ervation**. Seattle, Wash.: University of Washington Press, 1982. 831p. $131.00. LC 82-3464. ISBN 0-295-95899-5.

This work provides an excellent synthesis of information on 151 species of primates ving in the world today. It is arranged by family, with 2 to 10 page individual species ccounts. Taxonomic notes, geographic range, abundance and density, habitat, factors affect-g populations (including habitat alteration and human predation), and conservation action re all covered in each species account. A large distribution map for each species is also given s well as a table showing what countries the animal is found and references to where the formation was taken. No photos of species except for one opposite the title page. A lengthy 6-page bibliography provides further reading. Very useful to all groups, especially for onservation planning.

DENTATA

526. Montgomery, G. Gene, ed. **The Evolution and Ecology of Armadillos, Sloths, and ermilinguas.** Washington, D.C.: Smithsonian Institution Press, 1985. 451p. $49.95. LC 4-600292. ISBN 0-87474-649-3.

Containing the papers presented at the symposium "The Evolution and Ecology of enarthra" at the 1979 meeting of the American Society of Mammology, this publication ontains the current research and reviews of past research on this group of animals. Informa-on on the classification and nomenclature, evolution, anatomy, physiology, reproduction, iseases and parasites, and ecology are all contain in the various papers presented here. umerous black-and-white photos, drawings, maps, and figures illustrate the text. Unfortu-ately, there is no index; but the table of contents provides an outline for finding information this work.

IYSTICETA (Baleen Whales) and ODONTOCETA Smaller Whales, Dolphins, and Porpoises)

527. Bonner, Nigel. **Whales of the World**. New York: Facts on File, 1989. 191p. $24.95. C 88-33315. ISBN 0-8160-1734-4.

The title "Whales of the World" is used for three separate books listed in this section. his book is the most general of the three, providing a good overall review of the whales and olphins. The 10 chapters discuss the different types of whales in easy to read, nontechnical nguage. Color and black-and-white drawings are distributed throughout the work to illus-ate the text. An appendix provides a "Classification of the Order Cetacea," and a "Guide to urther Reading."

528. Coffey, David J. **Dolphins, Whales, and Porpoises: An Encyclopedia of Sea lammals**. 1st Am. ed. New York: Macmillan, 1977. 223p. LC 76-12469. ISBN 0-02-26660-8.

This work not only covers the whales, porpoises, and dolphins, it also includes the seals, a lions, walruses, dugongs, manatees, and sea otters. Although the title indicates it is an cyclopedia, it is included here because it is more of a handbook in nature. It is divided into ree sections. The first section describes the Cetacea (dolphins, whales, and porpoises), the cond the Pinnipedia (seals, sea lions, and walruses), and the third, the Sirenia (dugongs, sea ows, and manatees); plus, there is a chapter on the sea otter. Each section provides a long troduction to each group, describing behavior, anatomy, reproduction, captivity, distribu-on, and evolution. Following this is a dictionary of species where each species is briefly escribed in approximately one page. Color and black-and-white photos (some covering an tire page) and drawings illustrate the text. More general and condensed than *Handbook of arine Mammals* (see entry 1550). Especially useful to the amateur or student wanting easy ading with nice color photos.

1529. Ellis, Richard. **The Book of Whales**. New York: Knopf, 1985. 202p. $24.95. LC 80-7640. ISBN 0-394-73371-1.

This is a companion volume to Ellis' *Dolphins and Porpoises* (see entry 1530). Descrip tions of 35 species of the toothed and baleen whales include the beautiful drawings an paintings of the author. Each description is 5 to 20 (sperm whale) pages in length and include a black-and-white drawing of the species. A 30-page list of references is included and also bibliography for further reading. The nontechnical writing makes this good reading for th amateur, but it is complete enough for the specialist.

1530. Ellis, Richard. **Dolphins and Porpoises**. 1st paperback ed. New York: Knopf, 1989 270p. $24.95pa. LC 82-47823. ISBN 0-679-72286-6.

This work and its companion volume *The Book of Whales* (see entry 1529) wer originally intended to be one volume. Because of the size of the manuscript, it was publishe in two works. This work describes and pictures 43 species from the Platanistidae family (rive dolphins) and the Delphinidae family (the rest of the dolphins, pilot whales, and killer whales) It is written in a nontechnical style. The species accounts are arranged taxonomically and ar two to five pages in length. Each description usually includes a black-and-white drawing o the animal and/or the skull. A section of full-color drawings of nine selected species is inserte into the center of the text. A lengthy list of references cited in the text are arranged unde headings for the individual species. An index to common name, author, and subject complete the work. Although this is intended as a popular work, it is well documented and complet enough for the specialist.

1531. Food and Agriculture Organization of the United Nations. Working Party on Marin Mammals. **Mammals in the Seas**. Rome: FAO; New York: available from UNIPUB 1978-1982. $85.00 each. LC 79-320177. ISBN 92-5-100511-7 vol. 1; 92-5-100512-5 vol. 2 92-5-100513-3 vol. 3; 92-5-100514-1 vol. 4.

This four-volume work is the result of a Food and Agriculture Organization of the Unite Nations (FAO) report prepared by the Working Party on Marine Mammals. The report was response to public concern about the conservation status of marine mammals. These volume discuss the major groups of marine mammals, their status, management and conservation population biology, and current research. Volume one is titled *Report of the FAO Advisor Committee on Marine Resources Research, Working Party on Marine Mammals*; volume 2 *Pinniped Species Summaries and Report on Sirenians*; volume 3, *General Papers and Larg Cetaceans*; volume 4, *Small Cetaceans, Seals, Sirenians, and Otters*. Volume four is 53 pages in length. Information is also provided for some individual species (the Harbour Sea for example). For these species, the following information is provided: description, identifi cation, distribution, population size, reproduction, mortality, commercial killing, food, rela tion to man, current research, and references to further reading. A good source of conservatio information for these groups.

1532. Harrison, Richard, and M. M. Bryden, cons. eds. **Whales, Dolphins, and Porpoises** New York: Facts on File, 1988. 240p. $35.00. LC 88-11700. ISBN 0-8180-1977-0.

This is a very colorful, heavily illustrated work that will provide hours of entertainmen as well as good information. It is divided into three main sections (i.e., "Whales of the World," "The World of the Whale," and "Whales and People"). In the first section, each species i described and illustrated with beautiful color drawings. Longer discussions of each group the baleen whales, and the toothed whales, follow, illustrated with striking color photos tha often cover a full page. The second section provides information on their anatomy, senses reproduction, behavior, and intelligence. The last section includes a history of whaling whales in literature and art, captivity, relationships with humans and even a section of wha to do for whale strandings. Includes a bibliography and index. May be enjoyed for th beautiful illustrations or the easily readable text, at all levels of interest.

1532a. Leatherwood, Stephen, and Randall R. Reeves. **The Sierra Club Handbook of Whales and Dolphins**. San Francisco, Calif.: Sierra Club Books, 1983. 302p. $16.95pa. LC 83-388. ISBN 0-87156-341-X; 0-87156-340-1pa.

This is a guide to 75 species of whales, dolphins, and porpoises found worldwide. Concise descriptions and interesting text make this a good introduction to the Cetaceans. It provides background information on the group as a whole, including their classification and nomenclature, adaptations, thermoregulation, birth and care of young, migration and distribution and their conservation. The guide is arranged under headings for the baleen whales, and the toothed whales, dolphins and porpoises. The species descriptions are arranged under their respective families and are usually three to six pages in length. Species descriptions include information on the common and scientific names and the derivation of the name, the zone where it is found (eight zones are pictured in a map at the beginning of the guide), a description of the animal with distinctive features emphasized, the natural history, distribution and conservation status and a section for what other species it may be confused with. Each species description is illustrated with a colored painting and black-and-white photos. This is a companion volume to *The Sierra Club Handbook of Seals and Sirenians* (see entry 1549).

1533. Minasian, Stanley M., Kenneth C. Balcomb, III, and Larry Foster. **The World's Whales: The Complete Illustrated Guide**. Washington, D.C.: Smithsonian Books; distr. New York, W. W. Norton, 1984. 224p. LC 84-14142.

This colorful work covers information on the whales, dolphins, and porpoises of the world. The work features 160 beautiful full-color photos of almost all the whale, dolphin, and porpoise species found worldwide. The text gives introductory information on the species, then describes each family and provides page-long species accounts with a full-page photo on the facing page and two or more pages following. Facts on color, fins, length and weight, feeding, breathing and diving, mating, migration, distribution, and natural history are all provided along with the scientific name and common name. A black-and-white sketch of the animal stretches across the top of the page of description. Gives concise information and amazing photos.

1534. Ridgway, Sam H., and Richard J. Harrison, eds. **Handbook of Marine Mammals**. London, New York: Academic Press, 1981. 4 vols. $105.00 each. LC 80-42010. ISBN 0-12-588501-6 vol. 1; 0-12-588502-4 vol. 2; 0-12-588503-2 vol. 3; 0-12-588504-0 vol. 4.

Two volumes of this set, volumes three and four, cover the whales. (See entry 1550 for a complete annotation.)

1535. Tinker, Spencer Wilkie. **Whales of the World**. Leiden, Netherlands; New York: E. J. Brill; Honolulu: Published and distributed by Bess Press, 1988. 310p. $45.75. LC 87-70063. ISBN 90-04-08954-3.

This handbook emphasizes the anatomy and classification of the whales, dolphins, and porpoises. It is more detailed and technical than the other books of the same title reviewed in this section. In all, 77 species are described and pictured. It is divided into three parts, part one on the evolution and origin of whales, part two on the anatomy, and part three, the main text, on the classification and description of the living whales. This section is arranged in taxonomic order, first the toothed whales, then the baleen whales. Species accounts are one page or two in length and include brief information under the following headings: identifying features, size and shape, color, distribution, migration, fins, teeth, skeletal notes, food, swimming and diving, reproduction, abundance, economic importance, and miscellaneous notes. Most species accounts are illustrated with a black-and-white drawing of the animal, and two or three photos. Very thorough on anatomy and classification of the whales.

1536. Watson, Lyall. **Whales of the World**. Rev. pbk. ed. London: Hutchinson, 1985. 302p. LC 81-66869. ISBN 0-525-93202-X.

This is a survey of all the whales, dolphins, and porpoises living in the world today. This is the revised edition of the author's 1981 book titled *Sea Guide to Whales of the World*. It begins with introductory material on the study of whales, the evolution, natural history, status, and distribution. The guide section follows and includes two- to four-page descriptions of 76 species. Each species is pictured with a full-color drawing and a distribution map. Information on each species includes classification, common name, physical description, field identification, natural history, conservation status, distribution, and a short list of references at the end. The three appendices contain synonyms, record keeping and first aid for beached whales, and biographical sketches of 100 people prominent in whale studies. A good source of information for student and specialist.

CARNIVORA (Dogs, Cats, Seals, Sea Lions, Walruses, etc.)

1536a. Brown, Gary. **The Great Bear Almanac**. New York: Lyons & Burford, 1993. 325p. $30.00. LC 93-7686. ISBN 1-558121-210-8.

This popular book is packed with facts and lore on bears of the world. Easy-to-read text combined with many illustrations, photos, color plates, maps and charts will entertain and inform for hours. It is arranged into two parts and three appendices. Part one, "Bears in Their World," covers the evolution of bears, brief descriptions of eight species of bears found worldwide, their anatomy and physiology and bear behavior. Part two, "Bears in the Human World," covers the interactions of bears and man, and concludes with chapters on the conservation status of bears. Three appendices cover evolution, the names of bears found in CITES (Convention on International Trade in Endangered Species of Wild Fauna and Flora), and a list of organizations involved in bear studies or concerns. A bibliography and index complete the work. Written by Gary Brown after 31 years in the National Park Service, this work reflects his first-hand knowledge and dedication to bears gained during those years. For those seeking more information on individual bear species, see Terry Domico's *Bears of the World* (see entry 1539).

1537. Bueler, Lois E. **Wild Dogs of the World**. New York: Stein and Day, 1973. 274p. LC 72-96435. ISBN 0-8128-1568-6.

Providing interesting information on the lesser known relatives of the domestic dog, this book pictures and describes in depth 34 species of wild dogs. It is arranged by the major genera, *Canis* (wolves and jackals), *Vulpes* (foxes), *Dusicyon* (South American foxes), and includes additional, lesser-known species such as the raccoon dog and the dhole. Black-and-white photos and range maps are provided for each species. Species accounts are long, from 2 to 44 pages. In addition, it has four appendices covering anatomical charts of the domestic dog, the scientific and common names of all species of dogs, relationships between the members of the dog family, and genetic information. Good reading for the general reader to the specialist.

1538. **The Complete Dog Book: The Photograph, History, and Official Standard of Every Breed Admitted to AKC Registration, and the Selection, Training, Breeding, Care, and Feeding of Pure-Bred Dogs**. 18th ed. New York: Howell Book House; Toronto: Maxwell Macmillan Canada; New York: Maxwell Macmillan International, 1992. 724p. $27.50. LC 91-42714. ISBN 0-87605-464-5.

This work is a compilation of American Kennel Club standards by which purebred dogs are judged. In addition, it provides up-to-date information on how to select, train, care for, and breed the dogs. It was first published in 1929 and continues today as the most popular dog book available. After introductory information on dog sport and anatomy, the various groups of breeds are described and pictured. The breeds are grouped as sporting dogs, hounds, working dogs, terriers, toys, non-sporting dogs, and herding dogs. Each breed within these groups is described in four pages or so. The breed is pictured in a black-and-white photo, followed by information on the origin of the breed, when the breed was first registered by the American Kennel Club (AKC), and other interesting facts. The official AKC standard is then described in detail, covering the general appearance, size, head, neck and body, forequarters, hindquarters, color, gait, and temperament. A table with a scale of possible points is included at the end of the description along with disqualifications. The remainder of the book deals with care of the dog and its health, including first aid and training. A glossary of terms and an index finish this comprehensive work.

1539. Domico, Terry. **Bears of the World**. New York: Facts on File, 1988. 224p. $29.95. LC 88-3881. ISBN 0-9160-1536-8.

This popular work illustrates, in 160 beautiful color photos, and describes the common bears of the world. It also includes information on ursine development and the capabilities of bears. Individual chapters cover the black bear, brown bear, polar bear, Asian black bear, and the panda species. Human-bear interactions and how to avoid encounters are also discussed. Suggestions for further reading and an index complete this popular work for school and public libraries.

1540. Ewer, R. F. **The Carnivores**. Ithaca, N.Y.: Cornell University Press, 1973. Reprint, Ithaca, N.Y.: Comstock, 1985. 494p. (A Comstock Book Series). $47.50; $21.50. LC 85-12741. ISBN 0-8014-0745-1; 0-8014-9351-Xpa.

This classic work attempts to bring together information on the biology, social organization, systematics, and a little of the fossil history of the groups that make up the order Carnivora. These include the wild dogs, wolves, jackals, bears, racoons, pandas, weasels, badgers, skunks, otters, civets, hyenas, wild cats, ocelots, lions, and so on. The group made up of the seals, sea lions, and walruses (Pinnipedia) is not included in this work. There are chapters on the skeleton, anatomy, special senses, food and food finding, signals and social organization, reproduction, and fossil relatives. A final chapter is on the classification and distribution of the families. Black-and-white drawings are found occasionally in the text, with 16 pages of black-and-white photos of selected species found in the center of the text. A 37-page bibliography and indexes to author, species, and subject complete the work. Although dated, this work provides a foundation of information on the Carnivora that is expanded on in the other publications mentioned here, especially *Carnivore Behavior, Ecology, and Evolution* (see entry 1543).

1541. Food and Agriculture Organization of the United Nations. Working Party on Marine Mammals. **Mammals in the Seas**. Rome: FAO; New York: available from UNIPUB, 1978-1982. LC 79-320177. ISBN 92-5-100511-7 vol. 1; 92-5-100512-5 vol. 2; 92-5-100513-3 vol. 3; 92-5-100514-1 vol. 4.

Includes two volumes on seals, walrus, sea lions, and otters. (See entry 1531.)

1542. Fox, M. W., ed. **The Wild Canids: Their Systematics, Behavioral Ecology, and Evolution**. New York: Van Nostrand Reinhold, 1975. Reprint, New York: Krieger, 1984. 526p. $44.50. LC 83-268. ISBN 0-89874-619-1.

This work aims to provide comprehensive information on the canids, or wild dogs, in five main areas (i.e., taxonomy, behavioral ecology, social behavior, genetics and physiology, and behavioral evolution). Thirty papers contributed by individual experts in their field make up the text. A list and description of the genera and species of canids found worldwide is found in the first chapter. Black-and-white photos of selected species, and range maps are provided. Other chapters concentrate either on individual species, such as a review of the ecology of the gray fox, or provide a comparative approach such as a paper on the South American canids. A list of references for the entire work is found at the end of the text as well, as an author and subject index. Although this text is dated (1975), it still provides classic information by well-known researchers, and even includes an introduction by Konrad Lorenz. More for the specialist, but of interest and use to students and serious amateurs.

1543. Gittleman, John L., ed. **Carnivore Behavior, Ecology, and Evolution**. Ithaca, N.Y.: Comstock, 1989. 620p. $67.50; $25.00pa. LC 88-47725. ISBN 0-8014-2190-X; 0-8014-9525-3pa.

This recent work on the Carnivora expands on the work by R. F. Ewer, *The Carnivores* (see entry 1540), by concentrating on the behavior, ecology, and evolution of this group. Thirty contributors wrote chapters in three main areas from "Carnivore Group Living," "Adaptations for Aquatic Living by Carnivores," to the "Phylogency of the Recen Carnivora." Each chapter includes extensive references. Black-and-white photos, graphs maps, and figures illustrate each chapter. An appendix includes the "Classification of th Recent Carnivora," by W. Chris Wozencraft. When used in conjunction with R. F. Ewer' work, this forms a comprehensive source of information on the Carnivora.

1544. Green, Richard. **Wild Cat Species of the World**. Devon, England: Basset Publica tions, 1991. 163p. $60.00. LC gb92-2886. ISBN 0-946873-93-3.

This book describes 39 species of wild cats, including a description of the mysteriou onza cat. Each description is two to five pages in length and is arranged under identical topic for each species (i.e., vital statistics, description, distribution, habitat, habits, breeding development of young, status and systematics, and International Species Indexing Systen [ISIS] numbers). The vital statistics include measurements for length, height, weight, age a maturity, dental formula, chromosome count, karyotype, longevity, gestation period, numbe of young, weight at birth, and young weaned. Full-color photo plates of selected species ar inserted into the middle of the text. The information here is more concise than that o Guggisberg's work (see entry 1545) on wild cats and also includes the conservation status A 14-page bibliography of references and further reading, and an index to common nam complete this work.

1545. Guggisberg, C. A. W. **Wild Cats of the World**. New York: Taplinger, 1975. 328p LC 74-21020. ISBN 0-8008-8324-1.

After introductory material on the cat family, Felidae, and its origin, this book concen trates on describing each of the 38 species of wild cats found worldwide. Each cat is describe in one page to seven pages. The descriptions are easy to read and relate many of the author' personal accounts. The descriptions are, in most cases, more lengthy than those in Richar Green's book on wild cats (see entry 1544). In addition, basic characteristics, measurements distribution, and habits are detailed for each species. Many of the cats are pictured i black-and-white photos inserted into the center of the book. An eight-page bibliography an an index complete the work.

1546. Kelsey-Wood, Dennis. **The Atlas of Cats of the World, Domesticated and Wild** Neptune City, N.J.: T. F. H. Publications, 1989. $59.95. LC 90-158319. ISBN 0-86622-666-4

This large folio-sized work contains up-to-date information primarily on the domesti breeds of cats. It features striking full-color photos of all breeds of cats, plus other engagin photos showing behaviors, training techniques, care, and so on. It also covers the natura history of cats, domestication, anatomy, nutrition, care and breeding, as well as informatio on purchasing a cat and showing it. Health care is also discussed. The domestic cat breed are pictured and described in 120 pages of text. A bibliography of sources to read for mor information, and an index complete the work. A beautiful book for comprehensive informa tion on the domestic cat.

1547. King, Judith E. **Seals of the World**. 2d ed. Ithaca, N.Y.: Cornell Univ. Press, 1991 240p. $16.95pa. LC 91-7917. ISBN 0-19-858513-6; 0-8014-9953-4.

Now in its second edition, this work is an excellent introduction to the seals, and to se lions and walruses as well. It is arranged into three sections: one on classification an description of the species in this group; a section on pinniped biology, and a final section o appendices including a geographical index, origins of the scientific names found in the text references, further reading, and an index. The descriptions in the first section are arrange under groups such as the sea lions, the fur seals, walrus, northern phocids, and souther phocids. A chapter on fossil relations is included as well. Species descriptions are two or thre pages in length and include information on their distribution, a physical description, ecology breeding, predators, feeding, exploitation, and demography. The book features full-color photos of almost every species described as well as black-and-white drawings and distribution maps. A fine text for all interests.

1548. Kitchener, Andrew. **The Natural History of the Wild Cats**. Ithaca, N.Y.: Comstock, 1991. 280p. $27.50. ISBN 0-8014-2596-4.

This is a more comprehensive work on the wild cats than the other books on wild cats mentioned here (see entries 1544 and 1545). It covers their evolution, behavior, eating habits, domestication, and other studies. These topics the author has grouped under chapters such as "How to Become a Carnivore," "A Cat's Nine Lives," and "A Who's Who of Cats." In this last mentioned chapter, 37 species of wild cats are briefly described in about one page of text. Nice black-and-white drawings, graphs, maps, and numerous tables illustrate the text, along with eight pages of full-color photos. Appendices cover measurements and body weights, and a 23-page bibliography and an index complete this informative text.

1549. Reeves, Randall R., Brent S. Stewart, and Stephen Leatherwood. **The Sierra Club Handbook of Seals and Sirenians**. San Francisco: Sierra Club Books, 1992. 359p. $18.00. LC 92-946. ISBN 0-87156-656-7.

This handbook is a companion volume to *The Sierra Club Handbook of Whales and Dolphins* (see entry 1532a), and describes all other marine mammals. These include the pinnipeds (seals, sea lions, and walruses), the sirenians (manatees, dugongs, and sea cows), otters, and the polar bear. Altogether, it describes and pictures 42 marine mammal species.

After a general introduction to these four groups, the various species in each group are described in four to thirteen pages each. Each species account includes a note on the nomenclature, a physical description, the geographical distribution, the natural history, conservation status, and a list of references for further reading. A full-color painting by Pieter Folkens begins each species description, with black-and-white photos found throughout the text. The three appendices include a list of the common and scientific names of animals and plants mentioned in the book aside from the mammals in the text; a list of pinniped species and their dental formulas; and the principle breeding ranges in the major ocean areas.

This work and its companion volume provide concise, up-to-date information on these important species. A more thorough work is the four-volume *Handbook of Marine Mammals* (see entry 1550). However, for portability in the field and price, this two-volume set will give good service. Recommended for all public and academic libraries, and all school libraries from junior high school on up.

1550. Ridgway, Sam H., and Richard J. Harrison, eds. **Handbook of Marine Mammals**. London, New York: Academic Press, 1981. 4 vols. $105.00/vol. LC 80-42010. ISBN 0-12-588501-6 vol. 1; 0-12-588502-4 vol. 2; 0-12-588503-2 vol. 3; 0-12-588504-0 vol. 4.

This handbook treats all marine mammal types, providing both identification and basic information on individual species. The set is divided into four volumes. Volume 1, walrus, sea lions, fur seals, and sea otter; volume 2, seals; volume 3, sirenians and baleen whales; and volume 4, river dolphins and the larger toothed whales. Each chapter in each volume treats an individual species. The chapter covers the genus and species taxonomic information, distinguishing features and morphology, distribution and migration, life history and size of population in the world, internal anatomy and physiology, behavior, reproduction, morbidity and mortality, and a lengthy list of references for further reading. Black-and-white photos, drawings, and distribution maps enhance each chapter. Each chapter is up to 50 or so pages in length, providing good coverage of information on each species. Excellent, informative information that is useful to student, conservationist, researcher, and specialist.

1551. Riedman, Marianne. **The Pinnipeds: Seals, Sea Lions, and Walruses**. Berkeley, Calif.: University of California Press, 1990. $39.95; $16.95pa. 439p. LC 89-31542. ISBN 0-520-06497-6; 0-520-06498-4.

This handbook provides information on many aspects of pinniped biology. Classification and distribution of the pinnipeds is followed by information on ecology, predation, diet and food, reproduction, communication, and behavior. The book is well documented and laces the technical information with a readable text. Black-and-white photos (and a section of full-color plates) are found liberally placed throughout the work. A seven-page glossary and a 55-page bibliography extend the usefulness of this very worthwhile work.

1552. Sheldon, Jennifer W. **Wild Dogs: The Natural History of the Nondomestic Canidae**. San Diego, Calif.: Academic Press, 1992. 248p. $49.95. LC 91-14127. ISBN 0-12-639375-3.

This work describes all 35 species of wild dogs found worldwide. It is arranged in alphabetical order by genus (*Alopex* to *Vulpes*), with species arranged in alphabetical order within the genus. Common names are easily found in the index, table of contents, a table, and at section headings, so those unfamiliar with the scientific name will have no trouble finding information. Each species account is 3 to 10 pages in length and includes information under the following headings: distribution and habitat, physical characteristics, taxonomy, diet, activity, reproduction, social organization, and behavior. This allows for easy comparisons between species. Black-and-white photos of a few representative species are included. This work on wild dogs updates Lois E. Bueler's 1973 work (see entry 1537), but it does not include pictures for each species or range maps, and the descriptions in Bueler's work are longer. Sheldon's work is better documented, however, and includes a lengthy 35-page list of references cited in the work. This is a good, current reference work for all levels of interest.

LAGOMORPHA (Rabbits, Hares and Pikas)

1553. Chapman, Joseph A., and John E. C. Flux, comps. and eds. **Rabbits, Hares and Pikas: Status Survey and Conservation Action Plan**. Gland, Switzerland: International Union for the Conservation of Nature and Natural Resources, 1990. 168p. $25.00. ISBN 2-8317-0019-1.

This work reviews the current knowledge of the lagomorphs and provides an action plan for the worldwide conservation of this group. It is arranged into three sections. Section one introduces the group and provides a classification arrangement and a checklist of world lagomorphs. Section two gives species accounts of all known living rabbits, hares, and pikas. Each account is one to three pages in length and includes information on taxonomy, conservation status, a description, geographical distribution, habitat, ecology, behavior, and reproduction. A black-and-white range map is provided for each account and black-and-white photos illustrate many species or habitats. Section three provides the conservation action plan with specific recommendations for globally threatened species. A current, thorough review of this order, and at a reasonable price.

RODENTIA
(Rats, Mice, Squirrels, Gophers, Beavers, Porcupines, etc.)

1554. Ellerman, J. R. **The Families and Genera of Living Rodents**. London: Trustees of the British Museum, 1940-1949. 3 vols. LC 41-22525.

This classic work, based entirely on the collection of the British Museum, attempts to classify and describe genera and provide keys to genera for the living rodents. It is divided into two volumes with a separate third volume covering the additions and corrections to the main work, plus two papers. The two main volumes cover the rodents other than Muridae (Volume 1), and the Muridae (Volume 2). In volume one, 22 families of rodents, 151 valid genera with 2,773 named forms, are listed. Volume two, the Muridae family, lists 192 valid genera containing 3,600 named forms. After each family description, keys to genera are given (in most cases), then the genera are described, with species names listed after each genus. References to where first named and location where collected are provided for each species

Black-and-white drawings of skulls and dentition are included with each genera description. An index to scientific name is included at the end of each volume. Dated, but the only comprehensive classification work for the rodents.

1555. Hanney, Peter W. **Rodents: Their Lives and Habits**. New York: Taplinger, 1975. 224p. LC 75-10064. ISBN 0-8008-6804-8.

This introductory work uses nontechnical language to present a broad picture of the rodents, a very large group that accounts for nearly half of all mammal species. Rather than try to describe all the many families and species in this group, the author has written about the rodents in chapters such as "The Climbers," "Rodents of River and Swamp," and so on. Their social organization is explained in chapters such as "The Cosmopolitans," and "Social Life and Rodent Communities." Many black-and-white photos of selected species are found throughout the text as well as black-and-white drawings. A classification of the rodent families is found in an appendix at the end. A good introduction for students from high school on up, and for anyone interested in this diverse group.

1556. MacClintock, Dorcas. **Squirrels of North America**. New York: Van Nostrand Reinhold, 1970. 184p. (A California Academy of Sciences Book). LC 74-110059.

This work covers the squirrels and squirrel-like mammals of North America. These are in the suborder Sciuromorpha (one of three suborders in the Rodentia), which are dealt with in three sections in this book: the ground dwellers (woodchucks, marmots, prairie dogs, ground squirrels, rock squirrels, and chipmunks), the tree squirrels, and the gliders (flying squirrels). Species accounts are arranged within these three main groups. Each species account includes a physical description, its distribution, and behavior and ecology in very readable prose of one page or more. Distribution maps are provided for most species and charming pencil drawings (sometimes covering an entire page) are provided for many species. A long list of references is sectioned by group (i.e., chipmunks, gray squirrels, flying squirrels, etc.). No index. Interesting reading for students, amateurs, and specialists.

1557. Rowlands, I. W., and Barbara J. Weir. **The Biology of Hystricomorph Rodents**. New York: Published for the Zoological Society of London by Academic Press, 1974. 482p. (Symposia of the Zoological Society of London, no. 34). LC 74-5683. ISBN 0-12-6133334-4.

The Hystricomorpha comprise one of the three suborders of Rodentia, which includes the porcupines, guinea pigs, capybaras, agoutis, mole rats, and more. This work is the result of a symposium held at the Zoological Society of London on June 7th and 8th in 1973. Nineteen contributors presented papers on many aspects of this group. They ranged from "What is an Hystricomorph?" to papers on the evolution, ecology and behavior, vocalizations, reproductive physiology, embryology, and papers on individual species such as the guinea pig, gundis, and viscachas. Long lists of references are provided for each paper. Many black-and-white photos, graphs, maps, and figures are found throughout the text. An author and subject index completes the volume. For the specialist.

PROBOSCIDEA (Elephants)

1558. Eltringham, S. K., consultant. **The Illustrated Encyclopedia of Elephants: From Their Origins and Evolution to Their Ceremonial and Working Relationship with Man**. New York: Crescent Books, 1991. 188p. $19.99. ISBN 0-517-06136-8.

Lavishly illustrated, this work provides interesting information on this, the largest land mammal. It was written by seven authors and two consultants and includes information on evolution, anatomy, behavior, reproduction, ecology, disease, conservation, as well as the ivory trade and elephants as working animals. Full-color photos and drawings, some covering an entire page, illustrate the text. Good reading for junior and senior high school libraries, as well as public libraries, and at a very low price.

SIRENIA (Manatees and Dugongs)

1559. Reynolds, John E., III, and Daniel K. Odell. **Manatees and Dugongs**. New York: Facts on File, 1991. 192p. $24.95. LC 89-71399. ISBN 0-8160-2436-7.

Written in an easy-to-read style, this book describes the biology, behavior, and ecology of the living sirenian species: the Florida manatee, the Antillean manatee, the West African manatee, the Amazonian manatee, and the Dugong. It does this in light of conservation issues for this threatened group. Separate chapters of 30 pages each are written for each species, describing the anatomy, distribution, and behavior. Other chapters cover the relationship of the Sirenia to man and the need for more conservation measures. A list of organizations involved in sirenian conservation and a section on their legal status emphasizes the conservation issues. A list of selected readings arranged under headings for each species provides additional information on each species. The entire book is filled with excellent black-and-white drawings and two sections of color plates illustrating the manatees and dugongs in different settings and behaviors. A very good work for all levels of interest.

PERISSODACTYLA (Horses, Tapirs, Rhinoceroses)

1560. Evans, J. Warren, and others. **The Horse**. San Francisco: W. H. Freeman, 1977. 766p. (A Series of Books in Agricultural Science). $41.95. LC 76-22686. ISBN 0-7167-0491-9.

A well-used source on horses for many years, this work covers all the basic information and more. It includes the evolution of the horse, breeds in the United States, the biology, nutrition, reproduction, genetics and health of the horse, and even information on basic horsemanship and management of a horse farm. Each chapter includes a list of references cited. Black-and-white photos, graphs, and drawings are found throughout the text. A basic source on horses for all libraries.

1561. Goodall, Daphne Machin. **Horses of the World**. New and rev. ed. New York: Macmillan, 1973. 272p. LC 73-7480.

This book describes and illustrates horse and pony breeds that exist in the world today. It is arranged by geographical area, covering horses found in Europe, Asia, Africa, Australasia, and North and South America. Each species is described in a page or two, with black-and-white photos illustrating the horse. The written description contains information on locality where found, color and chacteristics, height, and a paragraph of interesting facts on the breed. Full-color photos accompany some descriptions.

1562. Groves, Colin P. **Horses, Asses, and Zebras in the Wild**. 1st U.S. ed. Hollywood, Fla.: Curtis Books, 1974. 192p. LC 74-79610. ISBN 0-88359-008-5.

This book provides a review of the wild horses and their relatives, the asses and zebras, that make up the family Equidae. This is not a treatment of domestic horses, which are dealt with in *Horses of the World* (see entry 1561). The individual wild species are described in nontechnical language, covering information on habitat, physical characteristics, and geographical distribution in 5 to 10 pages. Black-and-white photos are given for each species described. A good book for all libraries.

1563. Lydekker, Richard. **Catalogue of the Ungulate Mammals in the British Museum (Natural History).** London: Printed by order of the Trustees of the British Museum, 1913. Reprint, New York: Johnson Reprint, 1966. 5 vols. $150.00 set; $30.00/vol. ISBN 0-384-34486-0.

Although this serves as a catalogue of the ungulate mammals in the British Museum (Natural History), it also serves as a source for keys to identification, descriptions, and geographical range. Each volume covers specific groups: Volume 1, Artiodactyla, family Bovidae, subfamilies Bovinae to Ovibovinae (cattle, sheep, goats, chamois, serows, takin, musk-oxen, etc.); volume 2, Artiodactyla, family Bovidae, subfamilies Bubalinae to Reduncinae (hartebeests, gnus, duikers, dik-diks, klipspringers, reedbucks, waterbucks, etc.); volume 3,

Artiodactyla, family Bovidae, subfamilies Aepycerotinae to Tragelaphinae (pala, saiga, gazelles, oryx group, bushbucks, kudus, elands, etc.); volume 4, Artiodactyla, families Cervidae (deer), Tragulidae (chevrotains), Camelidae (camels and llamas), Suidae (pigs and peccaries), and Hippopotamidae (hippopotamuses); volume 5, Perissodactyla (horses, tapirs, rhinoceroses), Hyracoidea (hydraxes), Provoscidae (elephants). Each species account is usually one page in length with references to where first named, and other references to literature on the species, along with synonyms, the type specimen, a brief physical description, and geographical range. Places, dates, and collectors names are also provided for each specimen. Black-and-white photos and drawings of selected specimens are found occasionally in each volume. Each volume has its own index, and an index to the set is found at the end of volume 5.

1564. Mochi, Ugo, and T. Donald Carter. **Hoofed Mammals of the World**. New York: Charles Schribner's Sons, 1971. 268p. LC 75-169790. ISBN 0-6841-2382-7.

First published in 1953 at a size of 12 by 15 inches, this 9-inch edition features the remarkable silhouette art work of Ugo Mochi, with species accounts by T. Donald Carter. In all, more than 290 hoofed mammals of the world are depicted and described. The brief paragraph-long descriptions are arranged by major group, and listed by common name with scientific name provided for each illustration. The illustrations are silhouettes, cut from black paper, showing amazing detail and beauty in depicting each animal. Aside from the beauty of the illustrations, this is the only work that lists and describes the world's hoofed animals in one work.

1565. Willoughby, David P. **The Empire of Equus**. New York: A. S. Barnes, 1974. 475p. LC 72-5180. ISBN 0-498-01047-3.

This work provides a natural history of the living species of the horse. It thoroughly covers information on this group from nomenclature, evolution, ancestors, growth, relationship to the tapirs and rhinoceroses, the horse in art and history, information on selected breeds, coloration of coat, zebras, mules, and more. It is a very readable text that is liberally illustrated by black-and-white drawings, photos, figures, graphs, and tables. The appendices include detailed measurements and conformation diagrams, a phylogeny of the horse, and information on Baluchitherium, a fossil ancestor of the horse. A separate index to author's names, horses common and scientific names, and subjects provides adequate access to the wealth of information in this work.

ARTIODACTYLA (Pigs, Hippopotamus, Camels, Deer, Giraffes, Sheep, Cattle, etc.)

1566. Lydekker, Richard. **Catalogue of the Ungulate Mammals in the British Museum (Natural History).** London: Printed by order of the Trustees of the British Museum, 1913. Reprint, New York: Johnson Reprint, 1966. 5 vols. $150.00 set; $30.00/vol. ISBN 0-384-34486-0.

Volumes 1 through 4 of this set are devoted to the Artiodactyla. (See entry 1563 for a complete annotation.)

1567. Mochi, Ugo, and T. Donald Carter. **Hoofed Mammals of the World**. New York: Charles Schribner's Sons, 1971. 268p. LC 75-169790. ISBN 0-6841-2382-7.

The artiodactyls are included in this work. (See entry 1564 for a complete annotation.)

1568. Valdez, Raul. **The Wild Sheep of the World**. Mesilla, New Mexico: Wild Sheep and Goat International, 1982.

This work reviews general information and life histories of the wild (not domestic), sheep of the world. In addition, it provides species descriptions of two to four pages in length that give the taxonomy, type locality, distribution, synonymy, and physical description. A black-and-white photo of the species or the skull is provided with each account. A selective list of references and an index to name and subject complete the work. Written in semitechnical language, this work will be of use to a wide range of interests.

1569. Whitehead, G. Kenneth. **The Whitehead Encyclopedia of Deer**. Voyageur Press, 1993. 604p. $140.00. ISBN 1-85310-362-4.

This comprehensive work on deer is based on Whitehead's 1972 work *Deer of the World*. The book expands on the previous work by including not only descriptions of the 42 species of deer found worldwide, but also a glossary of terms and other entries on all aspects of deer studies. It also includes information on ancient and modern deer hunting practices, deer trophies, a dictionary of diseases and parasites affecting deer, and a reference section with deer classification and distribution information. The main part of the text is the descriptions, each one to six pages in length, of the 42 species of deer. Information on rare and endangered species is also included. Thirty-six pages of color plates illustrate the many species of deer, and black-and-white photos, range maps, charts, and reproductions of paintings, ceramics, and sculpture are found throughout the text. A 58-page bibliography attests to the thoroughness of this work. A comprehensive, scholarly work for all levels of interest.

FIELD GUIDES

Mammalia (General)

1570. Bang, Preben, and Preben Dahlstrom. **Collins Guide to Animal Tracks and Signs**. St. James's Place, London, 1974. 240p. LC 74-169035. ISBN 0-00-219633-6.

Primarily covering European mammals and birds, this field guide describes and pictures tracks and signs (droppings, pellets, homes, etc.) that can aid in identifying an animal. These tracks and signs were used originally by hunters in identifying and finding a deer, for instance, or beaver or raccoon for their dinner. But these signs and tracks are also used by naturalists, students, and wildlife specialists to simply aid in identifying species found in a particular area, even when they cannot be seen, as in the case of nocturnal animals. The book is arranged by type of track or sign. Tracks of animals with paws is one group, those with cloven hooves in another. Feeding signs, cones, nuts, and other fruits handled by animals, droppings, and pellets are other groups. For instance, if a tree has been gnawed at the base, the observer can turn to the section on bark gnawing to identify what animal may have caused this type of gnawing. The work is illustrated with over 600 color and black-and-white photos, drawings, and paintings. This is the English translation of a work published in Denmark in 1972, titled *Dyrespor*. The descriptions that go along with the illustrations are easy, informative reading for junior high school on up.

1571. Boitani, Luigi, and Stefania Bartoli. **Simon & Schuster's Guide to Mammals**. New York: Simon & Schuster, 1983. 511p. $14.00pa. LC 83-16222. ISBN 0-671-45447-1; 0-671-42805-5pa.

This guide pictures and describes 426 of the more common mammals found worldwide. It is arranged taxonomically from the Monotremata (platypus) to the Artiodactyla (pigs, deer, cattle, etc.). There are two species descriptions on each page, with the corresponding color photo on the opposite page. Each species description includes the classification (order and family) to which it belongs, a brief physical description, the geographical distribution, habitat, and behavior. A range map, drawing of distinguishing characteristics, and a graphic of the type of habitat it prefers (mountain graphic for mountain preference) are arranged next to the text description. Gives introductory information on the mammal group as a whole, a glossary of terms, a brief bibliography for further reading, and an index to common and scientific names. An easy-to-use guide to the world's more common mammals.

1572. Murie, Olaus J. **A Field Guide to Animal Tracks**. 2d ed. Boston: Houghton Mifflin, 1975. 375p. (Peterson Field Guide Series). $19.45; $14.95pa. LC 74-6294. ISBN 0-395-19978-6; 0-395-18323-5pa.

This work covers tracks and signs for mammals from North America, Mexico, and Central America, as well as some birds, reptiles, and a few insects. A key to tracks precedes the text and shows a drawing of a track with the name of the animal that makes it and the page number in the text for further description. The text is arranged by family groups. Each group and selected species are described for behavior and identifying tracks and signs. A black-and-white drawing of the animal is also provided as well as drawings of tracks and droppings of the more common species in the group. Although dated, this is the standard guide for animal tracks in use today.

Mammalia (Geographic Section)

AFRICA

1573. Cillié, Burger. **Mammals of Southern Africa: A Field Guide**. Sandton, South Africa: Frandsen, 1987. 182p. £13.95. LC 87-201687. ISBN 0-620-10367-1.

This work describes and pictures 83 species of mammals that are most likely to be seen by day. Arranged by order from Insectivora to Rodentia, selected species are described in one page with two striking full-color photos of the species on the facing page. The description is concise and includes information on sexual dimorphism, habitat, habits, voice, food, gestation period, breeding, number of young, weight, height, and life expectancy. A range map is also included as well as a drawing of the track made by the animal. A handy table is provided that tells which animals can be seen in one of the 28 game reserves and national parks in southern Africa. It also includes a unique illustrated index of the animals listed in this book and an illustrated guide to tracks. A clear and concise guide.

1574. Goss, Richard. **Maberly's Mammals of Southern Africa: A Popular Field Guide.** Craighall, South Africa: Delta Books, 1986. 346p. LC 87-137951. ISBN 0908387-63-6.

This is actually a revision of Charles Astley Maberly's *The Game Animals of Southern Africa* (Johannesburg: Nelson, 1967). It contains descriptions of 105 of the more common mammals of this area. The descriptions are longer, usually two to four pages, than Cillié's field guide (see entry 1573), and it describes more animals, but does not have as many color photos and the photos are grouped together in the center of the book, not with each description. Black-and-white drawings and range maps illustrate the text. An index to scientific, common, and Afrikaans names finish this guide. The text is bigger than most field guides and bound in a hard cover, so is not so portable in the field, but it includes good information on the more common mammals to be seen in this area.

1575. Haltenorth, Theodor, and Helmut Diller. **The Collins Field Guide to the Mammals of Africa including Madagascar**. Lexington, Mass.: Stephen Greene Press, 1988. 400p. LC 88-4791. ISBN 0-8289-0699-8.

A translation of *Shaugetiere Afrikas und Madagaskars* published in 1977, this guide lists and describes more than 300 species of the large and medium-sized mammals only. This was done to keep the guide down to a manageable size. The guide does not therefore cover the 200 species of bats or some of the smaller insectivore or rodent species. After a brief but good introduction to the mammals of Africa, the guide provides identification information for each order from Artiodactyla to Primates. Within the order, general information on genus and specific information on species are given. Species accounts average one page in length and are listed by common name followed by scientific name, the name's author and the date named. A detailed paragraph on identification of the species is also provided as well as the geographical distribution, habits, and reproduction. It includes 245 black-and-white distribution maps with plates of full-color drawings of most species placed at the end of the text. An easy-to-use pocket guide to African mammals, for the amateur and student.

1576. Smithers, Reay H. N. **Land Mammals of Southern Africa: A Field Guide**. Rev. ed. Halfway House, South Africa: Southern Book, 1992. 229p. $27.95. ISBN 1868-12-401-0.

Of the 291 land mammals of southern Africa, 197 species are described and pictured in this field guide by Reay Smithers, a specialist for many years in mammals of this region. Indeed, this guide should be usefully used in conjunction with Smithers more comprehensive work *The Mammals of the Southern African Subregion*, published in 1983 (see entry 1444). The guide is arranged taxonomically. Each species description is concisely done in one-half of a page, entered under the common name with the scientific and Afrikaans names found second. A black-and-white geographical range map is provided with each species description. A number is found at the end of each description that refers to the reference number in *The Mammals of the Southern African Subregion* for easy cross-reference to a fuller description. Beautiful color drawings of most species are found on 56 plates arranged throughout the guide. Drawings of tracks made by many of the mammals are also provided. This is an easy-to-use guide, especially useful to the student, amateur, or sightseer in the southern African region.

1577. Stuart, Chris, and Tilde Stuart. **Field Guide to the Mammals of Southern Africa**. London: New Holland, 1988. 282p. £9.95. LC 88-71277. ISBN 1-85368-015-X.

This is a guide to the mammals of the South African Subregion, an area south of the Zambezi and Cunene Rivers, which includes the countries of Zimbabwe, Namibia, Botswana, Mozambique, South Africa, Lesotho, and Swaziland. It includes one-page species descriptions with information on the physical description, distribution, habitat, behavior, food, reproduction, and general notes. On the page facing each species description is a color plate with two to four photos of the species. In addition to the land mammals, it includes the whales and dolphins and seals found just offshore. A separate section on tracks made by selected mammals is placed before the glossary and indexes to scientific, common, and Afrikaans names.

AUSTRALIA

1577a. Cronin, Leonard. **Key Guide to Australian Mammals**. Balgowlah, NSW, Australia: Reed, 1991. 190p. ISBN 0-7301-03552.

This is a guide to more than 170 species of Australian mammals, both terrestrial and marine, featuring striking full-color drawings of each species. A similar guide, by J. Mary Taylor (see entry 1578), describes all known genera native to Australia, rather than describing individual species. An easy picture key to the major groups of animals begins the guide, referring the reader to the pages in the guide where the individual species in the group are pictured and described. It is arranged under headings for "Monotremes," "Carnivorous Marsupials," "Bandicoots," "Wombats," "Koala," "Possums," "Kangaroos," "Bats," "Rodents," "Sea Mammals," and "Dingoes." The individual species are concisely described, two to a page, with a corresponding full-color drawing of the species on the facing page. Information for each species includes a brief physical description, measurements on size and weight, behavior, reproduction, food, habitat, and conservation status. A black-and-white range map accompanies each description. A page of suggested readings, and separate indexes to scientific and common names complete the guide.

1578. Taylor, J. Mary. **The Oxford Guide to Mammals of Australia**. Melbourne: Oxford University Press, 1984. 148p. LC 85-146114. ISBN 0-19-554584-2.

This is an easy-to-use guide to every genus of land and marine mammal native to Australia. Concise descriptions of each genus are written in nontechnical language. Information included in the page-long descriptions are the Latin and common names, body measurements, a physical description, reproduction, habits, habitat, and number and name of species in the genus. A black-and-white range map and a silhouette of the genus are included with each account. A full-color picture of a representative species would have been better for a field guide, especially the section on rats, but the silhouette works for the most part. A list of references, a glossary, and indexes to scientific and common names finish the work.

1578a. Watts, Dave. **Tasmanian Mammals: A Field Guide**. Text compiled by Margaret Grainer, Emma Gunn, and Dave Watts. Hobart, Tasmania: Tasmanian Conservation Trust, 1987. 111p. LC 88-169316. ISBN 0-9590500-4-3.

This is a field guide to 32 of Tasmania's native land mammals. Included here are the Tasmanian Devil and the Tasmanian Pademelon, animals not seen wild in any other area. It features excellent full-color photos of each species described. After information on the habitats where these mammals are found, the guide is arranged under headings for "Monotremes," "Marsupials," "Placental Mammals" and a section on "Introduced Mice and Rats." Each species account is a page long and includes the common and scientific name, the order and family to which it belongs and brief information on physical description, habits, habitat, food, breeding, geographical distribution and conservation status. A striking, full-color photo of the animal, faces the descriptive page. Three appendices include a footprint or track chart, a brief chart on identifying bats, and a table showing where each animal may be found in 13 of Tasmania's national parks. An index to common and scientific name finish the text.

BORNEO

1579. Payne, Junaidi, and Charles M. Francis. **A Field Guide to the Mammals of Borneo**. Sabah, Malaysia: Sabah Society with World Wildlife Fund Malaysia, 1985. 332p. LC 88-942095. ISBN 967-99947-1-6.

This field guide includes descriptions and color illustrations of all the mammals known to live in Borneo and its offshore islands and in surrounding waters. The text is arranged with the color drawings of species in the beginning of the text for easy identification. The reader is then referred to a page number where the species is described. Exact measurements, identifying characters, ecology and habit, as well as geographical distribution information are provided in each species description. Some black-and-white drawings of skulls, and a section on mammal tracks also illustrate the text. A gazetteer of place names mentioned in the text, a map of Borneo, a list of the English and local names for the animal, and a bibliography and index are found at the end of the work. A good guide for the specialist and the amateur.

GREAT BRITAIN and EUROPE

1580. Corbet, G. B. **The Mammals of Britain and Europe**. London: Collins, 1980. 253p. LC 81-161428. ISBN 0-00-219772-3; 0-00-219774-Xpa.

This is an easy-to-use pocket guide to the mammals of Britain, continental Europe, and the islands of Spitzbergen, Iceland, and those in the northeastern Atlantic. It illustrates 190 mammals in full color and has 194 distribution maps. Native and introduced species are included. After background information on the area, and finding and identifying mammals, the color plates of species, range maps, and tracks are found. Thereafter, the text is arranged by order, from marsupials to cetaceans. Descriptive information on each species includes geographic range, habitat, habits, and notes on identification. Additional distribution maps are inserted at the end of the text. Well illustrated and easy to use, a comprehensive guide.

1581. Lawrence, M. J., and R. W. Brown. **Mammals of Britain: Their Tracks, Trails and Signs**. Rev. ed. London: Blandford Press, 1973. 297p. LC 73-165999. ISBN 0713705884.

This is an interesting guide to identifying mammals by their tracks, trails, and signs. After introductory text on identifying tracks and making plaster casts, the book describes selected species and illustrates their tracks and other distinguishing marks left by the animal. The species are described in one paragraph to one page of text, with information on its geographical range, habits, and habitat. Black-and-white drawings of tracks as well as photos of trails and signs illustrate the text. For example, photos of otter trails on a river bank, on ice, and an otter "slide" on a bank are depicted as identifying marks of otters. Mammal identification by teeth, skulls, and other bones is also discussed with numerous drawings of skulls and teeth. Illustrated keys to skulls and lower jaws are provided for carnivores and other groups. A glossary of terms and a six-page bibliography finish the text. A detailed and useful guide to identification.

1581a. Macdonald, David, and Priscilla Barrett. **Mammals of Britain & Europe**. London, New York: HarperCollins, 1993. 312p. (Collins Field Guide). £14.99. ISBN 0-00-219779-0.

This field guide describes more than 200 species of mammals found in Europe and in the oceans around it. It is intended to be used with a companion volume on the natural history of mammals in this area. That volume was not available for review as of this writing. A set of color plates is inserted in the center of the guide illustrating 64 of the more common species found in Eastern Europe. Each illustration includes a drawing of the skull, track and sign as well as a drawing of the animal. Brief descriptions are found on the page opposite the drawing, with a page number provided to a more detailed description. The description is usually one page or more in length and includes information on the recognition, habitat, habits, breeding, life-span, measurements, and general notes. A distribution map accompanies each description. A glossary of terms used in the guide, a bibliography for further reading, and an index to scientific/common-name finishes the guide. This guide provides more textual description of each animal but less color illustrations than the guide by Corbet (see entry 1580).

MEXICO

1582. Wright, N. Pelham. **A Guide to Mexican Mammals & Reptiles**. México: Minutiae Mexicana, 1970. 112p. $6.50. ISBN 0-685-03696-0.

This small guide highlights the more common mammals and reptiles found in Mexico. It is written in easy, nontechnical language and is divided into two sections, one for mammals and the other for reptiles. Arranged by taxonomic groups, the mammal section begins with monkeys (Primates) and ends with pouched animals (marsupials). Almost every page contains a black-and-white or color photo or drawing of selected species. Very brief, but it does introduce the common native mammals found in this area.

NEOTROPICS

1583. Emmons, Louise H. **Neotropical Rainforest Mammals: A Field Guide**. Chicago: University of Chicago Press, 1990. 281p. $45.00; $19.95pa. LC 89-39353. ISBN 0-226-20716-1; 0-226-20718-8pa.

This is a field guide to the rainforest mammals found at elevations below 1,000 meters in Central and South America. It is arranged taxonomically from opossums to rodents. Each species account is under the common name, followed by the scientific name, with information on physical description, species variation, similar species, sounds, natural history, geographic range, conservation status, and local names. Black-and-white range maps are provided, and black-and-white or color drawings on 36 plates illustrate each species. The appendices add further information with a glossary of terms, keys to families and genera, a classification system, and notes on biogeography and conservation of neotropical rainforest mammals, tracks made by selected mammals, with black-and-white drawings, and a checklist and index to scientific names. Altogether, an excellent guide by an authority in this field.

NORTH AMERICA

1584. Alden, Peter. **Peterson First Guide to Mammals of North America**. Boston: Houghton Mifflin, 1987. 128p. $4.80. LC 86-27821. ISBN 0-395-42767-3.

Based on *A Field Guide to the Mammals of North America North of Mexico* (see entry 1585), this is a simpler guide, smaller and less detailed. It is for the amateur who is just starting out. It provides brief information and a full-color drawing of the most common species (and their tracks) found in North America. The species are arranged under the major groups (i.e., marsupials, insectivores, bats, carnivores, seals, rodents, rabbits, hoofed mammals, edentates, sirenians, and whales). Only common names are used; no scientific names are provided. A good beginning guide for junior and high school students and the amateur who is not yet ready for a more comprehensive guide.

1585. Burt, William Henry. **A Field Guide to the Mammals of North America North of Mexico**. 3d ed. Boston, Mass.: Houghton Mifflin, 1980. 289p. (The Peterson Field Guide Series; 5). $13.95pa. LC 75-26885. ISBN 0-395-24082-4, 0-395-240840pa.

Now in its third edition, this field guide provides an easy-to-use, yet informative text that describes and pictures 380 mammal species found in North America. It is arranged in taxonomic order from marsupials to whales. Each order is briefly described, then families, followed by descriptions of individual species in the family. Each species is described (in one page or less) by a physical description with sizes, color, shape, and so on, similar species are listed, the habitat and habits, the number of young, and the economic status (beneficial or not). Twenty-four color plates are inserted into the middle of the text, with color drawings of most species described in the text. Pages facing the plates give brief information plus the page numbers to where the description is found in the text, and a range map showing where the species are found in North America. Photos of skulls and a table for the dental formula of the groups are found at the end of the text. A list of general references and those for the individual states are helpful for finding more information on mammals. An index by scientific and common names finishes the volume. Black-and-white illustrations of the tracks that the more common mammals make are found at the beginning and end of the text. Very easy to use for student and amateur.

1586. Cockrum, E. Lendell. **Mammals of the Southwest**. Tucson, Ariz.: University of Arizona Press, 1982. 176p. LC 81-21834. ISBN 0-8165-0760-0; 0-8165-0759-7pa.

This guide describes and pictures 80 species found in the southwestern United States and northwestern Mexico. The species are grouped into the six major groups (i.e., hoofed animals, carnivores, rodents, hares and rabbits, bats, and insectivores). Each species account covers one page and includes the common and scientific names, identifying features, measurements, habitat, life habits, and related species. A black-and-white drawing and a map showing geographic range for each species are found on the opposite page from the species account. A short list of further readings and an index to common names completes this easy-to-use, yet authoritative guide.

1587. Farrand, John. **Familiar Mammals**. New York: Knopf, 1988. 192p. (Audubon Society Pocket Guides). $8.00. LC 87-46022. ISBN 0-394-75796-3.

This little guide to 110 of the most common mammals found in North America is distinctive for the full page color photos that accompanies each species description. It does not include as many species as the *Peterson First Guide to Mammals of North America* (see entry 1584), a similar guide, but the striking photos are helpful in identifying the individual animals in the field. Each species is described with information on the physical appearance, specific identifying characters, similar species, habitat, and geographical range. A range map and the track the animal makes are drawn on the same page as the description. Another good beginning guide for the student or amateur.

1587a. Katona, Steven K., Valerie Rough, and David T. Richardson. **A Field Guide to Whales, Porpoises, and Seals from Cape Cod to Newfoundland**. 4th ed. Washington, D.C.: Smithsonian Institution Press, 1993. $15.95. LC 93-84473. ISBN 1-56098-333-7.

Now in its fourth edition, this guide provides accounts of 22 species of whales and 7 species of seals that may be seen in the water or stranded on the beaches between Cape Cod and northern Newfoundland. It provides introductory information on how to observe whales (including information on avoiding or ameliorating seasickness) before going into the species accounts. It is arranged into three parts. Part I includes species accounts of whales, dolphins, and porpoises, part II provides a "bonus" section on spotting basking sharks, sunfish and leatherback turtles, and part III covers the seal species accounts. Each species account is several pages (six pages for the sperm whale) and is illustrated with a black-and-white drawing of the whale as well as several black-and-white photos. The descriptions emphasize field marks for identification. A lengthy bibliography of sources on whales, dolphins, and seals (including video and film sources) and an index complete this work.

1588. Palmer, Ralph S. **The Mammal Guide: Mammals of North America North of Mexico**. Garden City, N.Y.: Doubleday, 1954. 384p.

This guide is dated, but included here for the more lengthy descriptions given with each animal species than those found in other North American guides listed here. Most species descriptions are two pages long and cover information on habitat, reproduction, habits, and economic status, as well as a physical description. Black-and-white drawings of distinguishing characteristics or tracks are often found with the description. Range maps are provided for each species. Does not provide color photos or drawings like the other guides listed here, but gives better, interesting, prose descriptions.

1589. Whitaker, John O., and Robert Elman, text consultant. **The Audubon Society Field Guide to North American Mammals**. New York: Knopf, 1980. 743p. $12.50. LC 79-3525. ISBN 0-394-50762-2.

This guide is to all the land-dwelling or land-breeding mammals in North America, north of Mexico, and therefore excludes all ocean-dwelling species such as the whales and their kin. As with other Audubon field guides, this features striking full-color photos of most described species in the text. The guide is arranged to make identification easy by first turning to the "Silhouette and Thumb Tab Guide" that comes before the plates. Here the reader finds the silhouette that most closely matches the animal being identified. From the silhouette, the reader is referred to plate numbers where the full-color photo can be seen. From the color photo, the reader is referred to the text page where a written description is found, complete with range map and track drawing for final identification. Each species account covers a page or two and includes a physical description, similar species, sign (including scat, tracks, or other distinguishing sign), breeding, habitat, and geographical range. A glossary of terms and eastern and western range charts are found at the end of the text. Easy to use, with concise descriptions and good photos for quick identification.

1590. Zim, Herbert Spencer, and Donald F. Hoffmeister. **Mammals: A Guide to Familiar American Species**. New York: Golden Press; Racine, Wis.: Western, 1987. 160p. (A Golden Guide). $4.50. LC 61-8320. ISBN 0-307-24058-4.

A very basic, elementary guide to 218 familiar mammals found in the United States. Filled with colorful drawings of each mammal, the text provides interesting, readable facts for each species as well as a color range map. It is arranged taxonomically from the possums to the whales. The index is by common name, but the scientific names for each mammal are provided in a list. A good introduction to mammals for junior high school on up.

TEXTBOOKS

1591. DeBlase, Anthony F., and Robert E. Martin. **A Manual of Mammalogy with Keys to Families of the World**. 2d ed. Dubuque, Iowa: Wm. C. Brown, 1981. 436p. $30.00. LC79-53828. ISBN 0-697-04591-9.

Used primarily as a lab manual for upper-level and graduate courses in mammalogy, this text focuses on the identification of mammals of the world and on examining their evolutionary and ecological relationships. Keys to orders and to living families of each order are provided. Emphasis is on identification of North American mammals. Each order is broadly discussed in a paragraph or two, followed by a description of the main characteristics and the distribution of the order in the world. A key to the living families is provided as well as suggestions on identification of the group. Black-and-white drawings of skulls and the animal itself are given to aid in identification. The whole text is spiral bound so the text lies flat for easy use of the keys. In addition to the chapters on each order, chapters on age determination, diet analysis, identifying mammal sign, recording data, collecting, spacial distribution, and searching the literature are also provided. A useful text for students as well as for the interested amateur.

1592. Vaughan, Terry A. **Mammalogy**. 3d ed. Flagstaff, Ariz.: Northern Arizona University, 1986. 576p. $40.00. LC 85-10754. ISBN 0-03-058474-4.

This basic text in mammalogy is not exhaustive but covers the most important subjects in this area of study. Introductory chapters on mammalian characteristics, origins, and classification lead into chapters on the individual orders. Following these chapters are those on ecology, zoogeography, behavior, reproduction, metabolism, water regulation, acoustical orientation, and the impact of man on mammals. A good 44-page bibliography lists references from the text as well as some important references by the cited authors as well. The index is by scientific or common name as well as by subject. Each chapter is illustrated with black-and-white drawings of skulls and other identifying characteristics. Black-and-white photos are also found throughout the text. A good text for upper-level and graduate students as well as a good introduction for anyone interested in mammals.

JOURNALS AND SERIALS

1593. **Acta Theriologica**, Vols. 1, nos. 1- . Warsaw, Poland: Polish Academy of Sciences, Mammals Research Institute, 1934- . Published irregularly. Price varies. ISSN 0001-7051.

This is an international journal that publishes original research and review papers on all aspects of mammalian biology. Five or more book reviews are included in each issue. Each issue includes ten or so papers in English with Polish summaries. Although it publishes in all areas of mammalian biology, it has special emphasis on papers of the European bison and hare. It also publishes monographic issues such as the one in 1983 on the *Ecology of the Bank Vole*.

1594. **Acta Theriologica Sinica = Shou Lei Hsueh-Pao**. Vol. 1, nos. 1- . Beijing, China: Science Press for the Northwest Plateau Institute of Biology, 1981. Quarterly. $24.00/yr. ISSN 1000-1050.

Published in Chinese with English summaries, this journal contains articles of original research on all aspects of mammalian studies, with a concentration on mammals found in China. Each issue contains 10 or so articles as well as scientific notes and one or two book reviews. The last issue of each volume contains a keyword index and the table of contents of all issues in the volume.

1595. **American Journal of Primatology**. Vol. 1, nos. 1- . New York: Wiley-Liss, 1981- . Monthly. $756.00/yr. ISSN 0275-2565.

This is the official journal of the American Society of Primatologists. It publishes articles on original research and reviews in all aspects of primate studies. Three or four articles, a technical note, and brief report as well as a book review are published in each monthly issue. Author and subject indexes are included at the end of each volume.

1596. **Current Mammalogy**, Vols. 1- . New York: Plenum Press, 1987- . Published irregularly. Price varies. ISSN 0899-577X.

This irregular monographic serial is now up to volume 2 as of 1990. It publishes review articles as well as extended studies and accounts of important new discoveries on all aspects of mammalian studies. Papers are by invitation only by the editorial board, which review each paper before it is published. There are around 13 papers in each monographic issue and a comprehensive index for the entire work. Excellent for overviews of important past research and for insights on current and future research.

1597. **Folia Primatologica: International Journal of Primatology**. Vol. 1, nos. 1- . Basel, Switzerland: Karger, 1963- . 8 nos./yr. $171.00/yr. ISSN 0015-5713.

This international journal publishes papers in English, French, or German on primate studies. Usually three research papers (no more than 5,000 words each) and two brief reports (no more than 1,000 words each) are included in each issue. Book reviews are found in some issues.

1598. **International Journal of Primatology**. Vol. 1, nos. 1- . New York: Plenum Press, 1980- . Bimonthly. $275.00/yr. ISSN 0164-0291.

This is the official journal of the International Primatological Society and includes articles on laboratory and field studies in basic primatology. It publishes review as well as current research articles. Recently expanded in length in 1993, each issue now contains around 12 articles and two book reviews.

1599. **Journal of Mammalogy**, Vols. 1- , nos. 1- . Provo, Utah: Publication for the American Society of Mammalogists by Brigham Young Univ., 1919- . Quarterly. $33.00/yr. ISSN 0022-2372.

The official journal of the American Society of Mammlogists, this journal has been published since 1919. Each issue contains approximately 30 articles on all aspects of mammalian studies. Four or five mammalogy books are reviewed in each issue as well. Meeting announcements, information on student programs, fellowships and grants, and news items are found at the end of each issue.

1600. **Mammal Review**, Vols. 1- . Oxford, England: Published for the Mammal Society by Blackwell Scientific Publications, 1970- . Quarterly. $125.00/yr. ISSN 0305-1838.

This is the official publication of the Mammal Society, a society of both professionals and amateurs founded in 1954 to promote the study of mammals. As such it publishes the proceedings of the society meetings as well as review articles and reports, not original research papers. Each issue contains two to four of these review articles or reports. For example, one 1992 issue contained two papers, one on "The Endemic Mammals of Ethiopia" and the other on "World Distribution of the Rabbit *Oryctolagus cuniculus* on Islands." Both papers were lengthy, 30 to 50 pages, with detailed bibliographies. Each issue also has one or two book reviews.

1601. **Mammalia: Morphologie, Biologie, Systématique des Mammifères**. Vol. 1, nos. 1- . Paris: Muséum National D'Histoire Naturelle, 1936- . Quarterly. $160.00/yr. ISSN 0025-1461.

Including papers in English and French, this journal publishes original research on all aspects of mammalian biology, ecology, and systematics. Ten or so papers and four "notes" or brief (one- or two-page) research reports are published in each issue of approximately 200 pages. Several books are briefly reviewed, in French, in each issue.

1602. **Mammalian Species**. Nos. 1- . Provo, Utah: American Society of Mammalogists, c/o Dept. of Zoology, Brigham Young University, 1969- . Published irregularly. $10.00/yr. ISSN 0076-3519.

As the title suggests, each issue of this unique serial is devoted to a particular mammal species and is authored by an expert(s) on that species. Each issue follows the same format to promote uniformity of information for each species and for ease of comparisons with other species. The format begins with a list of references to where the species was first named and other references to type locality and taxonomy. The remaining headings are context and content, diagnosis, general characters, distribution, fossil record, form and function, ontogeny and reproduction, ecology, behavior, genetics, remarks, and literature cited. A black-and-white photo of the species and its skull, and a range map are provided for each species. Each species account is from 3 to 20 pages in length. Cumulative indexes to author and genera, and a systematic list are provided for every 100 species accounts. As of this writing, 443 species accounts had been published. A very valuable source for complete species descriptions and, using the lengthy list of references cited, for finding further sources of information.

1603. **Marine Mammal Science**. Vol. 1, nos. 1- . Lawrence, Kans.: Society for Marine Mammalogy, 1986- . Quarterly. $95.00/yr. (institution). ISSN 0824-0469.

This is the official journal of the Society for Marine Mammalogy and publishes primarily current research in all areas of marine mammal studies. Sea otters, dolphins, whales, seals, sea lions, manatees, and other marine mammals are all included. Usually five research articles and five notes (short articles on a very specific topic), plus a book review and a page of news on jobs or new publications or databases are included in each issue.

1604. **Primates: A Journal of Primatology**. Vol. 1, nos. 1- . Aichi, Japan: Japan Monkey Centre, 1957- . Quarterly. $198.00/yr. (institution). ISSN 0032-8332.

Although published in Japan, this international journal is entirely in English and includes articles on original research and review articles in all fields of primate studies. Nine or so articles and eight "Short Communications" are included in each issue.

TAXONOMIC KEYS

1605. Booth, Ernest S. **How to Know the Mammals**. 4th ed. Dubuque, Iowa: Wm. C. Brown, 1982. 198p. (Pictured Key Nature Series). LC 81-68878. ISBN 0-697-04781-4.

This is a very easy-to-use pictured key to all species of mammals in the United States and Canada, including Arctic Canada and Alaska, as well as some domestic animals and a few foreign species commonly found in zoos. Before the keys, introductory chapters are given on mammals as a group and how to study them. First a key to orders of North American mammals is provided, complete with black-and-white pictures of representative species from each order. Next are keys to family, then genus and species within each order. Black-and-white drawings of species and characteristic features, as well as distribution maps are given for each species. Brief information on the appearance or behavior of a genus and species is also given. The index serves also as a pictured glossary of terms, especially for names of individual parts of mammal skulls. An easy key for the interested amateur or student.

1606. DeBlase, Anthony F., and Robert E. Martin. **A Manual of Mammalogy with Keys to Families of the World**. 2d ed. Dubuque, Iowa: Wm. C. Brown, 1981. 436p. $30.00. LC79-53828. ISBN 0-697-04591-9.

Used primarily as a lab manual for upper-level and graduate courses in mammalogy, this text focuses on the identification of mammals of the world and on examining their evolutionary and ecological relationships. Keys to orders and to living families of each order are provided. Emphasis is on identification of North American mammals. See entry 1591 for full description.

1607. Fisler, George F. **Keys to Identification of the Orders and Families of Living Mammals of the World**. Los Angeles, Calif.: Los Angeles County Museum of Natural History, 1970. 29p. (Science Series 25, Zoology No. 12). $4.00. LC 72-196123.

This is a key to living mammals worldwide, but only to order and family. It is meant as an aid to the amateur and student in identifying a mammal's order and family, allowing the student to proceed to other identification aids, such as field guides, for identification to species. In all it provides keys to 19 orders and 123 families, following the classification scheme of George Simpson (1945) (see entry 1617). Twenty black-and-white line-drawings help to illustrate the keys, and a glossary of terms also aids the students in using the keys. Provides the basics in identifying the major groups of mammals.

1608. Jones, J. Knox, Jr., and Richard W. Manning. **Illustrated Key to Skulls of Genera of North American Land Mammals**. Lubbock, Tex.: Texas Tech University Press, 1992. 75p. $9.95. LC 92-18804. ISBN 0-89672-289-9.

This is an illustrated key to genera (not species) of native terrestrial mammals found in North America to the northern part of Mexico. After a key to the orders of land mammals, the book is arranged by order with keys to family and genera within the order. Each major group is introduced with a paragraph or more of information on the number of families, genera, and species, their distribution, and other taxonomic notes. Black-and-white drawings or photos by the second author illustrate most of the keys. A 10-page illustrated glossary aids in understanding the terms used in the keys. More technical than Roest's key (see entry 1610), but still intended for the use of mammalogy students, it is also useful for anyone trying to identify a mammal skull.

1609. Lawlor, Timothy E. **Handbook to the Orders and Families of Living Mammals**. 2d ed. Eureka, Calif.: Mad River Press, 1979. 327p. $20.00. ISBN 0-916422-16-X.

This handbook consists primarily of keys and diagnostic descriptions of mammalian orders and families worldwide. It is also listed in the Handbooks section of this chapter. For a full description, see entry 1432.

1610. Roest, Aryan I. **A Key-Guide to Mammal Skulls and Lower Jaws: A Nontechnical Introduction for Beginners**. Eureka, Calif.: Mad River Press, 1991. 39p. $5.95. ISBN 0-916422-71-2.

As the title suggests, this is a beginner's key to the identification of skulls of mammals found commonly in the United States and southern Canada. After basic introductory material and a five-page glossary of terms used in the key, the keys begin. There are two, one to skulls, and one to the lower jaw. An example of the simplicity of the keys is the beginning of the skull key where the first step in identifying the skull is a choice of either large, medium, small, or tiny skull (with measurements in inches and millimeters). If the skull to be identified is large, the reader is told to go to step 2 where other choices are given until the identity of the skull is derived. Common names of the animals are found in the keys, but a list of scientific names of the species are given at the end of the keys. A short description of cleaning skulls and a list of references for further reading complete this easy-to-use key. For a more detailed and technical key to skulls, the reader may try the one by J. Knox Jones and Richard Manning (see entry 1608).

AFRICA

1611. Meester, J., and H. W. Setzer, eds. **The Mammals of Africa: An Identification Manual**. Washington, D.C.: Smithsonian Institution Press, 1971. Loose-leaf. $105.00. LC 70-169904. ISBN 0-87474-116-5.

Composed primarily of keys and identification characters, this loose-leaf work was produced for the nontaxonomists, especially those ecologists and conservationists that have now taken such an interest in African mammals. It is detailed enough, however, for the specialist in mammology. This work is also consulted as a fine checklist or catalog of African mammals. It is arranged according to Simpson's classification of mammals (see entry 1617). Each chapter comprises either an order or family and is authored by an expert in that particular order or family. Detailed keys to genus are provided in each chapter. Species are listed under genus with the author and date when first named in the literature and geographical distribution. Subspecies are listed next, also with author and date when first named as well as a geographical range. Each chapter ends with a list of references mentioned in the chapter. A few black-and-white drawings are included. There is no comprehensive index to this loose-leaf work.

CHECKLISTS AND CLASSIFICATION SCHEMES

Mammalia (General)

1612. Corbet, G. B., and J. E. Hill. **A World List of Mammalian Species**. 3d ed. Oxford, New York: Oxford University Press, 1991. 243p. (Natural History Museum Publications). £30.00. LC 90-7790. ISBN 0-19-854017-5.

This is a list of all living species of mammals, including recently extinct species. The order names are arranged phylogenetically from Monotremata to Macroscelidea, with species listed alphabetically under genus names. No authors or dates are listed next to the name, for this information the reader is referred to J. H. Honacki's *Mammal Species of the World* (Allen Press: 1982, the second edition of which is reviewed here under entry 1618). The number of species is given with each order and family name as well as a geographical area where each group is found. The habitat preferred, such as forest, desert, or woodland, is also given. An indication of endangered status (R, rare; T, threatened; E, endangered) is included by a species name if found in the *1990 IUCN Red List of Threatened Animals* (IUCN: 1988) (see entry 79a). English common names are listed next to the scientific name if the name is distinctive and well established. Black-and-white line-drawings of representative species of each family decorate the margins of the list. An 18-page bibliography is arranged into three main groups: general works, geographical sources, and taxonomic sources, which is further grouped by order. An index by common and scientific names completes this basic reference source.

1613. Inskipp, Tim, and Jonathan Barzdo. **World Checklist of Threatened Mammals**. Peterborough: Nature Conservancy Council, 1987. 125p. £6.50. LC 88-112521. ISBN 0-86139-336-8.

This checklist is a combination of the lists found in Appendices I, II, and III to the Convention on International Trade in Endangered Species of Wild Fauna and Flora (CITES) and those included in the IUCN Red Data Books. Updates of these lists can be found in the *1990 IUCN Red List of Threatened Animals* (see entry 79a). The list is taxonomically arranged. Each name also includes the common name, and the countries where found, with numbers in brackets that refer to the entries in the bibliography at the end of the list where more information may be found. Beside the name is a column showing where the animal is listed in CITES, the status (rare, endangered, vulnerable, out of danger, insufficiently known, or indeterminate) as listed in the IUCN Data Books, and the major exploitation of the species (food, horn, teeth, ivory, etc., live animal trade, skins, trophies, wool), as well as further references numbers for further information in the bibliography. The references to the literature are especially useful.

1614. Luckett, W. Patrick, and Frederick S. Szalay. **Phylogeny of the Primates: A Multidisciplinary Approach**. New York: Plenum Press, 1975. 483p. LC 75-39714. ISBN 0-306-30852-5.

Although this work is primarily on the phylogeny of primates, it also contains several important chapters covering the entire class of Mammalia. Phylogenetic systematics studies the grouping of organisms for the purpose of classification, based on their evolutionary descent. The first of four main sections of this work contains background information on the whole mammalian group. A chapter by George Gaylord Simpson titled "Recent Advances in Methods of Phylogenetic Inference," one by Malcolm C. McKenna titled "Toward a Phylogenetic Classification of the Mammalia," and one by John F. Eisenberg titled "Phylogeny, Behavior, and Ecology in the Mammalia" are all important, frequently cited works. McKenna's paper presents a classification scheme for living and extinct mammals (to the ordinal level) that is more up to date than Simpson's (see entry 1617) and is frequently used as the basis for mammal classification today.

1615. Rice, Dale W. **A List of the Marine Mammals of the World**. 3d ed. Washington, D.C.: U.S. Dept. of Commerce, National Oceanic and Atmospheric Administration, National Marine Fisheries Service; for sale by the Supt. of Docs., U.S. Govt. Print. Off., 1977. 15p. (NOAA Technical Report NMFS SSRF-711). LC 77-603015.

This is a list of 116 species of recent (not fossil) marine mammals. These include the cetaceans, sirenians, pinnipeds, sea otters, and polar bears. Each name in the checklist gives the reference where first named and date, the common name, and the geographical distribution. Synonyms are listed at the end of the checklist along with a bibliography of literature cited.

1616. Seal, U. S., and Dale G. Makey. **ISIS: Mammalian Taxonomic Directory**. St. Paul, Minn.: Minnesota Zoological Garden, 1974. 1 vol. (loose-leaf).

ISIS is an acronym for International Species Inventory System. This system is used to give a number to each of the names for 3,986 species, 986 genera, 123 families and 19 orders of living mammals. The authors created a numbering system of 19 characters beginning at the level of kingdom. For example, the number for *Tachyglossus aculeatus acanthion* is 14010010010010003. The 14 is for the class Mammalia, the 01 is for the order Monotremata, the 001 is for the family Tachyglossidae, the next 001 is for the genus *Tachyglossus*, the next 001 is for the species *aculeatus*, and the 003 is for the subspecies *acanthion*. These numbers provide a specific identification number for a particular name. Listed with the number and scientific name is the common name and an indication of geographic distribution. Three appendices are also included. Appendix one is an outline of the classification numbering scheme; appendix two lists names from the *IUCN Mammal Red Data Book,* published in 1972; and appendix three lists endangered mammals from the U.S. list of Endangered Foreign Mammals from the *Federal Register* of 1970.

1617. Simpson, George Gaylord. **The Principles of Classification and a Classification of Mammals**. New York: American Museum of Natural History, 1945. 350p. (Bulletin of the American Museum of Natural History, vol. 85).

Although dated, this work is a classic and forms the basis of mammal classification today. It is divided into three sections. Section one discusses the principles of taxonomy; section two is the checklist of mammal species; and section three covers further information on the orders and families included in section two. Section two includes both living and extinct groups, 32 orders in all. Each name is followed by the author of the name and the year of publication, geographical distribution, or name of era if extinct, and common name. A good bibliography, especially useful for historical references, is found before the indexes to scientific and vernacular names.

1618. Wilson, Don E., and DeeAnn M. Reeder, eds. **Mammal Species of the World: A Taxonomic and Geographic Reference**. 2d ed. 1993. 1206p. $75.00. LC 92-22703. ISBN 1-56098-217-9.

This is the second edition of a checklist that has become the standard for mammalian taxonomy. It lists 4,629 (4,170 in the first edition) existing or recently extinct species from around the world. It is the most up-to-date list and also provides, for each name, the literature reference to where it was first named and described, the type locality, geographic distribution, conservation status, synonyms, and additional taxonomic notes. A lengthy 156-page bibliography of references cited in the work attests to the comprehensive research done to produce this checklist, and also provides suggestions for further reading.

The major difference between this edition and the first is the way it was compiled and edited. The first edition was the result of the reviews of approximately 160 specialists who contributed their individual views of valid names and references. It was decided that this approach was too free with dissenting opinions on taxonomic names and therefore this edition was authored by 20 specialists, who adopted the taxonomic arrangement they thought best synthesized the current literature for their area. The information for this edition was put into an electronic database, making future editions more easily updated and produced. Another difference is that the ISIS (International Species Inventory System) numbers, used in the first edition, were left out of this second edition. A table contained in the introduction summarizes the differences between the first and second editions in number of genera and species listed for each order and family (86 new names in the order Rodentia). It also shows rearrangements of taxons, especially the substitution of seven new orders for the old order Marsupialia.

A comparable checklist is *A World List of Mammalian Species*, third edition, by G. B. Corbet and J. E. Hill (see entry 1612). This does not contain the taxonomic references and is not as complete or up to date, but does contain and index the common names of each mammal, something *Mammal Species of the World* does not do. For its comprehensive, current, and authoritative work, *Mammal Species of the World* should be an essential purchase for all academic and large public libraries, and those special libraries with holdings in mammalian studies.

Mammalia (Geographic Section)

Many checklists are included in handbooks for the particular region. Check the Handbooks section of this chapter.

AFRICA

1619. Allen, Glover M. **A Checklist of African Mammals**. Cambridge, Mass.: Museum of Comparative Zoology at Harvard College, 1939. 763p. (Bulletin of the Museum of Comparative Zoology, vol. 83). LC a56-5747.

Though dated, this work is the main source for references on African recent mammal names. It covers the literature from 1758 to July 1939 and is updated by W. F. H. Ansell's work, *African Mammals, 1938-1988* (see entry 1620). It is arranged taxonomically with genera listed alphabetically under family, and species alphabetically under genera. The author and reference where first named is listed next to the name, giving volume, page, and date. The type locality follows the reference and geographical range is given for most species. An index to names follows the checklist.

1620. Ansell, W. F. H. **African Mammals, 1938-1988**. Cornwall, England: Trendrine Press, 1989. 77p. £10.65. LC gb89-30685.

This work updates Allen's 1939 book *A Checklist of African Mammals* (see entry 1619). It lists mammal names published since that publication up to December 31st 1988. References where the name was first published are in almost the same format as Allen's work except it places the date after the author instead of at the end of the reference. It also lists type locality and geographical location, with latitude and longitude. A short list of references and an index to names complete the work.

1621. Meester, J. A. J., and others. **Classification of Southern African Mammals**. Pretoria, Transvaal: Transvaal Museum, 1986. 359p. LC 87-173562. ISBN 0-907990-06-1.

Focusing on the area of Africa below the Zambezi and the Cunene River, this provides a classification of the mammals found in this area. It is arranged taxonomically by order from Insectivora to Lagomorpha. Keys to family, genus, and species are provided. Each species name includes the author and date when first named, geographical distribution, and references to where described. The common and Afrikaans names are also provided. Eight pages of selected references, and a detailed index complete the text.

1622. Meester, J., and H. W. Setzer, eds. **The Mammals of Africa: An Identification Manual**. Washington, D.C.: Smithsonian Institution Press, 1971. Loose-leaf. $105.00. LC 70-169904. ISBN 0-87474-116-5.

Although the title indicates this is primarily an identification manual, it is also serves as a checklist of mammals of Africa. See entry 1611 for a complete review.

AUSTRALIA

1623. Bannister, J. L., and others. **Zoological Catalogue of Australia. Mammalia,** Vol. 5. Canberra, Australia: Australian Government Publishing Service, 1988. 274p. $29.95. LC 85-147738. ISBN 0-644-05009-8.

This work, arranged by family group, lists 330 species. It includes species known at the time of European settlement and introduced species that are continuing as feral populations in Australia. Each family is authored by an authority on the group. After a brief paragraph or two on the family (including the number of species in Australia), each genus is listed with species underneath. The author and year of publication of the name of each taxon come before the title of the work, plus volume and page number where found. Type specimens and location as well as distribution in Australia are also provided. References to the biology of each species follow the information on geographical distribution. A taxonomic index finishes the catalogue.

NORTH AMERICA

1624. Jones, J. Knox, and others. **Revised Checklist of North American Mammals North of Mexico, 1991.** Lubbock, Tex.: Texas Tech Univ. Press, 1992. 23p. (Occasional Papers, The Museum, Texas Tech University, no. 146). $2.00.

This checklist, first published in 1973, has been revised for the fifth time and continues to be the main checklist for living mammals for this area. The list is introduced with commentaries on the main changes in the list. Following this, names of orders, families and genera are arranged in phylogenetic order, generally following Hall's *The Mammals of North America* (1981) (see entry 1477). Species are arranged alphabetically within each genus. Common names are listed opposite the scientific names. This checklist does not provide any further information such as references to where first named and author, or geographic distribution. The checklist ends with a list of references cited in the work. There is no index.

MEXICO

1625. Ramírez Pulido, José, and others. **Lista y Bibliografía Reciente de los Mamíferos de México**. México City: D.F., Universidad Autónoma Metropolitana, Unidad Iztapalapa, Dept. de Biología, 1983. 363p. ISBN 968-597-177-3.

This checklist includes 10 orders, 34 families, 148 genera, and 428 species. These are listed phylogenetically with species and subspecies listed alphabetically under genus. Each name includes the author, with reference to where first named in the literature, the type locality, and geographical distribution. References to other taxonomic works on the species are also provided. A long (55-page) bibliography of literature cited in the text follows the checklist. Synonyms are listed in an appendix. There is no index.

PALAEARCTIC REGION

1626. Corbet, G. B. **The Mammals of the Palaearctic Region: A Taxonomic Review**. London, Ithaca, N.Y.: Cornell Univ. Press, 1978. 314p. (British Museum (Natural History) Publication No. 788). $67.50. LC 77-90899. ISBN 0-8014-1171-8.

This work updates and supplements the *Checklist of Palaearctic and Indian Mammals 1758 to 1946* (see entry 1628), and provides identification keys, distribution, and concise taxonomic descriptions. It supplements this checklist by listing all names proposed from 1946 to 1972, and some to 1976. It is arranged from Marsupialia to Artiodactyla with range maps following the text. A comprehensive 27-page bibliography is useful for further study. The Palaearctic region includes approximately the continent of Europe, and Asia north of the

Himalayas, along with Africa north of the Sahara. References to where first named, the date and author, as well as a brief geographical range accompany each name. For more common species, range and remarks are more detailed. A supplement to this work was published in 1984 (see entry 1627).

1627. Corbet, G. B. **The Mammals of the Palaearctic Region: A Taxonomic Review. Supplement.** London: British Museum (Natural History), 1984. 45p. (British Museum [Natural History], Publication No. 944). $13.00. ISBN 0-565-00944-3.

Supplements Corbet's previous work (see entry 1626) and updates the literature to March 1984. In all, 113 new names are recorded.

1628. Ellerman, J. R., and T. C. S. Morrison-Scott. **Checklist of Palaearctic and Indian Mammals 1758 to 1946**. 2d ed. London: British Museum (Natural History), 1966. 810p.

This second edition is primarily a reprint of the first edition with only minor amendments. The work recognizes 809 species of mammals for this region. The list was compiled from the recent mammals named in the literature from the tenth edition of Linnaeus to the end of 1946 along with the study of the extensive British Museum collections. Domestic animals and extinct mammals are not included. Under each name are references where first named, and other references to the literature that includes pertinent descriptions and information. Brief geographical range information is also included. Updated by Corbet's work *The Mammals of the Palaearctic Region* (see entry 1626).

Mammalia (Systematic Section)

MARSUPIALIA (Opossums, Kangaroos, Wallabies, Koalas, etc.)

1629. Marshall, Larry G. **The Families and Genera of Marsupialia**. Chicago, Ill.: Field Museum of Natural History, 1981. 65p. (Fieldiana. Geology, New Series, no. 8). LC 81-65225.

This work is divided into two parts. Part one reviews the history of marsupial systematics. Part two provides a list of the current (1981) recognized families and genera names of living and fossil marsupials, along with their synonyms, references, and geologic or geographical ranges. A 23-page list of references cited, and an index to scientific and technical names completes the work.

1630. Oldfield, Thomas. **Catalogue of the Marsupialia and Monotremata in the Collection of the British Museum (Natural History)**. London: Printed by Order of the Trustees, 1888. 401p. LC 06-36238.

This work describes 151 species of Marsupials and three species of Monotremata, and is included here for its use as a checklist, although much of this information is outdated. This is reviewed in the Handbooks section (see entry 1506).

CHIROPTERA (Bats)

1631. Andersen, Knud C. **Catalogue of the Chiroptera in the Collection of the British Museum**. 2d ed. Vol. 1, **Megachiroptera**. London: Trustees of the British Museum (Natural History), 1912. Reprint, New York: Johnson Reprint, 1966. 854p. $72.00. ISBN 0-384-01395-3.

Included here in the checklist section for its listing of 186 species and 228 forms of the Megachiroptera, or large bats. More fully described in the Handbooks section (see entry 1508).

1632. Dobson, George Edward. **Catalogue of the Chiroptera in the Collection of the British Museum**. London: Printed by Order of the Trustees, 1878. Reprint, Forestburgh, N.Y.: Lubrecht and Cramer, 1966. 567p. $65.00. ISBN 3-7682-0300-X.

This work contains complete accounts of 400 species of bats. It is reviewed in the Handbooks section (see entry 1510), but is included here for its use as a checklist.

1633. Miller, Gerrit S., Jr. **The Families and Genera of Bats**. Washington, D.C.: Government Printing Office, 1907. Reprint, Forestburgh, N.Y.: Lubrecht & Cramer, 1966. 282p. $60.00. ISBN 3-7682-0534-7.

Still in print, this classic work lists and describes 36 families and subfamilies, and 173 genera of bats of the world. This work is reviewed in the Handbooks section (see entry 1512).

PRIMATES

1634. Gavan, James A. **A Classification of the Order Primates**. Columbia, Mo.: Museum of Anthropology, University of Missouri-Columbia, 1975. 36p. (Museum Brief, no.16). $1.90. ISBN 0-913134-15-5.

This brief treatise on the classification of the order Primates aims to discuss the biological traits that identify the suborders, infraorders, and superfamilies. It also provides a table of the classification of living Primates from suborder through genus. It gives historical information on the classification of Primates as well. The discussion of biological traits for each major group includes biological traits for the families and some genus groups. A seven-page glossary of terms used in the text, and a bibliography of references cited complete the work.

1635. Luckett, W. Patrick, and Frederick S. Szalay. **Phylogeny of the Primates: A Multidisciplinary Approach**. New York: Plenum Press, 1975. 483p. LC 75-39714. ISBN 0-306-30852-5.

Although this work is primarily on the phylogeny of primates, it also contains several important chapters covering the entire class of Mammalia. (See entry 1614.)

1636. Napier, P. H. **Catalogue of Primates in the British Museum (Natural History)**. London: British Museum (Natural History), 1976- . $55.00, pt. 1; $55.00 pt. 2; $65.00 pt. 3; $55.00 pt. 4; $49.95 pt. 5. (Publication nos. 744, 815, etc.). ISBN 0-565-00744-0 pt. 1; 0-565-00815-3 pt. 2; 0-565-00894-3 pt. 3; 0-565-01008-5 pt. 4; 0-565-01154-5 pt. 5.

These are catalogs of specimens held by the museum in its various collections. It is reviewed in the Handbooks section (see entry 1522), but is included here for its use as a checklist.

MYSTICETA and ODONTOCETA (Baleen Whales and Toothed Whales)

1637. Hershkovitz, Philip. **Catalog of Living Whales**. Washington, D.C.: Smithsonian Institution; for sale by the Superintendent of Documents, U.S. Govt. Print. Off., 1966. 259p. (United States National Museum Bulletin, no. 246).

This checklist of living whales of the world is a comprehensive work updated by Rice's work titled *A List of the Marine Mammals of the World* (see entry 1638). Each name in the list includes the author and reference where first published. References with further information and illustrations, such as those for the animal itself, or the skull or teeth, are also included. Brief geographical information and type locality are also provided for most species. A helpful glossary of common names with corresponding scientific names follows the list. A selective bibliography and an index to scientific name complete the work.

1638. Rice, Dale W. **A List of the Marine Mammals of the World**. 3d ed. Washington, D.C.: U.S. Dept. of Commerce, National Oceanic and Atmospheric Administration, National Marine Fisheries Service; for sale by the Supt. of Docs., U.S. Govt. Print. Off., 1977. 15p. (NOAA Technical Report NMFS SSRF-711). LC 77-603015.

This is a list of 116 species of recent (not fossil) marine mammals. These include the cetaceans (whales), sirenians, pinnipeds, sea otters, and polar bears. For a full review, see entry 1615.

ASSOCIATIONS

Mammalia (General)

1639. **American Society of Mammalogists**. Brigham Young University, Dept. of Zoology, 501 Widtsoe Bldg., Provo, UT 84602. (213) 548-6279.

1640. **Australian Mammal Society**. University of Tasmania, Dept. of Zoology, GPO Box 252C, Hobart, TAS 7001, Australia. 6-2509435.

1641. **Boone and Crockett Club**. 241 S. Fraley Blvd., Damfries, VA 22026. (703) 221-1888.

1642. **Mammal Society**. Bristol University, Dept. of Zoology, Woodland Rd., Bristol, Avon BS8 1UG, England. 272-2723000.

Mammalia (Systematic Section)

CHIROPTERA (Bats)

1643. **Bat Conservation International**. c/o Breckenridge Field Laboratory, University of Texas, Austin, TX 78712. (512) 499-0207.

PRIMATES (Monkeys, Gorillas, Chimpanzees, Lemurs, etc.)

1644. **American Society of Primatologists**. SUNY at Buffalo, Dept. of Anthropology, Buffalo, NY 14261. (916) 752-3273.

1645. **Francophone Primatological Society**. (Societe Francophone de Primatologie). Station Biologique de Paimpont, F-35380 Plelan le Grand, France. 99-078181. FAX 99-078089.

1646. **International Primatological Society**. Deutsches Primatenzentrum, Kellnerweg 4, W-3400 Gottingen, Germany. 551-38510.

1647. **Simian Society of America**. 1426 E. Weir, Phoenix, AZ 85040.

MYSTICETA and ODONTOCETA (Whales, Dolphins, and Porpoises)

1648. **American Cetacean Society**. P.O. Box 2639, San Pedro, CA 90731. (213) 548-6279.

1649. **Pacific Whale Foundation**. P.O. Box 1038, Kihei Azeka Pl., Suite 303, Maui, HI 96753. (808) 879-8811.

CARNIVORA

1650. **International Association for Bear Research and Management**. P.O. Box 3129, Station B, Calgary, AB, Canada T2M 4L7. (403) 264-7781.

1651. **North American Wolf Society**. 1409 Bentwood, Austin, TX 78722. (512) 459-5139.

PERISSODACTYLA

1652. **International Society for the Protection of Mustangs and Burros**. c/o Helen A. Reilly, 11790 Doedar Way, Reno, NV 89506. (702) 972-1989.

ARTIODACTYLA

1653. **Desert Bighorn Council**. c/o William C. Dunn, 245 Lakeshope Rd., Boulder City, NV 89005. (702) 564-1644.

1654. **Foundation for North American Wild Sheep**. 802 Canyon Ave., Cody, WY 82414. (307) 527-6441.

1655. **Society for the Conservation of Bighorn Sheep**. 3133 Mesalon Ln., Pasadena, CA 91107. (818) 797-1287.

APPENDIX
Internet Information Sources on the Animal Kingdom

The Internet has just recently become a significant source of information in the zoological sciences. Listed here are sources that may help in finding this information on animals and their environment. It is not a comprehensive list. Databases are constantly being created and access to them is constantly expanding. Not listed here are the many molecular, genetic, and agriculture sources. These sources may be found in the three specialized guides to the Internet by Una Smith, Wilfred Drew, and Ken Boschert listed below. Information on using the Internet will not be discussed here, but the following works may be consulted. They cover the subject in great detail.

Krol, E. **The Whole Internet: Catalog & User's Guide**. Sebastopol, Calif.: O'Reilly & Associates, 1992.

LaQuey, T. and J. C. Ryer. **The Internet Companion: A Beginner's Guide to Global Networking**. Reading, Mass.: Addison-Wesley, 1992.

Tennant, R., J. Ober and A.G. Lipow. **Crossing the Internet Threshold: An Instructional Handbook**. San Carlos, Calif.: Library Solution Press, 1993.

Three excellent guides to Internet resources in biology and agriculture are listed below. All can be accessed through Internet via Gopher (explained below) and doing a search using VERONICA on "Just Cows" for Wilfred Drew's guide, and on "Biology and Guide" for Una Smith's guide, and on "Electronic Zoo" for Ken Boschert's guide. FTP addresses are also supplied.

Boschert, Ken. **The Electronic Zoo: A List of Animal-Related Computer Resources**. 1993. 80p. Available via anonymous FTP from wuarchive.wustl.edu (128.252.135.4) in the subdirectory: /doc/techreports/wustl.edu/compmed/elec_zoo.x_x.

Drew, Wilfred. **Not Just Cows: A Guide to Internet/Bitnet Resources in Agriculture and Related Sciences**. 1992. 50 pages. Available via anonymous FTP from ftp.sura.net in directory pub/nic. File agricultural.list.

Smith, Una R. **A Biologist's Guide to Internet Resources**. 1994. 40 pages. Available via anonymous FTP from sunsite.unc.edu in file pub/academic/biology/ecology+evolution. Then use "get bioguide.faq" to copy the guide to your computer. Guide also posted on USENET sci.answers.

LISTSERV LISTS

The following are e-mail listserv discussion groups on many different topics of interest; birds and molluscs, for example. By subscribing to the listserv, the subscriber automatically receives e-mail from the group. One can reply to the group messages. Also, these messages are archived and can be searched. Much of the information on listservs listed below is from an excellent 40-page guide titled *A Biologist's Guide to Internet Resources* by Una Smith (see citation above).

To subscribe to the following lists, send e-mail to the listserv userid at the same node as the list. For example, to subscribe to bee-l send e-mail to the listserv@uacsc2.albany.edu and in the body of the message put

SUBSCRIBE BEE-L Your Name.

Internet Address	*Description*
aquarium@emuvm1.cc.emery.edu	Fish and Aquaria
batline@unmvma.umn.edu	Bat Discussion
bee-l@uacsc2.albany.edu	Bee Biology
biodiv-l@bdt.ftpt.ansp.br	Biodiversity Network
bird_rba@arizvm1.ccit.arizona.edu	National Birding Hotline
birdband@arizvm1.ccit.arizona.edu	Bird Bander's Forum
birdchat@arizvm1.ccit.arizona.edu	Birding Hotline(Chat)
birdcntr@arizvm1.ccit.arizona.edu	Birding Hotline (Central)
birdeast@arizvm1.ccit.arizona.edu	Birding Hotline (East)
birdwest@arizvm1.ccit.arizona.edu	Birding Hotline (West)
brine-l@uga.cc.uga.edu	Brine Shrimp Discussion
canine-l@psuvm.psu.edu	Canine Discussion
camel-l%sakfu00.bitnet@vtvm1.cc.vt.edu	Camel Discussion
cturtle@nervm.nerdc.ufl.edu	Sea Turtle Biology
conslink@sivm.si.edu	Biological Conservation
crust-l@sivm.si.edu	Crustacean Biology
deepsea@iubm.ucs.indiana.edu	Deep Sea and Hydrothermal Vent News
dis-l@iubvm.ucs.indiana.edu	Drosophila Studies
ecolog-l@umdd.umd.edu	Ecological Society of America
entomo-l@uoguelph.ca	Entomology Discussion
envst-l@brownvm.brown.edu	Environmental Studies
ethology@searn.sunet.se	Ethology
fish-ecology@searn.sunet.se	Fish Ecology
iapwild@ndsuvm1.bitnet	International Arctic Project Wildlife
icaeconet@miamiu.bitnet	Ecology and Environment
marmam@uvvm.uvic.ca	Marine Mammal Research
mollusca@ucmp1.berkeley.edu	Molluscan Studies
sbnbirds@indycms.iupui.edu	South Bend Area Birds
socinsct@albany.edu	Social Insect Biology
taxacom@harvarda.harvard.edu	Taxonomy and Systematics

NON-LISTSERV LISTS

These lists do not have automated subscriptions to their lists, as in the case of the listservs. Requests to join these non-listserv lists are read by the owner or manager, so to subscribe to one of these lists, send the e-mail message to the owner or manager at the e-mail address supplied.

American Society of Mammalogists.
To subscribe send message to editor at mnhvz049@sivm.si.edu.

Neotropical Birds.
To subscribe send message to Roberto Phillips at phillips@cipa.ec. Address of group is avifauna@rcp.pe.

Behavioral Ecology Digest.
To subscribe send e-mail message to b-e-requests@forager.unl.edu. Address of group is b-e-group@forager.unl.edu.

Bombus.
To subscribe to this bumblebee discussion list send e-mail message to bombus-request@csi.uottawa.ca. Address of group is bumbombus@csi.uottawa.ca.

Entomology Discussion.
To subscribe send e-mail message to Mark O'Brien at Mark.Obrien@um.cc.umich.edu. Address of group is ent-list@umich.edu.

Killiefish Discussion.
To subscribe send e-mail message to Killie-request@mejac.palo-alto.ca.us. Address of group is killie@mejac.palo-alto.ca.us.

Primate Talk.
To subscribe send e-mail message to primate-talk-request@primate.wisc.edu. Address of group is primate-talk@primate.wisc.edu.

International Marine Ornithologists' Network.
To subscribe send e-mail message to seabird@zoo.uct.ac.za. In subject field say SUBSCRIBE SEABIRD. Address of group is seabird@zoo.uct.ac.za.

Wildnet.
To subscribe to this discussion group on wildlife and fisheries issues send e-mail message to wildnet-request@access.usask.ca. Address of group is wildnet@access.uask.ca.

NEWS GROUPS ON USENET

Many of the above discussion groups are available on USENET. This is a system of ews groups that may be available on a computer system at a university or other facility with large enough system to accommodate USENET. Not all computer systems carry USENET ecause of the computer disk space and maintenance it requires. The advantage of accessing ese listservs on USENET is to avoid having mountains of e-mail messages delivered to your ddress. One can access the news group on USENET, rather than having the messages elivered via e-mail. The following are some USENET news groups that may be useful for oology topics. There are many more.

bionet.general	General Biology Discussion
bionet.software	Software for biology
bit.listserv.ecolog-l	Ecological Society of America
bit.listserv.ethology	Ethology
sci.bio	General Biology
sci.bio.ecology	Ecological Research

DATABASES

This is a listing of a few databases available on the Internet that apply specifically to information on the animal kingdom. Many more databases and documents can be accessed by Gopher, which is discussed in the Gopher section below.

AVES.

AVES is an archive of bird GIFs and bird-related information, including bird songs. It will be necessary to have GIF viewing software to see the bird images. It is accessible via anonymous FTP vitruvius.cecer.army.mil or through Gopher. The Gopher address is Gopher vitruvius.cecer.army.mil 70.

FORMIS.

FORMIS is a bibliography of ant literature containing more than 17,000 citations. It is searchable by taxon, subject or author. To access this database telnet to nerdc.ufl.edu. At the main menu select LUIS, then select "Indexes and Other Resources," then select "FORMIS."

GAIA: Survive Database.

This is an international database of multispecies survival information. It includes information on thousands of species lifespans under various conditions. It is assembled and maintained by the Gaia Multispecies Survival Project at the University of Texas System Center for High Performance Computing. Information on the database may be obtained by sending an e-mail message to gaia@chpc.utexas.edu.

Museum of Paleontology Invertebrate Database.

This database, compiled by the University of California, Berkeley, Museum of Paleontology, includes all the type specimens for invertebrates. It contains taxonomic, locality, and citation information for each type specimen. It can be reached through the Museum of Paleontology Gopher (see under Gophers on pp. 427-28), or via Gopher and doing a search using VERONICA on "Paleontology and Invertebrate."

Pacific Rim Biodiversity Catalog.

This database, presently being compiled by the University of California, Berkeley Museum of Paleontology with the support of the Pacific Rim Research Program, includes information from almost 200 natural history institutions throughout the world. The information includes their Pacific Rim zoological and paleontological holdings. So far, information is arranged from the 20 geographical regions of the Pacific Rim, with 16 taxonomic categories and 13 geological time units. The 16 taxonomic categories are Porifera, Cnidaria, Echinodermata, Brachiopoda, Annelida, Arthropoda, Mollusca, Other Invertebrata, Chondrichyes, Osteichthyes, Reptilia, Amphibia, Aves, Mammalia, Protista, Plants/Algae. The database can be searched by simple one-word searches or combined word searches using the words "and," "or," and "not," and with a wildcard (*). Available through the Museum of Paleontology Gopher (see pp. 427-28), or via Gopher and doing a search using VERONICA on "Pacific and Rim."

Research Results Database (RRDB).

The U.S. Department of Agriculture Extension Service offers this database that contains summaries of recent research from the USDA's Agricultural Research Service and Economic Research Service. Useful for information on many topics, but for the purposes of this guide, it is useful for literature on insects. To subscribe to the new titles added, send the message "subscribe usda.rrdb" to the address almanac@esusda.gov. Individual titles are also accessible via Gopher and VERONICA search using key words, "mites," for example.

Slater Museum of Natural History's Bird Collection.

The Bird Collection database from the Slater Museum of Natural History, at the University of Puget Sound, is available via the Internet. The 18,500+ records can be searched by genus, species, family, sex, plumage, country, state (or other geographical area), date, and preparation. Simple string searches are used to find information in the database. For information on searching and obtaining a password contact Gary Shugart or Dennis Paulson, Slater Museum of Natural History, University of Puget Sound, Tacoma, Washington 98416, or send e-mail message to slater@ups.edu. This database is also available through the Biodiversity and Biological Collection Gopher (see Gophers on pp. 427-28).

LIBRARY CATALOGS

Many college and university catalogs, as well as the Library of Congress, are available for searching through the Internet. These catalogs are especially useful for those who do not have access to a large library. By searching through the major collections, or in a university catalog that specializes in a particular area of zoology, several new sources may be found. Lists of these catalogs can be found through Gopher via VERONICA using the search words "library and catalogs." Search VERONICA using the word "LOCIS" for the Library of Congress Information System. LOCIS includes over 15 million catalog records and over 10 million records for other types of information. The telnet address is locis.loc.gov. Searching hours are restricted to certain hours of the day Eastern Standard Time.

ELECTRONIC JOURNALS AND NEWSLETTERS

A current list of electronic conferences, newsletters, and journals, compiled by Michael Strangelove, is available over the Internet. To obtain the directory send the following commands to listserv@uottawa.bitnet: GET EJOURNL1 DIRECTRY
GET EJOURNL2 DIRECTRY
Note the missing "a" in Journal and "o" in Directory.

The following is a list of e-mail journals and newsletters that pertain to the animal kingdom.

Animal Behavior Society Newsletter.

The editor of this newsletter is James C. Ha. To subscribe, send an e-mail note to jcha@u.washington.edu. The newsletter contains information of interest to society members and others interested in animal behavior studies.

BIOLOGUE: The Australian Biological Resources Study Newsletter.

This newsletter contains news of the Fauna and Flora of Australia Program. This includes information on their taxonomic and systematic research objectives, grant applications, and a summary of current research grants. Available through the Harvard Biodiversity

Gopher and the BioInformatics Gopher at the Australian National University (see Gophers on pp. 427-28).

Deep-Sea and Hydrothermal Vent News.

This is an e-mail bulletin and message service for biologists interested in deep-sea or hydrothermal vents. Specific areas are evolution, ecology, biogeography, paleontology, systematics, phylogenetics, and population genetics. It offers automatic distribution of e-mail and files to all members, keyword searching of archived messages, storage and access to such files as collection lists, as well as brief information on each member and their research. To join, send e-mail to listserv@yvvn.yvuc.ca. In the body of the letter say SUB DEEPSEA yourfirstname yourlastname.

Environmental Resources Information Network (ERIN) Newsletter.

See "ERIN" below in the Gopher section (pp. 427-28). This is available via Gopher and anonymous FTP from huh.harvard.edu.

Florida Entomologist.

The journal of the Florida Entomological Society titled *Florida Entomologist*, was being tested as of this writing to be available over the Internet as well as being published simultaneously in the printed version. Information on this may be found by contacting Tom Walker, Dept. of Entomology and Nematology, Univ. of Florida, Gainesville, Florida. 32611-0620. E-mail address tjw@ifasgnv or tjw@gnv.ifas.ufl.edu.

Laboratory Primate Newsletter.

This newsletter is available in a print edition as well as electronically through e-mail. It includes information on nonhuman primates and related studies such as care and breeding, conservation, requests for information or research materials, and miscellaneous information on what is happening in the world of primate research. It is published quarterly by the Psychology Dept. at Brown University. To receive issues of this newsletter (without graphics), send a message to listserv@brownvm.brown.edu. In the message say subscribe LPN-L yourfirstname yourlastname.

The Scientist.

This is available in paper or through e-mail. The Institute for Scientific Information and the NSF Network Service Center have provided for this publication through the Internet. Available through anonymous FTP on ds.internic.net, in pub/the-scientist or access via Gopher.

Starnet Echinoderm Newsletter.

STARNET is an electronic newsletter with information on many aspects of echinoderm biology. Posted here are pieces of information on meetings, book reviews, addresses of echinoderm biologists and more. Subscribers are from around the world. For information contact Win Hide at whide@matrix.bchs.uh.edu.

DIRECTORIES

Membership directories are being added to the Internet. For example, the membership directory of the American Society of Icthyologists and Herpetologists and a list of e-mail users at the Natural History Museum, London (including zoologists and entomologists) are currently available on the Internet. Both directories can be found on the Biodiversity and Biological Collections Gopher discussed on p. 427.

GOPHERS

A Gopher is a protocol and software package that can find information available on the Internet, then retrieve and display it. It "Gophers" through Internet-linked computers from around the world and finds this information for the user. The mother of all Gophers is found at the University of Minnesota where Gopher was created.

A very useful innovation for searching through "Gopherspace" is VERONICA (Very Easy Rodent-Oriented Net-Wide Index to Computerized Archives). VERONICA searches through Gopher holes to more easily find documents and software on the Internet. The searcher supplies simple word searches and VERONICA checks titles only (not contents) through Gopherspace and then retrieves and displays that document. For information on using Gophers, consult one of the books listed at the beginning of this appendix on the Internet. The following are Gophers with their addresses that relate to zoological topics. Again, the molecular and genetic Gophers are not included here.

Biodiversity and Biological Collections Gopher.

This Gopher, established by Harvard University, provides access to various types of systematics, biodiversity and biological collections. Included here are access to biological collections of Harvard and Cornell (fish and spider type catalogs) and Texas national history collection, fishes, as well as taxonomic authority files, directories of biologists, access to online journals such as the Erin Newsletter, and information about biodiversity projects such as NEODAT (the Inter-Institutional Database of Fish Biodiversity in the Neotropics). It also includes biodiversity authority files for ichthyology and malacology. The malacology authority file is from the University of Michigan's *Mollusk Reprint Database*. The Slater Museum of Natural History's bird collection is also available on this Gopher. Point the Gopher software to the address: huh.harvard.edu, or access via Gopher and do a search using VERONICA.

BioInformatics Gopher at the Australian National University.

This Gopher provides information under biodiversity, bioinformatics, and biocomputing, complex systems, neurosciences, and others. It provides online searching of databases such as *Australian Fauna*, bibliographies, and direct links to electronic news groups. To access the server, point the Gopher software to life.anu.edu.au, or via Gopher and do a search using VERONICA.

BOING (Bio Oriented INternet Gophers).

This Gopher offers searches through all the biology Internet Gophers in one search. To access this Gopher, point the Gopher client at merlot.welch.jhu.edu and select the following:

"Search Databases at Hopkins (Vectors, Promoters...)." Then select "Search BOING."

Once reaching BOING, just search for the topic using a single word or connecting topics with the words "and," "or," or "not." Wildcards (*) can also be used to truncate the search. For example, Bird* gets all records beginning with the word "bird," including "birds," "birdwatching," and so on.

Environmental Resources Information Network (ERIN).

This Gopher directory attempts to provide access to data and information on the distribution of endangered species, vegetation types, and other environmental information in Australia. The researcher can search through the *Biodiversity*, *Protected Areas*, *Terrestrial/Marine Environments* and other databases for information. ERIN's newsletter also available here (see Electronic Journals and Newsletters on pp. 425-26). Point the Gopher software to kaos.erin.gov.au, or access via Gopher and do a search using VERONICA.

MAILGopher for Biological Databases.

This is a Gopher that can be used with just the basic e-mail facilities. It was developed by the Biocomputing Research and User Support (BRUS) team of the National University of

Singapore (NUS). By using the electronic mail server address MailGopher@nusunix2.nus.sg, Gopher documents and directories can be retrieved and searches can be carried out.

Museum of Paleontology Gopher.

This Gopher is from the Museum of Paleontology, University of California, Berkeley. It includes Mollusca mailing list services, museum information forms and bulletin boards, the *Museum Invertebrate Type Specimen Catalog and Index*, and the *Museum Vertebrate Type Specimen Catalog and Index* as well as the *Pacific Rim Biodiversity Catalog and Index*. Point the Gopher software to ucmp1.berkeley.edu, or access via Gopher and do a search using VERONICA.

National Science Foundation Gopher.

This Gopher accesses the publications and awards of the National Science Foundation. This Gopher and other electronic access sources comprise the Science and Technology Information System (STIS). Many NSF documents and notices of grants, and deadlines for filing them can be found here. Especially useful are funded grant summaries with primary investigators and grant amounts listed. A simple search by topic will bring up all NSF-funded grants on that topic. Covers all aspects of science, including zoology. Point your Gopher software to stis.nsf.gov, or access via Gopher and do a search using VERONICA.

Primate Information Network.

This is a Gopher network covering many aspects of primate studies. It is maintained by the Wisconsin Regional Primate Research Center Library at the University of Wisconsin-Madison. Information that can be accessed here is Primate Taxonomy, the *Laboratory Primate Newsletter* (see Electronic Journals and Newsletters on pp. 425-26), bibliographies from the Primate Information Center, the Primate Supply Information Clearinghouse, an International Directory of Primatology, and the Primate Meetings Calendar. Connect to this Gopher server at Gopher.primate.wisc.edu, or access via Gopher and do a search using VERONICA.

Smithsonian Institution's Natural History Gopher.

This new Gopher contains information resources that are compiled and maintained by the Smithsonian's staff, including newletters and projects in the area of natural history studies. For zoology topics, this Gopher allows searching of the full text of Wilson and Reeder's *Mammal Species of the World* (see entry 1618). Here the researcher can search by, family, subfamily and genus of mammals found worldwide. An authority file on freshwater crayfish is also searchable here. Point the Gopher software to nmnhgoph.si.edu, or access via Gopher and do a VERONICA search.

Author/Title Index

Reference is to entry number. Those titles mentioned in annotations are designated by use of "n." Titles mentioned in places other than annotations are designated by the use of "p" for page number.

Subject Index

Reference is to entry number. Those subjects mentioned in annotations are designated by use of "n." In most instances, scientific and common names are both listed, with most scientific group names in the English form (Reptiles, not Reptilia). No attempt has been made to update outdated scientific names. Always check under the old and new scientific name (Coelenterates and Cnidarians). For geographic names, the entry is under the name of the country when the book was published (Ceylon or Sri Lanka).

Associations Index